Commando
A Boer Journal of the Boer War
Annotated and Illustrated

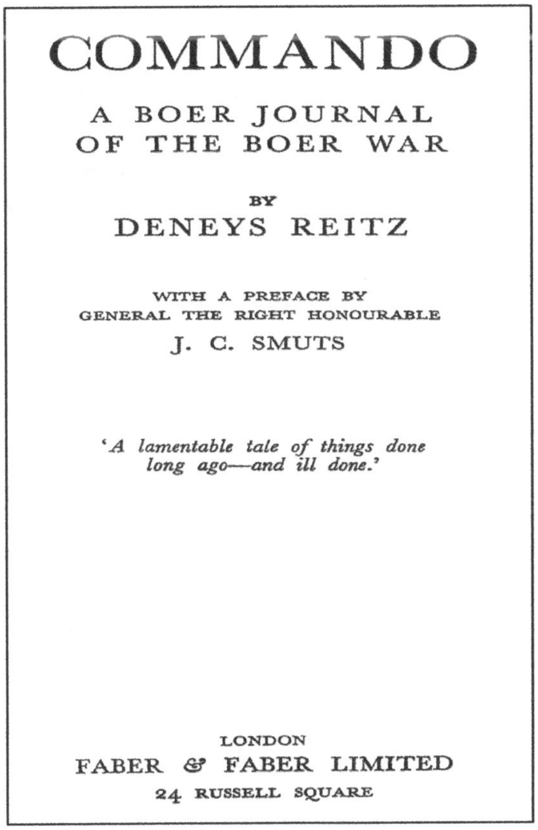
COMMANDO

A BOER JOURNAL
OF THE BOER WAR

BY
DENEYS REITZ

WITH A PREFACE BY
GENERAL THE RIGHT HONOURABLE
J. C. SMUTS

'A lamentable tale of things done long ago—and ill done.'

LONDON
FABER & FABER LIMITED
24 RUSSELL SQUARE

A Boer Journal of The Boer War
By Deneys Reitz
With A Preface By General the Right Honourable J. C. Smuts

'A lamentable tale of things done
long ago—and ill done'

Written in exile in Antanarive,[1] Madagascar in 1903.

Volume 4 of the series, *The Last Horse War*

Series compiled, annotated, illustrated and expanded by Payton Fireman, JD.

ISBN: 978-1-7320124-6-2

Front Cover Photo Credits:[1 2] Back Cover Photo Credits:[3 4 5] Title Page Image Credits:[6 7]

[1] Antananarivo, is the capital and largest city of Madagascar. Antananarivo - Wikipedia
[2] Please see the Bibliography for license links and other information regarding the material in this volume.

Brief Biographical Summary

Deneys Reitz (1882–1944), was the 17-year-old son of the Secretary of State for the Transvaal, Francis W. Reitz. Deneys enlisted at the outbreak of the war and soldiered, through it to its conclusion. After peace was concluded he and his family went into exile rather than accept the peace agreement.

After a period of exile in French Madagascar he returned to South Africa and under the tutelage of his former commander, Jan Smuts, studied law; eventually beginning his practice in Heilborn. Reitz was a supporter of the Botha/Smuts administration under the South African Party. During the Nationalist Party insurrection of early 1914 he was appointed military commandant of the Heilborn district. Later that year, in August, he took an active part in suppressing the Martiz rebellion.

In the First World War he fought for the Union of South Africa against the German Empire, as an officer in the British Army, where he was wounded in France before resuming command of the First Battalion of the Royal Scots Fusiliers.

In later life he was a South African politician and one of General Smuts' principal lieutenants, serving in many capacities.

In 1939, he became Minister of Native Affairs and Deputy Prime Minister until 1943, when he was appointed as South African High Commissioner to London, and served in that capacity until his death in 1944.

Contents

PF Preface 1
Maps & Tables Etc. 4
Dedication 17
Jan Smuts' Preface 18
I. Memory's Tower 21
II. On the Brink 25
III. To the Frontier 31
IV. We Invade Natal 34
V. A Battle 48
VI. Ups and Downs 58
VII. An Affair at Surprise Hill 67
VIII. A Visit to the Tugela Line 70
IX. The Battle of Spion Kop 79
X. The Rest of Our Corporalship is Destroyed 90
XI. A Campaign in the Free State 98
XII. The British Invade the Transvaal 112
XIII. Further Afield 125
XIV. New Conditions 132
XV. A Successful Affair, and After 139
XVI. From West to East 146
XVII. I Start for the Cape Colony 157
Middleburg Conference 164
Resume 172
XVIII. The Next Stage 180
XIX. Farther South 192
XX. We Go into the Cape Colony and Meet with a Warm Reception 207
XXI. Horses and Men 217
XXII. Moss-Trooping 235
XXIII. A Long Trail 250
XXIV. Calmer Waters 264
XXV. The Last Phase 278
XXVI. The Lost Cause 290
O'kiep & Cape Copper 306
Executions 309
Concluding Material 313
Costs of Second Boer War 318
Repatriation - Times History Vol 6 Chapter II 318
Epilogue 330
Family Biographical Sketches 336
Bibliography 344
Endnotes 347

PF Preface

Deneys Reitz[3] is first narrator in this series, whose participation in the Second Boer War extends from the beginning, in October of 1899, until the end, in May of 1902.

Through, dogged persistence and serendipity he was present at many significant events during the conflict.

I've tried to identify the dates of events from Deneys' and other sources as best I can. Sometimes I think he may have a few events out of chronological order. His narrative telescopes and then compresses time frames and it is sometimes difficult to tell what the date is. In some of his more wilderness adventures, he is not even sure what the month is when events occur and expresses that uncertainty.

He is the only memorialist in this series who has virtually overwhelmed my ability to annotate and illustrate his work. His recall of names, places and events is exceptional. He uses the language of the country around him and knows every bit of horse-craft. Despite this, I've made a concerted effort to include an identifying footnote associated with the first time an identifiable individual is mentioned.

It is surprising how few positive identifications I can make regarding his comrades. Many names simply have too many of the same or similar names to sort through in the records of primarily, Angloboerwar.com. Some progress can be made by looking at the dates individuals were taken prisoner and determining if it was before or after that time in the narrative.

Assembling the images, footnotes, references and other book elements has sometimes reminded me of playing Chess or Tetris. 'How does it fit? Where does it fit? etc.' With Deneys it is harder than usual because he is such a good explicator that I am not able to place the reference to a person, place or event, where it first occurs. I've had no end of trouble defining words and places when he first mentions them only to have him do a better job with an apt use or description in the upcoming sentence or paragraph.

Deneys' writing is so comprehensive that it is possible to simply cruise past my somewhat over enthusiastic use of footnotes; pretty much ignoring them completely. I write and include citations in my works as almost a legal brief and as a scholar's edition for my own purposes of veracity and accountability. It also leaves avenues for others to pursue their own scholarship.

Because the stories are synoptic, I end up using the same pictures and maps in several works. Since each work is intended to stand on its own, there is no reason not to do so.

Please see the South African Historical Background section of Volume 1; the Haldane, Frankland and Hofmeyr compilation, for an introduction to the time period and the events leading up to the conflict as well as for the series introduction and a discussion of the methodology for constructing this works in this series as well as the credits for those who have assisted me along the way.

I retain an aversion to italics as well as to blank pages, so very few are employed.

Once cited, maps and images may not be re-cited when portions are subsequently reused.

[3] Deneys Reitz (1882–1944) English Wikipedia: https://en.wikipedia.org/wiki/Deneys_Reitz Afrikaans Wikipedia: https://af.wikipedia.org/wiki/Deneys_Reitz

Orange Free State Commandants

Left: Dickson (State Aty.) Snawpoel, Steenkamp, Ferreira, Fick, Poigieter, Wessels Du Toit, Van der Mewe.
Middle: Albrecht,[4] Du Plessis, Nel, Pres. Steyn, J.H. Olivier,[5] M. J. Prinsloo, C.J. De Villiers.
Lower: Prinsloo, Vanyyl, Du Plooy, Lubbe, Naudi.[8] Photo.[9] Identifications from this cited copy.[10]

South African Republic Executive Council.[11]

[4] Major Albrecht, the commandant of the Free State Artillery, the man who turned his two hundred men into a model corps. By birth a German, after taking part in the Franco -Prussian War of 1870–71, Major Albrecht was charged by the Free State with the formation of a corps of artillery -men. He acquitted himself of his task with true German thoroughness: his corps was an example of bravery, strict performance of duty, and capacity. Major Albrecht was taken prisoner in General Cronjé's surrender at Paardenberg, on the 27th of February 1900. Heroes of the Boer War. Review of Reviews, 1903. P.144.

[5] Commandant J. H. Olivier, the hero of Stormberg; a great tactician. He and the Transvaal Commandant. Lemmer, succeeded, notwithstanding the occupation of Bloemfontein, in extricating the whole Boer force in the North of Cape Colony (21 March 1900), without losing a gun, waggon or horse. Heroes p. 133.

Original Book Summary

This unattributed book summary and review appeared in the first publication in 1929.

Deneys Reitz, who is now one of General Smuts's political lieutenants and a warm supporter of the British Commonwealth, fought through the South African War as our enemy. He enlisted on the outbreak of the war as a boy of seventeen and went through it to the bitter end. His father was President of the Orange Free State, and after peace was concluded he and his family went into exile. Later he returned to his own country; and later still fought in the Great War on the side of the Allies, first in West and South Africa, and then in France, where he was severely wounded, and where he came to command the First Battalion of the Royal Scots Fusiliers.

Commando abounds in interesting sidelights and portraits-Kruger, De Wet, Smuts, Kitchener. One of the most remarkable passages gives us a glimpse of the young Winston Churchill as prisoner of war in Pretoria. It is difficult to speak of this book in anything short of a string of superlatives. The spirit of it is magnificent; the uncomplaining courage of the boy-fighter rouses memories of all the stories down the ages about unrecking devotion to an ideal. And not one of them is a finer one than this.

A group of leaders at the beginning of the war.[12] (Identified right to left, Kock, Erasmus (Younger Brother), Kruger, Cronje and Coster.)

Maps & Tables Etc.

Multiple maps are provided of various areas. Each has details or nomenclature that are not reproduced in others.

SA Political and Railway

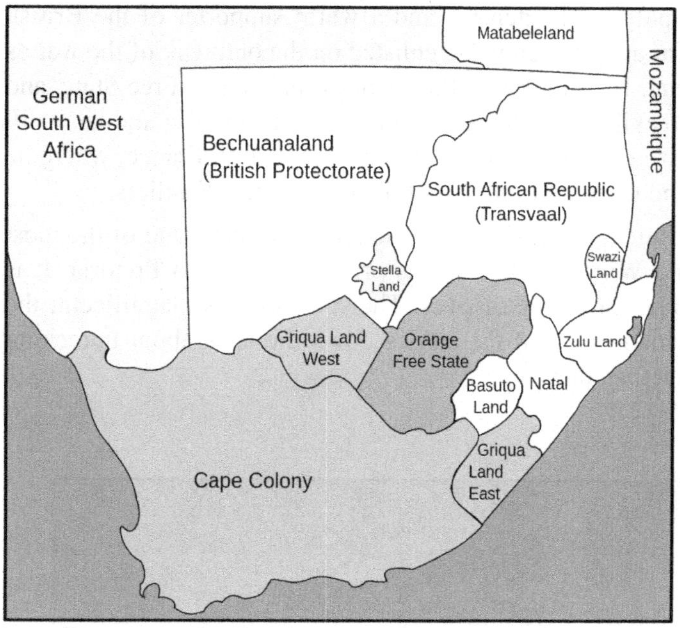

South Africa, 1885.[13 14]

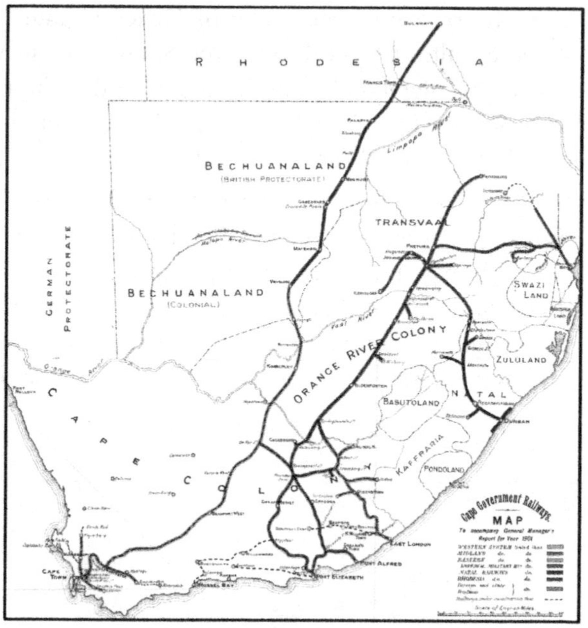

South African Railway System, 1901.[15]

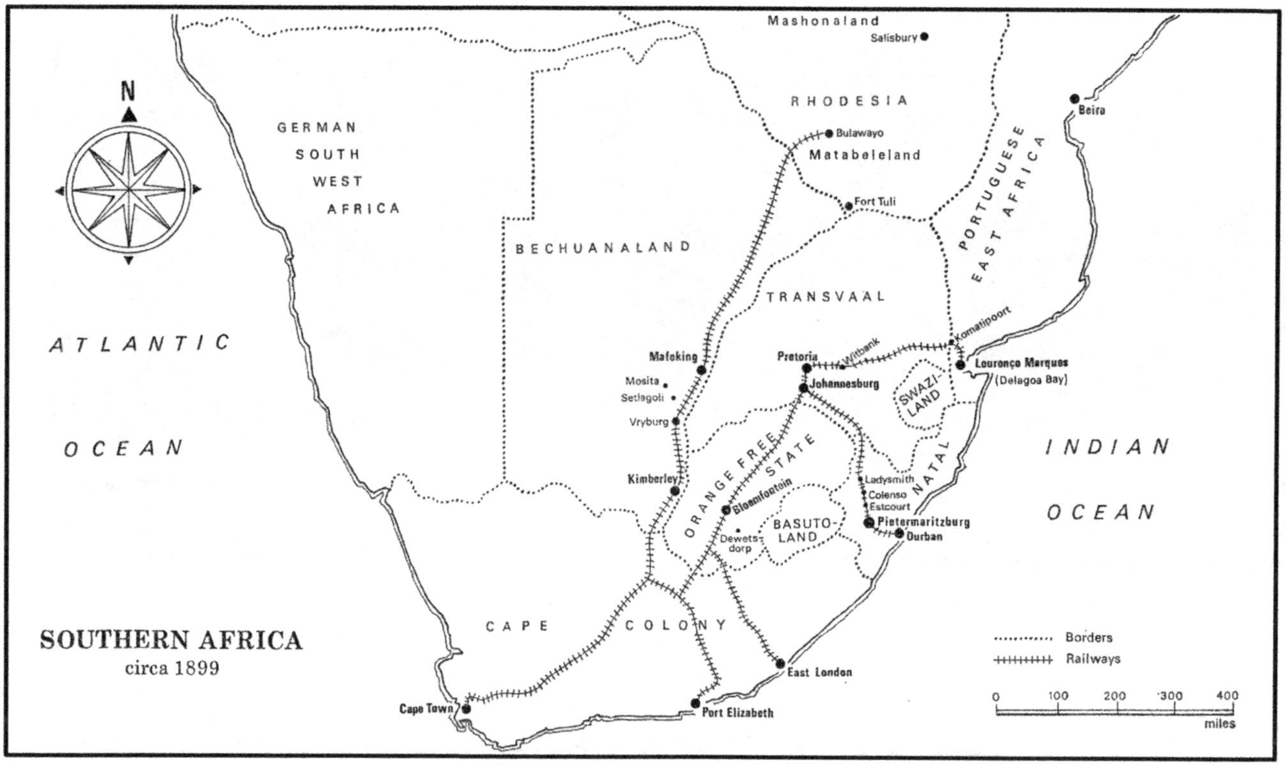

South Africa Railroad Network circa 1899.[16]

Topographical Map of South Africa

Topographical Map of South Africa showing the High Veld[17] of the Kaapvaal Craton.[18]

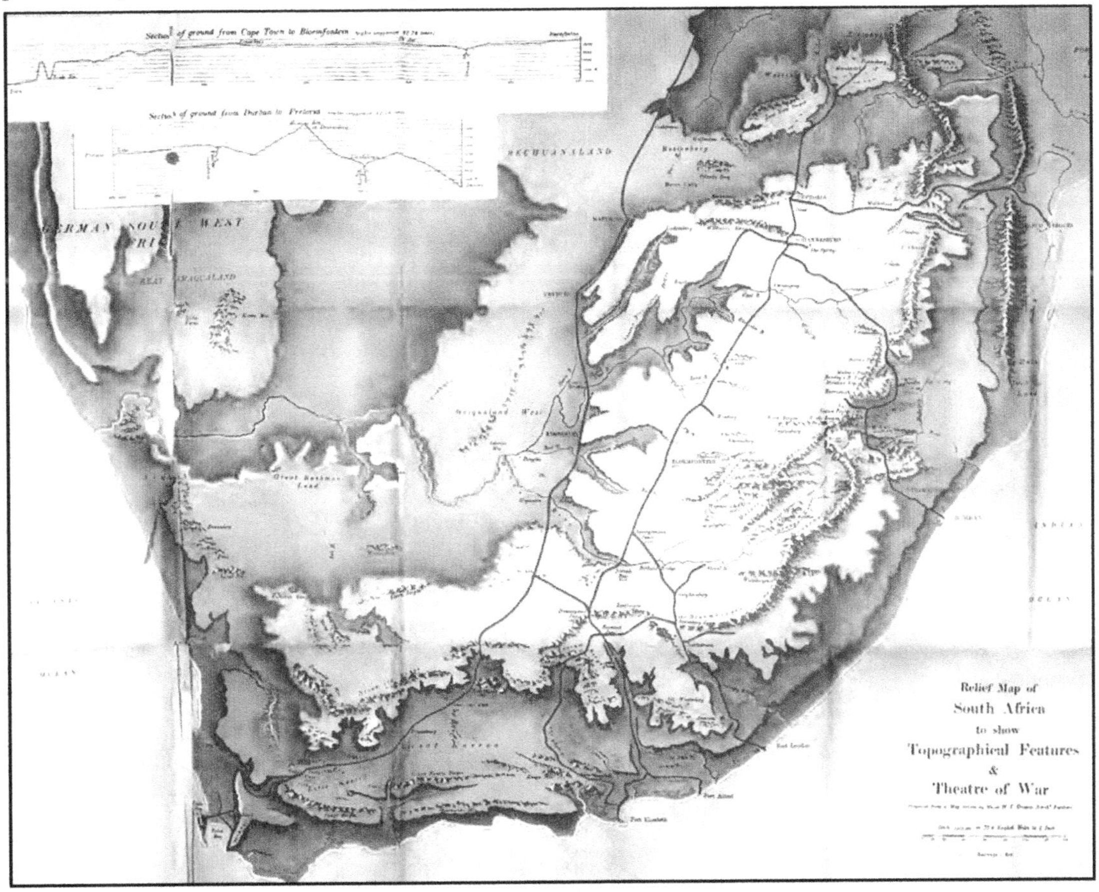

Topographical Map of South Africa.[19] Birch, 1899.

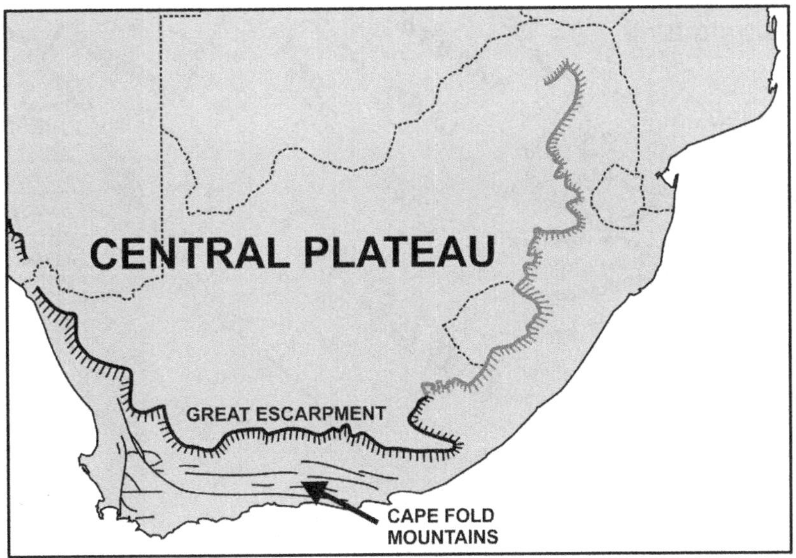

Great Escarpment.[20]

An Example of the Great Escarpment. Giants Castle in the Drakensberg Range.[21]

Theatre of Conflict

Theatre of Conflict.[22]

Northern Natal 1

Northern Natal[23] Note Volksrust and Majuba at top.

Northern Natal Detail #2

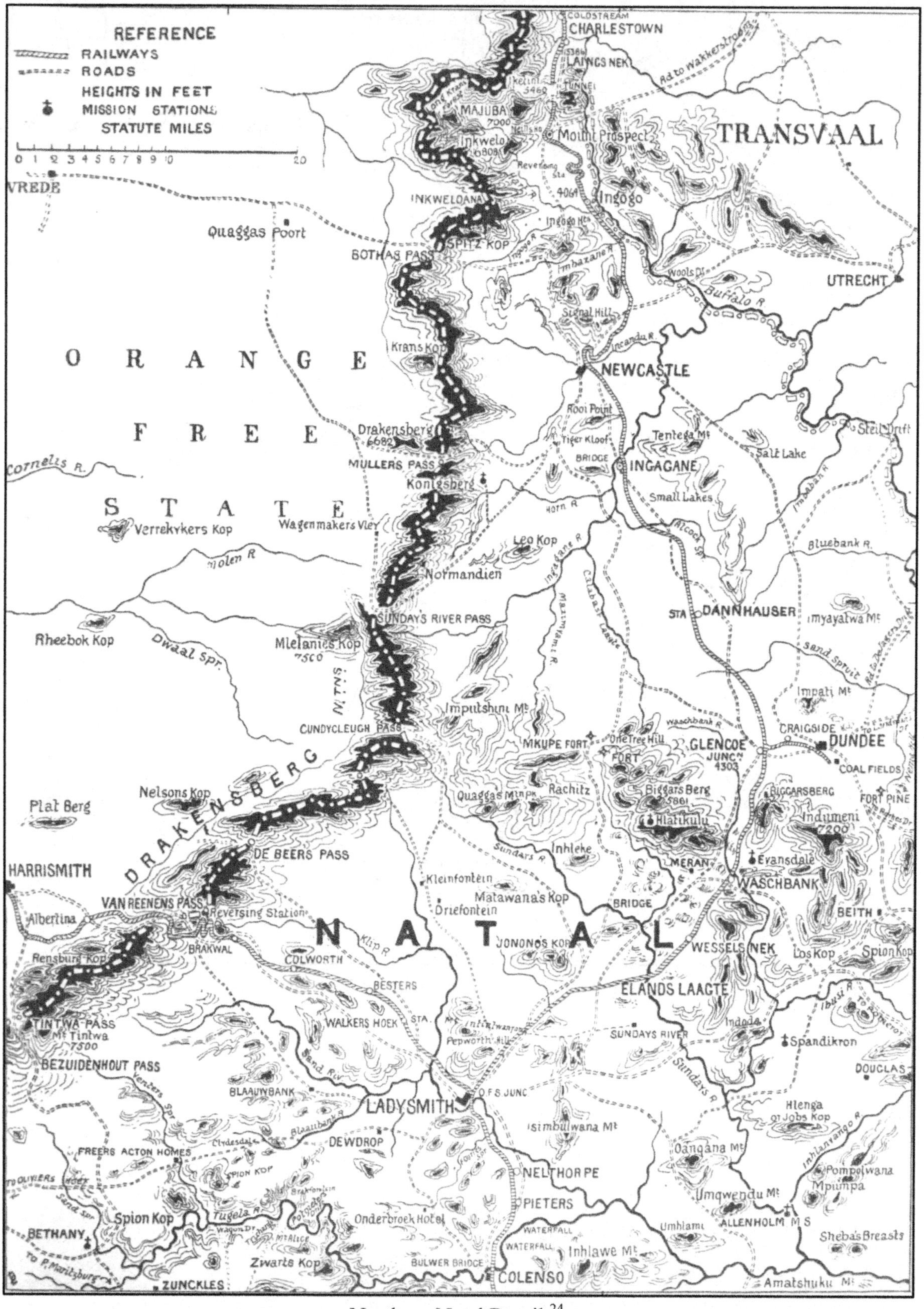

Northern Natal Detail.[24]

Northern Natal 3

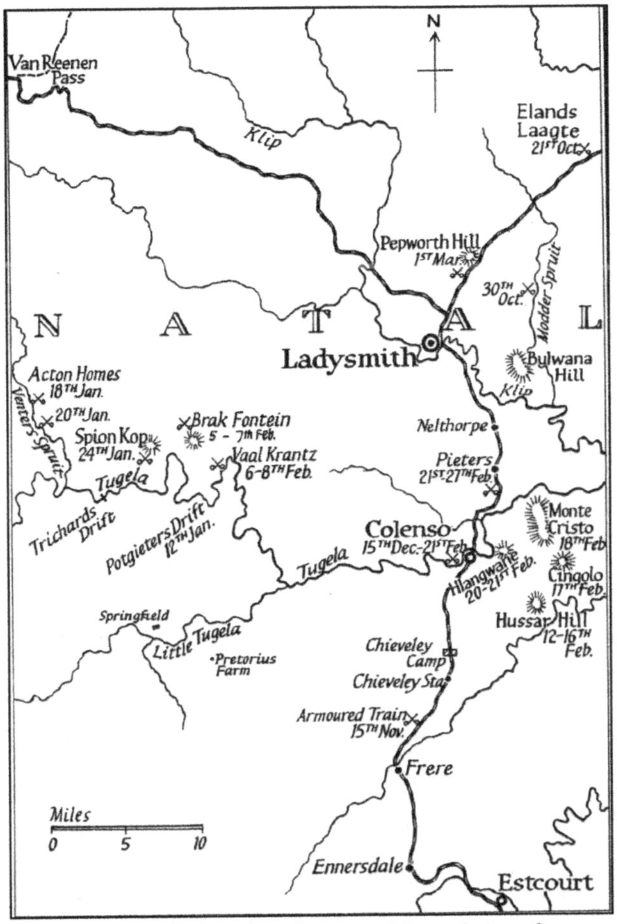

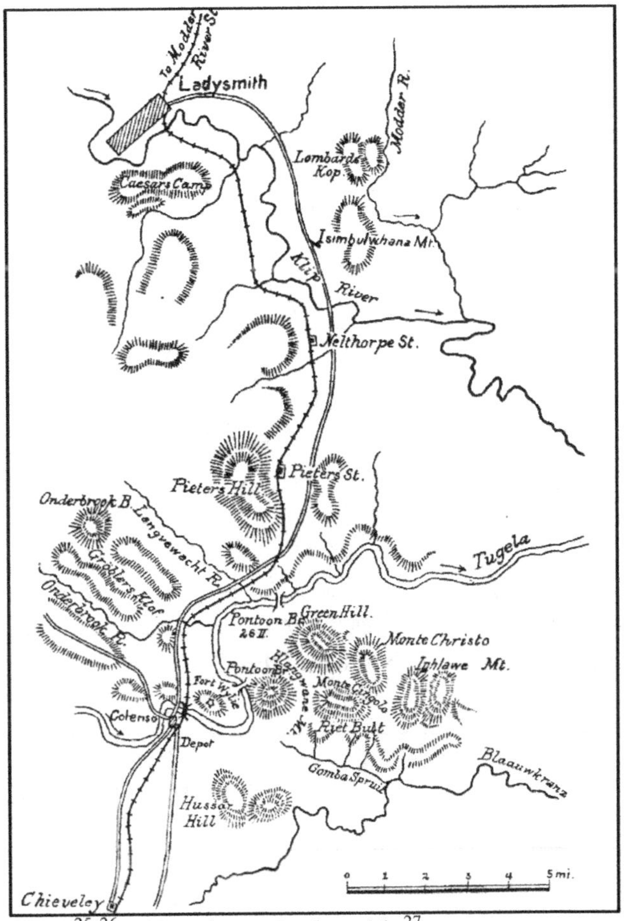

Northern Natal showing dates and locations of engagements.[25 26] Chieveley to Ladysmith.[27]

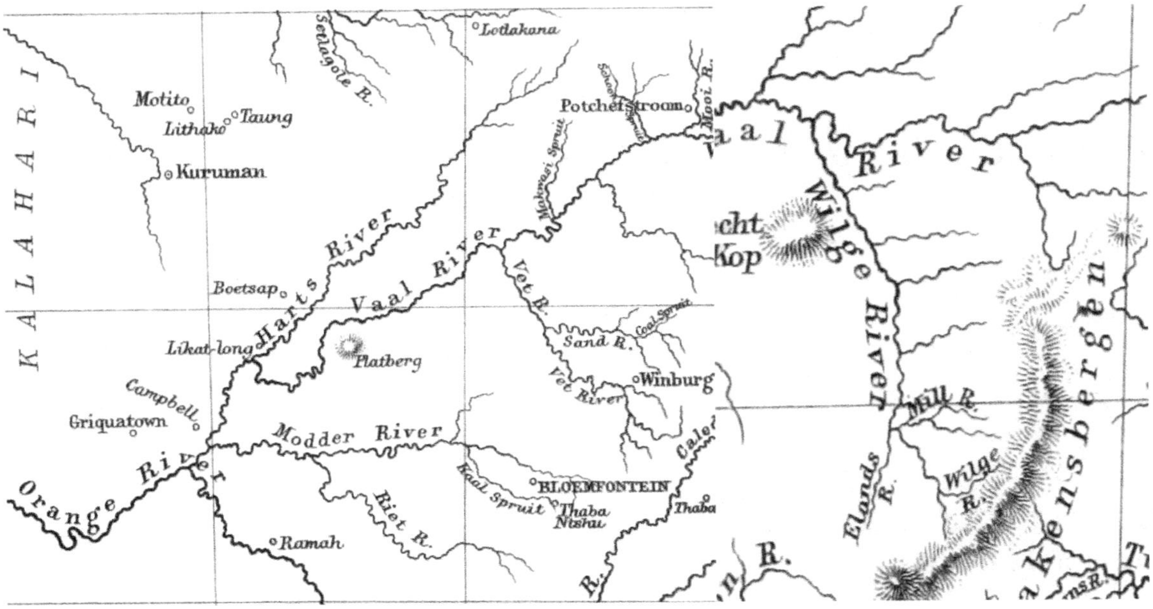

River System of the High Veld. A Composite of.[28 29].

Orange Free State Maps

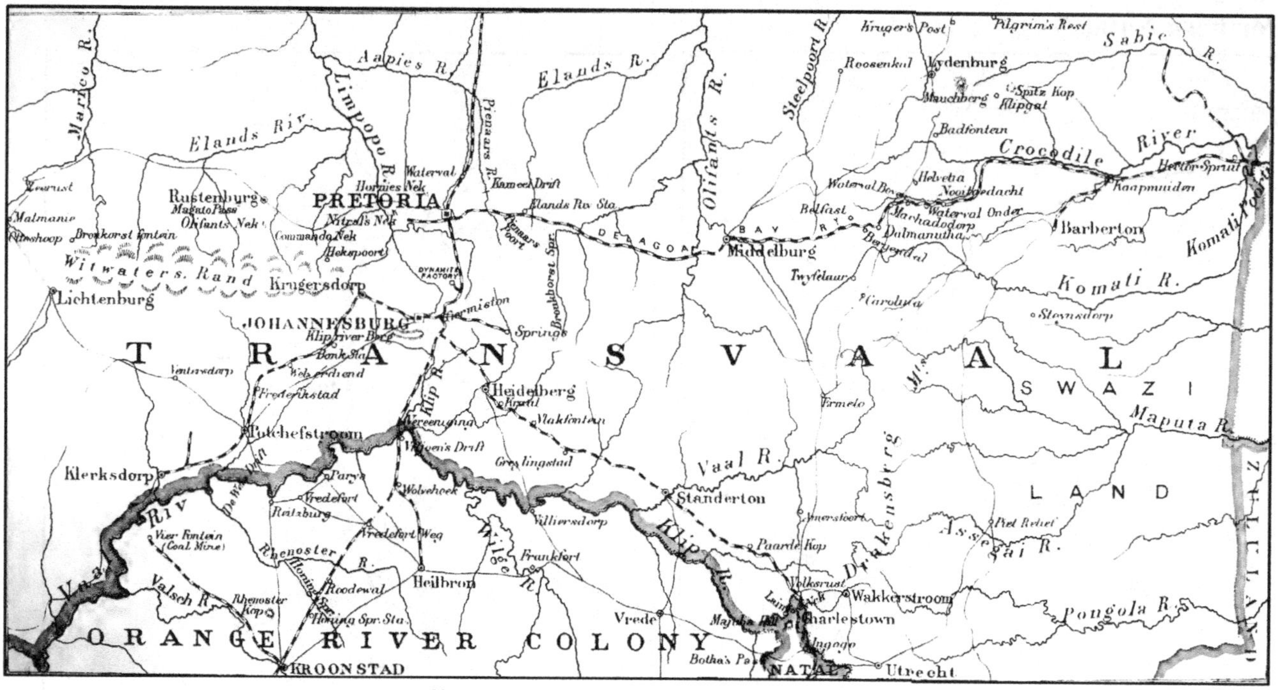

Southern Transvaal & Northern OFS.[30]

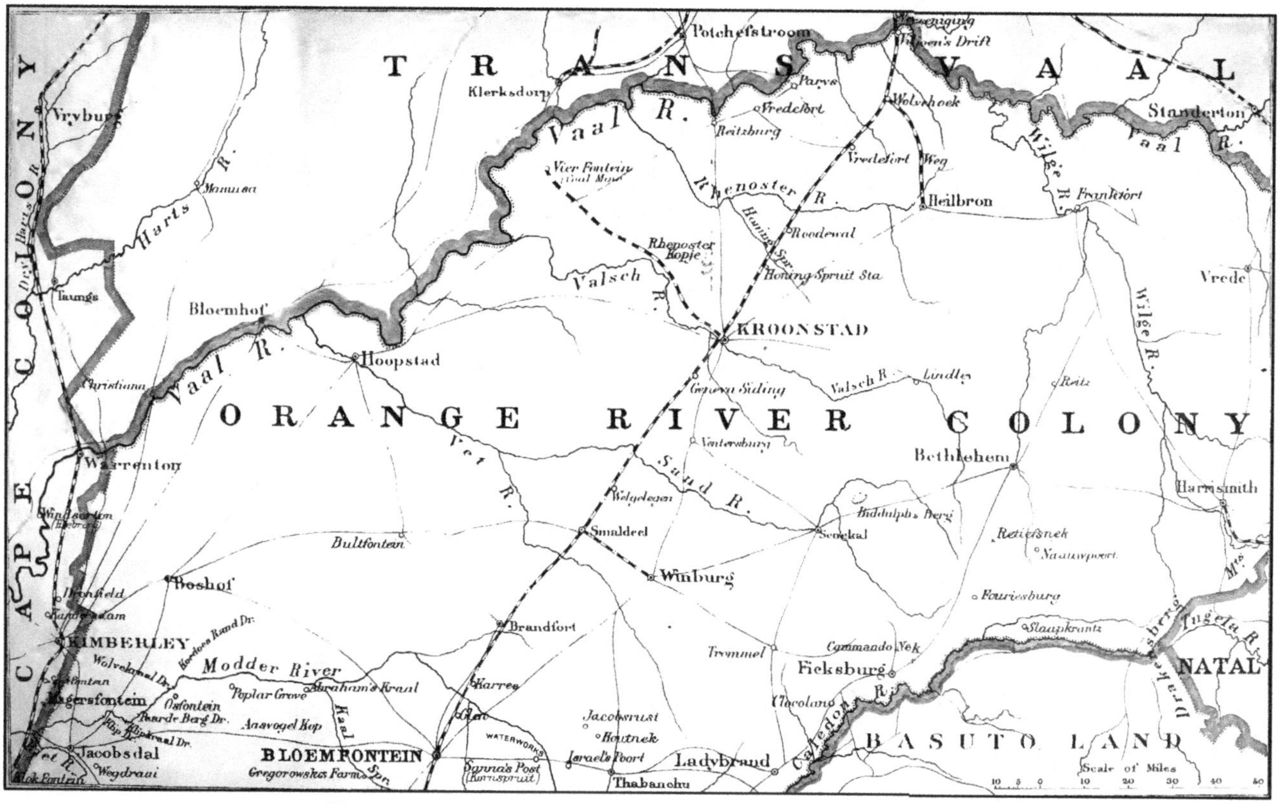

Orange River Colony North.[31]

Orange River Colony South.[32]

Central Orange Free State

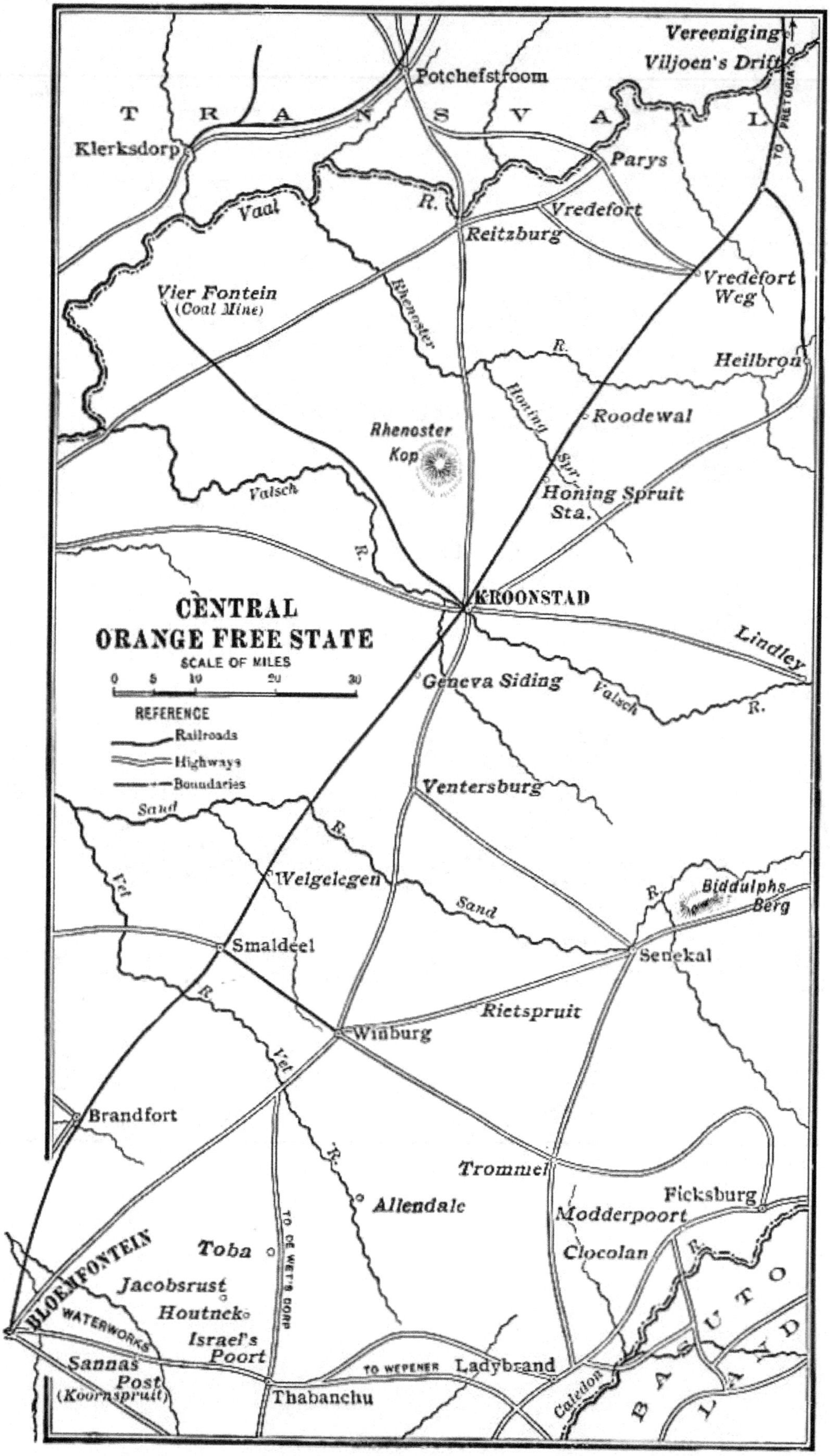

Central Orange Free State.[33]

Northeastern Cape Colony

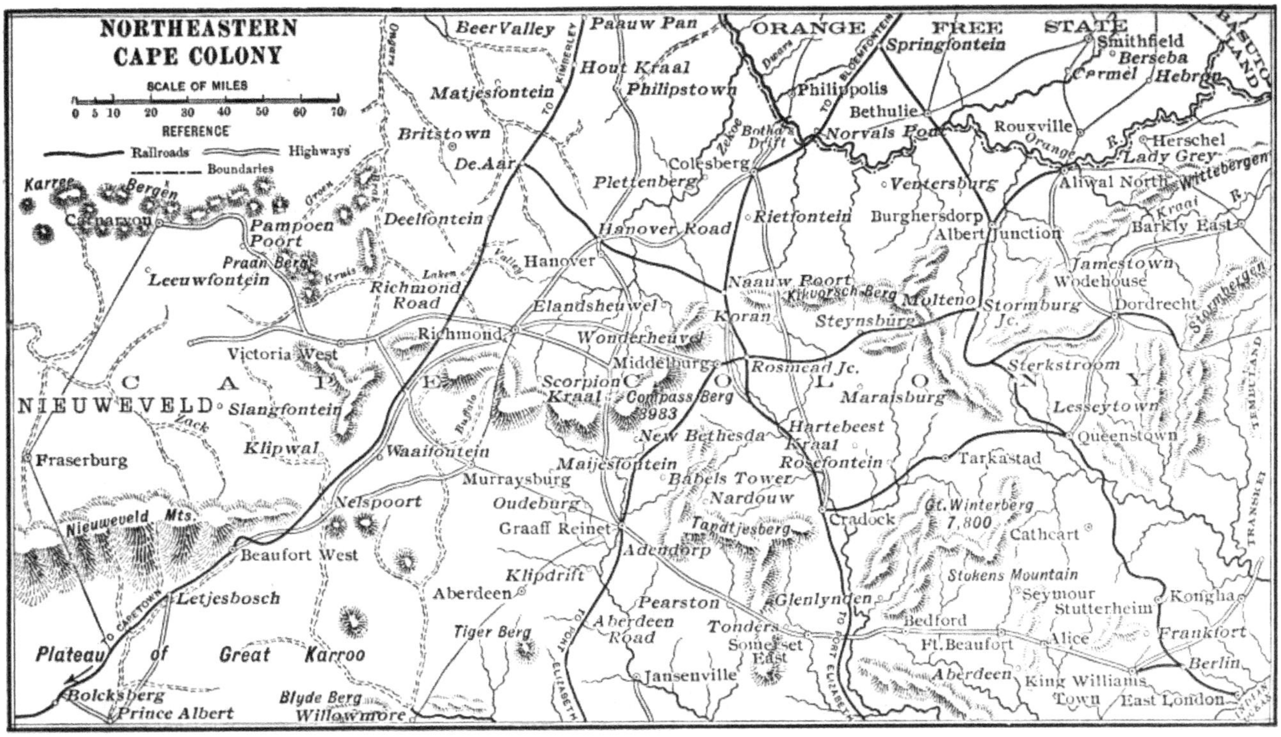

Northeastern Cape Colony.[34]

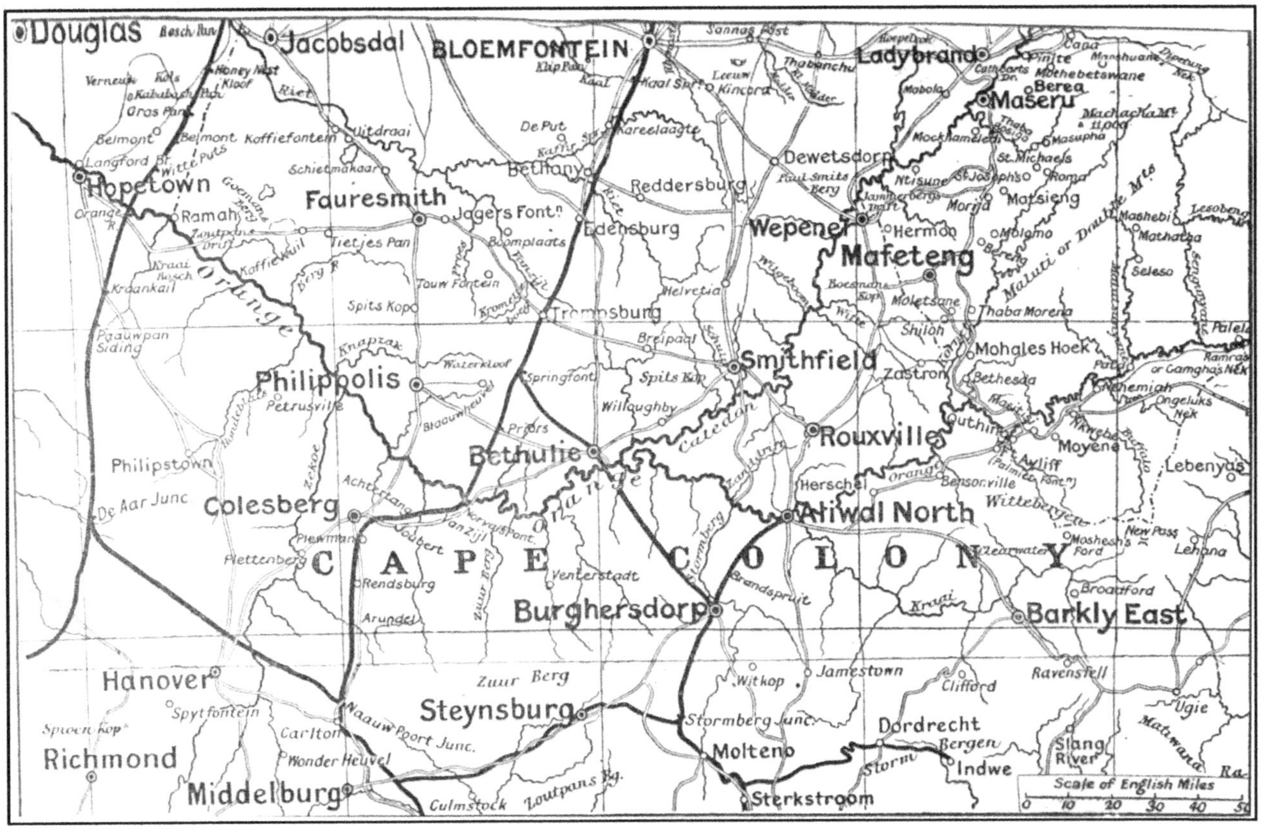

The Lines of Route of Methuen, Clements, Gatacre And Brabant.[35]

Modder and Reit Rivers

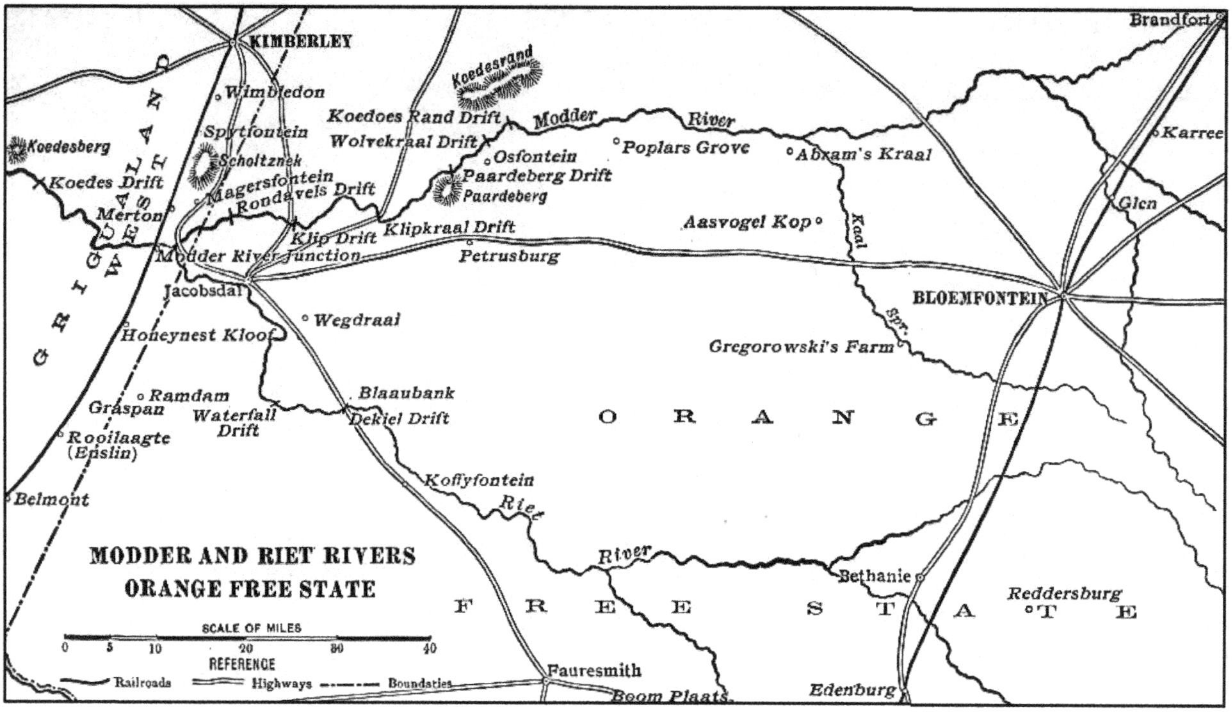

Modder and Reit Rivers.[36]

Commando Map

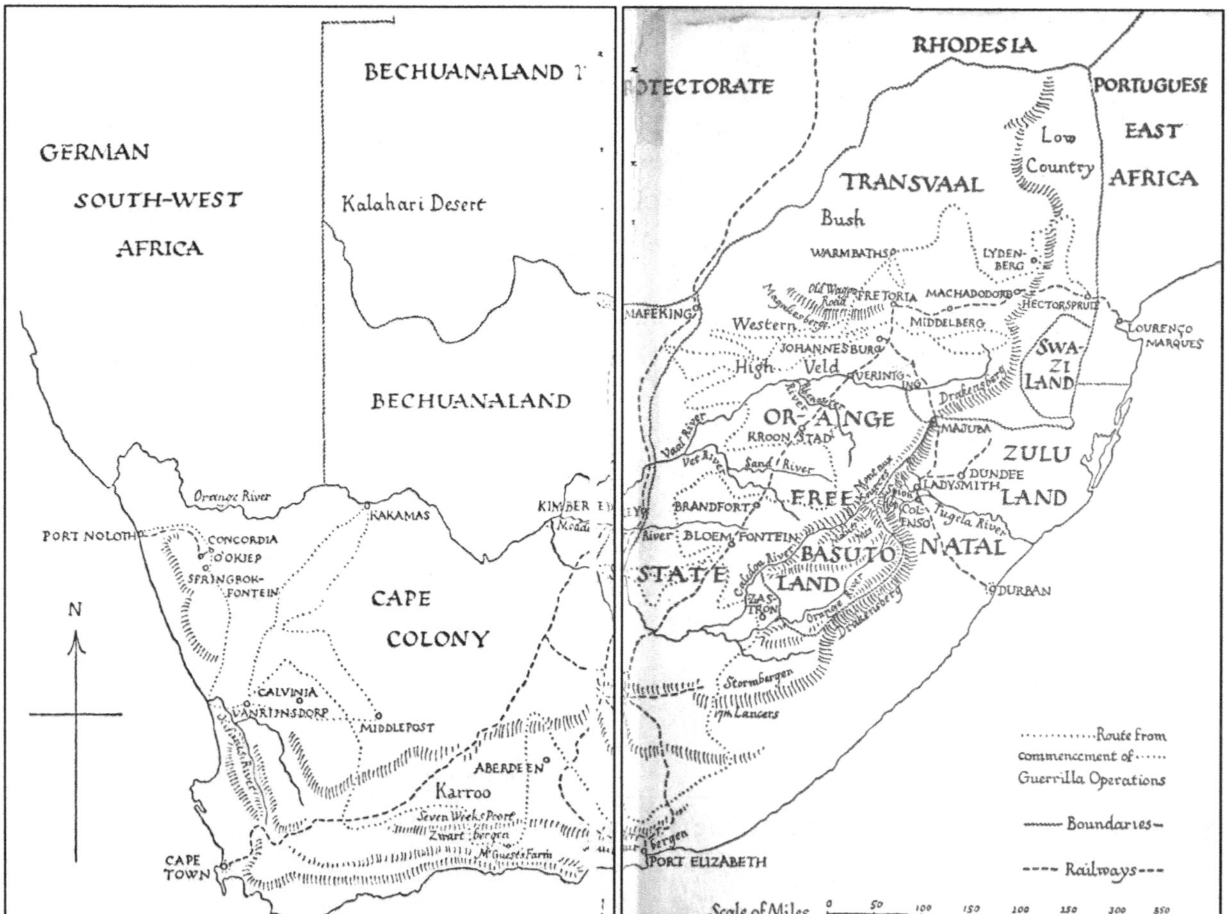

Map from C. Boni NY Publication of Commando in 1930.[37]

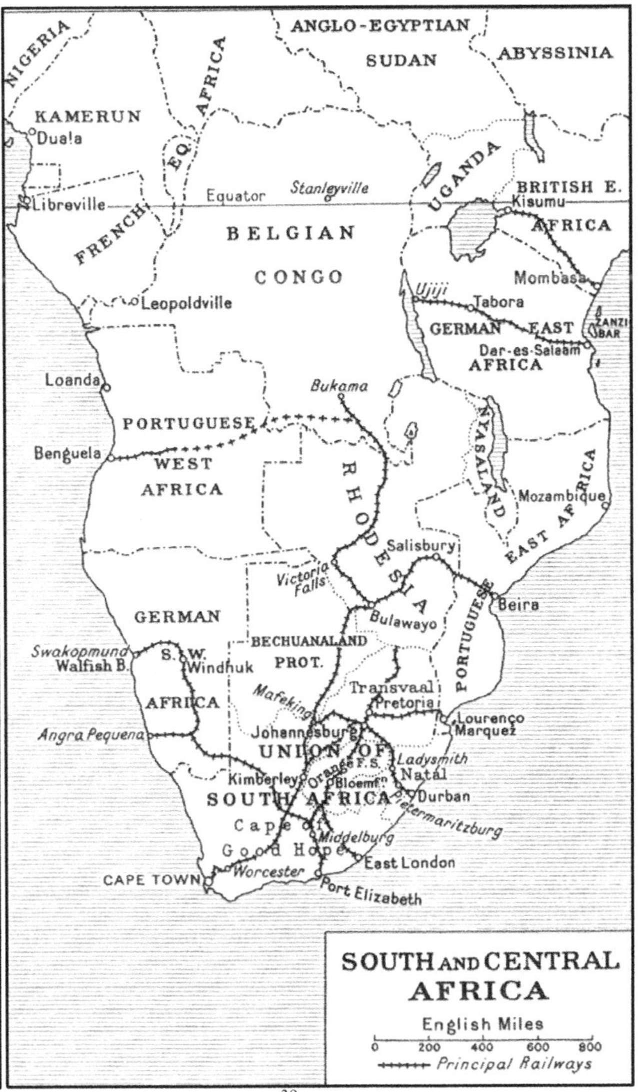

South West Africa.[38]

South & Central Africa.[39]

Dedication

This Book Is Dedicated To My Father Francis William Reitz
The Only Living President
of The Old South African Republics

Victrix causa
Diis placuit
Sed victa Catoni[6] [7]

Francis William Reitz, 1890.[40]

F. W. Reitz, Secretary of State of the Transvaal.[41]

[6] The victorious cause pleased the gods, but the conquered cause pleased Cato.

[7] PF: I believe he is referring to Cato The Younger. Marcus Porcius Cato (95 BC – April 46 BC), was an influential conservative Roman senator during the late Republic. A staunch advocate for liberty and the preservation of the Republic's principles, he dedicated himself to protecting the traditional Roman values he believed were in decline. A noted orator and a follower of Stoicism, his scrupulous honesty and professed respect for tradition gave him a political following which he mobilised against powerful generals of his day, including Julius Caesar and Pompey. https://en.wikipedia.org/wiki/Cato_the_Younger

Jan Smuts' Preface

When Colonel Reitz asked me to write a preface to his book of Boer War memories, I at first hesitated, as I feared that I might be introducing a book in which myself figured in some degree. However, on reading the manuscript, I find that I am only casually mentioned here and there, and have therefore no reason not to comply with his request.

It is a pleasure and a privilege to introduce this book to the reading public. To me it is a wonderful book—wonderful in its simplicity and realism, its calm intensity and absorbing human interest. Here is the book of the Boer War for which I have been waiting for the last twenty-five years and more. Many military books have been written on the Boer War— books full of interest and of valuable material for the future historian; but something else was wanted. The Boer War was other than most wars. It was a vast tragedy in the life of a people, whose human interest far surpassed its military value. A book was wanted which would give us some insight into the human side of this epic struggle between the smallest and the greatest of peoples.

Here we have it at last. There is no strategy and little tactics in this plain unvarnished tale. Wars pass, but the human soul endures; the interest is not so much in the war as in the human experience behind it. This book tells the simple straightforward story of what the Boer War meant to one participant in it. Colonel Reitz entered the war as a stripling of seventeen years, fought right through it to the end, and immediately after its conclusion wrote down these memories. Of military adventures there is of course full measure. He passed through as varied a record of exciting experiences as have ever fallen to the lot of a young man.

Indeed much of what is written in this book with such boyish simplicity may appear to the reader well-nigh incredible. But it is a true story, and the facts are often understated rather than exaggerated. The exciting incidents, the hairbreadth escapes, the daredevilry are literally true, and the dangers he passed through and courted are such as to make his unvarnished record read like one of pure romance. But there is here more than a record of war adventure. We have not only an unforgettable picture of mobile guerrilla warfare, but also an accurate description of life among the Boer forces. It is given, not in an abstract generalized form, but as the actual experience of one particular individual. As we read, we follow a true personal story which is often stranger than fiction.

The interest of the story deepens as it moves on from the heavy fighting in the Natal campaign under Botha, through the guerrilla warfare under De la Rey in the Western Transvaal, to the climax of marching and privation under me in the Cape Colony. The intimate picture gives us the inner truth of the war. We see how human beings react under the most terrible stresses to the passion of patriotism. We see how, under the influence of an ideal—in this case the ideal of freedom—the most ordinary human material rings true and rises superior to all danger and suffering and privation. And the effect is all the more striking because the story is so simple and unadorned and objective.

This book gives a wonderful personal record. But its wonder does not end with the book. The Boer boy who wrote this book was in the Wordsworthian sense the father of the man of after years. Let me add a few details to bring the account up to date. The boy left the country as an irreconcilable after the conclusion of the Boer War, as he and his family chose not to live under the British flag.

He drifted to Madagascar, where these memories were written in the intervals of malaria and transport riding.

There a letter from my wife[8] found him, urged him to return and pointed out to him that he was no better than her husband, and if the latter could afford to serve his people under the Union Jack surely his young friend could do the same.

The shaft went home; Reitz returned and was nursed back by her to health and peace of mind. He learnt to see Botha's great vision of a united South African people to whom the memories of the Boer War would mean no longer bitterness but only the richness and the inspiration of a spiritual experience. The loyalty of the Boer boy ripened into the broader loyalty of the South African. And I remember a night on the outbreak of the rebellion in

[8] Sybella (Isabelle or Isie) Margaretha Smuts (born Krige, Klein Libertas, Stellenbosch, Cape Colony, 22 December 1870 – farm Doornkloof, Irene, Transvaal, 25 February 1954) was the wife of the Union of South Africa's second Prime Minister, Gen. Jan Smuts, and a benefactor. Isie Smuts - Wikipedia

1914,[9] when Reitz once more appeared before me, this time a fugitive, not from the British, but from his own people in the Free State who had gone into rebellion.

Jan and Isie Smuts with one of their daughters, Santa.[42]

Such tricks does high fate play upon us poor humans. He did his duty in helping to suppress the rebellion, and thereafter he served on my staff in the German West campaign, just as he had done in the Boer War; in the German East campaign he rose to command a mounted regiment, and in the later stages of the Great War he commanded the First Royal Scots Fusiliers, one of the oldest regiments in the British Army. He was severely wounded early in 1918, but returned to France in time to lead his battalion in the fierce battles that closed the great drama and after the Armistice he led his men to the Rhine.

Since the War, he has taken an active part in the public life of his country. He has been a Cabinet Minister and still is a Member of Parliament in which capacity he is serving under me as loyally as he did in the sterner days of which he writes. This book is a romance of truth; but behind it is a greater personal romance, and behind that again is the even more wonderful romance of South Africa, to whom much should be forgiven for the splendour of her record during a period as difficult as any young nation has ever passed through.

J. C. Smuts

Pretoria

16th August, 1929

[9] The Maritz rebellion, was an armed insurrection in South Africa in 1914, at the start of World War I. It was led by Boers who supported the re-establishment of the South African Republic in the Transvaal. The rebellion failed, and its ringleaders received heavy fines and terms of imprisonment. Maritz rebellion - Wikipedia

Gen. Jan Christian Smuts and his cabinet in 1923. (All Left to Right) Front: Sir Thomas Watt, Sen. F.S. Malan, Smuts, Sir Thomas Smartt and Mr. Henry Burton. Back: Sen. Nicolaas de Wet, Col. Deneys Reitz, Sir Patrick Duncan, Mr. Deneys Reitz, John William Jagger and Col. Hendrik Mentz.[43]

Deneys Reitz in 1923.

Field Marshall Jan Smuts in 1942[44]

I. Memory's Tower

We lived in the Orange Free State.

My father was Chief Justice in Sir John Brand's[10] time and subsequently, in 1887, was himself elected President of the Republic.[11]

Our home was at Bloemfontein, the State capital, and here my brothers and I grew up. There were five of us, two older and two younger than myself, and we led a pleasant Tom sawyer like existence such as falls to the lot of few boys nowadays.

General view of Bloemfontein.[45]

Bloemfontein, the Capital of the Orange Free State. Population is 8,000.[46] The Presidency at Bloemfontein.[47]

We learned to ride, shoot, and swim almost as soon as we could walk, and there was a string of hardy Basuto ponies in the stables, on which we were often away for weeks at a time, riding over the game-covered plains by day, and sleeping under the stars at night, hunting, fishing and camping to our heart's content, and clattering home again when we had had our fill.

Sometimes my father took us with him on his long tours into the remoter districts, where there was more hunting and more camping, and great wapenshaws,[12] held by the Boer commandos to do him honour. Our small country was a model one. There were no political parties, nor until after the Jameson Raid of 1895 was there any bad blood between the Dutch and the English. We had no railways, and the noise of the outside world reached

[10] Sir Johannes Henricus Brand, GCMG (commonly known as Jan Brand and sometimes as John Henry Brand) (6 December 1823, Cape Town – 14 July 1888, Bloemfontein) was a South African lawyer and politician and the fourth State President of the Orange Free State, from 2 February 1864 until his death in 1888. His motto was "Everything will be all right if everyone does his duty." Jan Brand - Wikipedia Also English Wikipedia: Johannes Brand - Wikipedia

[11] Francis William Reitz, Jr. Francis William Reitz - Wikipedia Francis William Reitz, Jr. (5 October 1844 – 27 March 1934) was a South African lawyer, politician, statesman, publicist, and poet who was a member of parliament of the Cape Colony, Chief Justice and fifth State President of the Orange Free State, State Secretary of the South African Republic at the time of the Second Boer War, and the first president of the Senate of the Union of South Africa.

[12] "Weapons Show" A volunteer meeting or shooting competition. And An exhibition of weapons, according to the rank of the individual, by all persons bearing arms; formerly made at certain seasons in each district.

us but faintly, so that in our quiet way we were a contented community, isolated hundreds of miles from the seaboard.

The Market Square, Bloemfontein.[48]

Bloemfontein Post Office.[49]

Bloemfontein From The South.[50]

In 1894, when I was twelve years old, we were taken to Europe. It was a wonderful experience for inland-bred boys' to journey to the Coast, to cross the ocean in a ship, and to see the great crowds and cities of the old world. We went first of all to England, where we stayed for a while in London, marvelling at the things we saw. Thence to Amsterdam to visit the senior branch of our family, that had remained in Holland when our ancestors emigrated to South Africa long ago. The head of the old stock lived in a house on the Heerengracht; a wealthy man apparently, for he kept many servants and had fine paintings on his walls.

As our republic had taken its name from the House of Orange, my father was well received by the little Queen of the Netherlands,[13] and the Court and people made much of us. Next, we travelled to Paris to meet Casimir-Perier,[14] the newly elected President of the French. He took us to lay a wreath on the grave of Sadi Carnot,[15] his Predecessor, lately assassinated by an anarchist at Lyons. From there we went to Brussels to see Monsieur Jesslein, our Consul. His house stood in the rue de la Blanchisserie, and he told us it was the one in which the Duchess of Richmond had given her famous ball on the eve of Waterloo. We were presented to King Leopold,[16] an old man with a hooked nose and a long white beard, who extended only his little finger in greeting, perhaps because we belonged to a republic.

[13] Wilhelmina of the Netherlands - Wikipedia

[14] Jean Paul Pierre Casimir-Perier (French: (8 November 1847 – 11 March 1907) was a French politician who served as President of France from 1894 to 1895. Jean Casimir-Perier - Wikipedia

[15] Marie François Sadi Carnot (11 August 1837 – 25 June 1894) was a French statesman, who served as the President of France from 1887 until his assassination in 1894. Sadi Carnot (statesman) - Wikipedia

[16] Leopold II (9 April 1835 – 17 December 1909) was the second King of the Belgians from 1865 to 1909 and the self-made autocratic ruler of the Congo Free State from 1885 to 1908. Leopold II of Belgium - Wikipedia

From Belgium we went to Hamburg to take ship across the North Sea to Edinburgh, and from there to visit the Cathcarts[17] at Auchindrayne on the River Doon.[18] My father had studied law in Scotland and my grandfather before him had studied agriculture, and they had both spent much time at Auchindrayne, so my father wished his sons in turn to carry on the tradition of friendship which for clearly a hundred years linked the two families.

My grandfather [Francis Sr.] first went to Scotland in 1816. He met Walter Scott,[19] to whom he took a lion skin which the poet Thomas Pringle[20] had sent from Cape town, and he became intimate with the great writer. In later days in South Africa, he loved to tell of their meetings and of the banquet at which he was present when Scott for the first time admitted that he was the author of the Waverley Novels.[21] Both my grandfather and my father had returned to South Africa with a deep love of Scotland and Scotch literature, and at our home scarcely a night passed without a reading from Burns[22] or Scott, so that we felt as if we were among our own people.

From Auchindrayne we went to London to meet Sir George Grey,[23] who, as Governor of the Cape, had been a friend of my father many years before. My father used to say that if the English had sent more men like him to South Africa our history would have been a happier one, and although I was only a boy, and Sir George Grey a very old man, he made a deep impression upon me—a something of inward beauty not easily described, but which I have not yet forgotten.

From London we sailed for South Africa. On our return my brothers and I were received by our less fortunate playfellows like pilgrims safely returned from Mecca, so hazardous an undertaking did our journey seem to them in those days.

We took up our old carefree life once more, all unaware of the storm that was brewing between the white races in the Transvaal.

The Jameson Raid had not yet brought matters to a head, but there was trouble in the air. President Kruger[24] and the Commandant-General Piet Joubert,[25] came frequently to Bloemfontein on official visits to my father, and we eagerly questioned them and listened to their stories of hunting and of the wars against the natives and the British long ago.

Sir Henry Loch,[26] Governor of the Cape, also visited us, as did Cecil Rhodes,[27] a big florid man who cracked jokes with us boys, but on whose political aims my father looked askance. These two tried to prevent the Free State from entering into an alliance with the Transvaal, but they did not succeed, and a treaty was made with President Kruger wherein we agreed to stand by the Transvaal in case of war with England, a promise which the Free State loyally fulfilled.

My brothers and I did not understand the import of all this coming and going of noted men, and life ran on pleasantly enough, until in 1895 my father's health failed and he had to resign. We went to live at Claremont, a cramped suburb of Cape town, greatly missing our horses and the freedom of our wide Northern uplands.

[17] Earl Cathcart - Wikipedia

[18] The River Doon is a river in Ayrshire, Scotland. Its course is generally north-westerly, passing near to the town of Dalmellington, and through the villages of Patna, Dalrymple, and Alloway, birthplace of Robert Burns. River Doon - Wikipedia

[19] Walter Scott - Wikipedia

[20] Thomas Pringle (5 January 1789 – 5 December 1834) was a Scottish writer, poet and abolitionist. Known as the father of South African poetry, he was the first successful English language poet and author to describe South Africa's scenery, native peoples, and living conditions. Thomas Pringle - Wikipedia

[21] The Waverley Novels are a long series of novels by Sir Walter Scott (1771–1832). For nearly a century, they were among the most popular and widely read novels in Europe. Because Scott did not publicly acknowledge authorship until 1827, the series takes its name from Waverley, the first novel of the series, released in 1814. The later books bore the words "by the author of Waverley" on their title pages. Waverley Novels - Wikipedia And: Walter Scott - Wikipedia

[22] Robert Burns - Wikipedia

[23] Sir George Grey,(14 April 1812 – 19 September 1898) was a British soldier, explorer, colonial administrator and writer. He served in a succession of governing positions. Governor of South Australia, twice Governor of New Zealand, Governor of Cape Colony, and the 11th premier of New Zealand. He played a key role in the colonisation of New Zealand, and both the purchase and annexation of Māori land. George Grey - Wikipedia

[24] Paul Kruger - Wikipedia

[25] Piet Joubert - Wikipedia

[26] Henry Loch - Wikipedia

[27] Cecil Rhodes - Wikipedia

Panorama of Cape Town.[51]

Cape Town—Seat of Government, Cape Colony.[52]

Cape Town, Devils Peak, Table Mountain and Lion's Head from Table Bay.[53]

Cape Town docks.[54]

Birds Eye view of the Cape Town Docks, 1899.[55]

When my father recovered from his long illness [~1896–1897] we settled in the Transvaal, where he soon became Secretary of State under-President Paul Kruger. My eldest brother,[28] aged nineteen, was now sent to Europe to study law, and after a while the rest of us were put back to school at Bloemfontein until the middle of 1899.

During our absence at the Cape the ill-fated Jameson Raid had taken place, and we found on our return that feeling was running high between the English and the Dutch; and even in the Free State, where differences of this kind had hitherto been unknown, there was so much ill will that people openly talked of driving the English into the sea, whereas previously we had not given these matters a thought.

By July (1899) the situation had become so serious that my father ordered us up to Pretoria, as war with England seemed inevitable. We said goodbye to Bloemfontein the town where we had been born and bred and where we had spent such happy days, and journeyed north, leaving behind us the peace of boyhood, to face years of hardship, danger, and ultimate exile.

[28] Named Hjalmar, after a Norwegian uncle, Hjalmar Thesen his mother's brother. See infra, and: Hjalmar Reitz - Wikipedia

II. On the Brink

When we reached Pretoria, affairs were moving to a climax. Peremptory notes had been exchanged between the Transvaal and the British Government, and excitement was rising as each cable and its reply was published. Already the Transvaal capital was an armed camp. Batteries of artillery paraded the streets, commandos from the country districts rode through the town almost daily, bound for the Natal border, and the crack of rifles echoed from the surrounding hills where hundreds of men were having target practice. Crowded trains left for the coast with refugees flying from the coming storm, and business was at a standstill.

"A party of Boers at target practice."[56]

"A typical group of Burghers ready for Commando."[57]

"A Boer Commando passing through the streets of Johannesburg."[58]

A Burgher's farewell to his family, titled "Goodbye Daddy."[59]

Boers departing for the front.[60]

Looking back, I think that war was inevitable. I have no doubt that the British Government had made up its mind to force the issue, and was the chief culprit, but the Transvaalers were also spoiling for a fight, and, from what I saw in Pretoria during the few weeks that preceded the ultimatum I feel sure that the Boers would in any case have insisted on a rupture.

Transvaal Government Buildings and Dutch Reformed Church.[61] Supply Preparations—Despacthing Stores from Pretoria, October 1899.[62]

I myself had no hatred of the British people; from my father's side I come of Dutch and French Huguenot blood, whilst my mother (dead for many years) was a pure-bred Norwegian from the North Cape, so one race was much like another to me. Yet, as a South African, one had to fight for one's country, and for the rest I did not concern myself overmuch with the merits or demerits of the quarrel. I looked on the prospect of war and adventure with the eyes of youth, seeing only the glamour, but knowing nothing of the horror and the misery.

I was seventeen years old and thus too young to be enrolled as a burgher. President Kruger himself solved this difficulty for me. One morning when I was at the Government buildings, I met him and my father in the corridor and I told the President that the Field-Cornet's office had refused to enroll me for active service. The old man looked me up and down for a moment and growled: 'Piet Joubert says the English are three to one Sal jij mij drie rooi-nekke lever?' (Will you stand me good for three of them?) I answered boldly: 'President, if I get close enough I'm good for three with one shot.' He gave a hoarse chuckle at my youthful conceit and, turning to my father, asked how old I was. When he heard my age he said: 'Well then, Mr. State Secretary, the boy must go—I started fighting earlier than that,' and he took me straight to the Commandant-General's room close by, where Piet Joubert in person handed me a new Mauser carbine, and a bandolier of ammunition, with which I returned home pleased and proud.

General Joubert.[63] Mrs. Joubert.[64]

I saw a good deal of the President in these days as I used to go with my father to his house on the outskirts of the town, where they discussed State matters while I sat listening. The President had an uncouth, surly manner, and he was the ugliest man I have ever seen, but he had a strong rugged personality which impressed all with whom he came in contact. He was religious to a degree, and on Sundays he preached in the queer little Dopper[29] church he had built across the street, where I sometimes heard him.

President Kruger at his home.[65] President Kruger on his Veranda.[66]

There was Mrs. Kruger too, whom I often saw with her pails in the yard, for she kept dairy cows and sold milk to the neighbours. Once she brought us coffee while we were looking at a picture of the statue of her husband that was being set up on Church Square. The President was shown dressed like an elder of the Church in a top hat, and the old lady suggested that the hat should be hollowed out and filled with water, to serve as a drinking-fountain for the birds. My father and I laughed heartily on our way home at her simplicity, but we agreed that it was decent of her to have thought of such a thing.

Four generations of the Kruger family.[67] Mrs. Paul Kruger, Mrs. F.C. Eloff, Mrs. Louis Jacobz, and Baby Jacobz. President and Mrs. Kruger in their cottage in Pretoria, April, 1900.[68]

I also knew Piet Joubert, the Commandant-General, for, apart from his visits to Bloemfontein, his son Jan and I were friends, and I sometimes went home with him to talk about the coming war, and his father was generally

[29] Dopper: A member of the Reformed Churches in South Africa, a Calvinistic denomination.

there. He was a kindly, well-meaning old man who had done useful service in the smaller campaigns of the past, but he gave me the impression of being bewildered at the heavy responsibility now resting upon him, and I felt that he was unequal to the burden.

One afternoon he showed me a cable which he had received from a Russian society offering to equip an ambulance in case of war, and when I expressed my pleasure I was astonished to hear him say that he had refused the gift. He said: 'You see, my boy, we Boers don't hold with these new-fangled ideas; our herbal remedies (bossie-middels) are good enough.' Another time, when describing the festivities at the opening of the Delagoa Bay railway line, which he had attended as Commander-in-Chief, he told me that when the Portuguese paraded a thousand troops in his honour, he had gone down the ranks shaking hands with every one of the soldiers. I liked him very much personally, and to me he was always kind and fatherly, but I felt that he was unfit to lead armies, and it is a great pity that a younger man was not appointed in his place on the outbreak of the war.

And now the days were speeding by and in September of 1899 matters had come to such a pass that British troops were moving up to the western borders of the Transvaal and Free State, and other troops were on the water, while large Boer forces were mobilising on the various fronts. Committees and deputations from the Cape travelled up to make eleventh-hour attempts to avert the catastrophe of war, but it was clear that the die was cast and that neither side was in a mood for further parleying.

My eldest brother (named Hjalmar, after a Norwegian uncle Hjalmar Thesen) was away in Europe studying law, and my father had already cabled to him to return. My next brother, Joubert, named after the Commandant-General, was a year older than myself, and although he, too, was ineligible for burgher-rights, he intended volunteering for service, but the two younger ones were put back to school.

Joubert and I had made our preparations long before. Our horses were in good fettle and our saddlebags packed. My brother had a fine upstanding chestnut, and I had a strong little Basuto pony, and we were eager to be off.

Many of the country districts had been called up, but thus far no Pretoria men had gone forward. At last, on September fifth, the first batch from the town was ordered to entrain for the Natal border. The moment we heard of this we took our rifles, fetched our horses from the stable, within ten minutes had saddled up and mounted.

We said goodbye to our stepmother and her children, for my father had remarried years before, and rode up through the town to the Raadzaal to take leave of him.

Raadzaal, in Pretoria.[69]

Palace of Justice in Pretoria.[70]

We found him closeted with the President and members of the Executive Council, but we went in and, when we explained why we had come, all rose to shake us by the hand. The old President gave us a solemn blessing, and my father, who had not expected this sudden departure, bade us goodbye in a husky voice and said he knew we would do our best.

From the Government buildings we galloped to the station, where we found a great stir. Hundreds of friends and relatives- had come to see the contingent leave, but, in spite of the crowd on the platform and the loading of baggage and batteries, we were able to truck our animals after which we lent a hand with the stowing of the ammunition and other work.

When all was ready the train pulled out to the sound of the Transvaal National Anthem. There were enthusiastic cheers and the waving of hats and umbrellas by those remaining behind, and we were off to the front in good earnest.

Citizens of the Transvaal.[71]

Departure of Pretoria Commando for active service, October 1899.[72]

Commando with Johannesburg in the background Group of Mounted Boers probably with Johannesburg in the background.[73]

As for my brother and myself, we were not Transvaal burghers, nor had we been called out for service, but we automatically became soldiers of the Boer Army by virtue of having thrown our belongings through a carriage window and clambering aboard, little knowing on how long and difficult a trail this light-hearted enlistment was starting us.

Burgomaster Schutte of Johannesburg Witnessing the Departure of the Burghers: 4 November 1899.[74]

Loading Up Ammunition For The Transvaal Artillery.[75]

"The above picture represents a Train -load of the Wives of Boers who are still in Arms against us, and who have been Deported from the Towns we Occupy and sent by us to Rejoin their Belligerent Husbands. The Women are, apparently, as Hostile as the Men, and can Display their Rancour without Fear. It will be noticed that one has hoisted the Transvaal Boer Flag."[76] Also: And: Civil Prisoners from Dundee entering Pretoria, November 10th, 1899.[77]

III. To the Frontier

Our officer, or Field-Cornet, as he was called, was Mr. Zeederberg,[30] a coach contractor, and the rank and file were mostly young fellows from the Civil Service and the legal offices and shops in the town. Few of them had ever seen war, or undergone military training, but they were full of ardour, and in spite of cramped quarters and rough fare, we were like schoolboys as we clanked along.

After a monotonous journey of three days, often broken by interminable halts, we reached Saildspruit, a small station about ten miles front the Natal border, where we detrained. There were great numbers of burghers from the country districts already encamped on the plain, on either side of the railway line, and the veld on all sides was dotted with tents and wagon-laagers. On the left of the track stood a large marquee over which floated the vierkleter flag of the Transvaal, indicating General Joubert's headquarters. Both he and his wife were thus early on the scene, it being her invariable custom to accompany her husband in the field.

When we had detrained our horses, and helped to ground the guns, we moved away to where a halting-ground was assigned us. We off-saddled in the tall grass, and after building fires, and preparing supper, spent our first night in the open. For the next ten days we lay here enjoying the novelty of our surroundings, as if we were on a pleasure jaunt, rather than seriously awaiting the coming of war. One evening my brother [Joubert] and I received a pleasant surprise, for there arrived in camp an old native servant of ours, grinning from ear to ear at having found us. His name was Charley, a grandson of the famous Basuto chief, Moshesh.[31] He had been a family retainer ever since I can remember, first in the Free State and then in the Transvaal whither he had followed us. Latterly he had been on a visit to Umbandine,[32] King of the Swazis, but, learning that there was to be a war, he returned at once to Pretoria, and my father sent him on to us. He was more than welcome, for we could now turn over to him our cooking and the care of the horses, duties which we had been performing ourselves up to then; moreover, he had brought me a splendid roan[33] which my father had sent me, as he feared that the Basuto pony[34] would not be up to my weight.

A Basuto-Pony.[78]

Basuto Pony & Rider.[79]

Every morning my brother and I had our horses fetched from the grazing-ground and rode out to visit neighbouring camps and laagers, eager to see all that we could. We saw the stream of fresh contingents arriving daily by rail, or riding in from the adjacent countryside, and watched with never-ending interest the long columns of shaggy men on shaggy horses passing by.

[30] Probably Related to: Christiaan Hendrik (Doel) Zeederberg, b7c8 (1860 - 1907) - Genealogy (geni.com) No ABW reference found.

[31] Moshoeshoe I (c. 1776 – 11 March 1870) At the age of 34 Moshoeshoe formed his own clan and became a chief. He subsequently became the first King of Lesotho from 1822 to 1870. Moshoeshoe I - Wikipedia

[32] Either: Mbandzeni - Wikipedia or Ngwane V - Wikipedia

[33] Especially of a horse, having a coat of a dark base color with individual white hairs mixed in.

[34] The Basuto is a pony breed from Lesotho and South Africa. The Basuto is considered a small horse, since it possesses horse-like characteristics, such as an exceptionally long stride. Basutos have, long necks and long, straight backs, and straight shoulders. They have very tough legs and sound, very hard hooves. They can be up to 14.2 hands high (56.8 inches or ~142 cm). Basutos can be chestnut, brown, bay, gray or black. They are sure-footed, fast, fearless and are known for their stamina. Basuto pony - Wikipedia

At the end of the week there must have been nearly 15,000 horsemen collected here, ready to invade Natal, and we told ourselves that nothing could stop us from reaching the sea.

Our military organization was a rough one. Each commando was divided into two or more field-cornetcies, and these again were subdivided into corporalships. A field-cornetcy was supposed to contain 150 to 200 men, and a corporalship nominally consisted of 25, but there was no fixed rule about this, and a popular field-cornet or corporal might have twice as many men as an unpopular one, for a burgher could elect which officer he wished to serve under, and could even choose his own commando, although generally he would belong to one representing the town or district from which he came.

In the Pretoria commando, we divided ourselves into corporalships by a kind of selective process, friends from the same Government department or from the same part of the town pooling their resources in the way of cooking utensils, etc., and in this manner creating separate little groups that in course of time came to be recognized as military units. One of the number would be elected corporal, to act as the channel through which orders were transmitted from above, and much the same system held in all the other commandos. The commissariat arrangements were equally simple. Our Field-Cornet would know the approximate member of men under his command, and in order to maintain supplies all he needed to do was to send a party to the food depot, stacked beside the railway line, where they would break out as many bags of meal, sugar, and coffee as they considered necessary, load them on a wagon, and dump them in the middle of the camp for each corporalship to satisfy its requirements. The meat supply consisted of an immense herd of cattle on the hoof, from which every commando drew as many animals as wanted for slaughter purposes. This system, though somewhat wasteful, worked fairly well; the men were plainly but adequately fed on much the same diet as they were accustomed to at home, and there was little grumbling. Officers and men had to supply their own horses, rifles, clothing, and equipment, and nobody received any pay.

Ever since the Jameson Raid the Transvaal Government had been importing large quantities of Mauser rifles from Germany, which were sold to the burghers at a nominal figure, and as great stores of ammunition had likewise been accumulated, the commandos were very efficiently equipped.

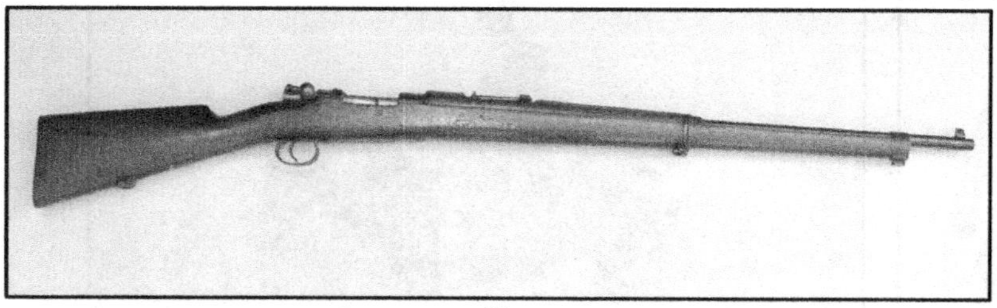

Mauser 1895 Bolt Action Mauser.[80]

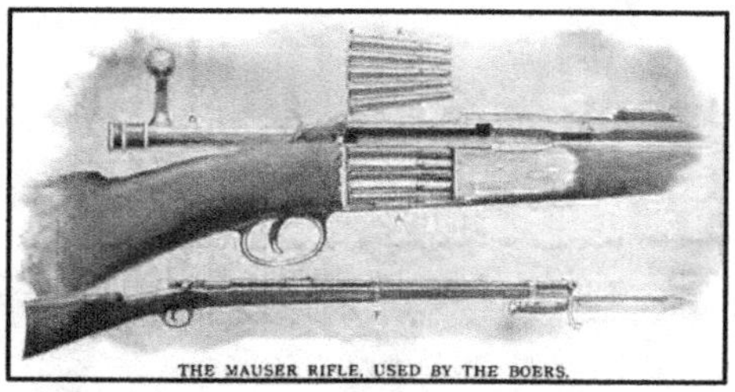

Mauser Rifle Detail.[81]

The two republics had mobilized between 60,000 and 70,000 horsemen, at this moment distributed west and east, ready to invade the Cape Colony and Natal at the given word. This great force, armed with modern weapons, was a formidable fighting machine which, had it been better led, might have made far other history than it did.

How many troops the British had in South Africa I do not know, but they were pouring reinforcements into the country, and I think our leaders underestimated the magnitude of the task on which they were embarked.

So far as our information went in regard to Natal, the nearest British troops lay at the town of Dundee, some fifty miles away. This force we subsequently found to be about 7,000 strong, and still farther south at Ladysmith they had another 6,000 or 7,000 men, but with fresh troops being landed every day it was difficult to say how soon the scales would dip against us.

On the 10th of October a great parade was held in honour of President Kruger's birthday. We mustered what was then probably the largest body of mounted men ever seen in South Africa. It was magnificent to see commando after commando file past the Commandant-General, each man brandishing hat or rifle according to his individual idea of a military salute. After the march-past we formed in mass, and galloped cheering up the slopes where Piet Joubert sat his horse beneath an embroidered banner. When we came to a halt he addressed us from the saddle. I was jammed among the horsemen, so could not get close enough to hear what he was saying, but soon word was passed that an ultimatum (written and signed by my father) had been sent to the British, giving them twenty-four hours in which to withdraw their troops from the borders of the Republic, failing which there was to be war.

The excitement that followed was immense. The great throng stood in its stirrups and shouted itself hoarse, and it was not until long after the Commandant-General and his retinue had fought their way through the crowd that the commandos began to disperse.

The jubilation continued far into the night, and as we sat around our fires discussing the coming struggle, we heard singing and shouting from the neighbouring camps until cock-crow.

Next day England accepted the challenge and the war began. Once more the excitement was unbounded. Fiery speeches were made, and General Joubert was received with tumultuous cheering as he rode through to address the men. Orders were issued for all commandos to be in readiness, and five days' rations of biltong[35] and meal were issued. Flying columns were to invade Natal, and all transport was to be left behind, so my brother and I were obliged to send our native boy to the central laager where the wagons were being parked until they would follow later.

My brother and I had joined hands with some friends from our Pretoria suburb of Sunnyside, and after a few days we had become merged ill a larger body, of which five brothers, named Malherbe, were the leading spirits. We chose Isaac Malherbe,[36] the eldest of them, to be our Corporal, and a better man I never met. We soon came to be known as 'Isaac Malherbe's Corporalship.' He was about thirty-five years old, dark complexioned, silent and moody, but we looked up to him because of the confidence which he inspired. His brothers were brave men, too, but he stood head and shoulders above us all. After his death on the Tugela[37] we found that he was a man of considerable means whose wife and two small daughters were left well provided for.

War was officially declared on October 11th. At dawn on the morning of the 12th, the assembled commandos moved off and we started on our first march.

As far as the eye could see the plain was alive with horsemen, guns, and cattle, all steadily going forward to the frontier. The scene was a stirring one, and I shall never forget riding to war with that great host. It has all ended in disaster, and I am writing this in a strange country,[38] but the memory of those first days will ever remain.

A Boer Encampment.[82]

Boer Camp at Elandslaagte.[83]

[35] Biltong is a form of dried, cured meat that originated in Southern African countries. It is related to beef jerky in that they are both spiced, dried meats. Biltong - Wikipedia

[36] Malherbe Isaac Diederick Johannes Prisoner number: 21182 Captured: Vaalkop 08 Jun 1901 Sent to: Bermuda, Darrell's Island Age: 16 Address: Leeuwfontein Source: Anglo Boer War Museum 2016 Anglo Boer War - Name search

[37] The vicinity of the Tugela river is where a great deal of combat will soon take place.

[38] Madagascar, where he is in exile in 1903.

IV. We Invade Natal

We reached the border village of Volksrust before noon, and here the entire force was halted for the day, the Pretoria men camping beside the monument erected to commemorate the Battle of Majuba, fought on the mountain nearby in 1881.

Camp of the Pretoria Commando. Majuba Hill is in the background.[84]

Arrival of a commando at Majuba Hill.[85]

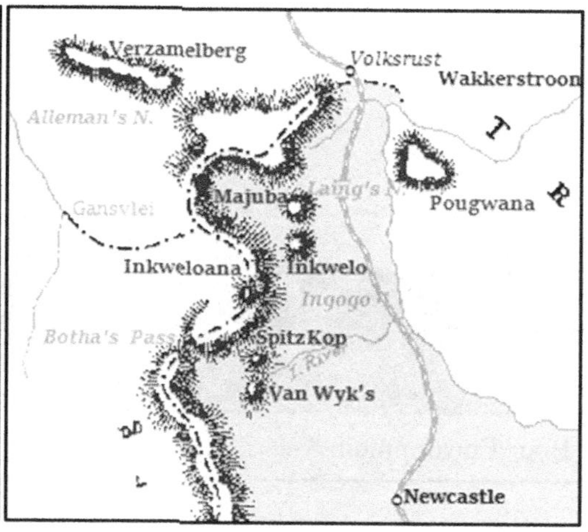

Northern Natal. Note Volksrust and Majuba at the top.[86]

They were tall, swarthy men, clad in black claw-hammer coats, and semi-top-hats, trimmed with crepe, a style of dress and headgear affected by so many Boer officers as virtually to amount to insignia of rank. General Maroola[39] had got his name during a recent native campaign in the Northern Transvaal, in the course of which he was said to have directed operations from behind a maroola[40] tree, while Swart Lawaai's was a tribute to his dark complexion and quarrelsome temper.

Several other forces, more or less equal to ours, were carved out of the main body, and in the afternoon each of these new commandos was assigned its route. We spent an unhappy night in the rain. We had neither tents nor over-coats, so we sat on ant-heaps, or lay in the mud, snatching what sleep we could. It was our first introduction to the real hardships of war, and our martial feelings were considerably damped by the time the downpour ceased at daybreak. When it was light we moved out, shivering and hungry, for it was too wet to build fires.

[39] Deneys mentions this person 25 times and yet never mentions his real name. It turns out it is Erasmus and his brother.

[40] The marula is a medium-large tree native to the miombo woodlands of southern Africa. Marula - Wikipedia

General Lucas Meyer[41] General Louis Botha and Commandant Daniel Jacobus Elardus Erasmus.[87] [42] Men of the Pretoria Commando.[88]

Our road lay between high mountains, and the rain came down again in torrents. Far away to our right and left we caught an occasional glimpse of other forces marching through the mist, also making slow progress over the heavy Country. We did not cross the border, but kept to a parallel road, and by dark we halted at a dismal spot, soon trampled into a quagmire by the multitude of horses. Again it rained all night, and again we could light no fires, and had to appease our hunger by munching biltong from our saddlebags.

It was a severe first test, for, in addition to the rain, a cold wind blew from the great Drakensberg range, cutting through us. Fortunately past troubles are soon forgotten, and when, towards sunrise, the weather lifted and we got under way, we were in good spirits again, with no thought of the unhappy night behind us.

Giants Castle in the Drakensberg Range.[89]

After a long ride we emerged into open country and there, winding across the plain ran the Buffalo River[43] with the green hills and pleasant valleys of Natal stretching beyond. With one accord the long files of horsemen reined in, and we gazed silently on the land of promise. General Maroola, with a quick eye to the occasion, faced round and made a speech telling us that Natal was a heritage filched from our forefathers, which must now be recovered from the usurper. Amid enthusiastic cries we began to ford the stream. It took nearly an hour for all to cross, and during this time the cheering and singing of the 'Volkslied'[44] were continuous, and we rode into the smiling land of Natal full of hope and courage.

[41] Lucas Johannes Meyer (19 November 1846 – 8 August 1902), was a Boer general, member of the Transvaal government and president of the Nieuwe Republiek.Lucas Johannes Meyer - Wikipedia

[42] Daniel Jacobus Elardus Erasmus (nickname Maroela, Pretoria, South African Republic, 4 March 1845 – Pretoria, 7 May 1914) was a Boer general during the Anglo Boer War (1899–1902). https://en.wikipedia.org/wiki/Daniel_Jacobus_Elardus_Erasmus
Daniel had two elder sisters and two younger brothers including Commander (Kommandant) Stephanus Petrus Erasmus (nicknamed Swart Lawaai, around 1850 – 1921).

[43] There are four Buffalo rivers in South Africa. This one is a tributary of the Tugela River. The source is the Majuba Mountain, the "mountain of pigeons" in Zulu. Buffalo River (KwaZulu-Natal) - Wikipedia

[44] The National Anthem of the Transvaal.

As soon as we were through the river, we spread out on a front of several miles, and went forward. Far away, on either side, we could see the other forces moving abreast of us. There was not a man who did not believe that we were heading straight for the coast, and it was as well that the future was hidden from us, and that we did not know how our strength and enthusiasm were to be frittered away in meaningless siege,[45] and in the holding of useless positions, when our only salvation lay in rapid advance. The nearest English soldiers were still many miles to the south, so that, beyond sending out vedettes,[46] the men went as they pleased- riding hither and thither to forage for supplies at the farm-houses we passed.

By nightfall we reached the hills near the village of Newcastle, and here we halted, our camp fires glowing far and near like the lights of a great city. Next morning we moved in columns through the streets, such of the inhabitants as sympathized with us waving encouragement, and the rest looking on in sullen resentment, as the long lines of horse-men went by. Beyond the village, a halt was called to allow the other forces to come abreast, and then, at dusk, all commandos started on a forced march that lasted throughout the night, except for an occasional halt to rest the horses.

By sunrise we had come within fifteen miles of the flat-topped mountain behind which lay the town of Dundee and the nearest English forces.

Contrary to expectation, we halted for the greater part of the day, and again, at sunset, were ordered to prepare for another night march. The commandos had by now closed in, and our fatal army was assembled practically in one body, making a brave show as we rode out. General Joubert had passed among us during the day, and we knew from him that he planned to surround the English troops that night. General Maroola, with the 1,500 men of who we of the Pretoria Commando[47] formed part, was to occupy the mountain overlooking the enemy's camp at Dundee while the other forces were to complete the pincers on the flanks and rear.

There was much excitement at the prospect of fighting, and even the heavy rain that set in after we started could not depress our spirits. The night was black and our route seemed to lie chiefly over an open mud-bound plain, varied at times by more broken country in the passage of which there was a good deal of confusion and intermingling between the different commandos; but for all that, steady progress was made, and towards dawn Maroola succeeded in disentangling his commando from the other columns.

Soon the frequent lightning revealed the steep side of a mountain, rising like a wall before us, and word was passed to commence the ascent, for this was the mountain from the top of which it was said one could look down on the English encampments on the other side. As there was almost certain to be a strong post holding the summit we climbed up in silence, expecting to be fired on at any moment, but when we reached the wide Plateau above we found it deserted.

This was so unlooked for that no one seemed to know what to do next, and, as it was still pitch dark, and the rain was coming down in torrents, we waited shivering in the cold for the coming of daybreak. When it grew light the rain ceased, but a mist enshrouded the mountain-top through which everything looked so ghostly and uncertain that we felt more at a loss than ever, and when Maroola was asked for orders he merely stood glowering into the fog without reply. We could not see fifty yards in any direction, but we knew that the English lines were immediately below us, for we could hear muffled shouts and the rumble of wagons, and we expected to be led down the face of the mountain to the attack. But General Maroola and his brother made no sign, and when President Kruger's son Caspar, who was serving with us as a private, and who for once in his life showed a little spirit, went up and implored them to march us to the enemy, Maroola curtly ordered him off.

He must have known what he was about though, for suddenly there came a violent cannonade, bringing us all to our feet as we listened to our first sound of battle.

We could see nothing, but heavy fighting had started close by, for the roar of the guns increased and at times we heard the rattle of small arms and Maxims.

None of the fire, however, was directed at us, and so far as we were concerned nothing happened, and we fretted at the thought of standing passively by when others were striking the first blow of the war. After perhaps an

[45] The Siege of Ladysmith. https://en.wikipedia.org/wiki/Siege_of_Ladysmith

[46] A patrol ship or lookout (person on guard) From veletta, influenced by vedere. vedetta - Wiktionary

[47] Membership List: **https://www.angloboerwar.com/?option=com_content&view=article&id=3061**

hour the sound died down, indicating, although we did not know it at the time, that the English had driven the Vryheid men[48],[Under General Lucas Meyer] from Talana Hill with heavy losses.

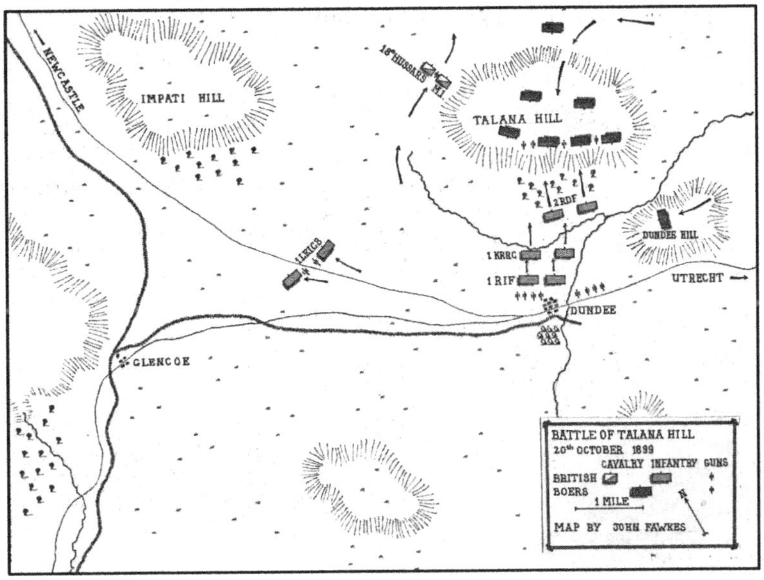

Battle of Talana Hill (Glencoe/Dundee)[90] October 20th, 1899. Erasmus is on Impate Mt.

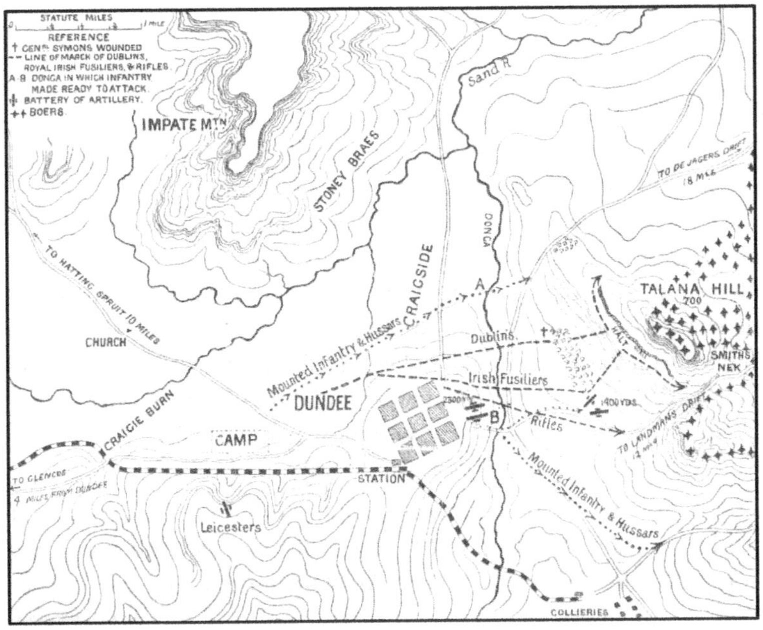

Battle of Talana Hill (Glencoe/Dundee).[91] Map with slightly more detail.

Towards midday the weather cleared somewhat, and while it still continued misty, patches of sunshine began to splash the plain behind us, across which we had approached the mountain overflight. And then, far down, into one of these sunlit spaces rode a troop of English horsemen about 300 strong.[49] [50] This was our first sight of the enemy, and we followed their course with close attention.

How this handful of men came to be right in the rear of due whole Boer Army I never heard, but they were on a desperate errand, for between them and their main body lay nearly 15,000 horsemen, and, now that the fog was lifting, their chance of regaining their base unobserved was gone. Already scattered Boer marksmen were appearing out of the mist, firing from the saddle as they came, and shepherding the soldiers still farther from their own people.

[48] Vryheid Commando Membership: https://www.angloboerwar.com/?option=com_content&view=article&id=3065

[49] At the beginning of the battle, anticipating victory in storming the hill, the cavalry had been detached to cut off the expected retreat. This failed as well because a large number were surrounded by a second force of 4,000 men under General Erasmus' men who arrived from another direction. They were late to the battle but just in time to capture the cavalry.

[50] This is Col. Colonel Moller who was sent to cut off the retreat of the Boer forces from Talana Hill.

Our men were by this time mostly crowding the forward edge of the mountain, hoping to catch sight of the English camp below, so that there were only a few of us who saw the troopers on the plain behind. Among these was our corporal, Isaac Malherbe, my brother [Joubert] and I, and five or six other Pretoria men, and, after watching the squadron below for a few seconds, we mounted our horses and rode down the mountain-side as fast as we could go. Arrived at the foot, we raced across the veld in the wake of the English troops, guided by the sound of dropping rifle-shots ahead of us, for we could no longer see our quarry, as they had disappeared for the time being among some low foot- hills. Following on, we soon came to the scene of action.

Boers watching the fight at Dundee.[92]

The English had gone to earth at a small homestead, and we were just in time to see the soldiers jumping from their horses, and running for cover to the walls of a stone cattle kraal, and among the rocks behind the farmhouse. Other burghers were flocking in, and soon the troops were completely surrounded. Across their front ran the dry bed of a spruit, and Isaac led us thither at once. This meant riding towards the enemy over the open, and now, for the first time in my life, I heard the sharp hiss of rifle-bullets about my ears, and for the first time I experienced the thrill of riding into action. My previous ideas of a battle had been different, for there was almost nothing to see here. The soldiers were hidden, and, except for an occasional helmet and the spurts of dust flicked up around us, there was nothing. We reached the spruit we were making for with one man wounded, and leaving him and our horses in the bed below, we climbed the bank and were soon blazing away our first shots in war.

The troops replied vigorously, but they were able to devote comparatively little attention to us, for by now the countryside was buzzing like an angry hive, with men arriving from every direction, and the end was but a question of time. After a few minutes a Creusot gun of the Transvaal Staats Artillery unlimbered and opened fire.

"Adjutant Roos and Gunners on the Tugela."[93] With 75 mm Creusot cannon. Boer Commandants reading a message from British officers after the fight at Dundee.[94]

The very first shell stampeded all the troop horses. The poor maddened brutes came tearing past us, and we leaped on our horses to head them off, but had to retreat to avoid being trampled down as they thundered by. I managed to hang on to the skirt of the mob, and, by seizing its flying reins, brought a fine black Waler[51] to a standstill. As I was looking over my prize I saw a white flag go up at the kraal and another from the farmhouse, so I hastened to be present at the surrender.

Australian light horsemen riding waler horses.[95] The first prisoners of war, captured near Dundee.[96]

By the time I got there the soldiers had thrown down their arms and were falling in under their officers. Their leader, Colonel Moller,[52] stood on the stoop looking pretty crestfallen, but the private soldiers seemed to take the turn of events more cheerfully. Officers and men were dressed in drab khaki uniforms, instead of the scarlet I had seen in England, and this somewhat disappointed me as it seemed to detract from the glamour of war; but worse still was the sight of the dead soldiers. These were the first men I had seen killed in anger; and their ashen faces and staring eyeballs came as a great shock, for I had pictured the dignity of death in battle, but I now saw that it was horrible to look upon.

Casualties In Action	Killed		Wounded		Missing and Prisoners			Forces Engaged		
	Officers	Men	Officers	Men	Officers	Men	Total	British	Boer	UK Loss
Talana Hill /Dundee	8	42	11	84	25	306	476	3,400	7,000	14.0%

Talana Hill / Dundee / Glencoe Casualties.[53]

[51] The Waler is an Australian breed of horse that was developed from horses that were brought to the Australian colonies in the 19th century. The name comes from their breeding origins in New South Wales; they were originally known as "New South Walers". https://en.wikipedia.org/wiki/Waler

[52] Bernhard Drysdale Moller Lieutenant Colonel QSA. Severe Wound, Van Wyks Vlei, 21 Aug 00. 18th (Victoria Mary, Prince of Wales's Own) Hussars. The regiment was at Glencoe under General Penn Symons when war broke out, and took part in the battle of Talana Hill (see 1st Leicestershire Regiment). In Sir G. White's dispatch of 2nd November 1899, para 12, he says: "Turning now to our cavalry, the 18th Hussars received orders at 5.40 am to get round the enemy's right flank and be ready to cut off his retreat. They were accompanied by a portion of the Mounted Infantry and a machine-gun. Making a wide turning movement, they gained the eastern side of Talana Hill. Here Lieutenant Colonel Moller halted with one squadron 18th Hussars, the machine-gun, and Mounted Infantry, sending his two other squadrons farther to the east. These two latter squadrons took part in the pursuit of the enemy, who retreated eastward, but Lieutenant Colonel Moller and the troops with him appear, so far as can be ascertained, to have pursued in a northerly direction, to have come in contact with superior forces not previously engaged, and to have been surrounded and forced to surrender while endeavouring to return to camp round the north of Impati Mountain" The Times historian (Leo Amery) states that Colonel Moller arrived at a strong defensible position from which he could have stampeded the whole of the Boers' ponies and commanded their line of retreat and that Major Knox begged to be allowed to fire on the ponies, but instead was ordered to advance with two squadrons right in the rear of the Boer position. This and the rest of Colonel Moller's proceedings are inexplicable. After exhibiting great rashness he seems to have become unnerved. Knox with difficulty got back with his two squadrons. The others were taken prisoners. In his evidence before the court of inquiry, Colonel Moller gave his reasons for taking the road he did, but Captain Lonsdale of the 2nd Dublin Fusiliers Mounted Infantry stated that he had informed the colonel he was taking the wrong road. The officers and men were exonerated, but Lord Roberts did not allow Colonel Moller to rejoin his regiment. https://www.angloboerwar.com/?option=com_content&view=article&id=508

[53] Doyle, Arthur Conan. The Great Boer War: Seventh Impression 1900 (September) London, Smith, Elder & Co. Pages 535-536. And With The Flag to Pretoria Vol 1 Preface page vi (Casualties) Forces engaged from various sources as a general guide, not exact figures.

I was too elated, however, at having taken part in our first success to be downcast for long, and I enjoyed the novelty of looking at the captured men and talk to such of them as were willing. After a final look round, those of us from Pretoria rode back towards the berg,[54] where we had left the rest of our commando, leaving the wounded to make their way as best they could to the nearest medical assistance.

It fell dark before we reached the foot of the mountain on which our men were, and, heavy rain setting in, we made for a deserted farmhouse, and here we spent a comfortable night, pitying our companions out in such weather.

At dawn the rain stopped, and before long we saw Maroola's force winding down the mountain, so we saddled our horses and made haste to rejoin them as they came down.

The men were cold and wet and hungry, and they looked with envy on our dry clothes and on the trophies in the way of swords and bayonets which we had brought with us. The weather at last was clearing and the sun came shining warmly through so that all were soon in a more cheerful frame of mind in spite of Maroola's telling us that the encircling movement had failed and that the English forces in Dundee had got away towards Ladysmith.

When his brother Swart Lawaai saw the horse I had captured the previous day, he made me hand it over to one of his Field-Cornets whose mount had gone lame, and I was fool enough to comply, for I still had some respect for authority, and that was the last I saw of my booty. This was the more unfortunate because the little Basuto pony that I had been using as a pack-animal had strayed during the night, and my brother and I were left without a led horse to carry our cooking-tins and supplies. We now felt our way round the corner of the mountain into Dundee town. As Maroola was not quite certain that the English evacuation was complete, he sent a patrol of ten men forward to investigate.

My brother and I were of the party, and, while we were riding ahead along the base of a kopje, we saw half a dozen English soldiers running up the slope about 500 yards off. We shouted to them to stop, but as they paid no heed, we sprang to the ground and fired, bringing two men down.

The others now halted, and, riding up, we found one dead and another badly wounded. The rest told us that they were a signalling party that had lost their way in the rains and mists of the preceding day, and they seemed greatly taken aback when they heard that their troops had evacuated Dundee.

We left the dead man lying where he fell and ordered the prisoners to carry their wounded companion into town, and as we were anxious to be first in, we left them and rode on.

By now Maroola's men were also making for Dundee, galloping hard behind us, but we were well in advance and easily got in before they came. They were not long, however, and soon 1,500 men were whooping through the streets, and behaving in a very undisciplined manner. Officers tried to stem the rush, but we were not to be denied, and we plundered shops and dwelling-houses, and did considerable damage before the Commandants and Field-Cornets were able to restore some semblance of order. It was not for what we got out of it, for we knew that we could carry little or nothing away with us, but the joy of ransacking other people's property is hard to resist, and we gave way to the impulse. My brother and I were hampered by the loss of our pony, but we brought away enough food for a royal feast, and after living on half-cured biltong for all these days, we made up for lost time.

There was not only the town to be looted, but there was a large military camp standing abandoned on the outskirts, and here were entire streets of tents, and great stacks of tinned and other foodstuffs, and, knowing the meagre way in which our men were fed and equipped, I was astonished at the numberless things an English army carried with it in the field. There were mountains of luxurious foods, comfortable camp stretchers and sleeping-bags, and there was even a gymnasium, and a profusion of other things too numerous to mention.

The looting of Dundee was the work of General Maroola's men, for the rest of the Boer forces were moving, east and west of the town in pursuit of the English garrison that was retreating to Ladysmith.

That afternoon Maroola managed to persuade his men to resume the advance, and he trekked off, but twenty men of our Pretoria commando were picked haphazard by him to remain behind to prevent further looting. I rode by just as he was making his selection and, his eye happening to fall on me, I was included in the party.

The old fellow probably thought he was doing me a good turn, but I was far from pleased, especially as my brother went on with the rest of the commando. After they were gone we searched around for one of the less-

[54] From German Berg or Norwegian berg, either term meaning "hill/mountain." (Impati Hill, in this case.)

damaged houses, as we had no idea how long we might be marooned here and had decided to be comfortable. Towards evening I rode up again to the English camp to have another look at it and, wandering- about, I came on the field hospital, flying the Geneva Cross. One of the tents was a large marquee for wounded officers, and here I saw General Penn Symons,[55] the Commander of the English troops. He was mortally wounded and the nurses told me that he could not last out the night. Next morning, as I was again on my way up to the camp, I met a bearer-party carrying his body, wrapped in a blanket, and I accompanied them to where they buried him behind the little English chapel.

Now that the commandos were gone, and we were left in sole charge, we looked like settling down for some time, but that afternoon a dishevelled horseman rode into town with tidings of the disaster that had overtaken the Johannesburg men at Elandslaagte. He was in a state of great excitement and gave us such an exaggerated account of the fight, probably to excuse his presence so far to the rear, that I decided to go off in search of our commando, as I could see that there would be other fighting. The temporary Corporal in charge of us, one Paul de Villiers,[56] forbade me to leave, as our instructions were to stay in Dundee until further orders, but I rode away within the hour. As it was late when I started, darkness came on before I had got very far, and I spent the night at Glencoe Junction, where I found other fugitives from the Elandslaagte battle, from whom I gathered that it had been a pretty serious affair, in which we had lost two guns and many men killed and wounded, besides some hundreds of prisoners.

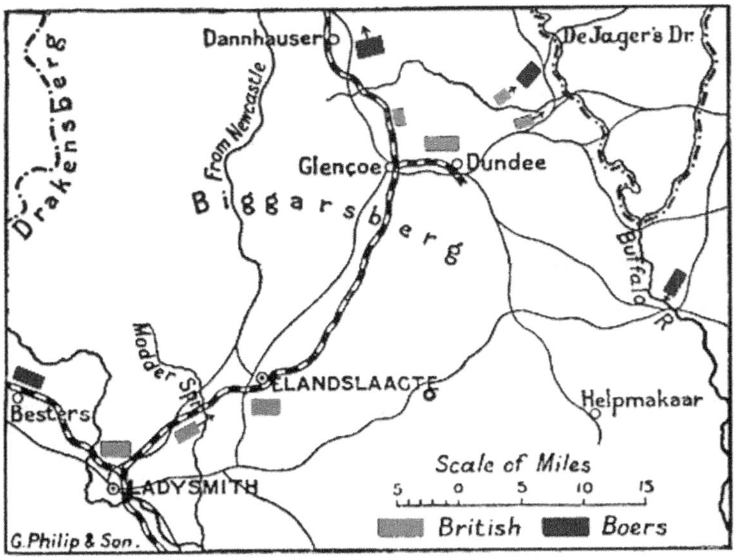

Area around the Battle of Elandslaagte.[97]

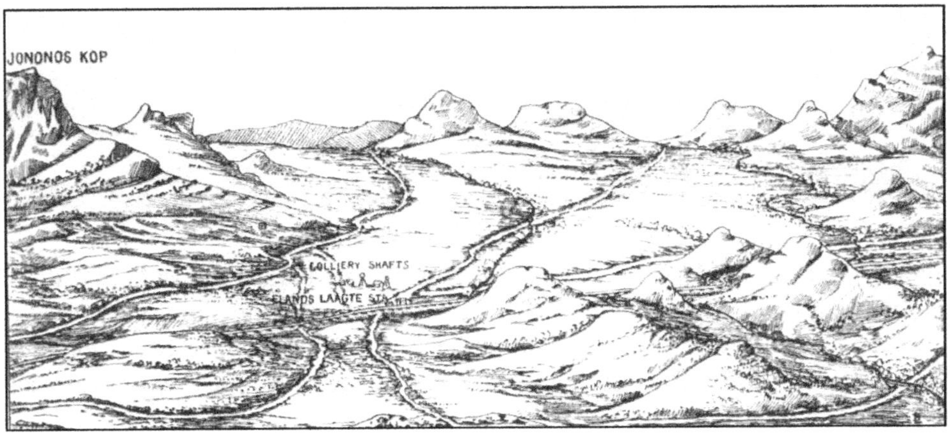

Landscape of Elandslaagte.[98]

[55] Major-General Sir William Penn Symons KCB (17 July 1843 – 23 October 1899) was a British Army officer who was mortally wounded as he commanded his forces at the Battle of Talana Hill during the Second Boer War. While his forces won the battle, by driving the Boer forces off of Talana Hill, they had to abandon their position and fall back to Ladysmith. Because the Boers had control of their water supply outside of town. Penn Symons - Wikipedia

[56] De Villiers Paul Hendrik Prisoner number: 11899 Captured: Harrismith Sent to: Ceylon, Diyatalawa Age: 20.

"Elandslaagte. The flat ground where the 5th Lancers charged the retreating Boers." [Note Coal mine shaft and works. Northern Natal is underlaid by coal which is one reason the British were obliged to defend it rather than retreating to the plains of Natal.][99]

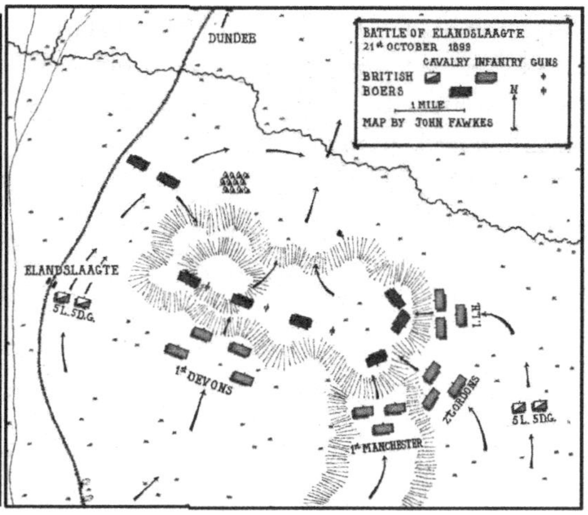

Gordon Highlanders advancing at Elandslaagte.[100] Where Haldane was wounded and Viljoen was shot in the foot and barely escaped. Battle of Elandslaagte.[101]

Imperial Light Horse.[102]

Charge of the 5th Lancers.[103]

At the station, too, I saw the wife of General Kock, who commanded the Johannesburg commando. [57] He had been badly wounded and taken into Ladysmith by the British, and the poor woman was hoping to get through their lines to see him. How she came to be here at all I do not know; probably she had accompanied the forces in the same way as the Commandant-General's wife, but I do not know whether she ever succeeded in reaching her husband, who died a few days later.

[57] Johannesburg Commando Records: **https://www.angloboerwar.com/?option=com_content&view=article&id=3052**

General Kock (Center) and his staff.[104]

Dr. ELSBURGER	Col. A. H. SCHIEL	JUDGE KOCK	COUNT ALBEDYLL	
		THE LATE GENERAL J. H. M. KOCK	COUNT ZEPPELIN	FIELD CORNET POTGIETER

General Kock and Staff identified.[105 58 59]

Just as the main Boer line began to wave white flags of surrender and the British ceased fire, General Kock led a counterattack, dressed in a top hat and his Sunday best. His party drove back the British in confusion, but they were rallied by the infantry commander, Ian Hamilton, and Kock and his assault force were killed or severely wounded. Kock was captured and died a week later while a prisoner in Ladysmith.

After this last counterattack the remaining Boers mounted their horses and began to retreat. A squadron of lancers and another of dragoons were able to get in amongst them with lances and sabers, inflicting severe casualties.

Carrying the wounded from the battlefield of Elandslaagte.[106] "Mrs. Otto Krantz, the wife of a hunter. Accompanied her husband to Natal and fought by his side at Elandslaagte, also in the battles on the Tugela, and after wars in the Free State. She is by no means the only Boer woman who has carried a rifle in the war, and used it."

[58] Johannes Hermanus Michiel Kock - Wikipedia

[59] "General Kock And His Staff : The Heroes Of Elandslaagte , 21 October 1899. J. H. M. Kock , the gallant old Transvaaler , Member of the Executive Raad , died at Ladysmith , on the 31st of October 1899, of inflammation of the lungs contracted on the field of battle, where he was wounded and lay long in the pouring rain before the English ambulance found him . His illness was increased by sorrow at his defeat. His son, Judge Kock, fell into the hands of the enemy, but was exchanged by after his father's death. General White Count Zeppelin was killed after firing off all his ammunition and eventually defending himself to the last against the charging Lancers with his whip. Colonel A. H. Schiel was till the end of the war a prisoner at St. Helena. Albedyll succeeded him in his command. With the exception of Potgieter and the lieutenant of the Transvaal Artillery , who did not take part in the battle, the others are all members of the German Corps, of which Dr. Elsburger stood at the head of the ambulance." Heroes of the Boer War. Review of Reviews, 1903.

Searching for wounded at night.[107]

Removing wounded after the Battle of Elandslaagte.[108 109]

"The German (Hollander) Corps. Before the Battle the of Elandslaagte (October 21st, 1899). The corps lost many killed and wounded in this battle."[110 60]

[60] The Battle of Elandslaagte (10/21/1899) was also notable for being the first and last battle of the volunteer Hollanderkorps. The Hollanderkorps was a group of ca. 150 Dutch volunteers which had been established a mere month earlier. During the battle the Hollanderkorps suffered 9 fatalities, including Herman Coster, along with fellow officer Cars Geerts de Jonge and seven soldiers: P.J. van den Broek, H. van Cittert, J.A. Lepeltak Kieft, Jan Moora, J.Th. Rummeling, M. Schaink, and F.W. Wagner. A further 35 of the Hollanders were taken prisoner by the British. Among the prisoners was Willem Frederik Mondriaan (brother of the Dutch artist Piet Mondrian) who was wounded in the battle. Although he was able to crawl away from the battlefield he was taken prisoner by the British shortly after. He was later sent to Saint Helena as a prisoner of war, returning to South Africa in 1903. Cornelis Vincent 'Cor' van Gogh, the brother of the Dutch artist Vincent van Gogh, also fought in the battle where he was wounded and taken prisoner. He died shortly after. The Boer generals were deeply unhappy with the Hollanderkorps' performance, and it was disbanded after the battle though several hundred Dutch volunteers continued to fight in Boer regiments. The names of the deceased Hollanders, including Coster, were inscribed at a monument at the location of the battle. The monument was destroyed by vandals in 2014. Battle of Elandslaagte - Wikipedia

The Hollander Corps leaving Pretoria for the front, 6 October 1899.[111]

Dutch Corps Monument.[112 113]

I rode on before daylight next morning, the memory of her tear-stained face giving me the first hint of what women suffer in time of war. I went down the Washbank Pass all by myself, and, although I was new to the country, I had no difficulty in finding my way, as the road was trampled and churned by the thousands of horsemen who had gone before.

During the two days while I was in Dundee, General Hubert had moved his whole army southwards, and was at this moment camped about ten miles from Ladysmith, where I rejoined after nightfall. As I approached I saw thousands of fires springing up on the hills, for the commandos were strung out over several miles, and it took me a long time to locate the Pretoria men in the dark. Stumbling about I came on General Maroola squatting by a fire, and, before I could slip away, he saw me, and testily demanded why I had left Dundee without his permission. I said I had come away to help him take Ladysmith, and, grinning sourly at my impudence, he packed me off. When at last I found my corporalship I was told that my brother [Joubert] had been missing since the day before, which troubled me, for we had promised my father to remain together if possible.

Next day our whole line moved- forward, and a fine sight it was, as the masses of horsemen breasted the green slopes towards the final hills from which we could look down to Ladysmith. The English were standing on the defensive, for during the morning we had not come in contact with them and we could now see them building forts and redoubts in the low kopjes surrounding the town.

Neither side made any attempt to get to grips that day, but by night our men had occupied Bulwana [Isimbulwhana] and Lombaards kop, two prominent heights, and they were also holding a line that stretched around by Pepworth Hill and the ridges lying towards Nicholson's Nek. The following day was spent in preparation for the coming battle, but I was riding from commando to commando in fruitless search for my brother.

Ladysmith with Lombard's Kop and Umbulwana noted.[114]

"Isandlwana Hill" [115] [(U)Bulwana on map].

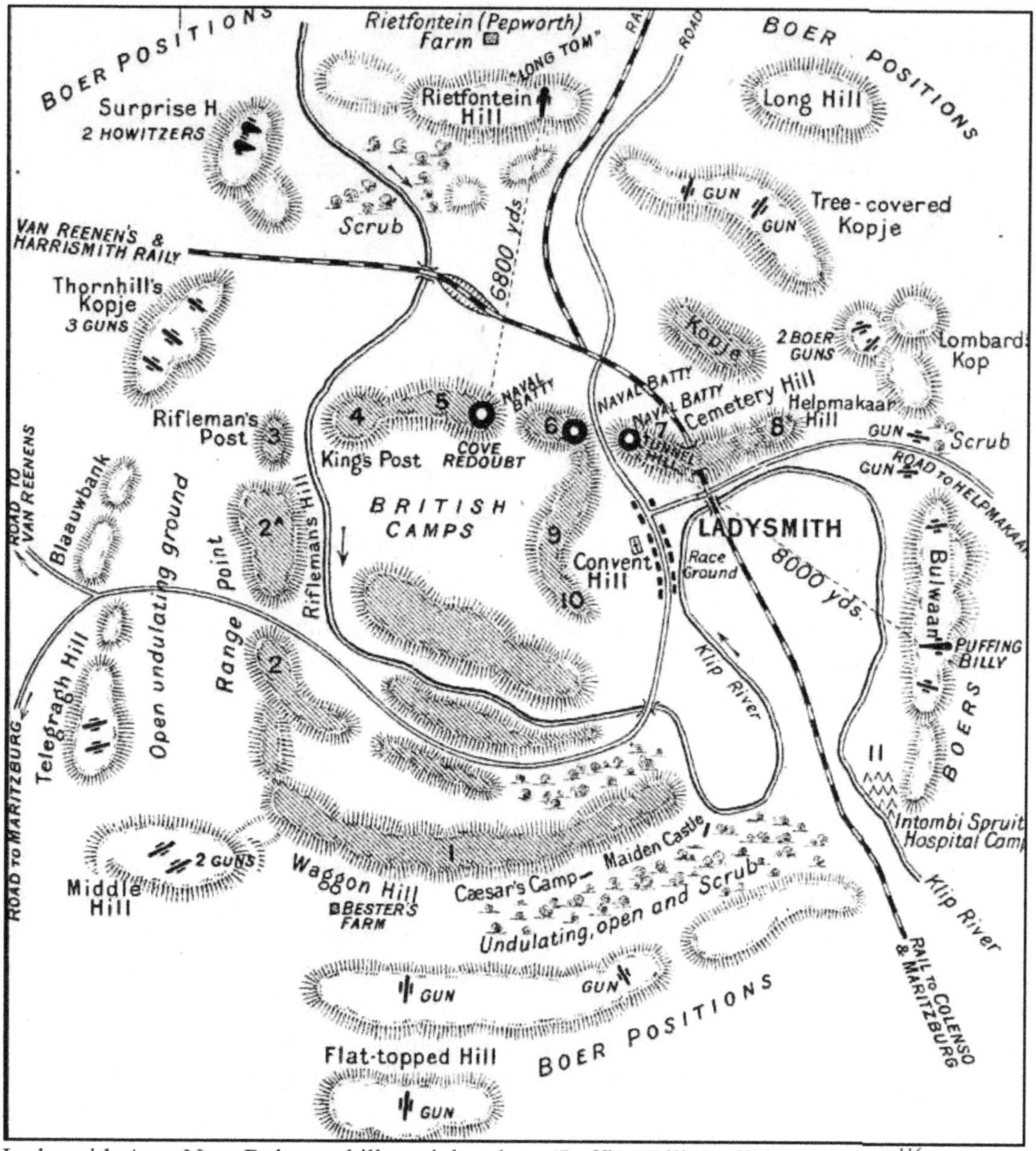

Ladysmith Area Note Bulwana hill on right where 'Puffing Billy' will be mounted.[116]

Towards eleven in the morning Piet Joubert, the Commandant-General, and his staff rode up to address us. He started by scolding the men for having looted a farm close by, and got so worked up that he forgot to tell us what the real object of his visit had been. We took his wigging in good part, but I am afraid no one treated our Commander in-Chief very seriously. His staff was jocularly known as the Royal Family, mostly relations. There was a story going round of a burgher who saw a man ordering people about. The burgher asked him if he was an officer and the man replied: 'Of course I'm an officer, I am the Commandant-General's nephew.' This tale might have been true, for some of the Royal Family put on airs, and we had no great love for any of them, although in fairness it must be said that none of the Commandant-General's own sons were on his staff, but were serving as ordinary privates.

After Piet Joubert had ridden off Mr. Zeederberg, our, Field-Cornet, set us to building breastworks in the line of hills forward of our halting-ground, and every man worked hard until sunset, for this was the position we were to hold when the enemy came against us.

A collision with the British was imminent. They had 10,000 to 12,000 men, including those who had retired from Dundee, and we had 14,000 or 15,000, so that something was bound to happen.

That night the Pretoria commando took up a suitable position to the right of Pepworth Hill, [aka Rietfontein Hill] where we lay on our arms till morning, expecting an attack at any moment, but sunrise came and all remained quiet, from which we gathered that another day was to pass before the dash.

General Joubert and staff at Modder spruit.[117] General Joubert with His Son-In-Law, Malan, and His Secretary, Bracht, in His Tent Before Ladysmith.[118]

Ladysmith, Natal.[119]

General View of Ladysmith.[120]

Casualties In Action	Killed		Wounded		Missing and Prisoners			Forces Engaged		
	Officers	Men	Officers	Men	Officers	Men	Total	British	Boer	UK Loss
Elandslaagte	5	50	30	169	0	4	258	3,500	2,000	7.4%

V. A Battle

When the men returned to the camp that evening after finishing their work, I was left on outpost duty with seven others, at the place where we had been building 'schans'[61] during the day.

In the schanzes near Ladysmith.[121]

The hill we were on formed the extreme right of the Boer line. From our station we looked over the plain leading towards Ladysmith, and to our right we looked across a broad valley, running between us and Nicholson's Nek, a flat-topped hill on the far side.

All remained quiet until three in the morning, when out of the darkness there came the sound of shots followed by confused shouting and trampling, but as the noise died down after a while we let things be. Shortly before daybreak, when it was growing light, two large mules came trotting up from below, their head-ropes trailing on the ground, and on bringing the animals to a halt we found that one of them carried on his back the barrel of a mountain gun, and the other a leathern box containing shell ammunition.

No.10 Mountain Battery in Natal, October, 1899.[62]

Close up of a British Mountain Battery.[122]

They had obviously come from the British lines, and we only learned afterwards that a strong force of infantry had marched out of Ladysmith intending to get into our rear under cover of darkness, but that their battery mules had stampeded, throwing the troops into such disorder that instead of achieving their object, they were obliged to take up a position on our flanks on Nicholson's Nek where, later in the day, I took part in the fighting that led to their capture.

Meanwhile it was sunrise, and we could now make out this force on the level top of Nicholson's Nek across the valley. The soldiers were working like ants, building sangars[63] of stone, and we could see a knot of officers, standing around what looked like an outspread map, while other men were pulling a tarpaulin over a tree for

[61] A sconce or a similar type of fortification, especially as built to defend a pass or ford. A platform or ramp used for ski jumpers (or any similar sport) to gain elevation during their jumps.

[62] Navy & Army Illustrated Vol 9 Photo by: T. H. Murray, Pieter Maritsburg. PF: Col. Frank Robert Carleton's column set out on the 29th accompanied by the 10th Mountain Battery, which required 200 mules to carry all the equipment and ammunition. While his troops were struggling up the hillside, which became progressively steep, some rocks were dislodged that rolled down amongst the mules and horses; alternatively, several mounted Boers rode down the hill and through the column, firing as they went, with the result that the artillery mules bolted back down the track through the troops, who thought they were under attack and fired on the mules. As the mules raced through the column they also stampeded the mules carrying the small arms ammunition reserve and disappeared into the night.

[63] A stone breastwork; a fortified niche or look-out post.

shade. At this time our commando, as well as the other commandos in the vicinity, was still peacefully at rest some 400 yards to the rear, out of sight behind a low ridge, and no alarm had yet been given, so we fired the first shots of the coming battle.

The range was too great for accurate shooting, but our volley had the effect of dispersing the officers, who hurriedly climbed up to join their troops on the hill-top above, where we could now no longer see any sign of life, as the men who had been working there had taken cover, and the plateau scorned deserted.

Our shots had the further effect of arousing the commandos and, before long, horsemen came hurrying from the different camps to occupy the forward crest, and within twenty minutes there were hundreds of riflemen in position, facing the plain that runs to Ladysmith.

Devonshire Regiment facing Pepworth Hill.[123]

As for the English force on Nicholson's Nek, orders were sent to the Free State commandos, coming up from the west, to attack them, and by seven o'clock we heard the popping of rifle shots from that direction, but for the next hour or two we could spare them no thought, for tall pillars of dust were rising from Ladysmith, and soon long columns of infantry debouched into the plain before us.

The Transvaal Staats Artillery had dragged a 6-inch Creusot gun 'Long Tom' up Pepworth Hill, a mile to the left, and they had installed several smaller guns there as well, and all these now began to fire on the approaching troops. I expected to see the shells blow great gaps and lanes in the enemy ranks, but instead of this our first shots were spent in finding the range, and by that time the columns had opened out, and, in place of the havoc which I had expected, the firing only caused smoke and local disturbances of earth, while the infantry came steadily on.

Long Tom showing breech and powder bag used along with ramrod/swab and water bucket.[124] Creusot Gun at Mafeking.[125]

And yet a great spectacle was developing.[64]

From where we held the sweep of the hills, we looked down as from an amphitheatre at every movement of the troops on the plain below, infantry hurrying forward in successive waves; guns being galloped up, and all the bustle and activity of a battle shaping before our eyes.

The soldiers, paying little heed to the shells that dropped amongst them, advanced without a halt, although many now fell dead and wounded while in the rear, battery after battery unlimbered. We saw the horse teams ridden back, and then, to cover the progress of their troops, heavy fire was opened and there came the sound, once heard never forgotten, of shells tearing towards us and exploding around us, and overhead, with deafening concussions.

12-pounder Naval Gun and its company.[126] Note: Manned by Navy bluejackets who were trained to use it onboard ships. Example of a 15-pounder field gun. The largest gun the British had before the arrival of the naval guns.[127]

Battery of the Royal Artillery heavy guns in action against the Boers. Natal.[128] A 4.7" Gun on Scott Carriage.[65]

Until Now the British did not have guns in the field that could match the range of the Boers guns.

The naval brigade from the HMS *Powerful* had reached Ladysmith by train about 9.30 AM to find the shells from "Long Tom" raining down on the town. Their 4.7 inch guns were unloaded and hitched to bullock teams and marched out to Limit Hill. This was too close to the enemy and they were ordered to retreat. Just as they started to retreat Long Tom was turned on to their position and the second shell broke the carriage of the leading gun, seriously wounding the crew. The other guns retreated and took up a position on the open ground just north of Ladysmith, and at about 12.30 PM opened fire on Pepworth at a range of 6,500 yards. In a very few rounds they effectively silenced Long Tom and the rest of the battery on Pepworth. To this point the British had been outranged by the Boer artillery. [The effect on morale] of the arrival of artillery with a range equal to that of the

[64] The Battle of Ladysmith: Battle of Ladysmith - Wikipedia

[65] Captain Percy Scott, of the "Terrible," who has been rewarded for his great service by receiving the C.B., is the now famous inventor of the wheel-mountings which have sent the 4.7- in. gun to do such splendid work far inland.

enemy's guns of position was significant. It went far to counteract the feeling of panic which had pervaded the town since it was apparent that the battle was not going successfully.[66]

"Naval 4.7 Inch Gun."[67] 4.7 Inch Naval Gun on improvised mounting. (From HMS *Terrible* with notation "Who sup with me require a devil of a long spoon.")

By now, what with the thunder of the British guns and of our own, the crash of bursting shells and the din of thousands of rifles, there was a volume of sound unheard in South Africa before. I was awed rather than frightened, and, once I had got over my first impression, I felt excited by all I saw and keenly joined in the firing. We were so successful that by the time the foremost infantrymen came within 1,200 yards of us, many fallen dotted the veld, and their advance wavered before the hail of bullets. They did not run away, but we saw them taking cover behind ant-heaps and such other shelter as the ground afforded. From there they directed a heavy fire on us, but their progress was definitely stayed, and our line held for the rest of the day.

British troops taking cover among rocks and ant heaps.[129] Ant mounds of "compass termites" oriented north/south.[130]

Our casualties were not heavy, for we were well protected, but the guns on Pepworth Hill close by were being severely punished. The English batteries concentrated upon silencing our pieces there and we could see that the people were making heavy weather on the hill, for its summit was covered in smoke and flame, and the roar of the bursting shells shook the ground even where we lay.

For the past few days I had been without word of my brother, but now an artilleryman came racing up on horse-back with a message from General Maroola to say that most of the guns were out of action. I recognized the man as an old acquaintance, and, when he saw me, he said my brother was at that moment serving as a volunteer with the crew of the big Creusot on Pepworth Hill.

[66] Adapted from Times History Vol 2 p. 236.

[67] See: Navy & Army Illustrated Vol 10 Photo by Cribb "Not only did Captain (Percy) Scott's ingenuity enable the long 12-pounder to be brought into use for fieldwork, but he also adapted a carriage by which the "four-point-seven" and 6-in guns could be utilised with troops in the field. This gun has a calibre and range equal to the heavy Creusot guns with which the Boers had supplied themselves."

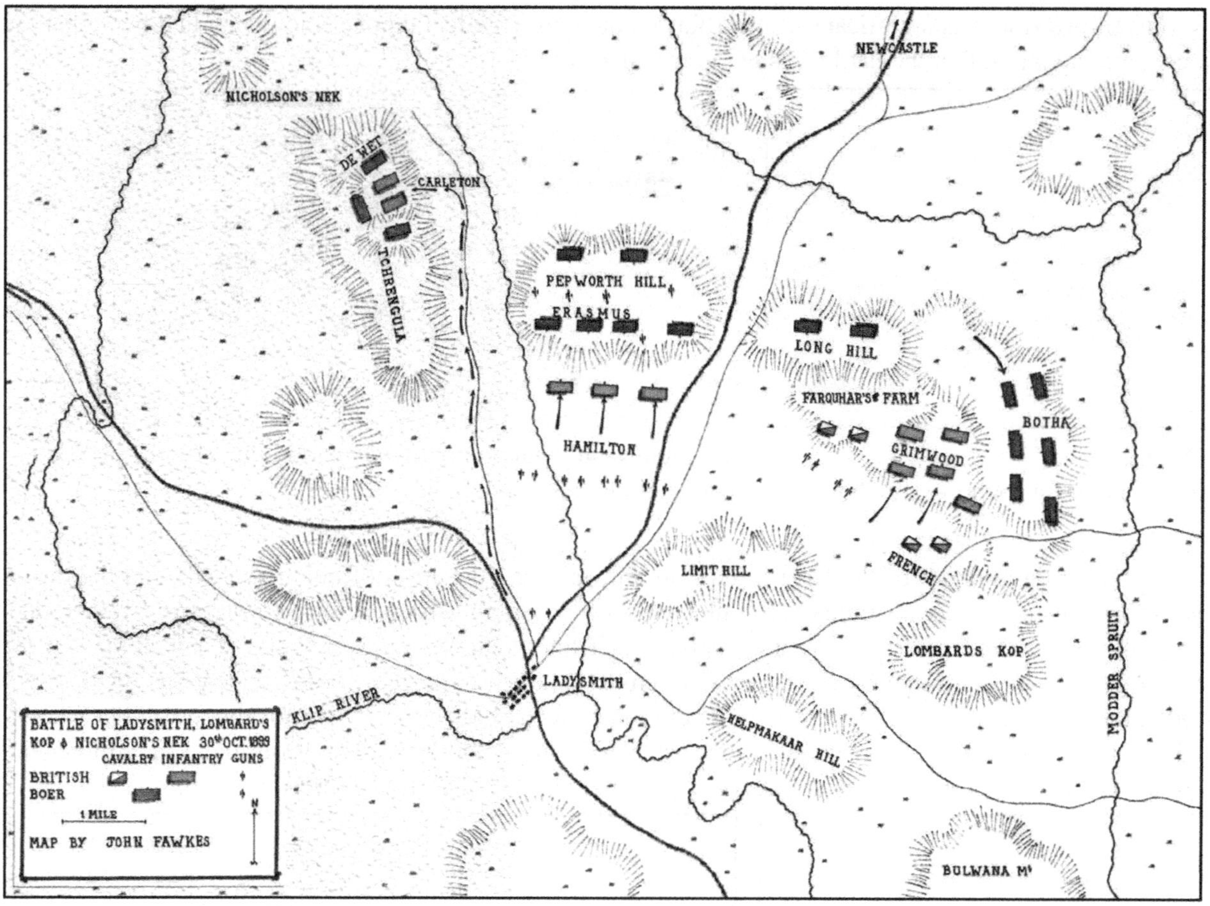

Ladysmith Battle Map.[131]

Concerned to hear that he was there, I ran back for my horse, which was standing under cover with the rest, and rode for the hill as fast as I could go, through shell and bullet fire I reached the rear of the hill in safety, and certainly the dispatch-rider had not overstated the case. Six or seven lead artillerymen, some horribly mutilated, were laid out on a square of canvas to which they had been carried from above, and Ferdinand Holz, the German military doctor, was attending to a number of wounded also brought down from the emplacements.

Boer Doctors.[132]

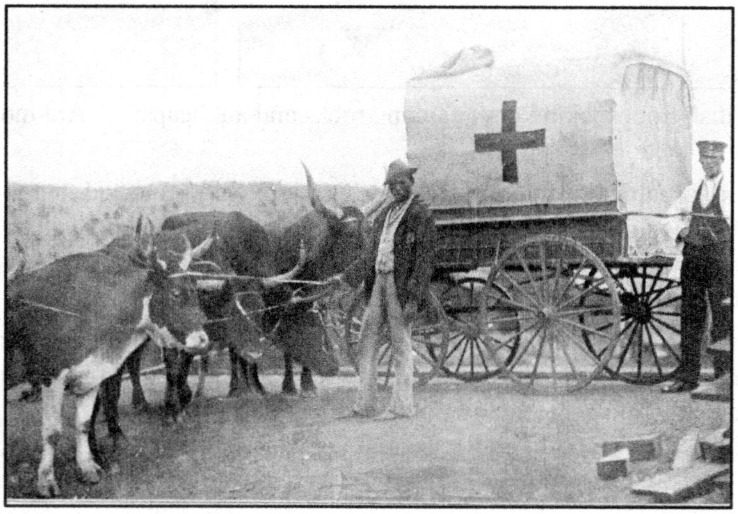

Red Cross Wagon.[68] With Driver and Attendant.

An ambulance-van was standing nearby with several of its mule team dead in their traces, and in the distance the native drivers were running wildly to the rear. At the guns above twenty to thirty shells at a time were bursting

[68] Webster Davis, John Bull's crime; or, Assaults on Republics. 1901 New York The Abbey Press

with terrific noise. I tied my horse to a tree and, after an anxious glance at the dead and wounded, to make sure that my brother was not of their number, I began to work my way up, crawling from stone to stone, and running forward whenever there was a lull in the fire, until I reached the sand-bagged ramparts. Here I found my brother, [Joubert] and the surviving artillerymen, crouching behind the wall, unable to serve their guns under the storm that was lashing the position. It was mostly shrapnel, so that none of the guns were actually destroyed, but all were silenced.

More dead lay about and wounded men were sheltering with the rest in the lee of the Parapet. I liked the spot so little that I tried to persuade my brother to return with me to our own commando, but, although he was somewhat shaken by this ordeal, he refused to come, and I had to admit that he was right. As there was no object in my remaining I bade him good-bye, and taking advantage of a slackening in the British gun-fire I made my way down. Below I found Dr. Holz, lying in a heap, struck dead by a shell while helping the wounded. A fresh ambulance wagon came up just then and I lent a hand at loading the casualties before I sought out my horse, fortunately unscratched, and made haste to get away.

I found the Pretoria men still holding the ridge on which I had left them, and on the plain before them the situation had not materially altered. The English troops showed no signs of advancing, and our commandos lay inactive while the guns pounded away, the battle resolving itself into an artillery duel, except that on the right we could hear the continuous crackle of musketry from Nicholson's Nek, where the Freestaters were engaged with the force that had come out of Ladysmith during the night.

Our Corporal, Isaac Malherbe, now suggested to those of us lying near that we should accompany him across the valley to Nicholson's Nek. About a dozen of us followed him to the horses, dodging among the rocks and trees to escape the shrapnel, and we rode down into the broad valley on the opposite side of which rose Nicholson's Nek.

We came under plunging rifle-fire as we moved across the floor of the valley, but we got under cover at the foot of the hill without loss. Here there was a party of Indian dhoolie-bearers, who had brought down some wounded English soldiers, our unexpected appearance among them causing considerable consternation. Ignoring them, we tethered our horses and started to climb the hill, going warily for fear of blundering into the wrong side of the fighting which we could hear above.

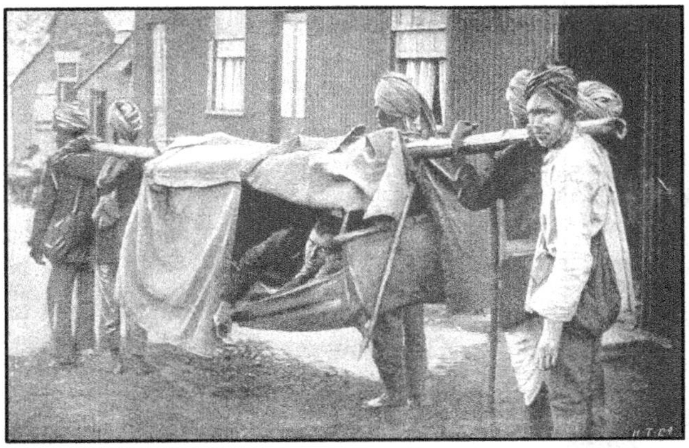

Carrying a soldier in a Dhoolie.[133]

The top of Nicholson's Nek is a broad level plateau dotted with outcrops of loose boulders, and, on breasting the rim, we were for a moment or two unable to gauge the situation. Rifles were cracking and bullets were whizzing, but there was no one in sight. Crawling forward, however, we came on small parties of Free State burghers lying behind rocks and other shelter in a rough line across the hill, and when we joined one of these groups, they pointed out to us where the English troops lay posted, behind similar cover, thirty or forty yards away.

Both sides were maintaining a vigorous short-range rifle contest, in which the soldiers were being badly worsted, for they were up against real old-fashioned Free State Boers for whom they were no match in sharpshooting of this kind. Time after time I saw soldiers looking over their defences to fire, and time after time I heard the thud of a bullet finding its mark, and could see the unfortunate man fall back out of sight, killed or wounded. We joined in the fight, and for the next hour we slowly but surely pressed the English to the far edge of the hill.

Infantry defending a position.[134]

As we gained ground we began to come on their dead and wounded, and realized what heavy losses we were inflicting, for behind almost every rock lay a dead or wounded man, and we knew that we should have possession of the hill before long.

Towards noon, as we were increasingly hustling our opponents, we heard a bugle ring clear above the rifle-fire, and at the same time a white flag went up.

Hundreds of khaki-clad figures rose from among the rocks and walked towards us, their rifles at the trail. We stood up to wait for them. The haul was a good one, for there were 1,100 prisoners, mostly Dublin Fusiliers. The commando responsible for this came from the district at Heilbron[69] in the Northern Free State. They were led by Commandant Mentz,[70] but the man who had chiefly urged on the fight was Field-Cornet Christian de Wet,[71] afterwards the redoubtable guerrilla leader. I saw him here for the first time as he made his way from point to point during the action, and I well remember his fierce eyes and keen determined face.

General Christian De Wet.[135] Christiaan Rudolf de Wet[72] Refer to: Colonel Carleton's surrender at Nicholson's Nek.[73]

[69] Helibron Commando: https://www.angloboerwar.com/?option=com_content&view=article&id=3024

[70] There are two possibilities: Mentz, Johannes Joachim, Veldkornet and Mentz Nicolaas Johannes, Veldkornet Heilbron Commando.

[71] Christiaan Rudolf de Wet (7 October 1854 – 3 February 1922) was a Boer general, rebel leader and politician. Christiaan de Wet - Wikipedia

[72] The photo was taken in Potchefstroom on the 27th of August 1900, six days after he and his commando of 246 men escaped from near encirclement north of the Magaliesberg File:Christiaan de Wet4.jpg - Wikimedia Commons

Shortly after the surrender I was talking to some of the captured officers when I heard one of them exclaim: 'My God; look there!' and turning round we saw the entire British force that had come out against us on the plain that morning in full retreat to Ladysmith. Great clouds of dust billowed over the veld as the troops withdrew, and the manner of their going had every appearance of a rout.

There were about 10,000 soldiers, but General Joubert had far more than that number of horsemen ready to his hand, and we fully looked to see him unleash them on the enemy, but to our surprise there was no pursuit. I heard Christian de Wet mutter: 'Los jou ruiters; Los jou ruiters' (Loose your horsemen—loose your horsemen), but the Commandant-General allowed this wonderful opportunity to go by, a failure that cost us dear in the days to come.

Judging by the disorderly appearance of the retreat, he could have driven the English clean through Ladysmith and out beyond, and he would have lost fewer men in the doing of it than we lost in the subsequent siege, but the English went hurrying back unmolested, save for an occasional shell from Pepworth Hill, where our guns had sprung into life again, and, with the whole Boer Army looking on, no attempt was made to exploit the victory that had been gained.

British forces retreating into Ladysmith after the battle.[136]

When we saw the troops falling back, and no chase given, we washed our hands of the business, and began to examine our own immediate surroundings on the hill-top. Dead and wounded soldiers lay all around, and the

[73] Carleton Frank Robert Crofton Lieutenant Colonel QSA. (Princess Victoria's) Royal Irish Fusiliers. At the battle of Ladysmith, General White sent a detachment consisting of the 1st Battalion, the Royal Irish Fusiliers and half of the 1st Battalion, the Gloucestershire Regiment, with Number 10 Mountain Battery (equipped with RML 2.5-inch Mountain Guns) to capture a pass known as Nicholson's Nek which lay about 3 miles (4.8 km) to the northwest of Pepworth Hill. The force was commanded by Lieutenant-Colonel Carleton of the Royal Irish Fusiliers. White intended that by seizing the pass, this detachment would prevent the Boer force from the Orange Free State reinforcing the Transvaal Boers on Pepworth Hill, and also prevent the defeated Boers retreating directly north. Carleton's men began preparing sangars on what they thought was the summit of Tchrengula. In fact the mile-long summit was divided into two peaks, and in the dark, the British had occupied only the southern, lower summit.

Christiaan De Wet quickly grasped the situation, and led some men on horseback to gallop to the northern end of the hill, where they dismounted and climbed the hill to occupy the undefended northern summit. From there, the Boers pressed forward along the summit, which was strewn with rocks that gave individual Boer riflemen ample cover. The British sangars were exposed and obvious targets to the Boers, and gave inadequate protection. Like the artillery, the British infantry relied on drill: firing volleys on the command of an officer. By the time the order had been given, the intended target was safe behind cover, while the British soldiers were exposed to fire.

The British infantry resisted for several hours, but the numbers of wounded and killed increased and ammunition ran out. Finally, as the main British force could be seen retreating into Ladysmith, Carleton gave the order to surrender. A British officer waved a white flag at the same time. Some of the British troops had seen no reason for the surrender and felt themselves humiliated by the order. Eight hundred soldiers were taken prisoner.

cries and groans of agony, and the dreadful sights haunted me for many a day, for though I had seen death by violence of late, there had been nothing to approach the horrors accumulated here.

Of our party under Isaac Malherbe not one had been hit, but the Free State men had eight or nine dead, and fifteen or twenty wounded. The English casualties were about two hundred killed and as many injured, the disparity being due to the fact that the English soldiers were no match for us in rifle-shooting. Whatever the defects of the commando system may be (and they are many) the Boer superiority in marksmanship was as great now as it had been in 1881. Having looked at the dreadful scenes on the plateau, we parted from the Freestaters and, returning down the hill to our horses, rode back through the valley to the Pretoria commando, which we found in the same place in the line where we had left them, the men still holding the ridge, and their horses still under cover behind.

There was on this day, and for long after, much acrimonious discussion regarding the Commandant-General's failure to pursue when the English turned back, and I was told by old Maroola himself, that when officers came up to implore Piet Joubert to follow; he quoted the Dutch saying: 'When God holds out a finger, don't take the whole hand,' meaning that the Almighty had sufficiently aided us for one day, and that it did not behove us to presume upon His bounty, a view which Isaac Malherbe said might be sound theology but no good in making war.

With that lack of vision that marred most of our doings in the earlier stages, we hailed the Ladysmith battle as a great victory, and we acted as if we had a broken and defeated enemy before us. It certainly was a notable success, but in the end it would have been better for us had the British smashed our line that day, for our leaders would then have followed a better plan of campaign than sitting down to a prolonged and ruinous siege.

Had the Boers made for the coast, instead of tying up their horsemen around towns that were of no value to them, the outcome of the war might have been different, but they sacrificed their one great advantage of superior mobility, and allowed splendid guerrilla fighters to stagnate and demoralize in the monotony of siege warfare, at a time when our only salvation lay in pushing to the sea.

Unloading British Prisoners from Train at Pretoria. [137]

More British Prisoners Arriving at Pretoria. [138]

British soldiers captured at Nicholson's Nek arriving in Pretoria. [139]

British prisoners from Nicholson's Nek on 30th October 1899 arriving at Pretoria Racecourse in the Boer War.[140] Below are British Casualties for the Battle.[74]

Casualties In Action	Killed		Wounded		Missing and Prisoners			Forces Engaged		
	Officers	Men	Officers	Men	Officers	Men	Total	British	Boer	UK Loss
Nicholson's Nek and Ladysmith aka Farquhar's Farm	6	57	9	244	43	925	1,284	10,000	20,000	12.8%

Last Journey. After the battle, October 30th, a soldier being taken from the hospital to the burial ground.[141]

[74] Doyle, Arthur Conan. The Great Boer War: Seventh Impression 1900 Pages 535–536.

VI. Ups and Downs

We rested on our laurels for a few days, each commando camping immediately behind the position which it had held during the battle.

Ladysmith was soon completely invested; on every hill and kook[75] lay a force of riflemen, and there must have been nearly ten thousand thus tied down who could have been put to far better use.

The breather remained fine, and we divided our time between watching the beleaguered garrison build forts and breastworks against our ultimate reception, and riding about the rear foraging for supplies.

After a while we of Pretoria received orders to move forward to occupy a hill named **Bell's Kop**, facing a murderous-looking English construction that had gone up within the last few days. We rode out before light one morning to take up our new position, which, we were informed, was to be held by us as part of the cordon that was being drawn round the town. When we heard that this was to be our permanent abode, all hands turned to building shelters against the weather, and generally making our camp habitable. Neighbouring farms were laid under tribute and plundered of everything that could be turned to account.

Boers on campaign during the Boer War.[142]

A Boer siege encampment.[143]

My brother rejoined us here. He had had enough of gunnery for the time being, and our native boy Charley also entered appearance, beaming with delight at having tracked us down. He travelled on foot all the way from the Transvaal border, and although several times arrested as a spy, he had talked and argued his way through until he found us. Needless to say, he was received with open arms, as we were once more able to turn over to him our duties of cooking, carrying water, horse-guard, etc., so my brother and I settled down to a life of ease, spending our time in sniping at the English outposts, or in riding to the neighbouring laagers.

Camp life was a pleasant existence. There were no drills or parades and, except for night picket and an occasional fatigue party to the railway depot to fetch supplies, there were no military duties. Our commando received many fresh drafts, and after a while varied from one thousand to fifteen hundred men, but discipline was slack, and there was a continual stream of burghers going home on self-granted leave, so that we never knew from day to day what strength we mustered.

Although we nominally stood under command of General Maroola and his brother, we seldom saw anything of them, for they lived at the head laager miles away, and to all intents and purposes we were a law unto ourselves.

Both Maroola and Swart Lawaai were incompetent as leaders, but they were true men nevertheless. They sacrificed all they possessed, and when, later in the war, they were deserted by their men and bereft of all authority, they remained in the field as private soldiers, until the end.

We settled down to a stationary life, and if occasionally time hung heavily on our hands, conditions were otherwise comfortable enough. Our boy Charley proved a capable freebooter, and thanks to his foraging expeditions into the hills among the Zulu kraals where, he made play with his descent from Chief Moshesh, our larder flourished and our mess was the envy of the rest of the camp.

After a week or two, tents were served out, and my brother and I shared one with the five good friends of our corporalship with whom we had kept company ever since leaving home. They were Charles Jeppe, a mine-

[75] Middle German: cake. In this case as to form of mound.

surveyor, Robert Reinecke, Walter de Vos,[76] Frank Roos and Samuel van Zijl,[77] Civil Servants, but they were all killed before long.

So quiet were things around Ladysmith, that, as time went on, many burghers got ox-wagons brought down from their farms, and some even had their wives and families with them, which tended further to increase the spirit of inactivity that was gaining on the commandos.

Preparing the telegraph apparatus for an observation balloon.[144]

War balloon at Ladysmith.[145]

Observation Balloon at Ladysmith.[146]

The Balloon at Ladysmith.[147]

[76] Possibly: De Vos Wouter Burger Heidelberg Commando.

[77] Possibly: Van Zijl Salomon C. Volunteer from Holland / Belgium serving with the Boer forces.

Ladysmith Town Hall.[148]

However, reports from all fronts were good, and we deluded ourselves into believing that everything was as it should be and, so far as my brother and I were concerned, we thoroughly enjoyed the business of besieging Ladysmith, and making regular excursions to see the guns fired into the town.

We went once every week to look up a Norwegian uncle of ours, named Theodor Thesell,[78] serving with a Free State commando on the west side of the cordon. He had settled near the Vaal River many years before, and had been commandeered for service very much against the grain, as he held that we were bound to be defeated. I was surprised to see in what good part the burghers took his outspoken comment on their chances, but the real fighting Boer is ever a tolerant, good-natured follow, and my uncle suffered no inconvenience for his views.

Once, while we were chatting to him on a kopje[79] occupied by a portion of the Kroonstad men beyond the Klip River, a considerable English mounted force moved forward to attack, or at any rate to demonstrate, before this position. Their guns unlimbered within range, and opened a cannonade, and their horsemen made as if to come at us. At the first alarm, the Kroonstad men ran from their tents, and directed such a hot fire that the cavalry wheeled round and galloped off; the guns were hooked up, and what looked like the promise of a lively battle fizzled out in a few minutes. Several shells had burst overhead, but as we were well sheltered behind rocks, we thought little of it until we found, when the affair was over, that three men had been killed and half a dozen wounded, quite close to us.

On another occasion, while my brother and I were returning from a call on the Commandant-General at the head laager, we were overtaken by darkness, and heavy rain coming on, we tied our horses in the shelter of some trees and, seeing a mule-wagon standing near with a canvas sheet drawn over it and pegged down on either side for a tent, we lifted a flap and crawled into a corner where we slept snug and warm all night. When we woke at daybreak a man on a camp stool with a mug of coffee in his hand was eyeing us appraisingly, so we unrolled ourselves and went forward to explain our use of his quarters. It was Mr. Smuts, the State Attorney, my father's colleague on the Executive Council.

[78] No Record Found for Theodor Thesell: Anglo Boer War - Name search

[79] A small hill or mound especially on the African veld.

"Mr. Smuts, Late State Attorney of the Transvaal."[149]

He was here on some Government business, and, after a hearty laugh at our impudence, he ordered coffee and breakfast for the pair of us. Afterwards he became a fighting General, and I was to see much hard service under him in the Cape Colony nearly two years later.

I also recovered my little Basuto pony that had been lost near Dundee. I found him tethered to General Joubert's own wagon in the head laager, and I had an angry scuffle with some dispatch-riders, before I was able to undo the head-rope and take him away with me. The Commandant-General was absent and they promised me all sorts of pains and penalties when he returned, but I knew that among the Boers, ownership of a horse is almost sacrosanct, so I carried him off in triumph.

When next I met General Joubert and asked him how he came by my horse, he said that the pony was found straying on the void by one of his staff; and that he had kept the little animal more as a pet than anything else.

The British now and then shelled our camp, upon which we would take cover at the first sign of trouble, and the damage was generally confined to a riddled tent or two, and perhaps a few horses injured. During one of these squalls our native boy [Charley] was hit, and for weeks afterwards he walked about proudly displaying a fragment of metal taken from his side.

At this time I was sent for to give evidence in Pretoria against a man who had stolen money and clothing, which my father had given him to take down to my brother and myself. I travelled up by goods train and the man was put in gaol. I was in Pretoria for only two days. Before returning to Natal, I walked with my father from his office in the morning and we touched at the State School where a number of captured British officers were confined, one of whom had asked for an interview.

The Staats Model School Officers Prison.[150] Model School (back view). The houses and gardens as well as the overhanging trees that gave cover for Winston's escape are just visible in the lower left.[151]

We passed through the sentries into a large classroom where he was playing games with his fellow-prisoners. His name was Winston Churchill, a son of Lord Randolph Churchill, of whom I had often heard. He said he was not a combatant but a war-correspondent, and asked to be released on that account.

My father, however, replied that he was carrying a Mauser pistol when taken, and so must remain where he was. Winston Churchill said that all war-correspondents in the Sudan had carried weapons for self-protection, and the comparison annoyed my father, who told him that the Boers were not in the habit of killing non-combatants.

Winston as prisoner in Pretoria, November 18th, 1899.[152]

Lord Randolph Churchill.[153]

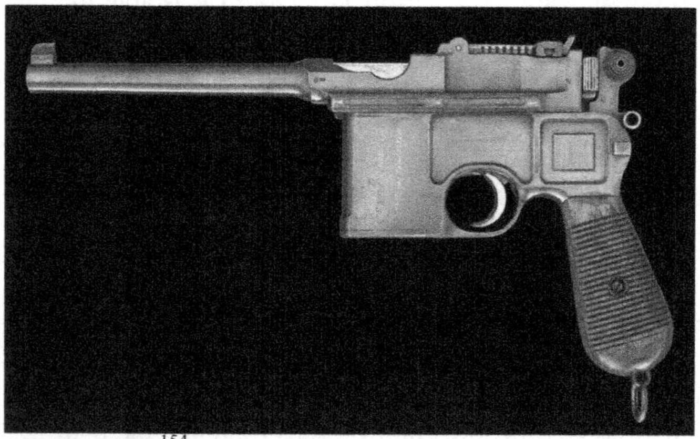

Mauser C96.[154]

In the end the young man asked my father to take with him some articles which he had written for a newspaper in England and if there was nothing wrong with them to send them on via Delagoa Bay. My father read portions of the articles to us at home that evening, and said that Churchill was a clever young man, in which he was not far wrong, for soon after the prisoner climbed over a wall and escaped out of the Transvaal— how, I never heard.[80]

When I got back to Ladysmith all went on as before, and only once during the first month of the siege did we indulge in any military operation. This was a movement against the fort opposite us, the same work that the English had built at the time of our arrival while we stood looking on. It lay eleven hundred yards away, across a level plain, and was known as the Red Fort,[81] and I had spent many moments sniping at it.

[80] PF: See Haldane et al. and London to Ladysmith.

[81] See Creswicke SA War map above where Red Hill is identified. Deneys may be at Rifleman's post just north of there.

Towards the end of November some three hundred men were ordered to attack it at dawn.

Our commando had of late been receiving reinforcements of inferior quality, mostly poor-whites from the burgher-right 'erven,' the slum quarters of Pretoria, a poverty-stricken class that had drifted in from the country districts after the great rinderpest[82] epidemic of 1896.

Cattle dead from rinderpest in South Africa, 1896.[155]

Shooting Rinderpest Oxen.[156]

They had become debased by town life, and had so little stomach for fighting, that their presence among us was a source of weakness rather than strength. The attacking force was drawn chiefly from these newcomers with Isaac Malherbe's corporalship, and one other of our better corporalships added to stiffen them. We left camp before daybreak, and assembled in a dry watercourse at the foot of Bell's Kop. From there we were to cross the intervening thousand yards of open, and rush the Red Fort at dawn.

Owing to various delays it grew light before we emerged from the spruit, and the soldiers must have observed some movement, for they opened fire before we could make a start. Assistant Field-Cornet de Jager[83] was in charge of the undertaking He was a poor-white himself, but a stout-hearted old warrior none the less, and seeing that we were discovered, he ordered us to advance, himself leading the way. Isaac Malherbe and his men followed, as did the other corporalship, but the remainder refused to budge from behind the safety of the bank. With de Jager at our head, about fifty of us ran forward, coming under heavy rifle-fire as we went. Visibility was still poor, so that we covered about four hundred yards without loss, until we reached a slight outcrop of rock, where we paused to take breath, and finding that we were not supported called the attack off. The net result of our efforts was that, having got thus far, we could go neither forward nor backward, as the sun was over the horizon by now, and to advance or retire over the level bullet-swept zone was too risky, so we made the best of it by lying in the blazing sun without food or water for the next ten hours.

The troops in the fort were very wide awake, forcing us to keep our noses glued to the ground to such an extent that we dared scarcely stir without bringing down a hail of bullets about our cars. A man named Anderson was shot through both legs, and Robert Reinecke of our tent pluckily carried him back to the spruit on his shoulders, the English firing all around him, until they realized that he was helping a wounded comrade, after which they left him to go in peace, and were even sporting enough to allow him to return to us without a shot fired.

We lay sweltering in the heat for the rest of the day, wishing ourselves well out of it, and the moment it began to get dusk we made a bolt, getting back over the plain without further casualties in spite of a hot fusillade. I had a lucky escape, for as we were nearing the dry course, I stopped for a moment to see whether anyone had been hit during the retirement, when a bullet grazed my throat, stinging like a hot poker and raising an angry weal. I was more startled that hurt, and quickly followed down into the spruit bed. Here we slaked our thirst from water-holes in the ground, after which we trudged home to find the balance of the assaulting column of that morning comfortably installed in camp, and I got the impression that they rather looked on us as fools for our pains.

The day after this typical piece of bungling I walked across to Surprise Hill, an isolated kopje standing about six hundred yards to the right of our camp. Here the State Artillery had a howitzer gun in position which I often

[82] Rinderpest (German 'cattle plague') was an infectious viral disease of cattle, and many other species of even-toed ungulates, including gaurs, buffaloes, large antelope, deer, giraffes, wildebeests, and warthogs. Death rates during outbreaks were usually extremely high, approaching 100% in immunologically naive populations. In 2010, the United Nations Food and Agriculture Organization (FAO) announced that field activities in the decades-long, worldwide campaign to eradicate the disease were ending, paving the way for a formal declaration in June 2011 of the global eradication of rinderpest. Rinderpest - Wikipedia

[83] Several possibilities under search: De Jager Ass. Veldkornet.

visited, to watch it bombarding the town. On this morning as I was returning through the bush-covered valley, I saw a man named Mike Hands walking ahead of me, leading a horse. As I was going forward to join him I was astonished to see a tall English soldier appear through the trees, his rifle at the ready. The soldier instantly fired at Hands, but hit the horse instead. The animal dropped and Mike fired. The soldier threw out his arms and fell back, and I found him dead with a bullet through his brain. How he came to be behind our lines was a mystery; possibly he had been scouting during the night, and had lost his way in the dark.

Boer Field Artillery[157] [At Ladysmith].

I was to know Surprise Hill better before long, but, in the meanwhile, camp life continued uneventfully. The besieged garrison contented themselves with shelling us at intervals, and, as they made no serious attacks, we assumed that they were marking time, for there was great activity on the south bank of the Tugela, twenty miles away, where the English Commander-in-Chief, General Buller, was massing troops to attempt the relief of Ladysmith.

Maxim machine gun and Krupp Howitzer at Ladysmith.[158] Lt Heinrich Grothaus of the ZAR Staatsartillerie, at the siege of Ladysmith. .[84] [159]

He was reported to have forty thousand men with great numbers of guns, and on our side reinforcements had been brought from the Transvaal and Free State, and there were about fifteen thousand Boers holding the north bank of the river, from a point below Colenso Bridge to Spion Kop, many miles upstream. As for us around Ladysmith, we felt secure in the belief that whatever might happen on the Tugela, the troops in the besieged town at any rate would make no move, so picket and other duties were carried out in a very perfunctory manner.

During the daytime no guards were set at all, as there was always a sufficient number of men on the hill above amusing themselves with sniping to make sure of an alarm being given in case of need; and at night, although we went on outpost so close to the English sentries that we could hear them challenge each other, and sometimes exchanged shouted pleasantries with them, we did not take our watches very seriously.

We used to go on foot after dark in parties of twenty or so, and, on reaching neutral ground on the plain between ourselves and the enemy line, two men at a time would walk forward a short distance. Here they stood or sat on

[84] Photo of Lt Heinrich Grothaus of the ZAR Staatsartillerie, at the siege of Ladysmith. The field gun is a 75mm Krupp QF. The medic on the right is Dr. Elsberger of the German corps. Note the tarpaulin over the epaulement under the muzzle of the Krupp. I understand this is to prevent a cloud of dust from rising up when the gun is fired, revealing its position. Image and quote from Rob Davidson: Anglo Boer War - Artillery and Ammunition - Page 15 - Boer War Forum

sentry-go[85] while the rest of us pulled off our boots, spread out blankets and went to sleep until it was the turn of the reliefs. At daybreak we collected our belongings, and tramped back to camp in time for morning coffee, and thus far no untoward incident had ever marred these tranquil doings.

But this happy state was disturbed on the night of December 9th,[86] for a detachment of two or three hundred soldiers came out of Ladysmith and, scaling Lombaards-kop, destroyed the big Creusot gun 'Long Tom' standing there. The kop lay six or seven miles from our camp, so we heard nothing till next morning, but when we did, it gave us food for reflection, for it looked as if the pleasant immunity of our night duties was a thing of the past.

"Long Tom in hospital: A Boer gun under repair at Pretoria."[160]

One of four Boer 'Long Tom' canons, damaged at Siege of Ladysmith, on its way to the Siege of Kimberley after it was repaired in Pretoria.[161]

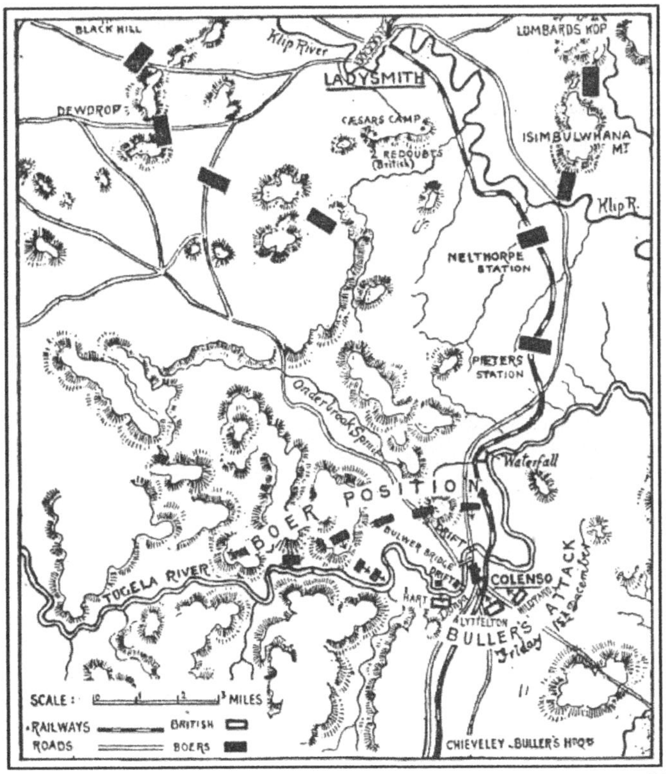

Boer fortifications around Ladysmith and near Colenso.[162]

Falls of the Tugela, noted as "waterfall on map."[163]

Colenso position

[85] Active military duty.

[86] Creswicke chronology cites: 12/8/1899 'British sortie from Ladysmith, Lombard's Kop being carried.'

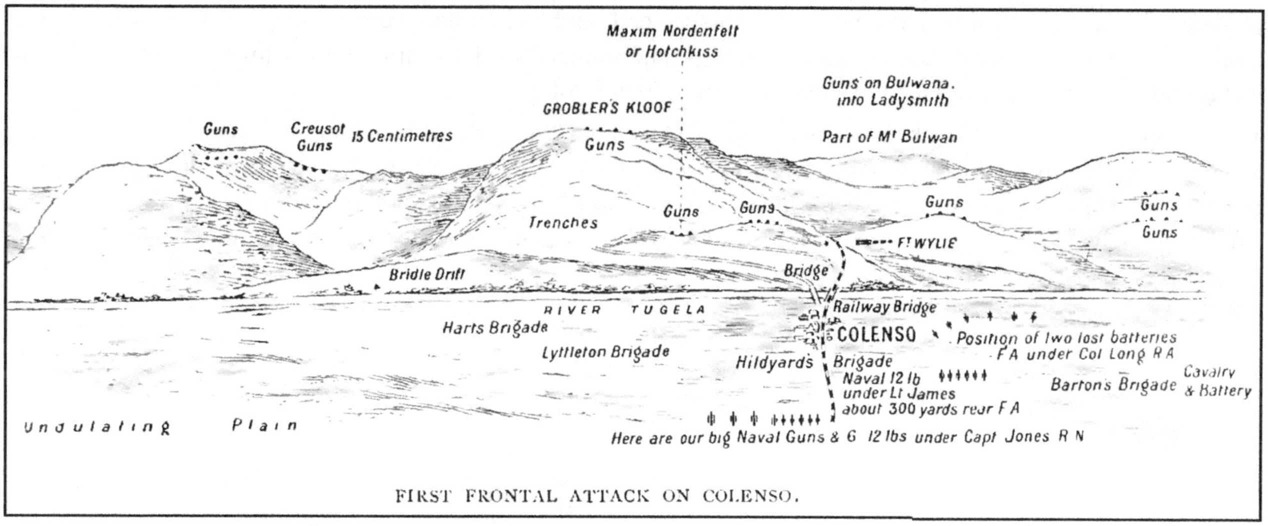

FIRST FRONTAL ATTACK ON COLENSO.

Boer positions at Colenso.[164]

Colenso Railway Bridge looking South from the Boer side of the river. Note tents in open area and people on island in the river. Colenso Bridge sometime after the battle of Colenso which took place on December 15th, 1899.[165] Note the change in the river's flow from dry to rainy season.

VII. An Affair at Surprise Hill

Two other corporalships went on duty at the same time. One, under Corporal Tossel, a former police detective, was posted at the foot of Surprise Hill and the other a long way to our left. My brother and myself and Samuel van Zijl were the only members of our tent who were present, the other four being absent with the carrying party. As we walked along in the dark behind Isaac Malherbe we discussed the previous night's attack on the Lombaardskop gun, and I remember poor Samuel saying he hoped our turn would not come next. But our turn did come next. When we reached the usual halting-place two men were sent forward according to custom, and the rest of us turned in. My time to go on duty was 1 a.m. At about half-past twelve I woke, and not thinking it worth while to fall asleep again, I lay on my blanket watching the stars.

After a while I distinctly heard the muffled sound of many footsteps in the direction of Surprise Hill, so I got up and walked forward to the two sentries to consult them. I found that they had also heard the noise, and the three of us listened for a few seconds to what were certainly men climbing the hill. We thought Corporal Tossel's men had taken fright at something, and were withdrawing up the slope towards the howitzer emplacement. This belief was rudely dispelled, for suddenly there broke from the summit of Surprise Hill a crash of musketry followed by wild bursts of cheering, and we realized that English troops were at the gun. As we stood undecidedly watching the hundreds of rifle flashes lighting up the hill-top, a vivid sheet of flame stabbed the darkness, followed by a tremendous roar, and we knew that our howitzer had been blown into the air.

Howitzer in use at Ladysmith.[166]

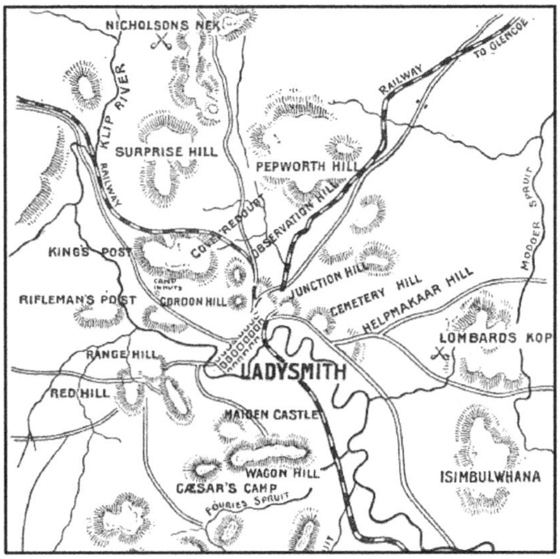

Surprise Hill is Upper Left.[167]

The two sentries and I rushed back to where our party were already on their feet. Isaac Malherbe now showed the stuff he was made of. Without a moment's hesitation he went straight for the danger-point, the continued cheering of the English soldiery and the volley-firing serving as a guide. His intention was to join hands with Tossel's corporalship if he could find them, and then to prevent or delay the troops from returning to Ladysmith, until the whole of the Pretoria commando could come from camp and so destroy or capture the intruders.[87]

As it turned out, Tossel's corporalship had bolted when they heard the English coming. They had not only given them a clear field but had fired no warning shot to alarm the unfortunate gunners up above, who were taken unawares and all bayoneted. And as for the remainder of our commando in camp, they stood to arms all night, but Field-Cornet Zeederberg refused to risk the confusion that he perhaps justly thought would ensue, if he tried to march his men in the dark to an unknown situation. So the twelve of us were left to our own devices.

As we approached we could hear by the firing and shouting that the main body of the attackers were still on the hill, but they had posted a string of pickets at the foot to secure their line of withdrawal back to the town, and before we had gone very far we ran into one of these. Isaac and I were a few yards in advance when a 'Halt! Who goes there?' was shouted at us from a few paces away. We simultaneously fired a shot apiece and ran forward. We came on a dead soldier, a sergeant, as I saw next morning from his badge, but the rest of the picket had run into the night.

[87] Creswicke Chronology cites: 12/9/1899 'Lieutenant-Colonel Metcalfe, 2nd Rifle Brigade, with 500 men from Ladysmith, captured Surprise Hill, destroying a howitzer.'

We went forward cautiously and soon we collided once more with another and stronger rear guard party. We were again challenged from close quarters, and, a heavy fire being aimed at us, we took shelter in a dry spruit bed that runs along the base of Surprise Hill. From here we returned the fire, until this outpost too gave us right of way, and we now began to file along the bed in order to seek out a convenient point from which to make a stand against the troops on the hill when they descended.

As we were going, a soldier, lying concealed in the grassy on the bank above us, thrust over the muzzle of his rifle and fired point-blank into us. My tent-mate, Samuel van Zijl, was walking immediately in front of me and I had my hand on his shoulder to steady myself on the uneven path. The bullet struck him full in the throat, and so near was the range that the discharge-scorched his face, and set fire to his beard, which flared up for a moment like a fusee. He staggered, and then dropped. He was still alive, but I could hear from his laboured breathing that he was badly wounded, so I made him as comfortable as I could by placing his blanket under his head, before hurrying up the spruit to rejoin the others. By now the troops were descending Surprise Hill, and we could hear them clattering down the slope towards us. Their officers were blowing whistles, and calling out: 'A Company here! This way B Company!' and so on, to collect their men. They seemed unaware that the road was to be disputed, for they made no attempt to conceal their progress, and there was laughter and repeated cries of 'Good old Rifle Brigade,' and here and there we caught the gleam of matches being struck and the glow- of cigarettes, to show how little they expected opposition.

In the meanwhile Isaac had selected a suitable spot on the bank, our faces towards Surprise Hill, our backs towards Ladysmith, and here we crouched, silently waiting for the oncoming troops.

From the sounds that reached us we judged them to be about three hundred strong and, with no sign of Tossers men or of help from the commando, it dawned on us that we were in a pretty tight corner.

While the soldiers were still some distance away I ran down to see how Samuel van Zijl was faring. He was only just alive, and he asked me in a faint voice to turn him on his side to ease the pain, but as I did so I felt his body stiffen, and then go limp in my arms, and when I laid him down-he was dead. I hurried back, for the laughter and talking were drawing very near. As I made up the course, a huge soldier, or he looked so in the dark, loomed up suddenly on the bank above. He lunged at me with his bayonet, but his insecure footing deflected the thrust and brought him stumbling against me. The man was at my mercy now, for I had my carbine against his side, but there came over me an aversion to shooting him down like a dog, so I ordered him to put up his hands instead, which he did at once, dropping his bayoneted rifle at my feet. I told him to sit down until I called him, a command which he so implicitly obeyed that I found him patiently waiting there next morning with a bullet through his leg from the cross-firing during the subsequent proceedings. This soldier must have come on ahead, for when I reached my companions, the main force was just reaching the foot of the hill and approaching us in a body.

Isaac whispered to us to hold our fire and each man peered into the darkness until, about fifteen yards away, we saw a black mass dimly outlined and then, at his word of command, we poured volley after volley into their closely packed ranks, shooting as fast as we could work the bolts of our rifles. When the blast struck them they thought they were being fired at by their own rearguard pickets, for there were cries of 'Rifle Brigade ! Rifle Brigade, don't fire!' but discovering their error a commanding voice called out 'Bayonets, bayonets,' and they came at us like a wall. In spite of our small number we delivered such a volume of fire that the head of the column swerved to the left and slanted across our front to make the spruit lower down, and, although we continued our volleys, we could not prevent their going by.

Several times, however, parties of soldiers who had lost their way in the dark walked in among us, and of these we shot some and took others prisoners.

A Captain named Geo. Paley[88] came up to where my brother and I knelt, firing over the edge of the bank, and as he failed to halt when called out, we both loosed a round and brought him toppling between us. Another time one of our men, Jan Luttig,[89] was seized by some soldiers within a few yards of us, and there was a hand-to-hand scuffle in which he was stabbed with a bayonet and clubbed on the head with a rifle-butt. In the dark we could not make out what was happening, but when we heard his cry for help and ran up, his assailants were gone. A moment later I made out three soldiers in the bed of the spruit behind me and slipped down towards

[88] Paley George Captain QSA (1) DoL Source: Ladysmith Siege Account and Medal Roll (Prince Consort's Own) Rifle Brigade.

[89] Several possibilities for Luttig. Anglo Boer War - Name search

them. One nearly spitted me with a vicious jab of his bayonet, which passed between my arm and body, but before he could repeat the thrust I had him covered and they surrendered. One was an army doctor with a bullet-wound in his foot, and the other two said they were helping him along. I ordered them to remain in the bed of the spruit, where I found them in the morning.

About this time we heard four or five shots in rapid succession, followed by groans, from the direction where the troops were still crossing the spruit twenty or thirty yards of. We did not know the meaning of this, and only at daybreak did we find that we had listened to the death-cry of some of our men who had come from the camp to our assistance.

It was now towards three in the morning, and we had nearly exhausted our ammunition, so we sat quietly watching the tail of the column vanish into the darkness beyond, on its way to Ladysmith.

When daylight came at last, a grim scene met our eyes. Before us, within a radius of less than twenty-five yards, lay over sixty dead and wounded English soldiers, and as we walked forward among them we came on the bodies of three of our men who had not been with us originally. Two were dead and dreadfully hacked with bayonets and the third was at his last gasp.

These had been away to the railway depot with the fatigue party, and on their return had gallantly attempted to make their way to us when they heard the firing. They nearly reached us, but ran into the withdrawing soldiers, and were bayoneted before they could fire more than a few shots. Behind us in the spruit bed lay poor Samuel van Zijl, and close by sat our prisoners, some twenty in number.

Dead, wounded, and prisoners, eleven of us had accounted for more than eighty opponents, an average of seven apiece.

Our work was now done, and shortly after sunrise Mr. Zeederberg and a large escort of men came scouting through the bush to see what was left of us, and they were surprised to find so many of us still alive.

We stood among the dead and wounded soldiers, the centre of an admiring crowd, and from now onwards Isaac Malherbe's corporalship was spoken of as the best in the commando.[90]

[90] Original Footnote 1: See Conan Doyle's 'History of The Great Boer War,' where he refers to me by name in connection with this affair. The Times "Story of The War" gives the following account:

The soldiers scrambled down Surprise Hill... . Suddenly a ring of flame blazed forth at their very feet. A party of some twenty Pretorians, mostly young lawyers and business men, had boldly come round the side of the hill and, regardless of the heavy fire, had lain down in a line across the slope to intercept the storming party of whose numbers they probably had no idea.

Disregarding the misleading orders shouted by the Boers, the men charged in grim silence. The gallant Pretorians emptied their magazines in a vain endeavour to stem the rush, and then the riflemen surged over them, killing or wounding several as they passed. The British casualties were fourteen killed and fifty wounded, not an excessive price to pay for so successful a feat.

VIII. A Visit to the Tugela Line

A few days after our fight at Surprise Hill we woke one morning to the rumble of distant gun-fire from the direction of the Tugela River and there was a buzz of excitement in the camps around Ladysmith when it became known that General Buller was launching his long-expected attempt to break through to the relief of the town. For a while there was no definite news, and we waited until sunset when word came that the British had been repulsed with heavy losses and that a number of guns had been taken.[91]

Boers with the captured guns after the battle.[92]

Boers at Ladysmith with one of Col. Long's captured guns.[168]

Hard on the success at Colenso arrived the news of victories on the other fronts. Piet Cronje had defeated and killed General Wauchope at Magersfontein,[93] and General Gatacre was beaten at Colesberg.[94] These tidings caused universal rejoicing, and there were few of us who did not believe that peace would soon follow, as it had after the Battle of Amajuba Hill [in] 1881.

[91] The Battle of Colenso took place on December 15th, 1899. See: Battle of Colenso - Wikipedia

[92] Apparently the breech blocks had not been removed and the guns were fully functional. Since the Boers still controlled the railroad between Colenso and the outskirts of beseiged Ladysmith they were able to use it to transport the weapons. Image source: britishbattles.com

[93] On December 11th, 1899. See: Battle of Magersfontein - Wikipedia

[94] December 10th, 1899 The Battle of Stormberg (Colesburg) was the first British defeat of Black Week, in which three successive British forces were defeated by Boer forces. Battle of Stormberg - Wikipedia

This time, however, their reverses seemed only to render the British more determined, but, the future being mercifully hidden from us, we confidently awaited the opening of peace negotiations and the surrender of Ladysmith, both of which events were expected to take place at any moment, and meanwhile camp life once more fell into the usual rut.

On Christmas Day another uncle of ours, Jan Mulder,[95] paid us a visit. He was a Hollander by birth and a rolling stone by nature after filling varied posts, and taking part in many native campaigns, he was Registrar of the Supreme Court in Swaziland when the war broke out and he enlisted as a private in the Swaziland Police,[96] with which force he was now serving on the Tugela.

As I was anxious to see the entrenchments there of which we had heard so much, I accompanied him on his return two days later.

We left the Pretoria camp at daybreak and rode round the east side of Ladysmith, touching at the laagers en route for a talk or a meal, and reaching the Tugela opposite Colenso by four in the afternoon. The Boer line roughly followed the ragged hills that fringe the north bank of the river, though here and there they were dug-in forward of the hills on the edge of the stream itself, and it was in one of these trench-lengths that we found the Swaziland Police, a small body of about one hundred strong.

1. Louis Botha's Commando
2. Boksburg Commando
3. Colenso
4. Krugersdorp Commando
5. Wakkerstrom Commando
6. Ermelo Commando
7. Swaziland Police
8. Ermelo Commando
9. British Camp, Cheively
10. Tugela River.

Battlefield of Colenso, December 15th, 1899.[169] 75 mm QF Creusot overlooking the battlefield of Colenso.[170]

The British Army lay across the river where their great tented camps were spread out on the plain. Judging by these and by the large numbers of troops being exercised every day, it was reckoned that there were forty thousand men facing us. I spent an interesting time riding up and down the line, and as things were quiet a party of us even rode through the river one morning to inspect the spot where the British guns had been captured and Lord Roberts's son killed in the recent battle. There were only decomposing horses and broken rifles lying about, and as the English gunners presently sent a brace of shrapnel over us we made haste to regain our own side.

[95] Likely: Mulder Johannes Henderik, Burger, Middelburg Commando. Deneys refers to him as his 'Hollander uncle.' Later his step-uncle, so a brother of his step mother Cornelia Mulder. Several other possibilities on ABW for this same name.

[96] Swaziland Commando: https://www.angloboerwar.com/?option=com_content&view=article&id=3063

Frere Camp.[171]

Our trench was shelled at intervals by a sixty-pound naval gun standing at Chieveley, seven miles away, but although some of the huge projectiles fell within a few feet of us, we suffered no damage. This was my first experience of lyddite, a newly invented explosive which went off with an appalling bang, emitting acrid green and yellow fumes that gave one a burning sensation in throat, and chest. I had read of its effect on the dervishes at Omdurman,[97] and the English newspapers had predicted equally terrible results for us; but the men made light of it and dubbed the shells 'little niggers' (klien kafferkies).[98] Whenever one of them came through the air there was a warning cry: 'Look out — a little nigger,' and everyone dived for cover behind the sandbags.

One of the 4.7's and a 12 pounder in action. Note the rope anchored to the ground to prevent recoil.[172] Photograph of naval ratings fitting primers / primer adaptors during the Tugela operations.[99]

97 The battle took place on 2 September 1898, just north of Omdurman in the Sudan. Battle of Omdurman - Wikipedia

98 This is unique and demonstrates that the two terms were distinct. During the Dutch and English Colonial Period the word Kaffir was used to describe monotheistic peoples (Nguni ethnic groups in particular) of South Africa, who were not of a Christian or of an Islamic religious background, without derogatory connotations, until the early twentieth century. It appears in many historical accounts by anthropologists, missionaries and other observers, as well as in academic writings. See Also: Kaffraria and British Kaffraria (Queen Adelaide's Province) were historical regions along the coast north or Port Elizabeth. See:File:Eastern Frontier, Cape of Good Hope, ca 1835.png - Wikimedia Commons Subsequently it became a racial slur. Kaffir (racial term) - Wikipedia See also the British military unit Kaffrarian rifles which were white and black units which were still organized as late as 1947. East London, 1 March 1947. King George VI inspecting 'Kaffrarian Rifles' guard of honour. - Atom site for DRISA And: Kaffrarian Rifles AKA: Buffalo Volunteer Rifles Buffalo Volunteer Rifles - Wikipedia Also: a historical region in Afghanistan and the setting of most of Rudyard Kipling's famous 1888 novella "The Man Who Would Be King." Kafiristan - Wikipedia And: an Arabic and Islamic term which, in the Islamic tradition, refers to a person who disbelieves in God as per Islam. Kafir - Wikipedia PF Note: I have no authority to change the wording used by original authors. There are instances where a pejorative or admiring modifier is used. While this is racist in and of itself, to identify someone by their complexion, in this narrative, identifiers such as Dutch, Boer, English, Uitlander, Indian, and many others are used as means of identification sometimes with pejorative or admiring modifiers.

99 For land service adaptors had to be inserted into the bases of the cartridges, enabling the electric primers to be replaced with percussion primers.

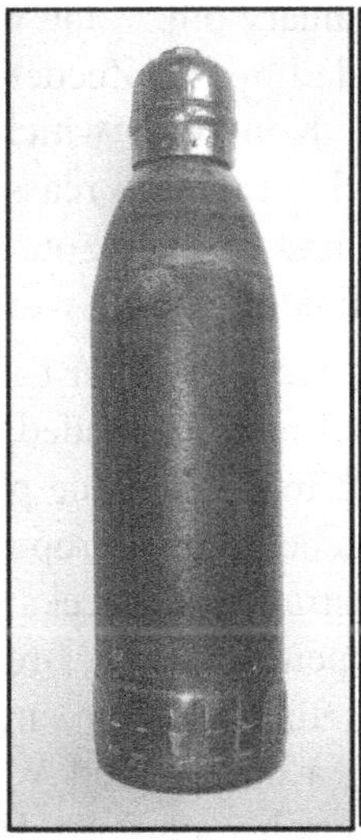

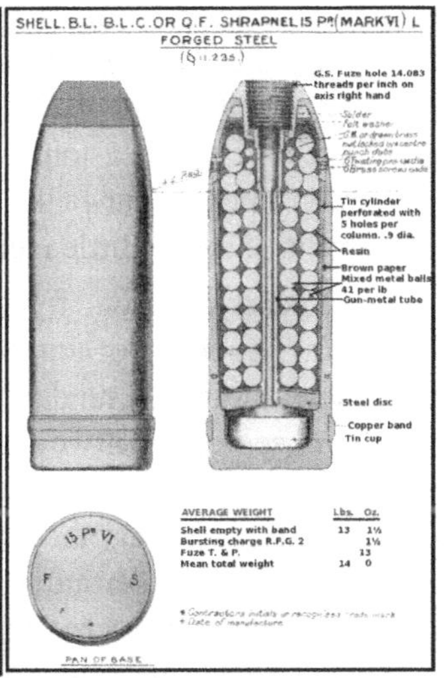

Varieties of Ammunition collected at Ladysmith.[173] 15 Lb. Shrapnel Shell with timed fuse on top.[174]
Diagram of Construction.[175]

I saw the New Year (1900) in with the Swazilanders, who held a sing-song in celebration, and three days later I said good-bye to my uncle, and started home.

This time I rode round the western circle of Ladysmith in order to visit the Free State commandos guarding that side of the cordon, as I had many friends and acquaintances there, and, after spending a pleasant day among them, I timed my journey to reach the Kroonstad camp by sunset, as I wished to put in a night with my Norwegian uncle.[100] I found him making ready to accompany a picket to the plain towards the English lines. His preparations for going on duty were unusual, for he inspanned[101] an American buck-board on which he piled a feather bed, blankets and pillows, and as soon as the rest of the Picket moved out after dark he told me to leave my roan horse in camp, and he and I drove in state to the place where we had to do guard for the night. The spot was not more than four hundred yards from the nearest English sentries, but this in no way disturbed my uncle's serenity. He unharnessed his two animals, hitched them to the wheels of the buggy, and giving one of the younger fry half a crown[102] to do his sentry-go, he spread his bedding and was soon comfortably snoring within earshot of the British. I shared his couch, and we both slept soundly until awakened shortly before daybreak in time to harness the horses and get back to camp before it grow light. I spent the day with the Kroonstad people and listened to a council-of-war.

For weeks there had been talk of an attack by the Free State forces against a loose-standing kop called Wagon Hill, considered to be the key of the Ladysmith defences. We had so often heard of the proposed attempt that by now we had ceased to believe in it, but this time it looked as if something was on foot at last, for Commandants and Field-Cornets came riding in from the neighbouring camps to attend the meeting. Here I learned that it had been decided to storm the hill within the next day or two. It was said that Piet Joubert was growing impatient at the delay of the Ladysmith garrison in surrendering, and hoped to help them to make up their minds by the capture of this commanding position. I had a good look at Wagon Hill, but I came away somewhat dubious of success, although the Freestaters were eager to have a shot at it.

Next morning I rode through the Klip River and past Nicholson's Nek back to the Pretoria camp, still wondering whether the attack would materialize. When I arrived at our tent I was told that it was in fact due for the very next morning, and that four hundred Pretoria men were to create a diversion by falling upon our old friend the Red Fort in order to draw off the enemy's attention from the main objective while the Freestaters attacked Wagon Hill.

Our corporalship was to take part, and Isaac Malherbe had already given the necessary orders.

[100] Theodore Thesell.

[101] To yoke (oxen).

[102] The British half crown was a denomination of sterling coinage worth two shillings and six pence, or 1/8 of one pound.

At 3 a.m. on the morning of January 6th,[103] the whistles blew for the attacking force to assemble, and soon we were marched off on foot, headed by Mr. Zeederberg and Assistant Field-Cornet de Jager. We collected in the same dry spruit below Belles Kop from which we had started on our previous abortive expedition. Our experience on that occasion did not tend to reassure us, for once more there was a strong dilution of burgher-right 'erven' men in our ranks, and we shook our heads when we saw them dejectedly standing behind the bank of the spruit, showing little inclination for the work in hand.

However, nearly half our force were of better calibre, and when, shortly before sunrise, the two Field-Cornets led the way, about two hundred men responded, the rest refusing to follow. There was no time to argue, so, leaving the shirkers behind, we made over the plain towards the enemy works. It was still dark and we went along un-observed until we reached the outcrop of rock halfway across, where some of us had spent such evil hours during the former demonstration six weeks ago. Here the English in the fort observed us, and although the dawn was just breaking, they opened a heavy fire in our direction. This brought us to a standstill, and for a few moments we were uncertain whether to go forward or to fall back. The Field-Cornets seemed equally undecided, but, while they were debating, a man named Willernse intervened. He was a member of President Kruger's bodyguard, a fine tall fellow in police uniform. He was a stranger to us, for he had been on a chance visit to a relative at our camp, and had volunteered when he heard of the attack. He made an eloquent appeal to them to proceed, his words creating such an impression that we started for the fort at a run. The place was built on the brow of a low stony kopje immediately beyond the Harrismith railway line and, except where its earthworks formed an embankment, there was no cover. We had about four hundred yards to go, but the light was still uncertain, so we gained the safety of the ramp with the loss of only one man. This railway causeway proved our undoing, for had it not been there the chances are we should have gone straight on into the fort, but the sheltering bank was too tempting, and with one accord the men halted behind it to recover their breath. In the circumstances the delay was fatal, for we lost the original impulse that had carried us thus far, and as the light was increasing and with it the fierce volleying, we became disinclined to leave cover, and, instead of resuming our advance, stayed where we were.

We were now close in under the walls of the fort and could plainly see the muzzles of the defenders' rifles as they stuck them through the loopholes to fire at us. Other soldiers were standing by to repel boarders, for above the breastworks their bayonets glinted in the rising sun and gave us still further food for thought.

After a while, Isaac Malherbe, my brother, and a few more of us crawled along the track and, jumping across the metals, ran up the slope to where fair-sized boulders gave good cover within twenty-five yards of the work. Our intention was to have a quick jumping-off point should the Field-Cornet give the order for a final rush, and from here we lay firing into the loopholes, waiting for the command to advance.

As we were now well ahead of where the rest of our force lay crouching behind the railway embankment, we were out of touch with them, although we kept looking back to see whether there was any sign of their coming on. Just as we were beginning to think that the attack had been called off we heard a shout, and saw Willemse leap out on to the track with about a dozen or fifteen men at his heels. At this we, too, sprang up to join in the charge, but almost as we rose, swift destruction overtook the storming party. A single volley flamed along the portholes, and before we had time to think the attack had withered away. We saw the men go down in a heap, leaving only one man erect. The rest were either dead or wounded or had flung themselves headlong for such cover as they could find.

Those of us with Isaac escaped annihilation because we were out of the direct line of fire and were able to regain protection before the soldiers had time to turn their attention to us, but we did not go quite unharmed, for Frank Roos, my tent-mate, fell dead among us with a bullet through his heart. The man who still faced the enemy was Willemse, and he, undeterred, ran up to the fort and tried to scale the wall. Bayonets were thrust at him which he parried with his rifle until a revolver was fired point-blank into him. He sank below the wall where he sat rocking to and fro, his head resting on his knees as if in great pain, and then another bullet found him, for he suddenly pitched forward on his face. We were so bewildered by the suddenness of everything that before we could collect our thoughts it was all over.

Willemse and six of his band lay dead, the survivors crouched behind cover against the driving bullets. Several of them were wounded, for they lay trying to bandage their injuries without exposing themselves from behind

[103] See Battle of Wagon Hill section of: Siege of Ladysmith - Wikipedia

their scant shelter, and it was painful to see their efforts without being able to assist; but it was suicide to venture out under the wall in the teeth of such heavy fire.

We did not know why this piecemeal sally had been made, but after waiting for some time we were satisfied that there was to be no further attempt on the fort, so we recrossed the railway line, one at a time, crawling over until we were all back behind the bank where the rest of the men were gathered. Mr. Zeederberg told me that the sortie had been made without his orders. Before he could stop them Willemse and the men were climbing up the slope to the disastrous end which we had witnessed.

There was now nothing else for us to do but to repeat our previous experience by staying where we were until the coming of night would enable us to retire. To renew the attack after what had happened was out of the question, nor was it feasible to withdraw in broad daylight from under the walls of the fort across a thousand yards of open veld.

It was as yet before eight in the morning, but we heard the sound of violent gun-and rifle-fire from Wagon Hill, three or four miles away, where the main battle was being joined. How the Freestaters were prospering we had no means of knowing, but we guessed that heavy fighting was in progress, for the throbbing of the guns increased, and the rattle of thousands of rifles. Our share in the proceedings, however, was over: caught like rats in a trap we had to confine ourselves to loosing an occasional shot and gazing longingly at the hill far to rearward where the balance of the Pretoria commando looked on comfortably at our plight.

As the hours dragged by, the sun blazed down unbearably, and the wounded moaned and cried for water; of which we had none. By scrambling across the track and crawling from rock to rock we managed to pull most of them into the shelter of the embankment until we had brought in all except a few who were lying too far up. While busy at this I took a hurried glance at the dead. One of them was Assistant Field-Cornet de Jager, the others were men I had seen about camp, but did not know personally. Then I lay reading an old newspaper I had in my pocket, and slept at intervals to pass the time. This brought me subsequent notoriety, for when I unfolded the paper those watching us from the rear thought that it was a white flag, and some busybody rushed down into the laager to spread the news that we were surrendering to the enemy. Before field-glasses could be procured to verify the report it was too late to overtake the rumour, and by nightfall every commando around Ladysmith had heard that we had been marched into the town.

In this manner the day wore through. Occasionally we put a shot into the loopholes above us, but otherwise we lay inactive behind the railway earthwork counting the hours still to elapse before darkness. Through it all we heard the battle on Wagon Hill ebb and flow. At times it flared up and then died away, until some new turn caused a further eruption. The condition of our wounded grew worse and worse as the heat increased, and there was nothing we could do for them. It was bad enough to be thirsty ourselves, but to hear them beg for water and to sit helpless by was terrible, so we averted our eyes and pretended not to hear them.

At length, towards five in the afternoon, when things were getting unbearable, relief came from an unexpected quarter. Almost without warning, black clouds raced across the sky and there broke upon us the most violent storm that I have over experienced. It leapt at us with a roar, and there fell a deluge which in an instant blotted all from view. The rolling thunder and the drumming of the waters were deafening. For perhaps a minute we waited and then we abandoned our dead and wounded and fled blindly through the hurricane in the direction of our camp, a mile away.

We found the whole place thrown into confusion by the storm. Our tents were down and our belongings were littered a mile and more along the valley, washed away by the flood. As we could neither light a fire nor dry our clothing, we spent a miserable night, made the more unhappy by the knowledge that we had selfishly left our wounded in the dark and rain and, to add to our discomfiture, came the tidings next morning that the Freestaters, despite magnificent courage, had failed to take Wagon Hill, and had lost three hundred men in the attempt.

Later in the day our dead were fetched in by wagon under flag of truce.

We carried poor Frank Roos to our tent, and for the second time in three weeks my companions and I sat by the body of a messmate, wondering whose turn was next.

Flag of Truce by Clive M. Dixon.[176]

Cartoons by Clive M. Dixon.[177] "The First Shell." "The Last Shell."

Cartoons by Clive Dixon.[178] Showing a Boer Commandant inspecting a position and then having his horse bolt when the gun is fired.

Battle of Wagon Hill (or Platrand)

The Boers around Ladysmith were also growing weak from lack of forage. With little action, many fighters took unauthorised leave or brought their families into the siege encampments. Eventually, with the Tugela in flood, preventing Buller from giving any support, some younger Boer leaders persuaded General Piet Joubert to order a storming attempt on the night of 5 January 1900, before another relief attempt could be made.

The British line south of Ladysmith ran along a ridge known as the Platrand. The occupying British troops had named its features Wagon Hill to the west and to the east Caesar's Camp (after features near Aldershot, well known to much of the British army). Under Ian Hamilton, they had constructed a line of forts, sangars and entrenchments on the reverse slope of the Platrand, of which the Boers were unaware.

In the early hours of January 6th 1900, Boer storming parties under General C. J. de Villiers began climbing Wagon Hill and Caesar's Camp. They were spotted and engaged by British working parties who were emplacing some guns. The Boers captured the edge of both features, but could not advance further. British counterattacks also failed.

At noon, de Villiers made another attack on Wagon Hill. Some exhausted defenders panicked and fled, but Hamilton led reserves to the spot and recaptured some empty gun pits. Late in the afternoon, a terrific rainstorm broke, and the Boers withdrew under cover of it.[104]

Imperial Light Horse.[179]

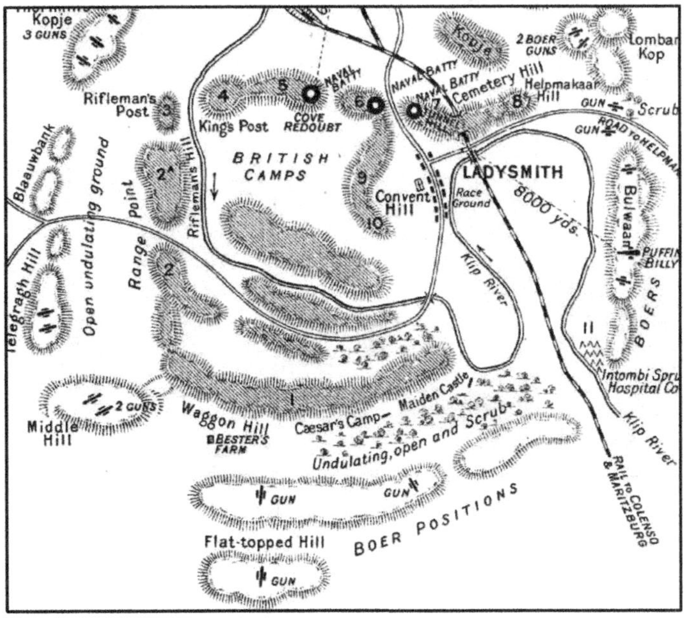

Caesars Camp Area.[180]

[104] Spiers, Edward, ed. (2010), Letters from Ladysmith: Eyewitness Accounts from the South African War (illustrated ed.), Frontline Books, p. 77–84, ISBN 9781848325944.

These are WSC's observations in London To Ladysmith.

> At two o'clock, according to Sir George White, the Boers in great numbers, evidently reinforced from Colenso, surprised the pickets and began a general attack on the outpost line round the town, particularly directing their efforts on Caesar's Camp and Waggon Hill. The fighting became very close, and the enemy, who had after all hardened their hearts, pushed the attack with extraordinary daring and vigour. Some of the trenches on Waggon Hill were actually taken three times by the assailants. But every time General Hamilton—the skilful Hamilton as he has been called—flung them out again by counterattacks. At one place, indeed, they succeeded in holding on all day, nor was it until the dusk of the evening, when the rain and thunderstorm which we saw hanging over Colenso broke on Ladysmith, that Colonel Park led forth the Devon Regiment—who, having had half their officers killed or wounded by a shell some days before, were probably spiteful—and drove the Dutchmen helter skelter at the point of the bayonet. So that by night the Boers were repulsed at every point, with necessarily great slaughter, greater at any rate than on our side. Their first experience of assaulting! Encore!

The British suffered 175 killed and 249 wounded. 52 dead Boers were left in the British positions, but their total casualties were not recorded.[105]

Casualties In Action	Killed		Wounded		Missing and Prisoners		
Ladysmith Seige & 1/6	Officers	Men	Officers	Men	Officers	Men	Total
Wagon Hill/ Platrand	14	164	33	287	0	2	500

Casualties per Conan Doyle's 'Great Boer War'.

Original Imperial Light Horse Monument.[181]

Imperial Light Horse Memorial in 2008.[106]

[105] By 4 PM Ian Hamilton surmised that the overall attack had failed and that the Boers would withdraw after dark but White insisted this costly attack be made. Pakenham p. 275.

[106] Memorial on Platrand outside Ladysmith to those members of the Imperial Light Horse that died during the Siege of Ladysmith. File:Platrand-Imperial Light Horse Memorial-001.jpg - Wikimedia Commons Author: NJR ZA. See Also: Herman Albrecht - Wikipedia and Robert Digby-Jones - Wikipedia, Both poshumously awarded the Victoria Cross for their actions during the engagement.

IX. The Battle of Spion Kop

About a week later, Mr. Zeederberg said I could go to Pretoria for a fortnight, so leaving my brother in charge of my horses, I once again boarded a north-bound goods train and travelled home in a cattle-truck, getting-there in three days' time. My father did not know I was coming, and although he gave me a warm welcome, he insisted on my returning at once, as he said the British were on the eve of delivering another great blow on the Tugela, and that my place was in Natal. I told him that the burghers thought there would be no more serious fighting, but he shook his head and said he only wished he could share our optimism.

I was disappointed, as I had been looking forward to the luxury of home life and good food for a little while, but I saw his point and started back on the second day.

I reached our camp at Ladysmith on **January 23rd** (1900) to find that volunteers were being called for to go to the Tugela, and I now heard that General Buller had moved the English army twenty-five miles upstream from Colenso in preparation for another big-scale attack in the vicinity of Spion Kop, a prominent hill forming part of the Boer line on the north bank of the Tugela. Already they were hammering away at different points, seeking a weak spot at which to thrust; so my father had been right, and, indeed, the situation was so critical that reinforcements were being sent from every commando lying around Ladysmith.

From the Pretoria laager fifty volunteers were asked for, and more than three times that number immediately offered themselves. The Field-Cornet made a selection which included Isaac Malherbe, my brother and myself and our three remaining tent-mates, Charles Jeppe, de Vos and Reinecke, as well as several more of our corporalship.

We set out within an hour of my arrival from Pretoria, and crossed the Klip River after dark, riding all night round by the west until we reached the rear of Spion Kop at day-break. As we rode, we could hear the sound of heavy gun-fire from the forward hills, and it never ceased for any length of time although we were still too far back to be in it, danger.

After a short halt to rest our horses and cook breakfast, we were ordered to the top of a steep ridge lying about a mile to the right, where we had to dig a reserve trench. A mule-wagon had accompanied us from Ladysmith carrying provisions, ammunition and a supply of pick-axes and shovels, with which unaccustomed tools we started up the slope, horses and all.

When we had dug for some time, Field-Cornet Zeederberg, who was always very kind to me, said that as I was the youngest I need not dig any more and could go down to where the wagon had been left for a rest. Nothing loth,[107] I made haste to reach the halting-place, and, leaving my horse in charge of the mule-drivers, I started out to see what was going on in the front positions, which were out of sight from where we had been digging.

Ever since sunrise there had come the unbroken boom of guns and the rattle of small arms, and now that I was free I decided to walk across the intervening hills to the firing-line. As I went, the gun and rifle-fire grew louder, and before long I reached a point from which I could see the Boer front strung out along the top of the next rise.

Black mushrooms of earth and smoke hung along the course of the positions from the heavy shells flung across the Tugela, and puff's of shrapnel flecked the air above. From the noise I judged that a battle was in full progress and, after some hesitation, I hurried on and reached the line in safety. The spectacle from here was a fine one. Far below on the plain the Tugela wound shining in the sun, and the bank beyond was alive with English foot and horse. From the wooded hills farther back came the flashes of the British guns, and in the din I asked myself more than once why I had been foolish enough to come.

During the preceding days the English had effected a lodgement at numerous points on our side of the river, and their troops were occupying such spurs and ridges running up from the water's edge as they had been able to seize. The Boers, on being pushed back, had reformed along the crest of the height, where they were now holding a series of hastily dug trenches and whatever natural cover they could find, and were stoutly resisting any further encroachments of the enemy, who in places were lying within a few hundred yards of us.

[107] He means nothing loath, as in not distasteful or unpleasant.

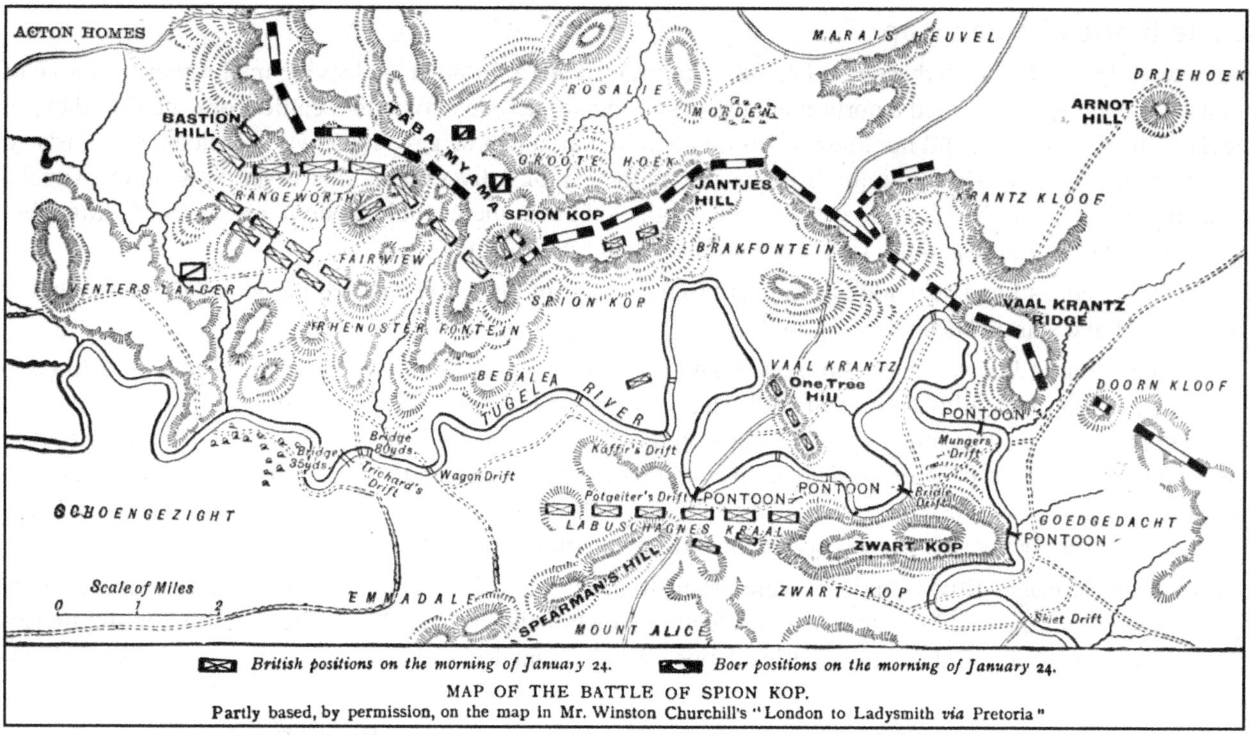

Boer Fortifications extending westward from Vaal Krantz Ridge to Acton Homes with Spion Kop centered.[182]

Spion Kop, 2008.[183]

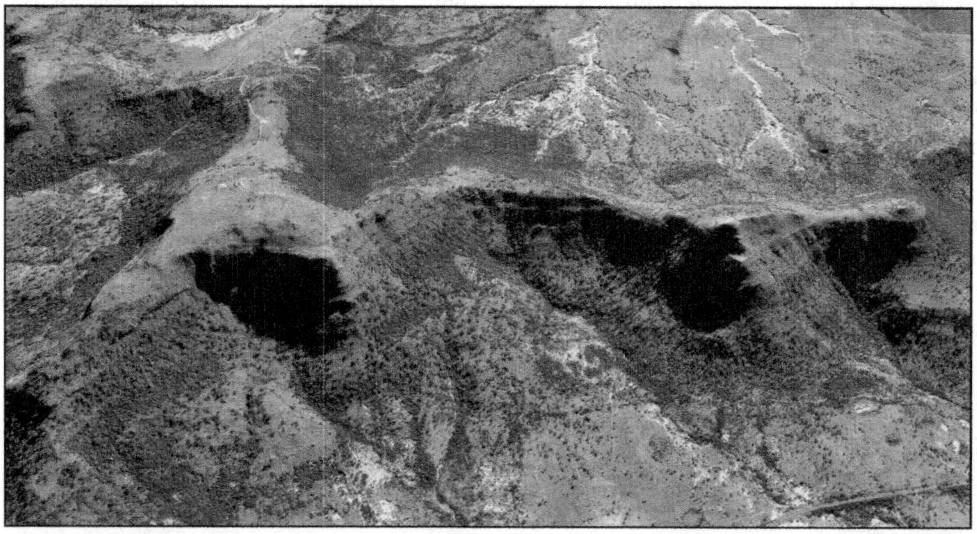

Spion Kop Google Image © 2023 Image Landsat/Copernicus.

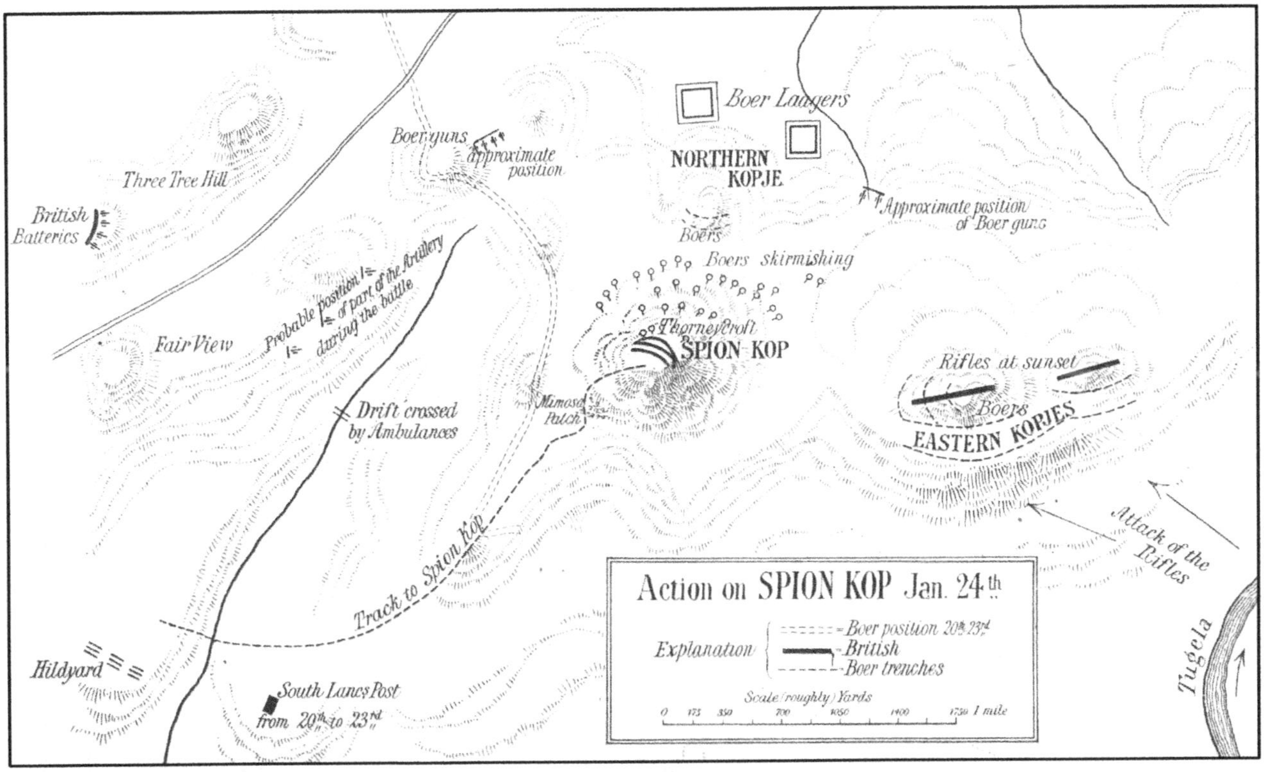

Spion Kop Battle Area.[184]

Royal Lancasters January 19th, 1900.[185] Cameronians Maxim Gun on Zwart Kop during Spion Kop Battle.[186]

The Positions here were held by Free State commandos while downstream lay the Transvaalers. There were probably ten or twelve thousand burghers in all on these hills, with the bastion of Spion Kop standing like a pivot in the centre. For the most part the men made slight reply to the fire in order to husband their ammunition, and our artillery kept silent for the same reason, although it was estimated that there were over two hundred guns firing at us, and I have heard that this was the heaviest concentration of gun fire that has been seen in any war up to the present.

Boer quick firing field gun and crew at Spion Kop.[187] Gun Emplacement.[188]

Bodies of British Soldiers on Spion Kop. [189] Note how much elevation the adjoining hill has overlooking their position.

Boer riflemen with their flag on Spion Kop.[190]

The casualties were considerable and I saw some men fearfully mutilated, including a father and son of the Frankfort commando who were torn to pieces by a howitzer shell, their rifles being sent spinning down the incline at the back of us.

It was a day of strain. Not only was there the horror of seeing men killed and maimed, but there was the long-drawn tension and fear of the approaching shells.

This tremendous volume of fire indicated an early attack, and throughout the day we looked to see the storm break at any moment, but, as it turned out, the bombardment was a feint, the real blow being delivered after midnight at a different point.

I was entitled to quit the line as my unit lay in the rear, but I did not like to go, and remained until things died down towards sunset, when I could return without loss of face. I found the Pretoria volunteers where I had left them digging that morning. They must have worked well, for they had completed quite a long trench.

I joined Isaac Malherbe and others sitting round the fires cooking their supper, and, watching the light fade away over the distant Drakensbergen, I chatted for a quiet hour with men who were mostly dead next morning.

Field-Cornet Zeederberg now ordered me down with him to the supply wagon. He said he was going to spend the night there, and, as he might require to send a message up to the trench, I was to come with him for I had good climbing legs.

When we got below, a tent had been pitched for him which I was allowed to share, and I was soon fast asleep. It rained at intervals during the night, and towards three in the morning we were waked by an angry stutter of rifle-fire coming from Spion Kop. We sat up listening, but as there was nothing we could do in the rain and darkness, and as after a while the firing died down, we fell asleep again.

At sunrise loud gun and rifle-fire broke out along the front on which I had been the day before; but, as it was no worse than it had then been, Mr. Zeederberg and I were not unduly perturbed and sat sipping our morning coffee in the lee of the wagon out of the way of the spent bullets that whined over our heads.

As we breakfasted one of our Pretoria men galloped up with a message from; Isaac Malherbe to say that the British had made a night attack and had captured Spion Kop. This was most serious, for if the hill went-the entire Tugela line would go with it, and we could hardly bring ourselves to believe the news. The man, however, assured us that it was true, but he said that a strong force of burghers was assembling below the hill and that Isaac Malherbe had ridden down by a short cut with all the men who were with him, so we shouted to the mule-drivers to saddle our horses, and filling up with ammunition from a box on the wagon we followed on the heels of our guide.

Heavy shells were lobbing over as we went but we had not far to go and in less than fifteen minutes reached the bottom of Spion Kop. Here stood hundreds of saddled horses in long rows, and we looked up at an arresting sight.

The Boer counter-attack had started shortly before. Eight or nine hundred riflemen were climbing up the steep side of the hill in face of a close-range fire from the English troops who had established themselves on the flat summit overnight. Many of our men dropped, but already the foremost were within a few yards of the rocky edge which marked the crest, and soldiers were rising from behind their cover to meet the final rush. For a moment or two there was confused hand-to-hand fighting, then the combatants surged over the rim on to the plateau beyond, where we could no longer see them. Spellbound, we watched until our men passed out of view, and then, recovering ourselves, dismounted, and tying our horses with the rest, hurried up in the wake of the attack.

Dead and dying men lay all along the way, and there was proof that the Pretoria men had gone by, for I soon came on the body of John Malherbe, our Corporal's brother, with a bullet between his eyes; a few paces farther lay two more dead men of our commando. Farther on I found my tent-mate, poor Robert Reinecke, shot through the head, and not far off L. de Villiers of our corporalship lay dead. Yet higher up was Krige,[108] another of Isaac's men, with a bullet through both lungs, still alive, and beyond him Walter de Vos of my tent, shot through the chest, but smiling cheerfully as we passed. Apart from the Pretoria men there were many other dead and wounded, mostly Carolina burghers from the eastern Transvaal, who formed the bulk of the assaulting column. Spion Kop, although steep, is not very high on the northern slope where we went up, and it did not take us long to reach the top. Here we found that the advance had got no farther than the fringe of loose rocks that runs like a girdle around the upper tableland. For the rest of the flat stretch beyond was still wholly in the hands of the British, who lay in a shallow trench behind a long low wall of stone about twenty yards away. From here came a vicious rifle-fire that made further progress impossible. It was marvellous that the Boers had got even thus far, for they had swarmed up the bare hillside in the face of a devastating fire, and they had pushed home the attack with such vigour that the narrow belt of rocks was thickly strewn with their dead.

I met my brother coming down in charge of captured soldiers and did not see him again as he had orders to escort them to Ladysmith, and he took no further part in the battle.

Giving him a hurried handshake, I went forward to the firing-line a few yards farther on. During the short delay I lost touch with Mr. Zeederberg, and when I inquired from the men crouching behind the rocks for Isaac Malherbe, I was told by Red Daniel Opperman,[109] the officer in command, that he had sent the Pretorians round to the ledge a few minutes earlier to rake the English flank. Working my way in that direction, I reached a spot where the out-crop of rocks came to a dead end. From here spread a patch of open ground until the ledge reappeared a hundred yards beyond.

One of the men holding this point told me that the Pretoria men had doubled across the gap shortly before and were now lying among the rocks on the far side, so I decided to follow; but the moment I left cover I drew so hot a fire that I was thankful to dive back for shelter and give up the attempt. Halfway across lay the huddled body of a dead man, and now that I had time to look more carefully at him I recognized **Charles Jeppe**, the last of my tent-mates. His death affected me keenly, for we had been particularly good friends. Outwardly he was a

[108] Krige Petrus Johannes, Burger, Pretoria Commando. 638 Records for the Commando: https://www.angloboerwar.com/?option=com_content&view=article&id=3061

[109] Opperman D J E Kommandant Source: Roll of DTD recipients.

surly man, but he had shown me many a kindness since first we messed together on the Natal border. As I was unable to find my Corporal, I now returned to where I had first reached the top and took my place in the firing-line.

During my absence about fifty soldiers had run forward to surrender, but otherwise things were going none too well. We were sustaining heavy casualties from the English 'schans' immediately in front of us, and the men grew restive under the galling point-blank fire, a thing not to be wondered at, for the moral effect of Lee-Metford volleys at twenty yards must be experienced to be appreciated. The English troops lay so near that one could have tossed a biscuit among them, and whilst the losses which they were causing us were only too evident, we on our side did not know that we were inflicting even greater damage upon them. Our own casualties lay hideously among us, but theirs were screened from view behind the breastwork, so that the comfort of knowing that we were giving worse than we received was denied us.

Fortunately, towards nine o'clock the situation eased, for the Transvaal artillerists got their guns into action on a commanding spur a mile away, and they began to fire over our heads into the troops crowded on the restricted space on the plateau before us. As the guns searched the hill-top the English fire slackened, and from then onward our losses were less. The position, however, remained unsatisfactory. The sun became hotter and hotter, and we had neither food nor water.

ZAR Staats Artillery using Maxim Pom Pom.[191]

Mountain Battery's 2.5-inch Mountain Gun.[192]

Around us lay scores of dead and wounded men, a depressing sight, and by midday a feeling of discouragement had gained ground that was only kept in check by Commandant Opperman's forceful personality and vigorous language to any man who seemed to be wavering. Had it not been for him the majority would have gone far sooner than they did, for the belief spread that we were being left in the lurch. We could see large numbers of horsemen collecting at the laagers on the plain behind, but no reinforcements reached us throughout the day. I repeatedly heard old Red Daniel assure the men that help would be forthcoming, but from the way he kept scanning the country below I could see that he was getting uneasy himself.

Top of Spion Kop 24 January 1900. Text "Lost [to] Artillery War 1899|00"[193]

Deceased soldier at Spion Kop.[194]

As the hours dragged on a trickle of men slipped down the hill, and in spite of his watchful eye this gradual wastage so depleted our strength that long before nightfall we were holding the blood-spattered ledge with a mere handful of rifles. I wanted to go too, but the thought of Isaac and my other friends saved me from deserting. No further attempt was made to press forward, and for the rest of this terrible day both sides stubbornly held their ground, and, although the battle remained stationary, the heavy close-range rifle-fire continued hours after hour, and the tale of losses mounted while we lay in the blazing heat.

I saw a strange incident during the morning. Near me was a German named von Brusewitz. He had been an officer in the German army, but the year before he had run a civilian through with his sword during some scuffle in a Berlin cafe. There was a great outcry over the incident, and to allay popular clamour the German Emperor broke him from his regiment. They say that in Germany the word 'Brusewitzerei' is still used to denote the arrogance of the officer caste. However that may be, von Brusewitz was now on top of Spion Kop, where he seemed bent on getting killed, for although we warned him not to expose himself too recklessly, he paid no heed, and repeatedly stood out from among the rocks to fire.

As the English soldier were so close to us this was sheer folly, and after he had tempted providence several times the inevitable happened. I saw him rise once more, and, lighting a cigarette, puff away careless of the flying bullets until we heard a thud, and he fell dead within a few feet of me, shot through the head. Not long after this something similar happened. An old Kaffir servant came whimpering up among us from below, looking for his master's body. I advised him to be careful as he went from rock to rock peering over to examine the dead men lying in the open, but he would not listen, and soon he too had a bullet through his brain.

The hours went by; we kept watch, peering over and firing whenever a helmet showed itself, and in reply the soldiers volleyed unremittingly. We were hungry, thirsty and tired; around us were the dead men covered with swarms of flies attracted by the smell of blood. We did not know the cruel losses that the English were suffering, and we believed that they were easily holding their own, so discouragement spread as the shadows lengthened.

Batches of men left the line, openly defying Red Daniel, who was impotent in the face of this wholesale defection, and when at last the sun set I do not think there were sixty men left on the ledge.

Darkness fell swiftly; the firing died away, and there was silence, save for a rare shot and the moans of the wounded. For a long time I remained at my post, staring into the night to where the enemy lay, so close that I could hear the cries of their wounded and the murmur of voices from behind their breastwork.

Afterwards my nerve began to go, and I thought I saw figures with bayonets stealing forward. When I tried to find the men who earlier in the evening had been beside me, they were gone. Almost in a panic I left my place and hastened along the fringe of rocks in search of company, and to my immense relief heard a gruff 'Wer da?' It was Commandant Opperman still in his place with about two dozen men. He told me to stay beside him, and we remained here until after ten o'clock, listening to the enemy who were talking and stumbling about in the darkness beyond.

At last Opperman decided to retreat, and we descended the hill by the way which he had climbed up nearly sixteen hours before, our feet striking sickeningly at times against the dead bodies in our path. When we reached the bottom most of the horses were gone, the men who had retired having taken their mounts and ridden away, but our own animals, and those belonging to the dead or wounded were still standing without food or water where they had been left at daybreak.

The first thing to do was to quench our raging thirst and that of our horses at a spring nearby. We then consulted as to our next move. Most of the wounded had been taken off in the course of the day, but we found a few serious cases that would not bear transport collected in charge of an old man, who, by the dim light of a lantern, was attending to their wants. We could get no coherent information and stood discussing what to do next, for we did not know that the English had also been fought to a standstill, and that they in turn were at that very moment retreating down their own side of Spion Kop. We fully believed that the morning would see them streaming through the breach to the relief of Ladysmith, and the rolling up of all our Tugela line.

While we were talking, Mr. Zeederberg came out of the dark. I had lost sight of him during most of the day, but he had been on the hill all the time, and had only come down shortly before us. He had seen nothing of Isaac Malherbe and the rest of our Pretoria men, and had no idea of what had become of them. A few more stragglers joined us and we agreed to lead our horses to the Carolina wagon-laager that, as we knew, lay not far off. We foraged for food in the saddlebags of such horses as were left, and then went off. When we reached the laager we found everything in a state of chaos. The wagons were being hurriedly packed, and the entire Carolina commando[110] was making ready to retire. They had borne the brunt of the day's battle and had fought bravely, but, now that the struggle was over, a reaction had set in and there was panic in the camp.

Fortunately, just as the foremost wagons moved away and the horsemen were getting ready to follow, there came the sound of galloping hoofs, and a man rode into our midst who shouted to them to halt. I could not see his face in the dark, but word went round that it was Louis Botha, the new Commandant-General, appointed in place of Piet Joubert who was seriously ill. He addressed the men from the saddle, telling them of the shame that would be theirs if they deserted their posts in this hour of danger; and so eloquent was his appeal that in a few minutes the men were filing off into the dark to reoccupy their positions on either side of the Spion Kop gap. I believe that he spent the rest of the night riding from commando to commando exhorting and threatening, until he persuaded the men to return to the line, thus averting a great disaster.

As for Commandant Opperman and our party, now that the Carolina burghers were returning we led our horses back to the foot of Spion Kop, to wait there.

We woke with the falling of the dew and, as the sky lightened, gazed eagerly at the dim outline of the hill above, but could make out no sign of life.

Gradually the dawn came and still there was no movement. Then to our utter surprise we saw two men on the top triumphantly waving their hats and holding their rifles aloft. They were Boers, and their presence there was proof that, almost unbelievably, defeat had turned to victory—the English were gone and the hill was still ours.

Leaving our horses to fend for themselves, We were soon hastening up the slope past the dead until we reached yesterday's bloody ledge. From here we hurried across to the English breastworks, to find them abandoned. On our side of the fighting-line there had been many casualties, but a worse sight met our eyes behind the English schanses.

In the shallow trenches where they had fought the soldiers lay dead in swathes, and in places they were piled three deep.

"The British trench in 'the murderous acre' on Spion Kop after the battle."[111] Spion Kop in 2005.[112]

The Boer guns in particular had wrought terrible havoc and some of the bodies were shockingly mutilated. There must have been six hundred dead men on this strip of earth, and there cannot have been many battlefields where there was such an accumulation of horrors within so small a compass.

Shortly after I reached the top, Isaac Malherbe and the remaining Pretoria men came up. They had spent the night somewhere below the kop, and like ourselves had come up the moment they realized that the English were gone. Isaac looked grim and worn, grieved at the death of his brother and of our other companions, but he was

[110] https://www.angloboerwar.com/unit-information/boer-units/177-transvaal-commandos/1954-carolina-commando 198 Records.

[111] This picture appears to be taken before internment. The soldiers appear to have been laid to rest. Image: File: Spion Kop Grave 1 and also: File:Spioenkop1.jpg - Wikipedia.

[112] File:Spioenkoploopgraf3.JPG - Wikipedia Spion Kop 2005: This photo shows a section of the British graves at the site of the Battle of Spioenkop. Many of the fallen soldiers were buried in the trenches where they died. These graves therefore give an indication of where the trenches were located at the time of the battle. Author: Renier Maritz. See Also: Spion Kop Battlefield Memorials - Wikipedia.

full of courage, and so were we all, for from where we stood we could look down on the Tugela River, and we were now able to grasp the full significance of our unexpected success.

Long columns of troops and long convoys of transport were re-crossing to the south bank, and everywhere the British were in full retreat from the positions which they had captured on this side of the streams and the clouds of dust rising on the Colenso road told us that General Buller's second great attempt to pierce the Tugela defences had failed.

Ambulance Corps crossing the Tugela.[195] "Bearers crossing the pontoon bridge at Trichard's [drift]."[196]

We spent the next hour or two helping the English Red Cross doctors and bearer parties that came up to bury their dead and carry away their wounded. By now hundreds of other burghers had arrived, mostly men who had retreated the day before, but like ourselves had loitered in neighbouring kloofs[113] and gullies to see if they could renew the fight.

Major Wright demanding possession of the wounded.[114] Searching for casualties after the Battle of Spion Kop.[197]

[113] A deep glen or ravine.

[114] "The Boers at first refused to allow the British volunteer ambulance men to carry off the wounded. Major Wright, who had been wounded at Elandslaagte and now serving under the Red Cross, persisted in his demand, claiming his right under the Geneva Convention, and finally carried his point in a personal interview with Commandant Botha." Image Source: With the Flag to Pretoria, H. W. Wilson Vol 1 1900 By R.B.M. Paxton, After a sketch by Ernest Prater.

Dawn on Spion Kop.[198]

Indian Bearers After the Battle of Spion Kop.[115]

Indian Stretcher-Bearers.[199]

Indian Ambulance waggon with Roberts' army.[200]

Another Work of Mercy. A New Ambulance Cart to be Used by the Native Indian Bearers.[201]

Battle of Spion Kop: Volunteer Ambulance Corps carrying wounded out of action.[202]

Casualties In Action	Killed		Wounded		Missing and Prisoners			Forces Engaged		
	Officers	Men	Officers	Men	Officers	Men	Total	British	Boer	UK Loss
Week of Battles and Spion Kop	27	245	53	1,050	7	351	1,733	17,500	12,000	9.9%

Casualty Figures.

[115] "Wilst the Boers treated our wounded with tenderness they did not hesitate to turn out the pockets of the dead and dying." Image Source: With the Flag to Pretoria, H. W. Wilson Vol 1. Credited to J. Greig.

Towards midday Isaac Malherbe ordered us to collect our Pretoria dead. We carried them down in blankets, and when the commando wagon came up we placed the bodies on board and escorted them to Ladysmith, whence they were sent to Pretoria for burial. So we rode behind the wagon which carried all that was left of our friends and companions, their horses trotting alongside with empty saddles.

I personally came home to a deserted tent, for within a few weeks four good friends had gone from it to their death, and our fifth messmate, de Vos, was lying dangerously wounded at some laager below Spion Kop. Only my brother and I were left, and he had been sent to Pretoria with the prisoners, so I was all alone, except for our faithful old native retainer, [Charley] who did what he could to cheer me up.

A contingent of Boer riflemen with Spion Kop in the background.[203]

British Casualties on Spion Kop.[204]

X. The Rest of Our Corporalship is Destroyed

For a while the success at Spion Kop went to our heads, and we thought that the English would be sure to make peace, but again the days came and went with no sign.

Indeed, we presently heard that General Buller was back at Colenso collecting an ever larger army to attack again, but we were confident that the Tugela defences would hold and we saw no shadow of the disasters that were soon to overtake us. After a week my brother Joubert returned from Pretoria. He said that he and the escort had been made much of by the townspeople, for, coming straight from the battle on the hill, they were the heroes of the hour, and they were marched down to shake hands with President Kruger, the nearest approach to battle honours that we ever attained.

He arrived in camp with my three remaining brothers. The eldest, Hjalmar, aged twenty, was studying law in Holland when the war broke out, and he had after considerable difficulty succeeded in reaching the Transvaal through Portuguese territory. The other two, Arnt and Jack, aged sixteen and twelve respectively, had been at school up to now; but they had at last persuaded my father to let them come to the war. Jack only remained for a few days, as old [General] Maroola on one of his visits to our laager caught sight of him and ordered him to be sent home, but at any rate from now onward there were four of us in our tent.

About the middle of February, a force of eight hundred mounted men was drawn from different commandos for the purpose of carrying out a raid into Zululand. The exact object of this expedition I never heard, but I think that it was intended to create a diversion, for the British were massing heavily at Colenso. My two newly arrived brothers were absent on a visit to the Tugela, for commando life was a novelty to them, and they spent much of their time riding about, but my brother Joubert and I joined the Zululand column. It assembled at the Commandant-General's headquarters behind Bulwana Hill, and we started the same morning under old Maroola. For several days we travelled due east through lovely mountain country everywhere dotted with picturesque native kraals. The Zulus showed no fear of us and refused to serve as guides, for they sided with the British, as they have always done.

Of English troops we saw no sign, and our journey was a restful interlude after the excitement of the past few weeks. Unfortunately it soon came to an end, for about the fifth day out a messenger came riding post-haste with orders for Maroola to return to Ladysmith at once.

We retraced our steps by forced marches, arriving back at the head laager on about the eighth day after leaving. During the last two days of the journey we heard a constant rumble of gun-fire coming from the direction of Colenso, and, on reaching our Pretoria camp, we were met with the disturbing news that the English had broken into our defences there to the extent of capturing Hlangwane Hill,[116] a commanding position that was considered the key to the Tugela line. Once more volunteers from every commando around the town were called for. My two brothers Hjalmar and Arnt had returned and gone off again on some jaunt, so Joubert and I, with the rest of Isaac Malherbe's corporalship, handed in our names. There was no difficulty in getting the required fifty men, and at dusk on the day after our return from Zululand we set out on a journey from which few were to come back.

We rode through most of the night in company with other units hurrying to the danger-point. Farther on we fell in with small parties riding in the contrary direction, who made no secret of having quitted the Tugela firing-line, and when we neared the river at daybreak we found a critical situation. Not only had Hlangwane been taken, but every Boer trench on the north bank for a distance of several miles had been evacuated, and, what was far more serious, there was a feeling of discouragement in the air that dismayed us.

Up to now, the prevailing note in Natal had been one of confidence in an early peace, but, almost in a night and without apparent cause, a wave of pessimism had set in and the nearer we came the more evidence was there of growing demoralization.

Hundreds of men were leaving the new line that had been formed in the hills behind the abandoned trenches. They were dispersing about the back area, some holding meetings, others making for the wagon-laagers in the rear, and from their talk and attitude we sniffed disaster, for though we knew before starting that Hlangwane had been taken, we did not know that the fighting spirit of the men had gone with it.

[116] 2/19/1900 Capture of Hlangwane by the Fusilier Brigade. The Boers evacuated the hill, and left a large camp behind them. Boer reinforcements were driven back.

Bad as it looked, however, things were not yet quite hopeless, for General Botha had taken up a new lane to the rear of the old one and was once more standing on the defensive with eight or nine thousand men. We were met by one of his officers, who allotted different points to the various fresh detachments. We of the Pretoria commando were told to leave our horses in charge of guards and march on foot down a rocky gorge which led towards the Tugela. Heavy Lyddite shells crashed against the cliffs on either side as we went down, but we emerged without loss at the lower exit of the canyon, to find ourselves in the bed of a dry spruit which ran across an open plain to where it joined the Tugela River about a mile and a half away.

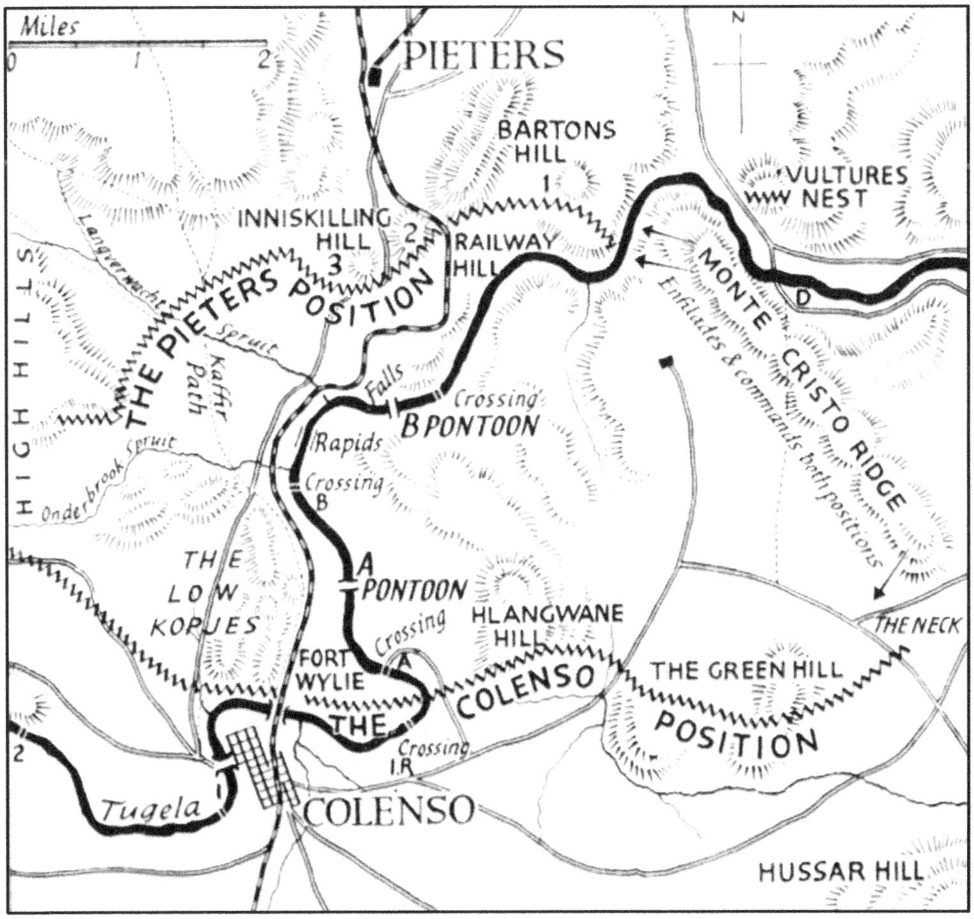

Hlangwane Hill Bottom Center and a view of the Monte Cristo position.[205]

This dry bed skirted along the base of the foothills for some distance before bending away across the plain, and, from the exit of the gorge to the point where it curved away, its bank had been converted into a sector of the new line. We were assigned a position near the top end, not far from where we had come out of the gorge, and we fell to at once, hacking firing steps, with pocket-knives and any other implements on which we could lay our hands.

A Boer Commando. "The boy third in the line is young Botha."[206]

The English infantry were eight hundred yards off, approximately along the line of our former trenches that had been abandoned two days before. From here they were maintaining a continuous rifle-fire, a multitude of bullets whizzing overhead or plugging into the ground in front.

The British infantry in firing line. Some soldiers are well dug in.[207]

They were also methodically shelling the course of the spruit with many guns, including high-angle howitzers, so that we were in a very warm and unpleasant locality.

The 5-inch Howitzer or Siege Gun.[208] 37th R.A. howitzer battery going to Maddox Hill to throw lyddite into the Boer Laager.[209]

The casualties were few, owing to the height of the bank, and not one of the Pretoria men was hit, although I got a shell splinter through my hat before I had been there ten minutes, and one or two of our men had scratches from flying earth as the shells exploded in the bed of the spruit below.

The British did not attack on our front, but later in the morning made a determined attempt to our left against some hills known as Pieters Heights, where the Bethal commando[117] was posted.

We saw waves of infantry going forward, but it was too far off for us to participate. At first the soldiers advanced in regular lines, but gradually their progress was stayed, and we could see the survivors crouching behind rocks and stones. They lost heavily, and it was soon evident that the attack had been repulsed.

[117] Bethal Commando: https://www.angloboerwar.com/?option=com_content&view=article&id=3041

View of Pieters Hill from south of the Tugela River showing lines of British infantry moving up the hill towards the Boer trenches on the crest.[210]

Nothing further happened that day-and by dark all was quiet. We spent an undisturbed night, and next morning (February 26th, 1900), the English sent in under flag of truce to fetch away their dead and wounded, so that there was no firing at all.

Having nothing to do, my brother and I walked down to watch the removal of the fallen soldiers and we spent some hours going over the ground discussing events with the English doctors and stretcher-bearers.

Ambulance Parties at Work.[211]

Returning, up the spruit on the way to our own quarters, we met my Hollander uncle, [Jan Mulder] still serving with the Swaziland Police, a chance encounter which probably saved the lives of both of us. My Uncle, seeing us pass, asked us to spend a day or two with him, and, falling in with his suggestion, we went to get our things and to tell Isaac Malherbe where we were going. He made no objection to our visiting the Swazilanders, seeing that they were close by, and he promised to send for us should we be wanted. As we were going off he looked at us with his quiet smile and said: 'Be sure and come quickly should I send word, for how shall I hold this bank without you two boys beside me?' He was referring to the night below Surprise Hill, but we never again saw him alive, nor any of the others.

Carrying our cooking-tins and blankets, we returned to the Swaziland Police. At sunrise next morning the English started in real earnest to bombard the spruit. All day long they shelled us with light and heavy guns, while we hugged the sheltering bank. Shrapnel and Lyddite crashed upon us, causing a great many casualties, and we suffered a terrible ordeal.

Considering the demoralisation that had set in, the men stood the bombardment well. There were some who retreated up the gorge, but there was no general desertion, and, when at sunset the fire died down, our forces were holding firm after one of the worst days of the war. This heavy shelling was the obvious preparation for an attack next day, so my brother and I, when the fire slackened, went up to find the Pretoria men. In the dusk we made our way along the bed of the spruit, past groups of burghers standing around their dead or wounded and other groups discussing the incidents of the day, but when we came to where we had left our companions the evening before, they were gone, and we were told by those nearby that they had marched up the gorge an hour ago for some other point of the line. We could not understand why Isaac had not let us know, but thought that his messenger had missed us in the failing light.

We were upset by his departure and decided to follow after at once, so we went back to tell my uncle of our intention. He said he was coming with us, too, as he had been thinking of a change for some time, and would like to join the Pretoria commando. Accordingly he told his Field-Cornet of his plans and the three of us set off through the gorge. It was pitch-dark by now, and we had a difficult climb up the boulder-strewn bed, stumbling and groping our way as best we could.

When at last we reached the top entrance, we were so weary with the heavy loads we were carrying, and there was so little chance of finding our corporalship at that time of night, that we flung ourselves down on the first piece of level ground and slept until daybreak. As soon as it was light we made out the tents of the horse-guards, where we found our saddles just as we had left them, and our horses grazing not far away. These guards were here on permanent duty to look after the animals of the burghers in the firing-line, and they told us that Isaac Malherbe and his party had spent the night with them, and had ridden off before dawn to join the Bethal commando at Pieters Heights some miles east of this, so we hurriedly cooked a meal and, saddling our horses, rode after him, following behind the range of hills that formed the Boer line.

As we went along, a bombardment more violent than that of yesterday broke out ahead of us, and, when we came to the rear of Pieters Heights, we saw the ridge on which lay the Bethal men (and our own) going up in smoke and flame. It was an alarming sight. The English batteries were so concentrating on the crest that it was almost invisible under the clouds of flying earth and fumes, while the volume of sound was beyond anything that I have ever heard. At intervals the curtain lifted, letting us catch a glimpse of the trenches above, but we could see no sign of movement, nor could we hear whether the men up there were still firing, for the din of the guns drowned all lesser sounds.[118]

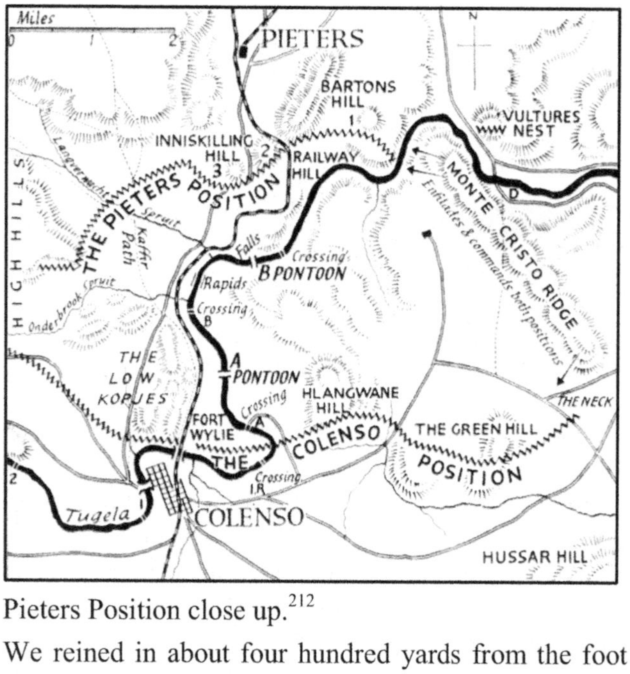

Pieters Position close up.[212]

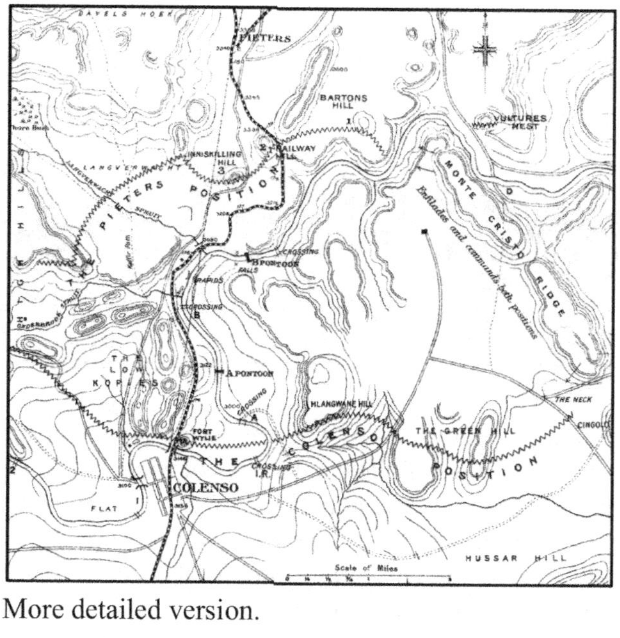

More detailed version.

We reined in about four hundred yards from the foot of the hill, at a loss what to do. To approach our men through that inferno was to court destruction, while not to try seemed like desertion. For a minute or two we debated and then, suddenly, the gun-fire ceased, and for a space we caught the fierce rattle of Mauser rifles followed by British infantry swamping over the skyline, their bayonets flashing in the sun. Shouts and cries reached us, and we could see men desperately thrusting and clubbing. Then a rout of burghers broke back from the hill, streaming towards us in disorderly flight. The soldiers fired into them, bringing, many down as they

[118] PF: From other accounts I've learned that Gen. Buller is using a creeping barrage to cover the assault on the position. A novel tactic for the time. After the summit is reached, he also shells the valley behind the position in order to prevent reinforcements.

made blindly past us, not looking to right or left. We went, too, for the troops, cheering loudly, came shooting and running down the slope.

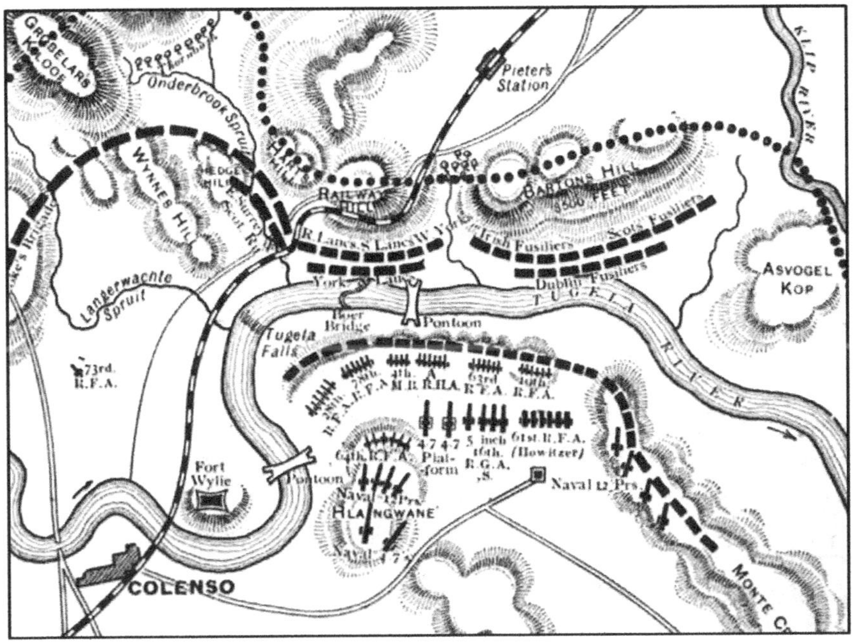

Pieters Positons.[213]

Of our Pretoria men who had been on the ridge not one came back. They had been holding an advanced position to the right of the Bethal section, and had been overwhelmed there. They stood their ground until the enemy was on them, and they were bayoneted or taken to the last man. Thus our corporalship was wiped out, with its leader, Isaac Malherbe, the bravest of them all, and their going at this calamitous time was scarcely noticed. For this day marked the beginning of the end in Natal. The British had blasted a gap through which the victorious Soldiery came pouring and, wherever we looked, Boer horsemen, wagons and guns went streaming to the rear in headlong retreat.

Receiving Boer prisoners after the Battle of Pieters Hill.[214] "Expanding bullets and other ammunition found at the bottom of Pieters Hill."[215] "

We followed the current, hemmed in by a great throng, all making for the various fords of the Klip River, and it was lucky indeed that the English sent no cavalry in pursuit, for the passages across the river were steep and narrow, and there was frightful confusion of men and wagons struggling to get past.

By nightfall my uncle and my brother and I had managed to cross, and as it started to rain we annexed a deserted tent behind Lombaards kop, picketed and fed our tired horses, and slept there till morning. We now resumed our journey as far as the head laager, where we spent a dismal hour or two watching the tide of defeat roll northward.

We knew that the siege of Ladysmith would have to be raised, and now came the news, while we were halted here, that Kimberley had also been relieved,[119] and that General Cronje had been captured at Paardeberg[120] with four thousand men, so that the whole universe seemed to be toppling about our cars.

Meeting of Lord Roberts & General Cronje. February 27th, 1900.[216]

Cronje at the Surrender.[217]

Cronje's infantry on the march to Klop drift under escort of a battalion of the Glouchesters.[218]

From the way in which the commandos were hurrying past, it looked that morning as if the Boer cause was going to pieces before our eyes, and it would have taken a bold man to prophesy that the war had still more than two long years to run.

We hung about the dismantled head laager till midday, after which the three of us rode on gloomily to the Pretoria camp, arriving towards five in the afternoon. Word of the disaster to our men on the Tugela had already preceded us, as is the way with evil tidings, and, as it was not known that my brother and I had escaped, our unexpected appearance caused a sensation, men running up from all sides to hear the truth. My other brothers had returned and their welcome was a warm one.

It was by now clear enough that the siege could no longer be maintained and, indeed, orders had already been received that all commandos were to evacuate their positions around Ladysmith after dark. Our wagons were standing packed, but at the last moment it was found that someone had levanted[121] with the transport mules, so everything had to be burned. It came on to rain heavily by nightfall.

Peals of thunder growled across the sky, and, wet to the skin, we stood huddled against the storm in depressed groups, awaiting our final orders. At last, long after dark, Field-Cornet Zeederberg gave the word that we were to move off to Elandslaagte some twenty miles to the rear. It was an inky night, with rain in torrents, through which we had to feel our way, and thus we turned our backs on Ladysmith for good and all.

[119] On February 15th, 1900 after a siege of 124 days. See: Siege of Kimberley - Wikipedia

[120] On February 27th. See Battle of Paardeberg - Wikipedia

[121] Literally: set the sail with the Levant, an easterly wind that blows in the Mediterranean Sea. To abscond or run away, especially to avoid paying money or debts.

Burning a camp..[219]

No march order was attempted; we were simply told to go, and it was left to each man to carry out his own retirement.

At the outset we travelled in company with many others, but, as I knew a short cut, threading the hills to the railway depot at Modderspruit, my brothers and I decided to go thither, for we saw no use in floundering about in mud and water, and the four of us, with my uncle and our native boy Charley, branched away by ourselves.

We reached the depot after two hours and found shelter until daybreak, after which we rode on.

The rain now stopped and the sun rose warm and bright, but it looked on a dismal scene. In all directions the plain was covered with a multitude of men, wagons, and guns ploughing across the sodden veld in the greatest disorder. Wherever a spruit or nullah barred the way there arose fierce quarrels between the frightened teamsters, each wanting to cross first, with the result that whole parks of vehicles got their wheels so interlocked that at times it seemed as if the bulk of the transport and artillery would have to be abandoned, for the mounted men pressed steadily on without concerning themselves with the convoys. Had the British fired a single gun at this surging mob everything on wheels would have fallen into their hands, but by great good luck there was no pursuit and towards afternoon the tangle gradually straightened itself out.

Our little family party remained behind with a number of others as a rearguard and we did not reach Elandslaagte until late that night. This place had been the chief supply centre for the Natal forces, and there were still huge quantities of stores that had been left to the enemy. These we burnt, lighting a conflagration that must have been visible for fifty miles around.

By now the bulk of the retreat had passed on, and next day we rode along leisurely, climbing up the Washbank valley to Glencoe near Dundee by the following evening. Here were stray remnants of almost every commando that had been in Natal, but things were in such confusion that most of the army had continued straight on, and there was scarcely a man who could tell us what had become of his officers, or what we were supposed to do next.

Mr. Zeederberg, however, was at Glencoe when we got there, and during the next few days he succeeded in collecting about three hundred Pretoria men, while more drifted back later on, as did stragglers from the other commandos, until after a week or ten days there was quite a respectable body of men numbering well over five thousand.

During this time my brothers and I with our uncle subsisted on what we could loot from the supply trains at the station, for there were practically no commissariat arrangements, but by raiding the trucks at night we did not do badly.

After a while General Botha reorganized everything, and a new line of defence was established along the forward slopes of the Biggarsbergen, to which all available men were marched. We of the Pretoria commando were assigned a post on the shoulder of the mountain to the right of where the Washbank valley reaches the plain below, and here we lay amongst pleasant scenery, from which we looked regretfully over the wide sweep of country to the south from which we had been driven, but we enjoyed the spell of peace and quiet after the turmoil of the past weeks.

XI. A Campaign in the Free State

The English Army, having forced the Tugela and relieved Ladysmith, was also resting.

Far down on the plain, large camps were springing up, but all through the month of March and half of April they made no move at all and the weeks went by unbroken, except for an occasional patrol when we would leave our horses below some kopje and climb up to watch the troops at exercise in the distance.

While matters went this easily in Natal, grave rumours reached us as to the situation in the Free State. Reliable information was scarce, but it was freely said that a powerful army under Lord Roberts had crossed the Free State border and was marching on Bloemfontein.

When my brothers and I heard this we felt we ought to go to our own country. We decided to go to Pretoria first and thence south by rail, for since the destruction of our corporalship we considered ourselves free to go where we pleased.

The day before we meant to leave we were ordered to take part in an attack on the English camps lying at Elandslaagte, halfway back to Ladysmith.

This much-criticized affair was to have been carried out by three thousand Transvaalers in conjunction with an equal force of Freestaters from the Drakensbergen. We assembled after midnight at the exit of the Washbank valley and reached the Elandslaagte hills as it grew light, but we looked in vain for the Free State commandos.

The Free State Commandant-General, Prinsloo (the same who surrendered so ignominiously with three thousand men a few months later), had telegraphed to General Botha at the last moment to say that he and his officers were attending a cattle sale at Harrismith on the day set for the attack, and were therefore unable to be present.

Prinsloo

General Prinsloo.[220] [122]

Surrender of Prinsloo's Boer forces in Brandwater basin, 1900.[221] [123]

[122] Prinsloo Marthinus General Born in 1838, he was serving as the Commandant of the Winburg Commando in 1867. He was a man of wealth and served as an elder of the church. At the start of the Boer War, he commanded Free State commandos in Natal. In the attack on Caesar's Camp on 6 Jan 1900, he was accused of indecision. He commanded the defensive line on the Tugela after the battle of Spion Kop. When Ladysmith was relieved and Buller pushed north, Prinsloo commanded the Free State commandos at Van Reenen's Pass. Now aged 62, he was seen as too old for combat and retired to his farm. He re-engaged and operated in the Free State. On 30 July 1900 he was captured near Fouriesburg in the Brandwater Basin with over 4,000 men, 2,800 cattle, 4,000 sheep, 5,000 horses and 2 million rounds. He died in 1903. See Also: Marthinus Prinsloo - Wikipedia

[123] Cape Mounted Rifles Article: https://www.angloboerwar.com/?option=com_content&view=article&id=316

Surrender of 5,000 Boers near Fouriesburg, to General Hunter. July 31st, 1900.[222]

In view of this, General Botha had to change his plans and content himself with a mere demonstration instead of serious business and the whole thing fizzled out in an artillery duel with heavy expenditure of ammunition and very little damage done. To me it was memorable only for the fact that while I was watching one of our Creusot guns being fired, the wind of a shell from the English batteries sent-me spinning yards away. It must have passed within a few inches of me and I shall not soon forget being blown head over heels.

We remained before the camps all day, subjected to severe shell-fire at times, and after dark we rode back to our stations in the Biggarsbergen, none the better for our outing.

The next day my brothers and I took leave of the Pretoria commando with which we had served for so long.

We rode up the valley to the train at Glencoe Junction, [Dundee] and it was months before we saw our old force again.

At the railway station we held up the first north-bound train, loaded our horses in one truck and ourselves in another, and steamed off, leaving Natal behind us for ever.

After the usual three days' journey and the usual delays at halts and stations, we arrived at Pretoria. My uncle, Jan Mulder, was not of our party, for he had remained with the Irish Brigade, a band of two hundred adventurers commanded by an American Colonel named Blake,[223] whose roystering habits and devil-may-care methods suited his own, so that we four brothers [Deneys, Hjalmar, Joubert, and Arnt] and Charley, our boy, formed our party.

Col. John Blake.[124 224]

John Blake alongside officers of the Irish Brigade. Blake stands third from left.[225]

[124] John Young Fillmore Blake, also known as John Y.F. Blake and J.Y.F. Blake was an Irish-American soldier and writer. He was born October 6, 1856, in Bolivar, Missouri in the United States, and died January 24, 1907, in New York City. Blake served as a foreign volunteer for the Boers of the South African Republic during the Second Boer War. John Blake (soldier) - Wikipedia and Irish commandos - Wikipedia

Francis Reitz had this to say about the Irish Brigade, as reported by Howard Hillegas:[226],[125]

My father did not know of our arrival until we came riding up to the front door, but when we told him we were for the Free State he approved. We gathered that the position was very bad, and he himself seemed aged and worn, for he bore great responsibilities upon his shoulders. He had signed the ultimatum to Great Britain, and in a large measure the policy which had led to the war had been his, so the gloomy military situation lay heavily upon his mind, as well as the personal anxiety of having four sons at the front. We spent the next few days in the luxury and comfort of home life, the last we were ever to enjoy.

I even visited Johannesburg for the first time. The city was practically deserted; the shops were boarded up and there was little or no life in the streets, but I remember the visit because that afternoon there came the roar of a great explosion and a column of smoke shot up into the sky a mile high.

Pritchard street before the crisis.[227] And as it is now, the effect of the war scare at Johannesburg.[228]

It was the Begbie foundry, where the Government was manufacturing shells and ammunition, that had gone into the air, by treachery it was said.

About thirty people were killed, but so fierce was the blaze that we could give no assistance and we had to look on helplessly while the fire burnt itself out.

Johannesburg, April 24th, 1900. Explosion accident at the Begbies munition works.[229]

On April 30th (1900) my three brothers and I, with our boy Charley, entrained for the Free State. We knew that the British had occupied Bloemfontein by now and were advancing towards the Transvaal, but no one seemed certain how far they had got, or for how far the railway line to the south was still open.

[125] When one of the four sons [probably Deneys] of State Secretary Reitz who were fighting with the Boer army asked his father for permission to join the Irish Brigade, the Secretary gave an excellent description of the organisation: "The members of the Irish Brigade do their work well, and they fight remarkably well, but, my son, they are not gentle in their manner.

Changing of the Guard at the Presidency, Bloemfontein.[230] [126]

Ambulance Train blown up at Kroonstad.[231]

The Red Cross hospital train wrecked.[232]

We crossed the Vaal River that night, and, after a slow journey over the rolling plains of the northern Free State, we reached a small station near the banks of the Vet River[127] by eleven o'clock next night. We were now within fifty miles of Bloemfontein and the train was going no farther, as the engine-driver told us that the British advance was at the next station but one down the line. On hearing this, we unslipped our horses and camped beside the track until daybreak, intending then to ride forward in search of the Boer forces that we knew must be somewhere ahead.

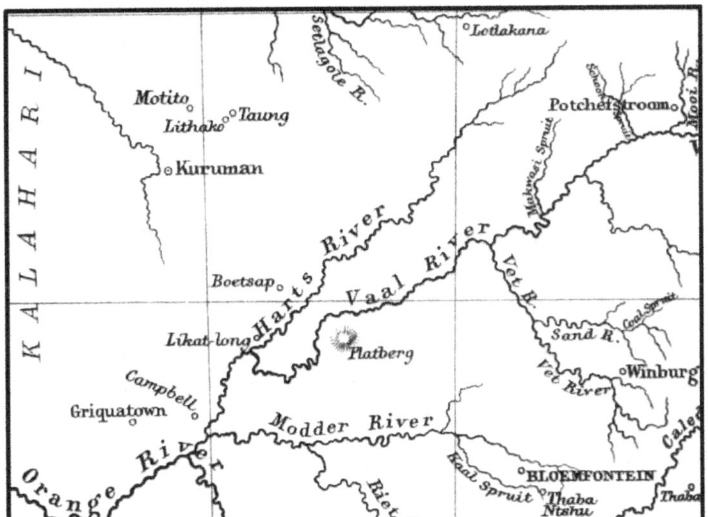

Rivers in the Orange Free State.[233] Note Vaal River, Vet River and Sand River west of Winburg.

[126] Roberts occupied Bloemfontein on March 13th, 1900. His despatch ran:—"The British flag now flies over the Presidency vacated last evening by Mr. Steyn, late President of the Orange Free State. The inhabitants gave the troops a cordial welcome."

[127] The Vet River is a westward-flowing tributary of the Vaal River, with the Sand river tributary to the north of it. Vet River - Wikipedia

As we were preparing to start next morning, another train steamed in from the north, carrying a hundred and fifty men under Commandant Malan,[128] a brother-in-law of the Commandant-General Piet Joubert, who had recently died.[129] Malan had collected a lot of young fellows whom he had formed into a flying column, the 'Afrikander Cavalry Corps,' and they were on their way to the nearest fighting.

They detrained here and we lost no time in enrolling ourselves as members of the 'A.C.C.' as it was called for short. Among them was my old schoolfellow Jan Joubert,[130] the Commandant General's son, with other friends and acquaintances.

We spent the morning getting ready and that afternoon we trekked off, riding south through the Vet River towards the sound of the distant gun-fire that we could now hear on the wind.

Darkness found us on the wide plain beyond the river, where we met hundreds of horsemen withdrawing, so they told us, to fresh positions, but whom we suspected of being on their way home for good. They said that great swarms of British troops were on the move, and that it was useless to think of fighting them in the open.

Nevertheless we rode on until midnight, when we came to where General de la Rey was halted with the Transvaal commandos. We found him squatted-by a small fire, a splendid-looking old man with a hawk-like nose and fierce black eyes. Beside him was his brother, nursing an arm shattered by a bullet that afternoon. He gave us a hurried account of the situation, which was very black. The British Army, after capturing Cronje and taking Bloemfontein, was now advancing on the Transvaal and, owing to the demoralized state of the commandos and the lack of defensive cover in this bare region, he saw little or no hope of stopping them.

Commandant De La Rey and his nephew.[234]

General Jacob Hercules 'Koos' De La Rey.[235]

He said he had about four thousand Transvaalers who had escaped the debacle at Paardeberg, but they were discouraged and were making the merest show of opposition.

The Free State commandos had disappeared altogether, although he believed that President Steyn and Christian de Wet were trying to reorganize them somewhere in the mountain country to the east, but for the time being they were out of action. The British were within a few miles of us, and would doubtless resume their advance in the morning.

So far as the 'A.C.C.' was concerned, he ordered us to ride forward for half an hour and halt till daybreak, after which we were to fit ourselves into the firing-line and act according to circumstances. We took leave of him and, mounting our horses, rode on for three or four miles.

Since we had come south the weather had turned bitterly cold, and we felt the change from the warmer climate of Natal.[131]

[128] Possibly: Herculaas Phillipus Malan, Commandant Rustenburg Commando.

[129] On March 28th 1900.

[130] Per Deneys, here and later. Jan Joubert is the son of Piet Joubert the late Commandant-General.

For months we were not to spend one really comfortable night until summer came round again, and on this particular night we sat, with our blankets wrapped around our shoulders, shivering till daybreak, for sleep was out of the question with the temperature below zero.

As soon as it grew light we were astir, anxiously scanning the ground before us, and soon we made out dense masses of English infantry on the plain. First came a screen of horsemen, and behind a multitude of infantry, guns and wagons throwing up huge clouds of dust.

A British mule train stirring up the dust as it toils up a steep incline.[236]

Column of the Royal Horse Artillery.[237]

We looked in dismay at the advancing host, for there were thirty thousand men approaching, whilst on our meagre front there may have been between three and four thousand Boer horsemen, strung out in a ragged line on the rising ground to right and left of us.

It was plain from the very way in which the men sat their horses that they would not stand, and indeed, on this bare veld and against such heavy odds, the task was manifestly beyond them.

The enemy forces came steadily on until their scouts were close to us. When we fired at these they fell back upon their regiments; the batteries unlimbered and in a few seconds shrapnel was bursting over us.

Our line gave way almost at once. The A.C.C. stayed as long as any, but we recognized the futility of remaining, so we went galloping back with field-gun and pom-pom shells besprinkling us as we rode. We had no casualties, but a number of men from other commandos were killed and wounded before we got clear, and after a hard ride we slowed down at a deserted farmhouse to breathe our winded animals.

English Vickers-Maxim Pom Pom.[238]

Pom Pom in action.[239]

English troops being mainly infantry, their progress was slow, and, although they were quickly at us once more, we were able to retire before them with very little loss for the rest of the day, firing on their scouts when they pressed us too nearly, and moving back in extended order to escape the shell-fire that came in gusts as the guns were brought forward. We were in the saddle until sunset, for we had to exercise ceaseless vigilance to keep the English horse from the wagons that were struggling to get away.

[131] Not only are they slightly farther South in the oncoming Southern Hemisphere's Winter but the High Veld plateau has a surface elevation of over 4,000 feet. Bloemfontein is situated at an elevation of 4,577 feet (1,395 m).

There must have been over a thousand of these, for, in addition to General de la Rey's transport, there were a great many vehicles belonging to the civilian population fleeing before the oncoming invasion.

By dark the English had pushed us right back to the Vet River, a distance of twenty miles or more, and next morning we had scarcely time to prepare a hasty breakfast before we could see the columns again advancing towards us.

General Louis Botha was standing at the drift as we rode through, for he had hurried round by rail from Natal to see for himself what was going on in the Free State.

He and General de la Rey disposed such commandos as were available along the river from the railway bridge to a point about four miles down, with orders to take a stand. The 'A.C.C.' was allotted the extreme right, so we searched out a suitable spot on the river bank beyond the next farthest commando, and, leaving our horses in the bed below, took up our posts. As we expected trouble, we sent the native boy Charley to the rear with our spare kit loaded on my Basuto Pony (which we had brought from Natal as a pack-horse).

The British were by now feeling their way over the plain that ran down to the river, and before long were volleying at us from tall grass in which we could only just see where they lay. Then their guns came forward and started shelling us. Our horses were safe in the river bed behind, but owing to the thorn trees fringing the bank our view was impeded, and we had to crawl to the outside edge of the bush to see the enemy, with the result that we had practically no cover and had casualties almost at once. Several men were killed and wounded close to me, and altogether it was a beastly day.

The shelling was not confined to our portion of the line, but ran up and down, like a piano, as far as the bridge and back again, and at times was as heavy as that on the Tugela. It continued until far into the afternoon and it was only at three o'clock that we saw the infantry preparing to charge.

We had six killed and about fifteen wounded by then. The dead had been laid on the sand in the river bed, and the wounded had been placed on their horses and told to get away as best they could, while the rest of us stayed on in such holes and hollows as we had been able to scrabble in the soil. The shell-fire and the casualties had shaken us, and when the British rose to their feet, and from their rear some three hundred cavalry came riding, sword in hand, we rose to fire a few wavering shots, and then broke for the river behind, tumbling down pell-mell to get our horses. Leaving our dead, we rode up the opposite bank and went racing across the open to the hills a mile to rearward. We were heavily fired on, but reached sanctuary with only two or three men down and a few wounded, including my younger brother Arnt with a scalp-wound from a glancing rifle bullet. The English horsemen that had been the chief cause of our flight came no farther than the river, but the infantry were breaking through at several places, and we could see them well on our side of the stream, the rest of the commandos also retreating wildly.

The Boer commando guarding Van Reenan's Pass.[240] Vet River Bridge before repairs.[241]

Before long we were again being shelled, and near sunset the soldiers came on to drive us out of the hills to which we had fled. As we were holding the extreme end of the Boer line, we were soon outflanked. We saw a regiment change direction and before we could stop them they were climbing into the very hills on which we ourselves were at a point about fifteen hundred yards off, from which they began to work their way towards us.

We fronted round to meet them, but the sinking sun was straight in our eyes, making accurate shooting difficult, and, when the soldiers came swarming towards us at short range, another stampede took place. The whole of what remained of the 'A.C.C.' rushed for their horses and made haste to be off. By the time I was in the saddle, the nearest infantrymen were so close that I could see their faces and the brass buttons on their tunics, but they were blown with running and their aim was poor, and although several horses were hit, none of the men were injured. My eldest brother's pony was shot through the body from saddle-flap to saddle-flap, but the plucky little animal carried him a thousand yards before he fell dead. Riderless horses were careering all over the place, and by cornering one of them we succeeded in providing my brother with another mount. Bullets and shells were striking everywhere, so we hurriedly transferred his belongings to the new horse and followed the rout now fast vanishing over the next rise. In the darkness we galloped on to catch up with the press, for all positions along the river had been given up and a wholesale retreat was in progress.[132]

The 'A.C.C.' had over thirty men killed and wounded during the day, and if this was a fair average the total Boer casualties must have been heavy.

We trekked on till after midnight, our only crumb of comfort being our native boy [Charley] awaiting us beside the road, his voice quavering with emotion at seeing all four of us still alive.

The withdrawal was continued next morning without waiting for the enemy, and by midday we were across the Sand River, thirty miles and more from the scene of yesterday's encounter.

The wagon convoys were far away by now, so we were free of that trouble, and we had moved to the rear so rapidly that it was twenty-four hours before we saw the English on the skyline once more. It was towards evening of the next day that they came— a small advance guard of two hundred horsemen with a gun, riding so fast that before we could stand to arms they were on the south bank of the Sand River, and killed one of the Irishmen preparing to dynamite the bridge.

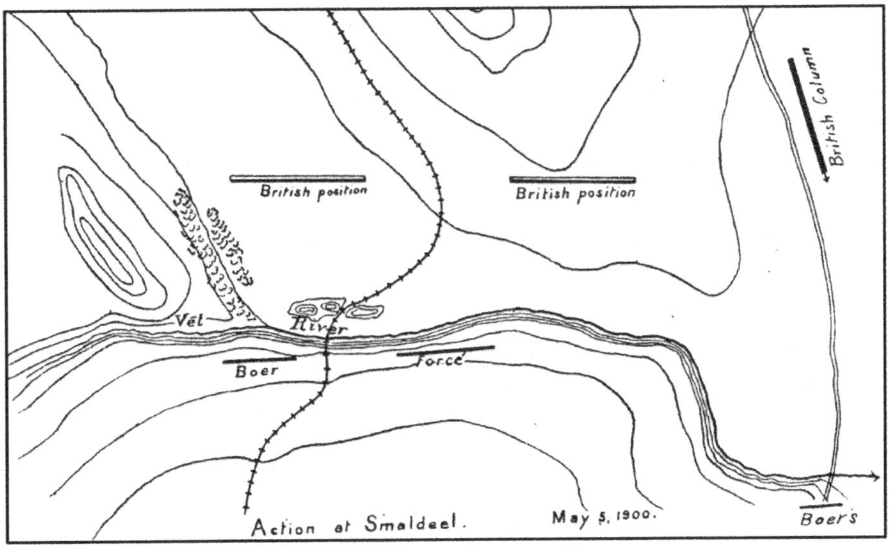

"DYNAMITE DICK."

Action on Vet River at Smaldeel. May 5th, 1900.[242] The Irish Brigade were renowned for their ability to destroy infrastructure. See John H. King,[133] better known as "Dynamite Dick," Boer Scout, and American soldier of fortune. He survived the war returned to Chicago.

[132] Sunday, May 6, 1900 The Vet River passed and Smaldeel occupied.

[133] Pennsylvanian John H King was a miner-turned-adventurer. King was nicknamed 'Dynamite Dick' for his expertise in blowing up trains and bridges. King earned a reputation for bravery as well, when he and another man retrieved a seriously wounded comrade under heavy fire at the battle of Vaal kranz. The rescued Boer was Alec Brand, the son of Johannes Henricus Brand, a former President of the Orange Free State (Farwell, 1976, p. 92). King fought at Spion Kop in January 1900, where he captured an old friend who had joined the British side. According to a report by Howard Hillegas in the New York World, King and his compatriot had a brief conversation, shook hands, and then King shot his unnamed friend to death. There are other anecdotes from participants claiming that Americans on each side took pains to identify and target their fellow countrymen.
The brigade was augmented by the so-called Chicago Ambulance Corps raised among Irish-Americans by Finnna-Gael, who landed about 50-strong at Beira in Mozambique in April 1900 (Lee, 1985, pp 53–4). The Portuguese authorities allowed them to entrain for the Transvaal, where they divested themselves of their medical guise, armed themselves and, under a Captain Patrick O'Connor, fell in with the Irish Brigade, helping to swell its ranks. The ambulance service itself, including five doctors and two nurses, accompanied the corps and provided professional support for its volunteers as well as for Boer commandos in general.

The South African Military History Society -Military History Journal Vol 15 No 6 - December 2012 Americans Who Fought In The Anglo-Boer War By Benjamin N Brown South African Military History Society - Journal - Americans who fought in the Anglo-Boer War (samilitaryhistory.org)

The commandos were some distance back, but the 'A.C.C.' was halted nearby, so we ran for our horses and recrossed the river lower down to outflank the intruders. When they saw that they might be cut off, they fired a few shells at us and fell back on their main troops, masses of whom we could now see on the far horizon. We got near enough to use our rifles, bringing down two troopers from their horses. I rode up to have a look at them. They were both Canadians, badly wounded, one of whom told me that many thousands of their people, as well as Australians and other Colonial forces, had volunteered for the war, as if the odds against us were not heavy enough already.

Partial Chronology

Monday, April 30, 1900: Deneys, Joubert, Arnt and Chaley arrive in theatre.

Saturday, May 5, 1900: British occupation of Brandfort.[134] Roberts's further advances North to the Vet River.

Sunday, May 6, 1900: The Vet River passed and Smaldeel occupied

Monday, May 7, 1900: General Hunter occupied Fourteen Streams

Tuesday, May 8, 1900: Ladybrand[135] deserted by the Boers.

Wednesday, May 9, 1900: Capture of Welgelegen.

Thursday, May 10, 1900: Battle of Zand River. Occupation of Ventersburg.

Saturday, May 12, 1900: Roberts occupied Kroonstad without resistance.

It was getting dark, so we left the wounded men to be fetched in by their own side, and for the next two days we were not disturbed, the English apparently resting their men before resuming the advance. On the morning after this, Commandant Malan sent me with a message for General Botha, whom I found, after a fifteen-mile ride, camped beside the railway line. Having delivered my dispatch, **I started back for the Sand River** where I had left the 'A.C.C.,' only to find them gone. Men in the neighbourhood told me they had ridden away hours before, going west on some unknown errand. It was growing dark and bitterly cold, and as it was useless to try and follow, with no idea of where they had gone, I broke enough timber from the pumping station at the railway bridge to make a bonfire, and spent the night in solitary comfort within, my old roan horse snug beside me.

As soon as it grew light next morning, high dust-clouds rising south of the river showed that the English were coming, so I saddled and hastened to fall back to where a force of six hundred burghers was approaching under the personal command of General Louis Botha.

Louis Botha.[243]

Boer Artillery unit. Photo by 'L. Botha.'[244]

[134] Brandfort, officially renamed Winnie Mandela in 2021, is a small agricultural town in the central Free State province of South Africa, about 60 km northeast of Bloemfontein Brandfort - Wikipedia See also for Concentration camps located there and Brandfort Commando - Wikipedia

[135] Ladybrand is a small agricultural town in the Free State province of South Africa, situated 18 km from Maseru, the capital of Lesotho. Ladybrand - Wikipedia And: Ladybrand Commando - Wikipedia

Another view of the apparently the same artillery unit. Note the heliograph set up on the right.[245] Photo by "L. Botha."

He ordered us to open out, and each man stood before his horse awaiting developments. These were not long in coming. Soon the English scouts had crossed the river, and infantry were coming down to the drifts, while the country behind was black with more troops and transport columns. Batteries came into play, and, as the ground was devoid of cover, we mounted and sullenly retired. Now began anew the long-drawn humiliation of retreat. All day we were driven relentlessly; the British herded us like sheep to the incessant shriek of shells and the whizz of bullets, and by evening we were a demoralized rabble fleeing blindly across the veld.

After sunset the pursuit tailed of I spent a cold night on a kopje with a few other stragglers, and next morning we rode into the town of Kroonstad.[136] Here we found President Steyn addressing a crowd of burghers from a market-table in the square. He had succeeded my father as President of the Free State in 1896, a burly, heavily bearded man, not brilliant, but possessed of dogged courage.

President Kruger and President Steijn Addressing the Boers in the Market Hall at Kroonstad after the occupation of Bloemfontein.[246] President Marthinus Theunis Steyn.[247]

For the moment his words fell on deaf ears, for so universal was the discouragement that few paid any heed to his appeals. His audience consisted mainly of Transvaal burghers, more concerned with getting back home than with forlorn hopes, so there was little response, but I stopped to listen, and was so carried away by what he said that I rode back the way I had come with a dozen more until we reached the scattered hills a few miles south of the town.

[136] Kroonstad is a town in the northern part of the Free State and was founded in 1855. Kroonstad was named after a horse, called Kroon, which was owned by Sarel Cilliers. The horse was injured in the Crown Creek. Kroonstad - Wikipedia

President Steyn Exhorting His Burghers at Smaldeel (Orange Free State).[248]

From here we could see the English columns advancing towards us, and after some time several hundred other men rode up, coming across the plain from various directions. By midday the troops were within range and began to shell us, but made no attempt to push home an attack. In the afternoon my brother Arnt made his appearance from the rear. He brought me bad news of the 'A.C.C.' He said that after I had left them on the Sand River, they had received orders to go west, to watch the movements of a cavalry force in that direction.

In the neighbourhood of Kopje-Alleen[137] they were ridden down by a regiment of horse. It was a case of every man for himself, and he had escaped by hard riding, but he could not tell whether our remaining two brothers [Hjalmar and Joubert] and the native boy [Charley] had got away.

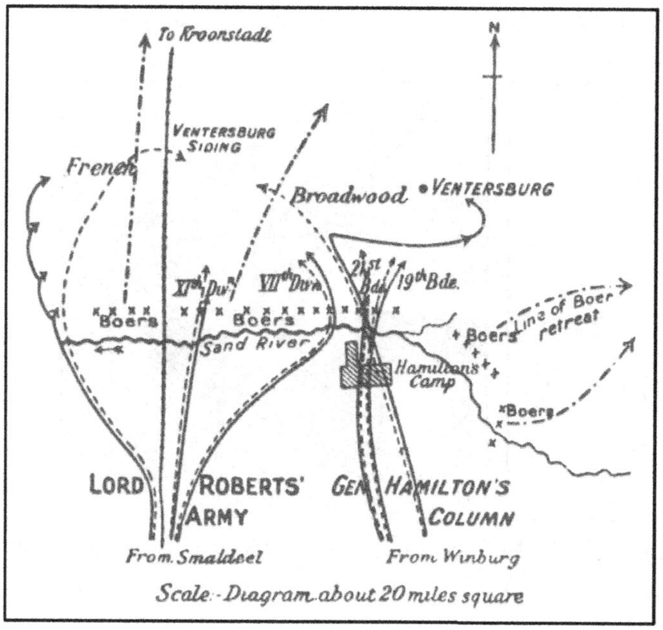

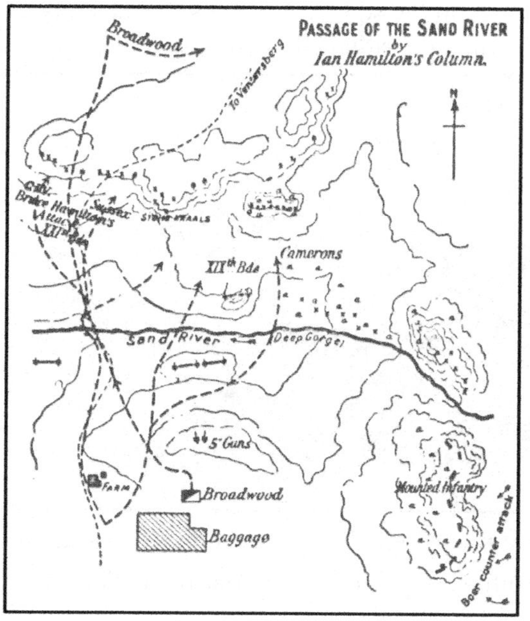

Diagram to Explain the Passage of the Sand River,[138] May 10th, 1900. The dotted lines show what was proposed, the continuous lines show what was done. The crosses indicate the Boers. Scale is 4.47* 4.47 Miles for 20 Square Miles.[249] Ian Hamilton's Action at the Sand River, May 10, 1900.[250]

[137] Riebeeckstad Riebeeckstad - Wikipedia

[138] Sand River (Free State) - Wikipedia

Reporters and photographers await the arrival of Roberts in Kroonstad on Saturday, May 12, 1900.[251] The Lace Diamond Mines near Kroonstad.[252]

This was serious news, and when a retreat was called after nightfall we rode back into Kroonstad greatly troubled. The town was in darkness, and we went right on for twenty miles without a halt until we overtook the commandos.

Next day the retreat was continued [Northwards] to the Rhenoster River. In the course of the morning we came on our boy Charley, who had not only escaped but had brought out the Basuto pony as well, thereby saving much of our gear, but he had no word of the other two.

An hour later we saw my eldest brother [Hjalmar] ambling along in company with the rest of the 'A.C.C.' They had got off better than we thought, having lost only twelve men in the fight. Of my brother Joubert they could tell us nothing; no one remembered having seen him and it was a long time before we heard of him again.

The 'A.C.C.' had now lost, in less than a week's fighting, nearly a third of its strength, with very little to show for it, but at any rate we were still a coherent unit, which was more than could be said for most of the other commandos, as the process of disintegration was gaining so rapidly that the major portion of the burghers riding in the retreat were no longer members of any recognized force, but merely individuals on their way home.

At the north bank of the Rhenoster River, we lay for nearly a week without sign of the British, during which time the Boer Army melted still further until General Botha had a bare handful of men left.

One morning a dozen of the 'A.C.C.,' of whom I was one, were ordered to go back in the direction of Kroonstad to reinforce a small body of Scouts under Captain Daniel Theron,[139] who was keeping touch with the enemy.

Rhenoster River Area.[253]

[139] Daniel (Danie) Johannes Stephanus Theron(9 May 1872– 5 September 1900) was a master scout in the Second Boer War. Danie Theron - Wikipedia "He was without a doubt one of the best scouts the Boer nation had produced. He repeatedly infiltrated our lines and obtained valuable information. Over and over he cut off our scouts and patrols, looted our supplies, and performed all sorts of dazzling military services for his people." Frederick Russell Burnham, Chief Scout for the British Army in the Second Boer War (1900).

We recrossed the [Rhenoster] river, [Southwards] and after a forty-mile ride south over the plains, we found him and his men on a hill over-looking the English camps that had sprung up around Kroonstad. Captain Theron had gained considerable notice before the war for thrashing Mr. [W. F.] Moneypenny, the well-known journalist,[140] and had of late added to his reputation by his daring at the time when Cronje was surrounded at Paardeberg. He was a light, wiry man of about twenty-six, dark complexioned and short-tempered, and although I never once saw him really affable, his men swore by him for his courage and gift of leadership.

Daniel (Danie) Theron.[254] Danie Theron's Scouts: at Dinner.[255]

For two days we watched the camps, and then one morning pillars of dust slowly rising and troops marching on every road showed that once more trouble was afoot. We made immediate preparations to depart, for we were only there as an observation post. Before we started a companion and I went down to a farmhouse to fill our saddlebags with meal and biltong from a supply stored there, and in returning we rode into a troop of English horse that unexpectedly appeared through the trees. We whipped up and raced away towards where we could see Theron and his men retreating in the distance, but the patrol was hard on our heels, and galloped after us for over a mile, firing about our ears, until at last our men, seeing the danger, came to the rescue. We fell back slowly until dark, and spent the night in view of their camp fires, and next morning, when we reached the north bank of the Rhenoster River, we found it deserted. A solitary member of the 'A.C.C.,' left behind for the purpose, told us that General Botha was summoned to Pretoria and that General de la Rey had taken the remainder of the forces towards the Transvaal border. We were to follow after. [Northwards towards the Transvaal.]

A large group of horsemen of the Imperial Yeomanry galloping over a plain.[256] A typical Imperial Yeoman on campaign.[257] 42nd (Hertfordshire) Company, Imperial Yeomanry.[141] [142]

[140] Over an article Moneypenny had written that was chauvinistic towards women.

[141] 42nd (Hertfordshire) Company was a unit of the Imperial Yeomanry formed to supplement the British Army in the Second Boer War. Raised by the part-time Hertfordshire Yeomanry and largely drawn from its ranks, the company was engaged at the Rhenoster River, in the Brandwater Basin 42nd (Hertfordshire) Company, Imperial Yeomanry - Wikipedia

Rhenoster River Engagement

When the British marched north from Bloemfontein towards the Vaal River there were still scattered parties of Boers active in the area and it had to deploy advance and flank guards at all times.

The British crossed the Vaal on June 1st, and was about 15 miles (24 km) from Johannesburg when news arrived that the Boers under Christiaan de Wet had overrun several garrisons along the railway and destroyed the Rhenoster bridge.

Roberts sent his chief of staff, Major-General Kitchener back to organise the defence of the line of communications. Kitchener moved south with a column including 42nd and 44th Companies Imperial Yoemanry and brought the Boers to action near the Rhenoster river. 42nd Company was involved in two-hour fight to clear De Wet from his position.

Then early next morning the company was roused from its camp to ride 2 miles (3.2 km) to where the Boers were attacking an engineers' train. Later that morning they drove the Boers off some hills, but had to return to the camp which had been attacked.

A yeoman standing by his horse.[258]

Rhenoster Bridge with just cribbing supporting it.[259]

Rhenoster bridge midway through reconstruction.[260]

[142] Chronology states that on Wednesday, May 23, 1900 the Rhenoster position was turned.

XII. The British Invade the Transvaal

Travelling on [Northwards] through empty abandoned country, we overtook some sort of a rearguard by next evening at Viljoen's drift,[143] and in their company we crossed over [the Vaal River] that night into the Transvaal at Vereeniging. Here there were only a few Irishmen of the dynamite squad, who told us that the 'A.C.C.' was camped a few miles on, and that some of de la Rey's men under General Lemmer[144] were ten miles down the river, but that for the rest the Boers had vanished.

Goods Shed at Viljoen's Drift, 1898.[261]

Boer commando in fighting trim.[145] This is the Winburgers Commando (Winburg, OFS). At the time the photo was taken [in Viljoen's Drift], they had already been on active service in Natal and were being sent down to the assistance of the forces around Bloemfontein.

Captain (Danie) Theron asked me to remain with him, but I refused as I wished to rejoin my brothers, so I said good-bye next morning, and went in search of the 'A.C.C.' This was the last I saw of him, for he was killed a few miles from here; a man who would have made a name for himself had he lived.

The 'A.C.C.' now decided to join General Lemmer's men, and, after a long ride, we came up with them just in time to see a strong body of English cavalry crossing the Vaal River, cheering loudly at setting foot in the Transvaal again, for it was twenty years since a British soldier had trodden its soil.[146]

[143] Viljoensdrif is a town in the far North of the Free State, just south of the Vaal River and close to Vereeniging in Gauteng. Viljoensdrif - Wikipedia

[144] Unable to identify this individual.

[145] Anglo Boer War - Images from 'On the Transvaal border' published by W H Gill, 1900 - Boer War Forum Credit Dr. David Biggins.

Hermanus Richard Lemmer.[147] [148] Refugee engines captured at Viljoen's Drift.[149]

Mounted Infantry Crossing the Vaal River.[262] Thursday, May 24, 1900 British Army entered the Transvaal, crossing the Vaal near Parys, unopposed. First camp inside the Transvaal.[263]

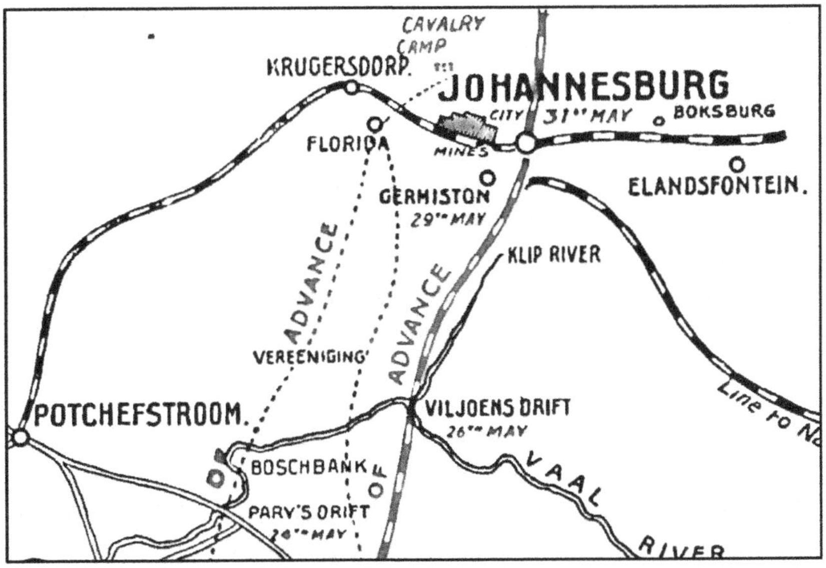

Note Parys Lower Left and Florida near Johannesburg.

[146] Chronology, Thursday, May 24, 1900 British Army entered the Transvaal, crossing the Vaal near Parys, unopposed.

[147] General Hermanus Richard Lemmer was born on 29 June 1849, at the place Hartebeestfontein, distr. Potchefstroom. Not before his thirty-second year was his Calm life as a cattle farmer, disturbed by extraordinary events. It was two years after the first War of Independence, he volunteered to enlist in the war of Moshette vs. Mochua. Here he served as a Field Cornet over a division of the volunteers and was badly injured. Later, Lemmer represented his district in the First People's Council until the last war broke out. He made the battle of Kraaipan as but was later appointed as a Fighting General and in the beginning operated in Colesberg, then on the Natal border and then at various places in the Free State and Transvaal, in numerous battles and smaller battles. In a bloody battle between Lichtenburg and Marico, during the last days of 1900, he lost his life, while he led a storm-attack at the head of only a small number of civilians. Two of his sons were later killed as Field Cornets.

[148] Our [Transvaal] Military Officers See Bibliography for full citation.

[149] These engines were sent across the border to the Transvaal. Note that all the connecting rods have been removed so they can be towed by a front engine. Anglo Boer War - Images from 'On the Transvaal border' published by W H Gill, 1900 - Page 3 - Boer War Forum Credit Dr. David Biggins.

They had batteries posted on the Free State side, and, as we had ridden too close up, Lemmer lost three men killed and several wounded, with no corresponding advantage to himself. He thereupon drew off into rougher country where we halted for the night, and next morning, seeing the British troops advancing from the Vaal, we retreated in the direction of Johannesburg, thirty miles away.

By noon we found General de la Rey, with nearly a thousand men, holding some low hills within sight of the mine-stacks. I was surprised that he had managed to keep so many with him, considering the way in which things were going to pieces, but he had more control over men than any officer who I had thus far seen.

Commando with Johannesburg and smokestacks in the background.[264] Gold mine near Johannesburg.[265]

At four o'clock the advance was on us again. Armstrong guns were unlimbered, and we were severely handled. The position we held was a strong one, however, and despite casualties we stood our ground until dark, by which time word came through that we must fall back on the Klip River, a small stream on the outskirts of Johannesburg.[150]

We groped our way through the night with hundreds of other men all jostling each other on the narrow road, and, having forded the river, we slept till daybreak. We were now practically backed right up against the city, so close that sightseers and even women came out in cabs and on foot to view the proceedings, and soon after dawn the English came pouring over the ground to the south of the Klip River with horse, foot and guns. As we were watching them, Commandant Gravett,[151] of the Boksburg commando,[152] came riding by asking for volunteers to accompany him to a low ridge just beyond, from which he said we could make a showing against some English cavalry that had crossed the river[153] and were approaching in our direction.

The men of the 'A.C.C.' hung in the wind, for they were sulking over some gibe Gravett had flung at us when we elbowed his people off the road the night before, so only a man lamed Jack Borrius[154] and I went. We rode rapidly forward, reaching the hills just in time to forestall the English horse-men from getting there first. We brought down three, whereupon the rest galloped back through the river, but soon returned reinforced, and they came at us so determinedly that we loosed only a few shots before running for our lives. We came under brisk fire without any casualties, but my roan horse had a piece clipped from his ear by a rifle bullet before I got back to the 'A.C.C.' in the rear.

My two brothers had been absent since earlier in the morning. For several days the younger one [Arnt] had been ailing, and now he was so ill that Hjalmar and our boy Charley had taken him to Johannesburg, one riding on either side to hold him on his horse. We did not know what was wrong with him, but it subsequently turned out to be typhoid fever from which he barely escaped with his life.

[150] Monday, May 28, 1900 The Battle of Biddulph's Berg. Also: Orange Free State formally annexed under the title of Orange River Colony.

[151] Gravett Johan David, Burger. He is the only 'Gravett' listed in the Boksburg Commando. https://www.angloboerwar.com/?option=com_content&view=article&id=3043

[152] Boksburg Commando 108 Records: https://www.angloboerwar.com/?option=com_content&view=article&id=3043

[153] Possibly: the Jukskei River, just west of Johannesburg. It is the southernmost river in the Crocodile River (West) basin. Jukskei River - Wikipedia

[154] Borrius Johannes Phiiiippus Commandant Commando: General Smuts. Awarded the DTD (Dekoratie voor Trouwe Dienst). Anglo Boer War - Borrius, Johannes Phiiiippus. Commandant - Boer War Forum

They had not yet returned when I rejoined the 'A.C.C.' and in the meanwhile we were kept busy enough, for the British troops were by now crossing the Klip River[155] in large numbers, deploying on the open ground between us, and before long shelling had commenced. No doubt they knew by now that Johannesburg was theirs for the taking, and they ran no risks with their infantry, confining themselves to most unpleasant gun-fire.

For the first time for many days we, too, had guns in action, and there were several batteries of Creusots blazing away from close by. The gunners suffered terribly, and I counted seven artillerists killed in less than fifteen minutes during one particularly violent burst. We of the 'A.C.C.' were snugly tucked away in a kopje where the shelter was so good that we did not lose a man or a horse, and we passed most of the day idly watching the scene. Shortly before sunset we saw activity away to our right, and there came line upon line of infantry, with guns roaring General de la Rey had his Lichtenburg men[156] there, but although they are reputed the best fighting men in the Transvaal, they were overborne by weight of numbers and were soon riding back in full retreat. This was the last effort to defend Johannesburg. When the line gave, all was over, and during the night de la Rey drew off to the west to his own country, where the-doughty old warrior was to fight many another battle in days to come.

"Group of Boer Commandants On the Klip River, Near Johannesburg. This picture was taken just before the great battle on 28 May 1900, when all the forces united under Lord Roberts were needed to capture the positions held by 2,000 Boers. Lord Roberts succeeded, but only after he had encountered the most determined resistance. His losses were very heavy. Johannesburg was occupied by him on the 31st of May."[266 157 267]

155 The Klip River is the main river draining the portion of Johannesburg south of the Witwatersrand, and its basin includes the Johannesburg and Soweto. The mouth of the river is at Vereeniging where it empties into the Vaal River. Klip River - Wikipedia

156 Lichtenburg Commando, 312 records: https://www.angloboerwar.com/?option=com_content&view=article&id=3055

157 The Doornkop area is also the site of the third day (29 May 1900) of the Battle of Johannesburg that took place nearby from 27 to 29 May 1900 during the Second Boer War. The ridge was taken by cavalry under command of Lt Gen John French while a nearby ridge, now covered by several Soweto suburbs, was taken by seven infantry battalions.

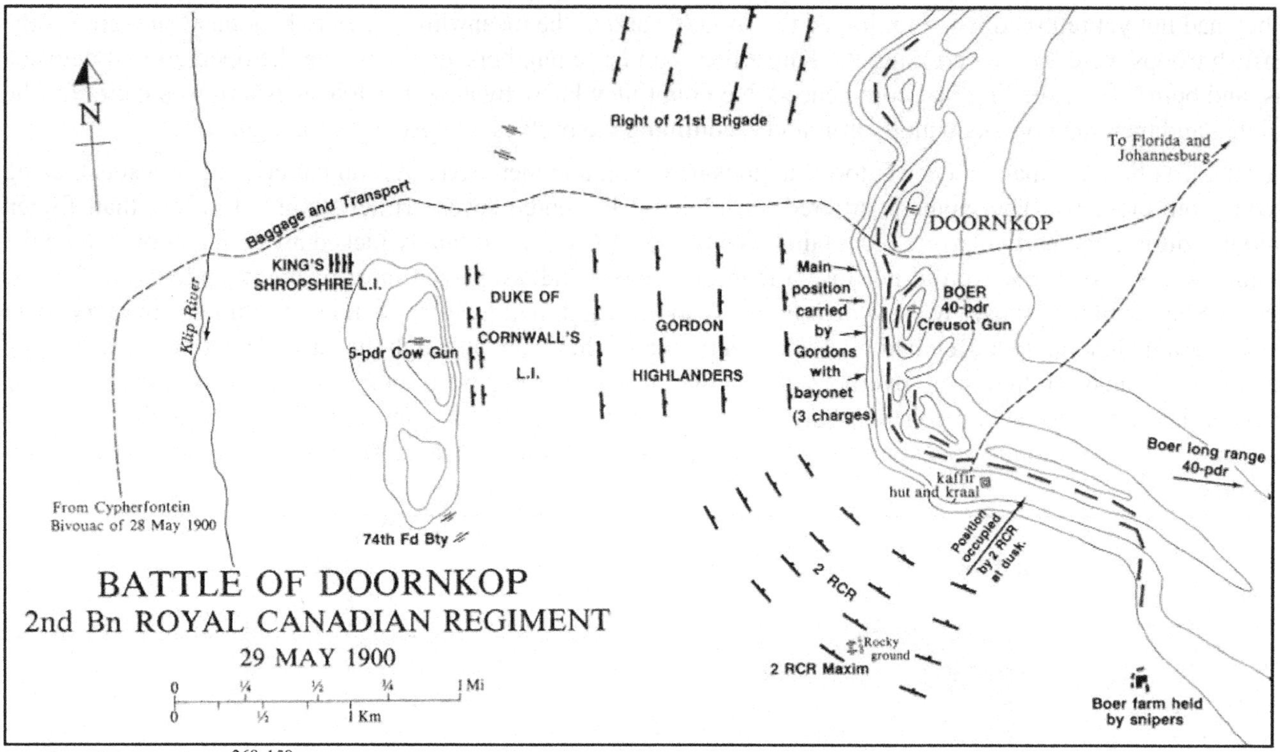

Battle of Doornkop.[268] [158]

All semblance of order or resistance now disappeared. Wherever one looked, men were departing wholesale, and the universal cry was: 'Huis-toe,' the war is over. Several of our 'A.C.C.' men deserted at this juncture, but most of them remained, and we fell back that evening to Langlaagte, a suburb of Johannesburg, where we spent the night. My brother Hjalmar and our boy were waiting for me. They had reached Johannesburg railway station, and had succeeded, in spite of the disorder, in getting Arnt on board a goods train for Pretoria. They said that all trains were crammed with fugitives, but that they had left him in the care of a man who promised to deliver him into my father's hands. With this they had to be satisfied, but they returned towards the firing very worried as to the outcome, for by now he was delirious. Next morning (it must have been about the 1st or 2nd of June) we saw the British feeling their way into Johannesburg, so we followed the drift of retreating men going round by the eastern side of the town. As we passed the gold-mines that lay on our route, there was a small column of cavalry drawn up not far off watching us go by, who made no attempt to interfere with us, probably thinking that we were refugees and not worth bothering about.

Gold Mines at Johannesburg 1899.[269]

When we got to the main road leading to Pretoria, we found it crowded, with mounted men, wagons and herds of cattle, and we had to make our way through dreadful confusion. To the right was another British column

[158] Battle of Doornkop - Wikipedia

moving parallel with us, which caused our native Charley to remark, 'Baas, those English people don't know the road to Pretoria, so they are coming along with us to make sure,' and, indeed, I believe that the English could have ridden in amongst us that day without firing a shot, so strong was the conviction that our army was disbanded and the war at an end.

By sunset the 'A.C.C.' worked itself out of the throng and halted at Six-mile Spruit,[159] a rivulet that distance from Pretoria. Commandant Malan intended to wait here until next morning, but my brother and I pushed on as we were anxious to get home and see our father.

We reached Pretoria by ten o'clock, and rode through the deserted streets to our home in the Sunnyside suburb. Here disappointment awaited us, for the place was in darkness and the house was empty. We went to several neighbours to make inquiries. They seemed to think that the enemy was upon them, for it was only after we had tried at several doors that at last a shrinking figure appeared in response to our knocking with rifle-butts, and, seeing who we were, curtly told us that President Kruger and my father had run away, and that Pretoria was to be surrendered to the British in the morning, after which the door was slammed in our faces. We knew the President and my father too well to believe that they had ignominiously run away, and the fact that they had left Pretoria together was proof to us that they had gone to carry on the war, so we returned home, and after stabling and feeding our weary horses, broke open one of the doors and went inside.

We made a roaring fire in the kitchen, at which we cooked a dinner with supplies from the pantry, and then slept in comfortable beds, a change after the freezing nights we had endured of late.

It was nevertheless a dismal homecoming. Our younger brother [Arnt] had been left stranded in a cattle-truck weak and ill, amid the chaos of a general retreat, our other brother [Joubert] was missing, and for all we knew dead, while my father was gone and our home was deserted.

We only heard later that my stepmother [Cornelia Maria Theresa Mulder [160]] and the younger children had been sent to Delagoa Bay and thence by sea up the East Coast of Africa to Holland, where they still are.

Park Station, Johannesburg.[270] President Kruger's private car used as the capital after evacuation of Pretoria.[271]

Early next morning we set about making plans for the future. First we saddled our horses and rode uptown to find out what was happening. The streets were swarming with leaderless men, knowing even less of the situation than ourselves. Of the 'A.C.C.' there was no trace, and all was utter confusion with looting of shops and supply depots, and a great deal of criticism of our leaders.

After commandeering provisions for our future requirements, we returned home. The British by now were shelling the forts outside the town, and an occasional 'over' fell in our vicinity, but we were accustomed to gun-fire by now, and remained quietly resting until the afternoon.

Towards three o'clock a gaunt figure appeared before us. It was our missing brother Joubert, whom we had given up for lost. He said that his horse had been killed when the 'A.C.C.' were rushed at Kopje-Alleen a

[159] 9/2/258/0162 Wierda Bridge, Six Mile Spruit, Centurion, Pretoria the national road to Johannesburg crosses the Six Miles Spruit ten km south of Pretoria. List of heritage sites in Gauteng - Wikipedia

[160] Francis married Cornelia Maria Theresa Mulder on December 11th, 1889. At the time of her marriage Mulder was acting director of the Eunice Ladies' Institute at Bloemfontein. With his second wife, he had six sons and one daughter. She was born in Delft, Netherlands 25 December 1863; died in Cape Town on 2 January 1935.

fortnight before, but he had succeeded in escaping on foot. After tramping it for many days, he reached Johannesburg in time to board the last outgoing train, which had just brought him to Pretoria. As burghers now came galloping past, shouting that the English were entering by the road above the railway station, I hurried back on horseback to the centre of the town, where I annexed a saddled-horse, from among several standing before a shop that was being looted, and absconded with this remount for my brother. We now prepared to leave, though as a matter of fact the English only occupied Pretoria next day, but, as we did not know that the rumour was premature, we thought it safer to get away in good time.

In the circumstances it seemed best to leave our faithful old native boy [Charley] behind, as we felt that with the increasing difficulty of securing horses and food we could no longer indulge in the luxury of a servant, and besides, we needed his animal as an additional pack-horse. The poor fellow piteously entreated us to keep him, but we had to harden our hearts, and, having no money to give him, we allowed him to take from the house as many blankets and other articles as he could carry, and so parted from him after an affecting scene.[161] Our arrangements were easily made. We loaded what we needed from pantry and wardrobes on to our pack-horses, and after a last look round at our home we rode away on the main road leading east, along which many other fugitives were already hurrying.

[161] PF: Charley was in his hometown; he was well known and respected as member of the Reitz household. He had friends there and probably had family too. He could get a pass to engage in work. (See picture infra.) He was out of the War and would not have to endure the danger and privations that Deneys and his brothers will encounter.

The Occupation of Pretoria

Boer Retreat from Pretoria.[272]

Boer Refugees.[273]

First Troops to Enter Pretoria.[274]

Natives procuring passes in Pretoria.[275]

Waterval after the liberation of the [British] prisoners.[276]

"Boer Prisoners in the Prison Prepared for the British Officers Just Outside Pretoria."[277] Note Dutch hats behind the wire.

Roberts riding in to Pretoria.[278] Roberts is riding the white horse. The entry of Roberts into Pretoria.[279] Drawn by Sidney Paget from a photograph taken by the Earl of Rosslyn.

For three hours the broad river of steel and khaki flowed unceasingly, and the townsfolk gazed in awe and wonder at those majestic soldiers whose discipline neither perils nor hardships had disturbed, whose relentless march no obstacles could prevent.[280]

THE ENTRY INTO PRETORIA

This picture shows the arrival of the Duke of Marlborough (on the grey horse), his cousin, Mr. Winston Churchill, and a few other officers in Pretoria, shortly before the formal entry of Lord Roberts

This is the caption that appeared for a reprint of the photograph in The Sphere on July 21st, 1900.[281]

Regimental band of the Gordon Highlanders parading through Church Square in Pretoria.[282] Gordon Highlanders parading through Church Square in Pretoria.[283] Note Kilts.

The scene in the square on June 5th, [1900] when the flag was raised on the Raadzaal.[284]

Pretoria Square occupied by the British, June 5th 1900.[285] Hoisting the British flag on the Raadzaal, Pretoria.[286]

Colonials "Marching Past" at the formal Occupation of Pretoria, June 5th 1900.[287] "Lord Roberts and staff cheering the Queen--Raising Union Jack on occupation of Pretoria, June 5th, S.A."[288]

Resume

By dark we had got as far as the big distillery[162] eight or nine miles distant, where we spent the night. By morning so many other horsemen had arrived that there must have been nearly fifteen hundred, few of whom were under officers, and none of whom seemed to know what to do next. My brothers and I rode about, looking for the 'A.C.C.,' but, although we found no trace of them, it did not worry us overmuch, and we agreed to remain on our own until we fell in with them again or until we had made further plans. In going round we met Mr. Smuts, the State Attorney, off-saddled under a tree with his brother-in-law, [Petrus] Krige, who had been one of Isaac Malherbe's men and had been seriously wounded at Spion Kop. I had not seen him since, for he had only just left hospital to avoid being captured in Pretoria by the British.

Commandant Smuts.[289]

As Mr. Smuts was a member of the Government, we persuaded him to tell us where my father and the President had gone to, and what the general position was. He said that the President and my father were at Machadodorp,[163] a small village on the Pretoria [to] Delagoa-Bay railway line, at which place they had set up a new capital. So far from making peace, they were determined to carry on the war by means of guerrilla tactics. They hoped to stop the rot that had set in, and Mr. Smuts himself was starting immediately for the Western Transvaal to reorganize that area, while similar steps would be taken elsewhere; and in the Free State, President Steyn and Christian de Wet had undertaken to pull things together.

The Commandant-General, Louis Botha, was lying not many miles away, collecting as many burghers as he could to form the nucleus of a fresh army and everyone was to be directed thither. All this was better news than we had heard for a long time, and already we could see, from the animated way in which the men were standing around their fires talking and laughing, that there was a more hopeful feeling in the air.

My eldest brother and I decided that, before joining a commando, we should seek out my father at Machadodorp, partly to find out his views, and partly to hear whether he knew what had become of our younger brother Arnt. My brother Joubert refused to accompany us and rode away to look for General Botha, so that once more we lost sight of him for many days. Machadodorp lay a hundred and seventy miles due east, and Hjalmar and I set out on the long ride without delay. We did the ninety miles to Middleburg in two days, and here we were lucky enough to get a lift by goods train for the rest of the journey, arriving at Machadodorp by the following morning.

This village was for the time being the capital of the Transvaal. Long rows of railway coaches constituted the Government Buildings, where such officials as had not preferred surrender made a show of carrying on the

[162] Per Haldane: The Hatherly Distillery.

[163] Machadodorp, also known by its official name eNtokozweni, is a small town situated near the edge of the escarpment in the Mpumalanga province, South Africa. The Elands River runs through the town. Machadodorp - Wikipedia

public business of the country. In one of these coaches we found my father installed, and his welcome was a warm one, for he had received no news of us since we had left him in April to go south into the Free State. We were greatly relieved to hear that our brother Arnt was at that moment lying in the Russian ambulance at Watervalonder, in the low country, forty miles down the line. He had arrived delirious some days before, but there was hope of his recovery.

My father confirmed what Mr. Smuts had told us of the military position, and he said that guerrilla war was better suited to the genius of the Boer people than regular field operations. He spoke of George Washington and Valley Forge, and of other seemingly lost causes that had triumphed in the end, and although we did not altogether share his optimism (for we had the memory of demoralized and flying columns fresh in our minds), yet his faith cheered us tremendously.

When we asked after President Kruger we were told that he too was down at Watervalonder, for he was an old and feeble man in these days and was unable to stand the bitter cold up here.[164]

Before returning west in search of General Botha, Hjalmar and I took train down the mountain to see our sick brother.[Arnt] We found him with many wounded men in a hospital improvised by that very Russian ambulance corps[165] which General Joubert had refused to accept, but which had nevertheless come to our assistance.

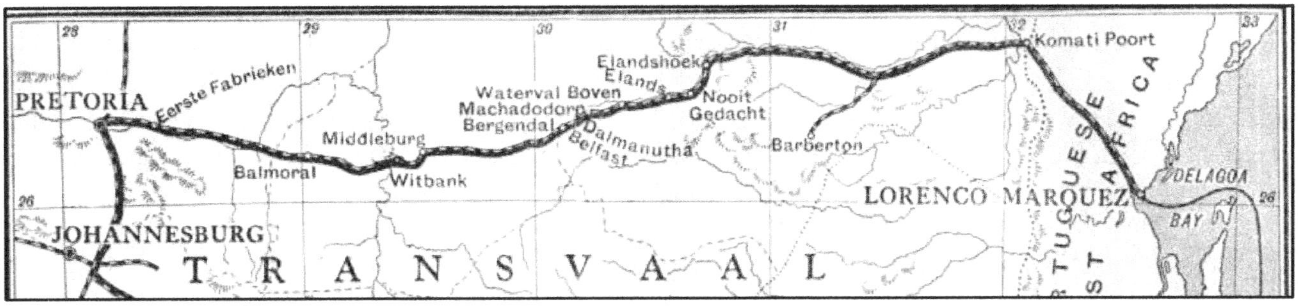

General Map of SA.[290]

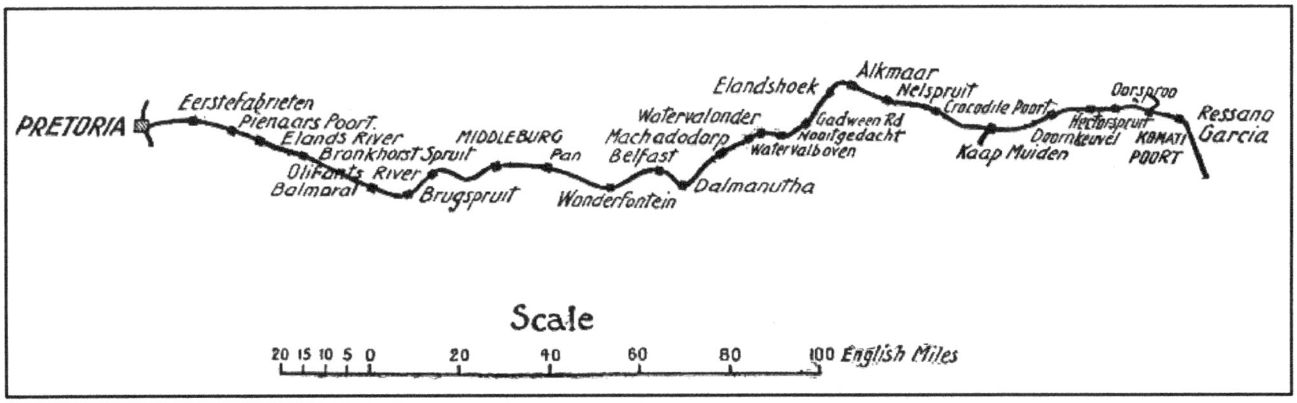

Pretoria to Komati Poort Railway.[291] Machadodorp/Belfast[166] is in the middle.

[164] On the high plateau.

[165] Russian Volunteers. 99 Records: https://www.angloboerwar.com/unit-information/boer-units/179-foreign-vols/3246-russian-volunteers

[166] Belfast (also known as eMakhazeni) (Dutch Berg-en-dal (Hill and Dale) is a small town in Mpumalanga Province, South Africa. Belfast, Mpumalanga - Wikipedia Spoiler alert: During the Anglo-Boer War several battles and skirmishes took place in and around the town.

The Battle of Leliefontein November 7th, 1900, took place 30 km south of Belfast at the Komati river, an engagement for which several Victoria Crosses were awarded to Canadian soldiers. The British built a concentration camp here during the Boer War to house Boer women and children. Several Victoria Crosses were awarded for action at Monument Hill which is on the edge of the town. During the war Meyer de Kock and his family resided in town. He was a member of the local peace committee and visited the Boer commandos to persuade them to surrender. He was convicted by them of high treason on the grounds that he was not communicating with the commander(s) of the unit and was instead making his appeal to rank and file Burghers and soldiers. He was executed by firing squad on 12 February 1901. Battle of Leliefontein - Wikipedia

Cog Wheel Railway.[292] From Waterval-Boven to Watervalonder the line drops at a gradient so steep that the engines used a cog-wheel rack railroad.[293] Waterval-Boven, circa 1901. Note steep incline at top left.[294]

He was conscious when we got there, and the Russian nurses said that he had turned the corner, although still in grave danger.

At Watervalonder we had our last sight of President Kruger. He was seated at a table in a railway saloon, with a large Bible open before him, a lonely, tired man. We stood gazing at him through the window, but as he was bowed in thought, we made no attempt to speak to him. He left for Portuguese territory not long after, and I never saw him again, for he was taken to Holland on a Dutch man-of-war, and he is still an exile. (He died in Switzerland in 1904.)

We now returned up the Berg to Machadodorp, where we said good-bye to my father, and travelled back to Middleburg by rail to get our horses, which we had left in charge of one of the townspeople.

Here there was a small contingent of German volunteers,[167] about sixty strong, under an Austrian, Baron von Goldeck,[168] whom we had known in Natal. As we had no idea where the 'A.C.C.' had gone, and as one commando was as good as another, we obtained admission to the 'German Corps,' as it was somewhat grandiloquently called. Von Goldeck was preparing to ride his men west to scout for General Botha, so we trekked away the next day, going via Balmoral Station, until in three days' time we gained contact with the British patrols on the outskirts of Pretoria.

Lord Roberts was resting his army around the capital, so we spent the next ten days skirmishing over the uneven country to watch his movements. We had several exciting encounters, in the course of which we lost five Germans, but it was an enjoyable time. We lived on what we could forage, and, what with scouting to within sight of Pretoria and raising alarms in the big camps, there was not a dull moment. Then the country got too hot to hold us and we fell back twenty or thirty miles to where General Botha was busy collecting as many men as he could get together.[169]

We found him halted near the old battlefield of Bronkhorstspruit, where Colonel Anstruther's force was cut up in the war of 1880.[170] He lay by his saddle on the open veld, and save for a few dispatch-riders and some pack-horses, there was nothing to distinguish his headquarters from any of the other groups of burghers dotted about. He said we had done well and could now take a holiday, so we rode to a deserted farm some distance off, and remained there quietly for some days.

[167] German volunteers 863 Records: https://www.angloboerwar.com/?option=com_content&view=article&id=3243

[168] Von Goldegg Und Lindenburg Anton Austrian and Hungarian volunteers: 77 Records: https://www.angloboerwar.com/?option=com_content&view=article&id=3245 See Also: German Volunteers: 863 Records: https://www.angloboerwar.com/?option=com_content&view=article&id=3243

[169] PF: Given the amount of time it has taken Deneys to travel East and then back West, it is unlikely the he participated in the battle of Diamond Hill on June 11th and 12th. Battle of Diamond Hill - Wikipedia He is describing harassing operations subsequent to that general engagement.

[170] The Battle of Bronkhorstspruit was the first major engagement of the First Boer War. It took place on 20 December 1880. In an action lasting just 15 minutes, 156 British soldiers were killed or wounded, with the rest taken prisoner. Reported Boer casualties were two killed and five wounded. Battle of Bronkhorstspruit - Wikipedia

XIII. Further Afield

It was by now becoming generally known that General Botha was assembling a new army, and the more spirited elements were gradually coming back from their homes and elsewhere in order to rejoin him.

Already he had three thousand men, and more were riding in daily. Things were, therefore, shaping better and there was, at any rate, a visible recovery from the deplorable conditions that had held during the retreat through the Free State.

At this stage, remnants of the Natal commandos began to come through. The forces we had left there had been driven out, much as we had been hustled from tile Free State, and had endured the same humiliations. Thousands of the weaker men had surrendered or gone home, and in many cases entire commandos had melted like snow, whilst those that did remain were mere skeleton formations. The remainder, however, were seasoned fighting men, and nearly two thousand of them joined General Botha here, bringing his total force up to five thousand or more.

One evening I saw a Natal contingent approaching over the hills. When they reached us the men turned out to be the remnant of our old Pretoria commando, now dwindled to about one hundred and fifty, amongst whom were many old friends of the Ladysmith days. The burgher-right men had deserted in a body, and the commando had lost heavily in killed and wounded since we had left them. Mr. Zeederberg, the Field-Cornet, had broken down in health; and one Max Theunissen,[171] a young fellow of twenty-five, was in command.

Old General Maroola, they said, and his brother Swart Lawaai, had been deposed, but they had remained in the field and were serving as privates with some other commando.

I had got on well enough with von Goldeck and the Germans, but I decided to return to the Pretoria men. My brother Hjalmar preferred to stay where he was, so I said good-bye, and rode away with my roan horse and the little Basuto pony to join my former unit.

The morning after we moved off, going south over the plains for two days and then bending round so as to make for the railway line between Pretoria and Johannesburg, as Max Theunissen was under orders to destroy the English communications. There were, however, so many English soldiers [were] guarding the track that we never got within five miles of it, and for the next fortnight we wandered about searching for an opportunity that did not come. The English troops clamped about Pretoria occasionally flung a few long-range shells at us, but otherwise there was nothing of importance.

While we were operating in this quarter I had unexpected word of the A.C.C. We were taking cover one morning in a hill from the shell-fire of a column that had appeared, when two burghers galloped up to seek shelter. They belonged to the A.C.C., and when I asked after the others they pointed to a farm far out on the plain, where they said Commandant Malan and the rest were off-saddled, so when they rode away I went along with them, partly to see my old companions, and partly to explain to Commandant Malan about my brothers and myself.

To reach the farm we had to ride across the front of the English column, but they let us go in peace, and I was soon shaking hands with Malan and his men. A little later, the English force moved nearer, and began to drop Lyddite shells from a howitzer posted on a distant rise. This caused the worst incident I saw in the war. For a time the shells burst harmlessly on open ground, but as the aim grew truer, Commandant Malan ordered us to take cover, and we distributed ourselves behind the garden wall and behind the wall of a small dam that stood near the dwelling-house.

By the wall of the dam was a huge willow tree, in the rear of which seven men of the A.C.C. took station. Suddenly a shell hit the bole[172] about two feet from the ground and, passing right through the tree, exploded as it emerged on the other side. The result was terrible, for the seven unfortunate men were blown to pieces which strewed the ground for thirty yards beyond. They were so mangled that, when the English gun ceased firing, their remains had to be collected with a shovel, a most sickening spectacle. And a further trial was in store for the A.C.C. After the howitzer had gone, Commandant Malan mounted his horse and with a few men rode up towards the English troops to fire the intervening grass.

[171] Theunissen Max Jeremias Vorster Veldkornet Awarded the DTD. Also: Prisoner number: 26496 Captured: Driebad 05 Dec 1901.

[172] The trunk or stem of a tree.

Half an hour later, as I was returning to the Pretoria commando, I saw by the way in which his men were standing that something was wrong, so I galloped up to find him lying with a rifle bullet in his throat, and he died in a few minutes, and any lingering idea I may have had of re-joining the A.C.C. now vanished, for they seemed under an unlucky star. For another week or two the general military situation remained quiet; British columns came from Pretoria now and then, but they were only on the prowl, and there was no serious fighting.

This spell of calm was invaluable, for it gave the Boer leaders breathing space in which to reorganize their scattered forces, and it gave the men time to recover from the demoralizing effect of the long retreats, with the result that when Lord Roberts resumed operations in the middle of July a far better spirit prevailed.[173]

The British advance started early one morning on a broad front. We of the Pretoria commando took up a position in the nearest kopjes, but before long were so heavily shelled that we withdraw, the other commandos also falling back. The direction in which the English were moving was east along both sides of the Delagoa Bay rail-way line. Up to now this way to the Portuguese port had been open and the Transvaal Government had been importing supplies on a large scale, but apparently Lord Roberts intended to close this final loophole and cut us off completely from the outer world.

Accordingly great numbers of troops, apparently more than thirty thousand, moved to right and left of the railway, sweeping us easily before them. They were spread fanwise over a distance of fifteen miles, and as the usual curtain of scouts approached, followed by shell-fire from the guns moving up behind, we pursued the same methods that we had employed in the Free State, fading back from hill to hill and rise to rise, firing when opportunity offered, but not really fighting.[174]

In this manner we retreated for four or five days, by which time the English had pushed us along the railway line through the town of Middelburg,[175] and right up to Belfast village forty miles beyond. On the night we reached this place there was such a press on the road that in the dark I lost the Pretoria men. By next morning they were nowhere to be seen, but, as I was getting accustomed to quick changes of commando, I joined a body of Boksburgers who happened to be passing. They were part of a larger force under that same Commandant Gravett with whom I had gone down to the river the day before Johannesburg was taken.

Boksburg is a small mining village on the Reef, and the Boksburg men for some reason or other were known as 'Gravett's Guinea Fowls' (Gravettse tarantaal-koppe), a title in which they took great pride. Gravett was of English extraction, a fine big man, greatly liked and trusted by all. He was killed a month or two later, and I was with him when he died.

Gerhardus Hendrik Gravett.[176 295]

[173] Saturday, July 21, 1900 Advance begun from Pretoria east, along Delagoa Bay Railway.

[174] Wednesday, July 25, 1900 Roberts's force reached Balmoral on the way to Middelburg. French's Cavalry and Hutton's Mounted Infantry put Boers to flight six miles south of Balmoral.

[175] Friday, July 27, 1900, Occupation of Middelburg by advance guard of Roberts without opposition.

[176] Military Officer Album (Direct Translation from Dutch) General Born at Alexandria, Cape Colony, on June 28, 1858, Gerhardus Hendrik Gravett, when the war broke out, living at Elandstfontein (Germiston), where he was born in 1888 and which place its origin was almost chiefly due to his zeal and work force. Was chosen as Field Cornet and took then took an active part in the fighting near Colesberg. At Norval's Pont he was elected Commandant by his citizens and the success of the masterly retreat from the Cape Colony, through the eastern districts of the Free State, past Bloemfontein - already in the hands of the enemy - At Lijdenburg he was appointed as Fighter general. At Witpoort, Mapochsgronden, with a handful of burgers fighting valiantly, Gravett was knocked next to a Nordenfeldt Maxim by a shrapnel seriously wounded, **and died on 26 October 1900 near Roossenekal**. Was buried the cemetery at Germiston for the last time. Commandant General Botha was present on this occasion and testified of Gravett: "He was a man who loved his people dearly; In the darkest days, he was always a source of courage and hope for all who came in contact with him; of the civilian ranks he was ascended to the highest position that a brave man could hold in our country; He was faithful unto death and did what he could.

In the company of the 'Guinea Fowls' I trekked from Belfast to Dalmanutha, another forty miles up the line. After being marched about for some days over mountainous country, we were allotted a position on the edge of the escarpment near Machadodorp, where General Botha intended making a stand athwart the Delagoa railway.

Along this crest he was going to fight a last pitched battle before taking to guerrilla war.[177] All through the retreat we knew that sooner or later it was planned to break away from before the advance and scatter in smaller bands, and this knowledge had kept the men in good heart. Although they had been badly harried, there was no tendency to dissolve as had been the case in the Free State, and when they were called upon to make a final stand they were willing enough.

The position General Botha had selected was a natural fortress. Between us and the enemy there stretched a level plain that could be swept by rifle-fire, and immediately behind us the ground fell steeply into a valley, giving excellent cover for men and horses. We were practically on the farthest rim of the high veld, for a few miles back the country drops into the malarial lowlands that lie towards the Portuguese border. The British had come to a temporary halt at Belfast, and for a week there was no sign of them. Towards the end of that time, while I was working on the defences, I saw the Pretoria commando come riding up, and with them were my two elder brothers Hjalmar and Joubert. Needless to say we were all three glad to be together again, and I at once took leave of the Boksburgers in order to join them.

The Pretoria men were given a portion of the line close by, and at sunrise on the next day heavy dust-clouds arose in the distance, and before long masses of British infantry appeared on the skyline.

They were calculated to be thirty-six thousand strong (but it is difficult to count infantry on the march), and within an hour their skirmishers were firing, and their batteries were unlimbering almost within rifle-range of us.

By ten a heavy bombardment was in full swing, although no actual advance was attempted, as they evidently intended first to batter down our works.

This lasted until sunset, but our cover was so good that the casualties were nowhere heavy, and the Pretoria commando went scot-free. By dark it had all died down, and we passed a quiet night lying around our fires. Next day the programme was repeated. We were shelled to such an extent that one dared scarcely look over the edge of the breastworks for the whirring of metal and the whizzing of bullets. Several of our men were wounded, and my brother Hjalmar was shot below the eye. My other brother led him down into the valley, for he was partially blinded, and there placed him on his horse and rode with him for the nearest medical assistance.

This second day of the bombardment was a crowded one. Shortly after my brothers had left there was an earth-quake, the first I had ever experienced. It came with a loud rumbling, and the ground rocked beneath us like a ship, while stone fell from the works, causing much alarm, for disturbances of this kind are practically unknown in South Africa.[178] We thus suffered a bombardment from above and an earthquake from below at one and the same time, and this remained a topic for wondering discussion months afterwards. When it was over, a Lyddite shell from a howitzer dropped almost on top of me. It was like another earthquake. I was stunned for several minutes, and, after that, lay for a while in a semi-conscious state, hardly knowing whether I was dead or alive.

In the afternoon a detachment of infantry came down a defile on our left. We saw them in time to drive them back, killing and wounding about fifteen, but owing to the cross-fire we could not reach the fallen men until some time after dark, when we groped our way to get their rifles and equipment. The night was so cold that we found only three soldiers alive, some wounded, who might otherwise have survived, having died of exposure. We carried the three men back, and laid them by a fire, where one more died before morning.

As soon as it grew light on the third day, the bombardment recommenced more furiously than ever, but, instead of being spread all over our front, it was concentrated on that section held by the Johannesburg Police,[179] a mile to our right. Tremendous gun-fire was poured on them, and from the massing of the infantry columns we knew that a crisis was at hand. The police behaved splendidly. Twice they threw back the attacks, and hung on

[177] The Battle of Bergendal (also known as the Battle of Belfast or Battle of Dalmanutha) August 21–27 took place along the railway line east of the town Belfast in what was then the South African Republic. It lasted from 21 to 27 August 1900. The 7,000 Boers were under the command of General Louis Botha, whose headquarters were at a railway station called Dalmanutha. The 19,000 British Empire forces were under the overall command of Lord Roberts. General Sir Redvers Buller led the final assault against the Boer position on the farm Berg-en-dal (Hill-and-dale). Battle of Bergendal - Wikipedia

[178] August 26th, 1900.

[179] The South African Republic Police (Dutch: Zuid-Afrikaansche Republiek Politie; ZARP) South African Republic Police - Wikipedia

doggedly under some of the fiercest pounding of the war. While this was going on the rest of us could do little, and for the most part we sat perched on our scanses[180] watching the struggle. By sunset the police were all but annihilated, and in the dusk we saw the English infantry break into their positions. Here and there a hunted man went running down the slope behind, but the majority of the defenders were killed. Our line being broken, we had to give way too, and after dark General Botha ordered a withdrawal. We fetched our horses from the valley below, and fell back for two or three miles before halting for the night.

"The Johannesburg Police (Van Dam's Commando) [Possibly Gerard Mari Johan van Dam, See MOA] Off To The Front, 26 October 1899. No commando behaved more gallantly in the war than the Johannesburg Police, which, with the other police and the State Artillery, formed the standing army of the country. Originally the Johannesburg Police numbered over a thousand men. After the Battle of Dalmanutha (23–27 August 1900) their strength was reduced to less than 200."[181] PF: These men are the much maligned Zarps in the British narratives. They fought bravely at Dalmanutha.

Next morning we made for Machadodorp, long-range shell-fire accompanying us. We found the village deserted, the movable capital having left the day before for Watervalonder below the berg.[182]

Beyond Machadodorp a single road climbs the last range, and from here one can look down upon the low country. As this was the only avenue of retreat, we soon found ourselves travelling among a medley of burghers, guns, wagons, and a great crowd of civilian refugees fleeing with their flocks and herds and chattels. It was pitiful to see this exodus, for the English brought their guns up with great speed and the road was heavily shelled over at times, as the wagons with women and children came under fire, but on the whole their behaviour was good, and in the end the shelling proved more unpleasant than dangerous.

After a while the Transvaal Artillery managed to get a battery of Creusot guns into action, which held up the advance sufficiently long to enable the non-combatants with their wagons, carts, and animals to get out of range, after which we too moved slowly up the mountain. I had a narrow escape when we halted to rest our horses. I was sitting on an ant-heap reading a book, when someone called to me that my roan horse was in trouble. He was plucking grass some distance off and had got his legs entangled in the reins, so I went to free him. While I was away a shell burst on the ant-heap, blowing holes in my book, and tearing up the mound itself.

We spent the night at Helvetia[183] on the mountain, and next morning rode down the pass to Watervalonder, a drop of two thousand feet in less than three miles. The British appeared above the berg an hour or two later, but did not follow us down, so we wended our way at leisure.[184]

At Watervalonder the railway comes out of a tunnel from the high country, and here, standing beside the line, were my father and my brother Hjalmar. They had come from Machadodorp by train the day before and were halted to let the engine get up steam, after which they were to be hauled farther down with other officials and a

[180] A crescent-shaped structure of stones built to afford cover in battle.

[181] Heroes of the Boer War. Review of Reviews, 1903.

[182] Tuesday, August 28, 1900 General Buller's troops occupied Machadodorp. Bergendal occupied.

[183] Later the site of the Battle of Helvetia. December 28th, 1900. Battle of Helvetia | Endless streams and forests (wordpress.com)

[184] Wednesday, August 29, 1900 Kruger fled to Nelspruit. The Boers evacuated Helvetia, which was occupied by General Buller.

number of wounded. Hjalmar had a bloody bandage over his eye like a pirate, but otherwise he made light of his hurt. Our remaining two brothers [Arnt & Joubert] were missing, but we were so continuously losing and finding each other nowadays that their absence caused us no undue anxiety, as we felt they would turn up sooner or later.

After a while the engine-driver ran up to say that he was going off, so my father and brother had to climb on board, and I continued down the road behind the retreat which was crowding the long valley that runs towards the Godwan River.[185]

By dark I caught up with the Pretoria commando, and we spent the night pleasantly encamped beside a stream. Since morning we had at one stride descended from the bleak highlands to the warmer climate of the low country, and I for one spent the first comfortable night for many weeks.

Next day the Boer forces retired still farther down the valley to Nooitgedacht, where about two thousand English prisoners were confined in a camp. They were lining the barbed-wire enclosure beside the railway line to watch us go by, and were in high spirits, for they knew that they were to be liberated that day.[186] They exchanged good-natured banter with us as we passed, although one of them, less amiable than the rest, said to me: 'Call this a retreat? I call it a bloody rout!' I must say it looked like it, for by now the English advance was on our heels once more, and the narrow valley road was thronged with horsemen, wagons and cattle, all moving rearward in chaos. With the Boers, however, appearances are often deceptive—what might seem to be a mob of fugitives one day, might well prove to be a formidable fighting force on the next, and the soldier who spoke to me little thought that the men pouring by in disorderly flight were yet to test the endurance and the patience of Great Britain to its utmost.

We now headed down along the banks of the Godwan River for another twenty-five miles, by which time the pursuit had slackened, and for the next three days we lay in peace amid beautiful surroundings of mountain and forest. The English probably thought that General Botha meant to cross over into Portuguese territory for internment, rather than surrender, so they lay on their arms and rested. But General Botha intended far otherwise. One morning my father came up by special train from Nelspruit,[187] where another temporary capital had been established, and he told us that the Commandant-General was going to strike north into the wilds. He would then make for the mountains beyond Lydenburg[188] [189] where the forces were to be reorganised for the carrying on of guerrilla warfare.

Long Tom Pass and Monument.[296] [190]

[185] Unable to identify by this name. By location near Nelspruit it is the Crocodile River. Crocodile River (Mpumalanga) - Wikipedia

[186] Thursday, August 30, 1900 British occupation of Waterval Boven. Release of about 2,000 British prisoners at Nooitgedacht.

[187] Nelspruit - Mbombela - Wikipedia

[188] Lydenburg, is a town on the Mpumalanga highveld. It is situated on the Sterkspruit/Dorps River tributary of the Lepelle River at the summit of the Long Tom Pass. Lydenburg - Wikipedia

[189] Long Tom Pass is a mountain pass on the Great Escarpment, between Lydenburg and Sabie (South Africa). It is named after the Long Tom cannon. A monument commemorating the last use of the Boer 155 mm Creusot Long Tom guns during the Second Boer War Long Tom Pass - Wikipedia And 155 mm Creusot Long Tom - Wikipedia

[190] Tuesday, September 4, 1900 General Buller and Botha engaged at Lydenburg.
Thursday, September 6, 1900 British occupied Lydenburg, Botha retreated.
Tuesday, September 11, 1900 Kruger, having fled from the Transvaal, arrived in Portuguese territory, and proceeded to Lorenzo Marques.
Thursday, September 13, 1900 General French occupied Barberton.

My father had really come up to see how my brothers and I were getting on, but I was the only one there. Hjalmar was at an ambulance recovering from his wound—Joubert had not returned since he had ridden back to the fighting at Dalmanutha, and the youngest was wherever the Russian ambulance wagons had gone, so that at the moment we were a pretty scattered family.

My father remained with me for the day, and then steamed back down the valley. Unfortunately the engine that brought him ran over and killed my poor little Basuto pony. Besides having served me faithfully since the first day of the war, he was an intimate link with our old home life, for he had come with us from the Free State as a foal, and the loss of this loyal companion was a great blow to me. However, there was not much time for repining,[191] as the English resumed their advance early next morning. The commandos fell back at once; some retired down the valley, but others slipped into the mountains.

Among the latter was the Pretoria commando, which climbed up to a place called Devil's Kantoor,[192] where we arrived that night.

The Kaapsehoop heights seen from below.[297] Avoca Bridge on the Kaapmuiden to Barberton line, 1895[193]

The Pretoria men had for some days been discussing the question of doubling back to the high veld. They felt drawn towards their own part of the world, and wished to return to where they might get occasional word of their families left behind in Pretoria: so they now held a meeting and decided to break away along bridle-paths through the mountains.

I refused to accompany them, as my father and brothers were ahead, so I said good-bye, and soon they were riding off into the dark. I did not see them again, but they succeeded in making their way back to their own district, where such of them as were not killed or captured remained in the field to the end of the war. I myself joined a party of burghers whom I found on the mountain, and in their company I rode all night, until dawn found us far down on the plain that runs by Barberton towards Kaapmuiden. Beyond this we got into rugged country and, travelling slowly, reached Hectorspruit[194] in three days' time.

This was the last railway station before the Portuguese border, and here lay nearly five thousand horsemen awaiting General Botha's orders. We had shaken off the English army by fifty or sixty miles, and I found all three of my brothers there before me. Hjalmar's wound was better and Arnt had so far recovered that he could sit a horse, while Joubert was none the worse for his wanderings during the retreat. My father was missing, but he arrived two days later. He had abandoned his railway coach owing to a stoppage on the line, and had come across the difficult area above Crocodilepoort[195] on horseback, so our family was united again for the first time after several months.

Sunday, September 16, 1900 British occupied Nelspruit. British occupation of Kaap Muiden.

Monday, September 24, 1900 Arrival of the British at the Portuguese frontier. Evacuation of all the Boer positions near the frontier.

[191] The act of fretting or feeling discontent or of murmuring.

[192] Kaapschehoop (the Devil's Office) is situated 1640 m (5380 ft.) above sea level on the Drakensberg Highveld escarpment, about 28 km from Nelspruit. Kaapsehoop - Wikipedia

[193] Kaap River - Wikipedia File:Avoca bridge, South Africa.jpg - Wikimedia Commons

[194] Hectorspruit, officially Emjejane, is situated between Kaapmuiden and Komatipoort on a southern tributary of the Crocodile River. Hectorspruit - Wikipedia

[195] Very near or actually, Komatipoort is a town situated at the confluence of the Crocodile and Komati Rivers. The town is 8 km from the Crocodile Bridge Gate into the Kruger Park, and just 5 km from the Mozambique border. Komatipoort - Wikipedia

General Botha now got everything ready. Surplus guns were destroyed or thrown into the Crocodile River, and the sick and wounded were sent over the Portuguese border, while such stores as had been accumulated were distributed among the men or else burnt. Then, on a morning early in September (1900)[196] he led the way into the uncharted bush to begin a new phase of the war.

"Wonderboom Tom" (a.k.a. "Puffing Billy," of Umbulwana fame). And in UK as War Trophy exhibited outside Glasgow Art Gallery.[298] [197]

Remains of a Boer big gun, Waterval.[299]

Flour abandoned on Delagoa Bay Railway.[300]

[196] Monday, September 24, 1900 Arrival of the British at the Portuguese frontier. Evacuation of all the Boer positions near the frontier. Tuesday, September 25, 1900 Surrender of Boers to the Portuguese.

[197] Wonderboom Tom" (a.k.a. "Puffing Billy", of Umbulwana fame) photographed outside Glasgow Art Gallery, where it stood for the duration of the Glasgow International Exhibition (2nd May to 9th November 1901).
The description on the plaque reads: "GUN CAPTURED from the BOERS / Loaned by Field Marshal Earl ROBERTS / This "Long Tom" was one of the four 95 pounders used by / the Boers against us from the commencement of the War. / It was first brought to bear on the 11th Division at / Pienaars Poort to the East of Pretoria two days / before the fight at Donker Hoek (now called / Diamond Hill) and subsequently at Belfast. / Its remains were taken on Sept 24th 1900 by the / 11th Division on the hill to the East of Komati Poort".

XIV. New Conditions

Our road ran through the Sabi low country teeming with big game of all descriptions. By day great herds of zebra, wildebeest, and sable stood fearlessly gazing at us, and at night lions prowled roaring around our camps. Of hunting we had our fill, and to me this journey through a strange and remote region was full of fascination, for we were passing through country as untouched as that upon which the old pioneers had looked when first they came north in the days of the Great Trek.

Black Wildebeest.[301]

Adult male Sable antelope.[302]

After long marches we reached the foot of the mighty range that runs north as far as the eye can see. From here General Botha sent the commandos on, with instructions to follow the base of the escarpment to the Murchison Range,[198] two hundred miles away. He himself, escorted by the remnant of the Johannesburg Police and accompanied by the handful of officials who still remained, climbed up by a hunting-path to make for a lost village called Ohrigstad[199] in the Steelpoort Gorge, our way leading by Gras Kop and Pilgrim's Rest[200] amid mountains and forests and gorges more beautiful than any I know of in South Africa.

Will Barter, John Hygelsen, Kloprer.[303] The Treasury Department in the Field. William Barter, chief of the Money Printing Division, on the left.[304] Coins of the SAR.[305]

Now that President Kruger had left the Country, the Transvaal Government consisted of my father and Schalk Burger who was the Vice-President, together with a few other heads of departments, and they naturally followed General Botha up the mountain, as did my brothers and I. We remained at Ohrigstad for a week, but when

198 In this case the meaning is an area of open, often unfenced, grazing land.

199 Ohrigstad is a town to the north of Lydenburg in the Limpopo province Ohrigstad - Wikipedia

200 Pilgrim's Rest was the location of an emergency mint during the Second Boer War. This mint struck the famous and extremely rare Veld Pond. Pilgrim's Rest, South Africa - Wikipedia

malaria broke out the Government moved to within a few miles of Lydenburg village on higher ground. General Botha now went off to plan guerrilla operations.

General Schalk W. Burger.[201] A large male greater kudu.[202]

He did this so effectively that in a few months' time he had mobile forces in every quarter of the Transvaal harrying and worrying the British far and near, a reawakening that must have been particularly galling to them, considering how close we had been to a general collapse.

All this, however, was still to come, and in the mean-while my brothers and I lay in the Government laager, fretting at our enforced idleness. We endured it for two weeks, after which we made up our minds to leave. My youngest brother was still so weak from his illness that he had to remain, but the other three of us made ready. My brother Hjalmar, who had a queer bent of his own, preferred to go off by himself, and although we tried to keep him with us he rode away towards the eastern Transvaal.

He was subsequently captured and sent to a prison camp in India, and I did not see him again.[203] My brother Joubert and I decided to go west. We had heard from a passing burgher that General Beyers[204] was collecting a force at the Warm Baths,[205] two hundred miles distant, so we agreed to ride thither. Our preparations were soon made, for they were simple enough. We shot a koodoo[206] in the mountain, and made biltong, and we collected a supply of mealies from a neighbouring field. This was all the commissariat we had, for now that we were cut off from the outer world through losing the Delagoa Bay line, we were on Spartan diet, and for the next two years such luxuries as sugar, coffee, tea, bread, and soap were only to be had on rare occasions by capture from the enemy.

We said good-bye to all, and set out.

[201] General Schalk W. Burger. Vice-President S. African Republic, 1900 Schalk Willem Burger - Wikipedia Image: Times History War in S.A Vol 3 Photo by Steger.

[202] Often called Africa's grey ghost, the Greater Kudu is remarkable in its abilities to disappear. They are able to filter back into the surrounding bush and thickets, as silently as they appeared. File:Kudu (17418137574).jpg - Wikimedia Commons Credit StomSignal.

[203] Served: Pretoria. Received the Wound Ribbon (Lint voor Verwonding). No Prisoner Record Anglo Boer War - Reitz, Hjalmar. Burger - Boer War Forum

[204] Christiaan Frederik Beyers (23 September 1869 – 8 December 1914) was a Boer general during the Second Boer War. When responsible government was granted to the Transvaal, Beyers became speaker of the Lower House. He showed in the speaker's chair remarkable gifts. He was acute, tolerant and rigidly impartial, thus making a deep impression upon English-speaking South Africans, who would have supported his claims to be the first speaker of the first South African House of Assembly, had they been pressed by Louis Botha, the first Prime Minister. Instead, Beyers was made commandant general of the Citizen Forces of the Union Defence Force of South Africa, and in that capacity paid a visit to Great Britain, Germany, Switzerland and the Netherlands in 1912.

When World War I broke out, he set himself in almost open opposition to the policy of the Botha government. Beyers resigned his post as commandant general in a letter addressed to General Smuts, then Minister of Defence, and published in Het Volk, an anti-government journal. General Smuts replied in a stern letter declaring that the war was a test of the loyalty to their pledged word of the Dutch-speaking people, and accepting Beyers' resignation.

Beyers took the field as a leader of the Boer Rebels against the government, only to be overwhelmed by the government troops under the command of General Botha, to be driven from pillar to post as a fugitive, and to be wounded and consequently drowned on 8 December 1914, while trying to escape from his pursuers by crossing the Vaal River. With his death the Martiz rebellion was brought to an end. Christiaan Frederik Beyers - Wikipedia

[205] Bela-Bela is a town in the Limpopo Province of South Africa. Deriving its name from the geothermic hot springs around which the town was built, it was called Warmbaths, until 2002 Bela-Bela - Wikipedia

[206] The kudus are two species of antelope of the genus Tragelaphus: Kudu - Wikipedia

General Christiaan Beyers.[306]

Christiaan Frederik Beyers. [307] .[207]

I did not see my father again for twenty months.[208] He remained in the field until the end of the war with the Government laager, handling his rifle like a private soldier when they came under fire, which was quite often, and doing much to keep up the spirits of the fighting men by his poems and his personal example.

It was a long ride to the Warm Baths. Our way went through bush country, with neither track nor road to guide us, for the region we traversed was untenanted save by native tribes and wild animals. I was on my splendid old roan who had carried me from the start, and my brother rode the horse which we had taken back from our native boy at Pretoria.

On the third day of our journey we had a mournful encounter. We came on Commandant Gravett dying of wounds in the bush.[209] He had been hit by a shell ten days before, and his men had brought him here to prevent his falling into the hands of the British. He knew his end was approaching, but he bore his sufferings without complaint, and spoke of his coming death with resignation. He called us to him a few minutes before the end to tell us of his friendship with my father in former days, and a little while after he lapsed into unconsciousness from which he never recovered. We helped to bury him under a tree, and rode saddened on our way.

For ten days we journeyed, straying far north beyond the Olifants River,[210] misled by natives who told us of a commando in that direction. When we came up with this force, it was only a patrol on their way to Secocoeni's country[211] to inquire into fighting that had broken out between two native tribes. Among the patrol were old General Maroola and his brother Swart Lawaai, now serving in the ranks.

They were dressed as in the Natal days, only shabbier from long exposure on the veld, but they greeted us as if nothing had changed since. At length we turned south, and eventually reached Warm Baths, where we found General Beyers with a thousand men. He had been a lawyer before the war, and was now in command of the North-West under General Botha's new scheme of reorganization. He was a brave man, but I never liked him.

Here we were surprised to find our old A.C.C. commando almost intact. After that unfortunate day, when Commandant Malan and seven men were killed, they had left the high veld and had made straight for these

[207] Born in Banhoek, Stellenbos, on September 21, 1869, Christiaan F. Beyers left Victoria College in 1889 and settled in Transvaal, first in Pretoria, where he passed his procureurs exams five years later. Established at Boksburg, his name soon became favourably known; As a participant in the commandos that resisted the Jameson invasion with such success, he obtained full Transvaal voting rights in 1896. He also distinguished himself on the sports field as a fearless footballer. Deployed to war as a private citizen, he rose to the rank of Ass. Commander-General within twelve months and, with varying degrees of luck, confronted the enemy's superior forces in the Waterberg and Zoutpansberg districts. Delegated by his audience to the Peace Conference, he was elected Chairman and therefore played an important role in the relevant negotiations. Our [Transvaal] Military Officers: Dutch:

[208] After the peace in June of 1902.

[209] Gravett was knocked next to a Nordenfeldt Maxim by a shrapnel seriously wounded, and died on 26 October 1900 near Roossenekal.

[210] The Olifants River is a river in South Africa and Mozambique, a tributary of the Limpopo River Olifants River (Limpopo) - Wikipedia

[211] Sekukuniland is a natural region in north-east South Africa, located in the Transvaal Province, also known as Bopedi (meaning "land of Bapedi"). The region named after the 19th century King, Sekhukhune I. Sekhukhuneland - Wikipedia

parts. This was why we had never come across them during the retreat to the Portuguese border. They were still about sixty strong, under command of a newcomer, a young officer named **Lodi Krause**,[212] who invited us to rejoin, which we did at once. We remained in the neighbourhood of the Warm Baths for the better part of a month while General Beyers was raising more men, and the time passed pleasantly enough. We hunted a good deal and several times rode out on patrol to Pienaars River, twenty-five miles towards Pretoria, to watch the doings of a large English camp there. On one of these occasions we had a narrow escape. Six of us were approaching a kopje overlooking the camp, when suddenly a strong body of horsemen tried to cut off our retreat. We turned and made a dash for it, passing so near to some of the troopers that I heard them shouting at us to halt, while their bullets whipped around us as they galloped through the bush. We shook them off, however, without getting a scratch.

Except for these patrols and hunting for the pot, we spent a quiet time. Late in November [1900] my younger brother Arnt arrived at the Baths looking fit and well, having ridden hundreds of miles to find us, for he too had found life in the Government laager too slow once he was restored to health. We were very glad to see him, and the three of us built a weatherproof hut and fared well as food was plentiful. Rations were chiefly game and mealie meal, last of which General Beyers had accumulated a large supply from the Waterberg farms. He himself was camped close to us, a dark moody man who lost no opportunity of holding prayer meetings. With him was the Reverend Mr. Kriel, a Dutch Reformed parson, equally zealous; so between the two we were continually bidden to religious services, and they even went the length of ordering all the younger men to attend Bible classes. When my brothers and I ignored the order, General Beyers and Mr. Kriel rode over in person to expostulate with us, and even threatened to turn us out of the commando, but we stuck to our guns and heard nothing further of it; in fact we rather gained in reputation, for the Boers, although a religious people, are not intolerant in matters of faith.

Rev. Abraham Paul Kriel.[213]

I did not care for Beyers, but I liked the old predikant[214] [Kriel] for all his narrowness, and afterwards in the Cape Colony I grew to admire him as a steadfast man.

On about the 7th of December (1900) we were told to be in readiness to move south next day. My brother Joubert now surprised us by saying he was not coming. He said that he was going to be a gunner with a Creusot gun, one of the few still left, and was leaving at once with his new unit. He had always wished to be an

[212] Krause Ludvlg Emil Commandant Commando: General Beyers. Awarded the DTD (Dekoratie voor Trouwe Dienst).

[213] Rev. Rev. Abraham Paul Kriel was ordained as a minister in 1880 and had already served for several years the Witwatersrand was working when the last war broke out. Initially joined His Rev. joined the Krugersdorp commando, which moved to Dalmanutha after the battle Zoutpansberg and from there to the western districts. Remained under General Beyers However, a small part of this commando is in the Zoutpansbergse behind, and Among these was also Rev. Bantam. The battle of Nooitgedacht, the march around Pretoria and he attended the later battles on the Hogeveld and was known for his dedication and calm helpfulness in moments of danger. As a Field Chaplain and Patient Comforter he served General Beyers' command until the end of the war. Returning to Langlaagte, Rev. Kriel there, with the help of numerous friends, a blessed orphanage has been established, in which no fewer than 260 orphans and the children of needy parents from the last war are raised.

[214] A minister in the Dutch Reformed Church, or allied denomination.

artilleryman, and his obstinacy defeated our arguments. He rode away that same afternoon and I have not seen him since, for he was captured and sent to Bermuda, where he still is.[215]

Boer Prisoners awaiting transportation.[308]

Arriving at the wharf.[309]

Boer Prisoners at Bermuda."[310]

"Singing class of Boer prisoners on Burt's Island, Bermuda."[216]

General Beyers left a few hundred men and his guns behind to oppose the English columns that were preparing to invade the northern Transvaal, and with the rest, about eight hundred strong, he marched off. His purpose was to make south over the Magaliesbergen to the high veld beyond, in order to carry on guerrilla warfare there.

Our first day's trek brought us to the big native stad[217] of a local chief named Koos Mamogali.

A native town, this one near Mafeking.[311]

From here we looked across a valley, twelve miles wide, towards the Magaliesbergen. The mountain range stretches east to west for a hundred miles and more and, in this area, forms the dividing line between the high veld and the bush country from which we were emerging. We crossed the valley that night, and by daybreak reached the foot of a disused pass known as 'the old wagon road.' Here there was evidence that General de la Rey had lately been in the vicinity for we found a burning convoy of fifty or sixty English supply wagons, and

[215] Reitz Joubert Prisoner number: 21388 Captured: Majatje 02 Jun 1901 Sent to: Bermuda, Burts Island Age: 20 Source: Anglo Boer War Museum 2016.

[216] Over 3.000 Boer prisoners are quartered in the five healthful coral Bermuda Islands. At Darrell's Island are 851 mostly irreconcilables and strictly guarded. At Morgan's Island are 884, including 27 officers; at Tuckers Island, 809, at Burt's Island, 607; and at Port's (the hospital) Island, 35.

The prisoners employ their time in camp work, under the direction of the Industrial Association connected with each camp, in making toys for disposal by the Boer Toy Agency, and in recreative games. Each camp has a singing class, and every night towards dusk the members may be heard singing their evening prayer. On Darrell's Island they have clubbed together to purchase a piano, and those who care to be taught got to Captain Mosert for musical instruction. The prisoners, with few exceptions, appear contented with their treatment, which certainly possesses compensating features." H.A. Wilson After Pretoria: The Guerilla War Vol 3 Jan-April 1901 p.521.

[217] From Old Danish stath, ("place, city") Originally the same word as sted ("place").

one of his men who came riding by told us that they had ambushed the wagons in a stiff fight and had taken many prisoners.[218]

We looked on this as a good omen after the unbroken run of ill luck that had dogged us for so long, for it showed that General Botha's reorganization was beginning to take effect.

After a rest we climbed the pass, reaching the top by four that afternoon. From there, southwards, we had a wide view over the grass-covered plains, and far away on the skyline we could even see the smoke-stacks and goldmines of the Witwatersrand. The sight raised our spirits, and the men crowded the edge of the cliff animatedly discussing the improved outlook, for we were back from the wilds to within hail of Pretoria and Johannesburg.

Magaliesberg Range.[312 219]

Then we came down the south side, and by dark were off-saddled among the gardens and orchards below. We had had no sleep the night before and were looking forward to a good night's rest, but this was not to be. As we were preparing our supper, General de la Rey came riding amongst us on his famous little white-faced pony, and word went round that we were to attack an English force camped at the foot of the mountain not far away.

Looking West along the Magliesberg in the Skeerpoort Area.[313]

[218] Possibly: 12/26/1900 General Charles Knox engaged with De Wet near Leeuw Kop.

[219] The Magaliesberg of northern South Africa, is a modest but well-defined mountain range. It rises at a point south of the Pilanesberg National Park to form a curved prominence that intersects suburban Pretoria before it peters out some 50 km (31 mi) to the east, just south of Bronkhorstspruit. The highest point of the Magaliesberg is reached at Nooitgedacht, about 1,852 metres (6,076 ft.) above sea level. Magaliesberg - Wikipedia

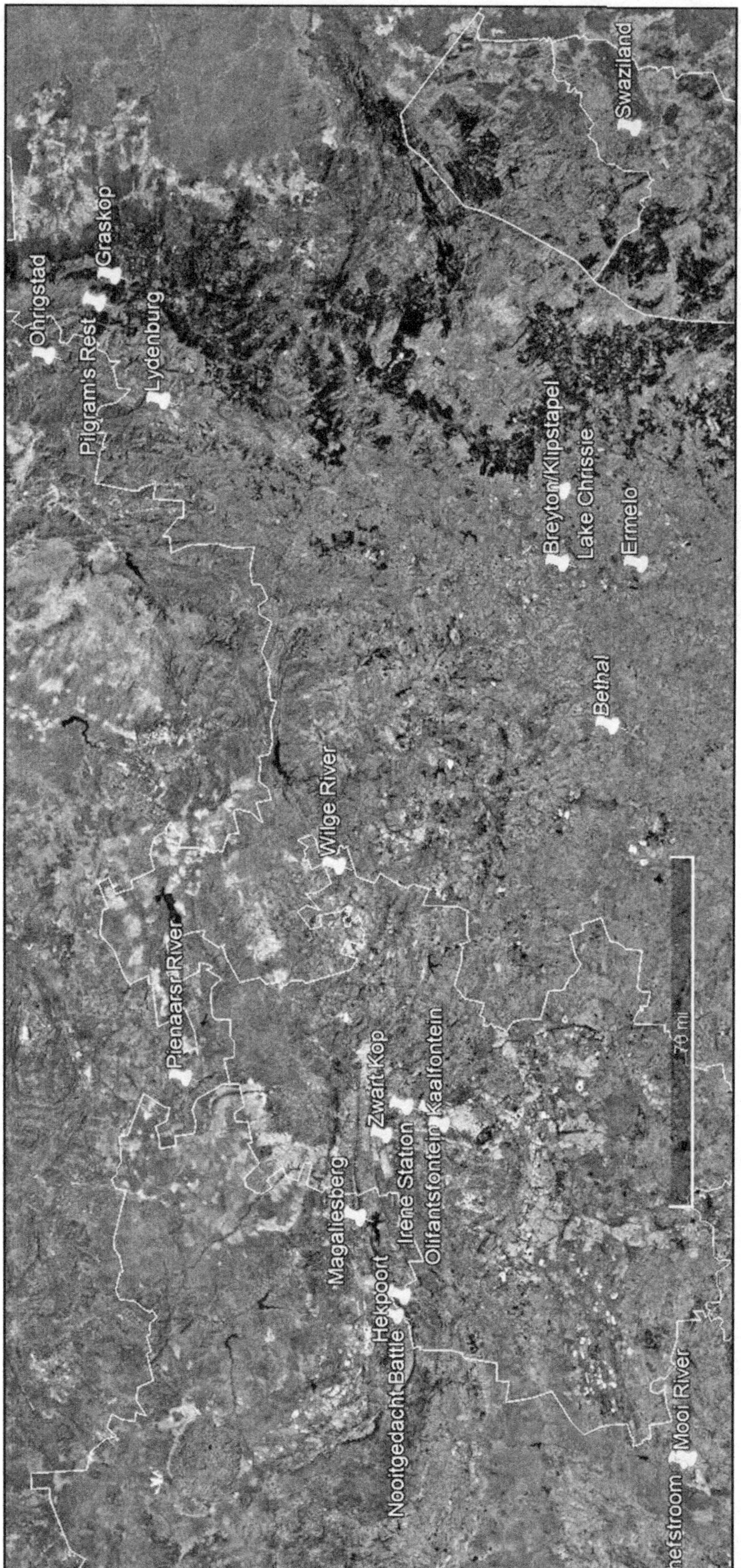

Map data: Google Image© 2023 Image Landsat/Copernicus.

XV. A Successful Affair, and After

I had last seen General de la Rey during our great retreat through the Free State. Since then he had been busy in the western Transvaal raising fresh commandos, and infusing new spirit into the fighting men by his ceaseless activity, and by the great affection they had for this wonderful old man. Only two days before he had fallen upon and captured the British convoy that we had seen on fire beyond the berg, and now he had crossed over to this side with four hundred men, to strike another blow.

An English commander, General Clements,[220] was camped round a bend of the mountain about nine miles off, with many troops, wagons, and guns, and when Beyers arrived General de la Rey was quick to seize the opportunity-for combined attack.[221]

A plan of action was soon arranged. De la Rey with his horsemen was to rush the English camp at dawn from the landward side, while we were to return up the pass and make our way along the mountain-top under cover of night, until we could fire down on the enemy from the cliffs.

Having completed his dispositions, de la Rey rode away in the dark to find his own men, and soon the whistles blew for us to saddle our horses. General Beyers led us up the road down which we had come an hour or two before, but when we reached the head of the pass we changed direction, picking our way eastwards along the boulder-strewn crest; of the range. We had to lead our animals, for riding was out of the question over the uneven surface and progress was slow.

Towards morning, tired and sleepy, we were halted for an hour, to enable de la Rey's guides who had accompanied us to reconnoitre. When they returned they reported that **General Clements's** camp stood at the base of the mountain and was held by about five hundred soldiers. On the cliffs above, close to where we now lay, was a force of equal strength entrenched behind sangars and breastworks in case of overhead attack. These men we were to deal with while de la Rey's men fell upon those below.

General Beyers, whatever his faults, was a bold and resourceful leader, and he made immediate preparations for the assault. Ordering all horses to be left behind, he passed word that we were to advance on foot. We knew no drill, so it was difficult to keep alignment in the semi-darkness, but we got ourselves sorted out into some kind of extended order and moved forward in a long ragged front.

We of the A.C.C. were on the extreme right at the edge of the cliff, with a drop of five or six hundred feet below us. Beyers was with us, and to our left walked the Waterbergers,[222] and beyond them the Zoutpansberg[223] men. Before we had gone far, dawn lit the mountain-tops, and with it came a fierce rifle-fire from the enemy schanses some distance ahead.

We had gone without sleep for two days and two nights, so that our spirits were low, and our advance came at once to a halt. Our line fell down behind rocks, and whatever shelter was to be had, leaving General Beyers walking alone, his revolver in one hand and a riding-switch in the other, imploring us to go on, but we hugged our cover against the hail of bullets lashing around us.

From where I lay on the tops of the crags, I could look straight down into the English camp hundreds of feet below. I could almost have dropped a pebble upon the running soldiers and the white-tented streets and the long lines of picketed horses.

As I looked down on the plain, from behind a jutting shoulder of the mountain came winging into view a force of mounted men who galloped hard for the English camp. It was General de la Rey timing his attack to synchronize with our own. They closed in, and for a moment it seemed as if they would overwhelm the British, but then the soldiers rushed to their posts opening heavy fire. The plain became dotted with fallen men and horses, and the attack wavered and broke. The survivors turned towards the shelter of the buttress whence they had come, and in less than ten minutes the assault was over.

The troops facing us on the mountain now made a mistake.

[220] R.A. P. Clements Local Major General Ralph Arthur Penrhyn Clements - Wikipedia

[221] Thursday, December 13, 1900.

[222] Waterberg Commando, 155 Records: https://www.angloboerwar.com/?option=com_content&view=article&id=3067

[223] Zoutpansberg Commando, 137 Records: https://www.angloboerwar.com/?option=com_content&view=article&id=3068

Like ourselves, they were able to look down at the attack, and when they saw our men retire in confusion they set up loud shouts of triumph. Stung by their cries, our whole force, on some sudden impulse, started to its feet and went pouring forward. There was no stopping us now, and we swept on shouting and yelling, men dropping freely as we went.

Almost before we knew it, we were swarming over the walls, shooting and clubbing in hand-to-hand conflict. It was sharp work. I have a confused recollection of fending bayonet thrusts and firing point-blank into men's faces; then of soldiers running to the rear or putting up their hands, and, as we stood panting and excited within the barricades, we could scarcely realize that the fight was won.

Our losses were severe. On the ground across which we had charged lay a trail of dead and wounded, and yet more by the schanses.

In all we had about twenty-five men killed and some seventy wounded, and we shot down nearly a hundred of the English, besides taking as many more prisoners. But it was a heavy price to pay for success, even in this strange affair in which our men who had been cowering disorganized behind the rocks suddenly flung themselves upon a fortified enemy with a furious desire to silence their shouts of triumph.[224]

We had now taken the main defences, but scattered rifle shots were coming from a nest of granite boulders to the rear, and General Beyers ordered Krause, Commandant of the A.C.C., to clear the place. Krause took a dozen of us, and we worked our way forward in short rushes. But he grew impatient and told us to close in more quickly. The result was disastrous, for as we rose a salvo rang out which brought down four of our men. Having fired this parting volley, the soldiers, of whom there were only six, went running towards the mouth of the ravine which led down a cleft to their camp below.

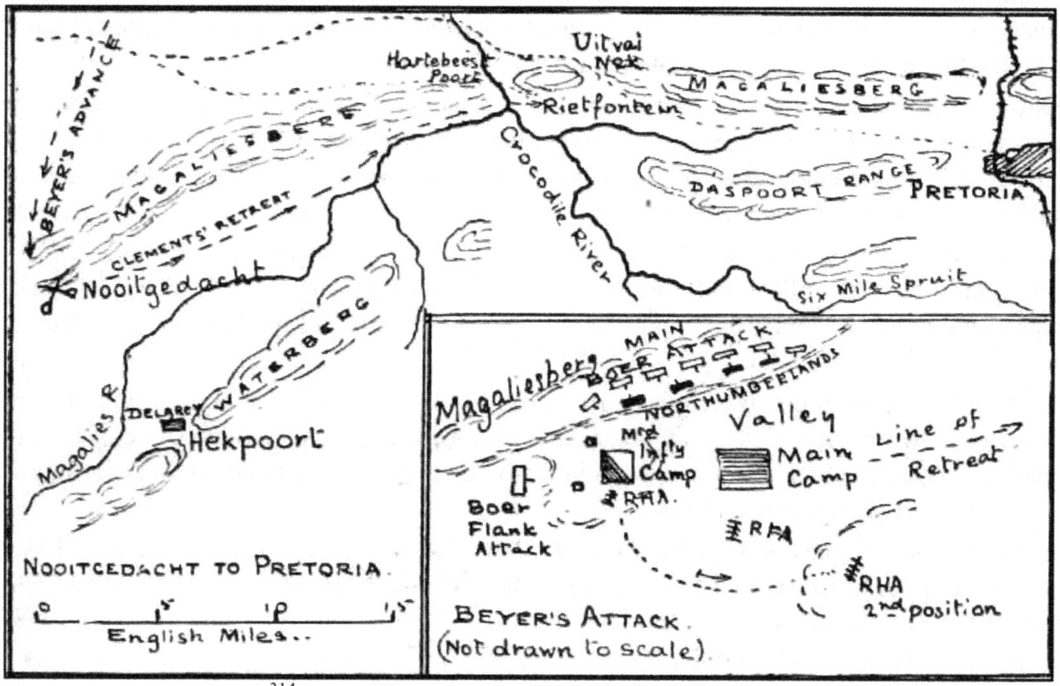

Battle of Nooitgedacht.[314] [Magaliesberg]

I hit one of them through the thigh, and Krause shot another dead, but the rest escaped. We walked back to see the extent of the damage, and it was bad enough. My old school fellow Jan Joubert, son of Piet Joubert the late Commandant-General, had a bad chest wound, and the other three men were dead. Two of them were young brothers named Koekemoer, about eighteen and nineteen years old, who had been with the A.C.C. since the Free State days. I also went to see the soldier whom I had shot. He had a nasty wound, but he was bandaging it himself with the first-aid pad which they all carried, and he said he could manage. He was a typical Cockney,

[224] Original Footnote: Of this attack The Times History of the War gives the following account:
In a bold reckless fashion for which there was no precedent since the attack on Wagon Hill a year earlier, and which was a startling novelty to the troops present, the Boers rushed forward on foot, cheering and shouting as they ran. Indifferent to loss, with great skill and dash . . . they shot down most of the British officers, rolled up the pickets and, in spite of a stout resistance, mastered the position after killing and wounding ninety-seven officers and men.

and bore me so little ill will that he brought out a portrait of his wife and children, and told me about them. I made him comfortable, and left him cheerfully smoking a cigarette. **Jan Hubert** was badly hurt, a portion of his rifle-stock having been blown into his lung, so Krause asked me to take a water-bottle from one of the dead soldiers, and go down into the ravine in search of water. I got a flask and went down the slope to the mouth of the gorge. Unknown to us there was a path to the English camp, along which reinforcements were climbing up to dislodge us. I saw twenty or thirty soldiers already near the top, standing in a group not a stone's throw from me, while many more were coming on behind in single file.

I fired at once and dropped a man, the remainder disappearing amongst the trees. From here they opened fire on me, and I in turn had to take cover, dodging from rock to rock to get back to Krause. On hearing my news he took a number of men, and we ran down just in time to see the path crowded with soldiers. We lost no time in pouring close-range volleys into their midst. In less than a minute only dead and wounded were left; more than twenty men of the Imperial Yeomanry of London lying in the space of a few yards.

This was the final clearing, and we now had the camp below at our mercy, for we were able to fire into it without opposition. Soon we could see the occupants retiring with their guns, and we descended the ravine and entered the camp.

In passing by the intake of the gorge, I found the soldier whom I had killed. I was horrified to see that my bullet had blown half his head away, the explanation being that during one of our patrols near the Warm Baths I had found a few explosive Mauser cartridges at a deserted trading-station, and had taken them for shooting game.[225] I kept them in a separate pocket of my bandolier, but in my excitement had rammed one of them into the magazine of my rifle without noticing it. I was distressed at my mistake, but there is not a great deal of difference between killing a man with an explosive bullet, and smashing him with a Lyddite shell, although I would not knowingly have used this type of ammunition. I flung the remainder into the brook that ran by, now red with the blood of dead men lying in the water.

Having sent back for our horses, we hastened down the path into the camp.

On my way down the gorge I found two wounded officers beside the track, one with his thumb shot away and the other with a broken arm. As I came up I heard one of them remark: 'Here comes a typical young Boer for you,' and they asked me whether I understood English. I told them 'Yes,' and the man with the thumb said: 'Then will you tell me why you fellows are continuing the war, because you are bound to lose?' I replied: 'Oh, well, you see, we're like Mr. Micawber, we are waiting for something to turn up.' They burst out laughing and the one said: 'Didn't I tell you this is a funny country, and now here's your typical young Boer quoting Dickens.'[226]

The camp was filled with supplies of all kinds, and such a smashing of cases, and ransacking of tents and wagons, had not been seen since we looted the Dundee camp long before. While we were at this, General Beyers came riding among us in a rage, and ordered us to follow the enemy, but we thought otherwise. We considered that the object of the attack was to capture supplies, and not soldiers, as soldiers would have to be liberated for want of somewhere to keep them, and besides, if we went off, we might return to find the camp already looted during our absence. So we attended to the matter in hand, more especially as de la Rey's horsemen had recovered from their setback earlier in the morning, and could be seen stringing out across the veld towards where Clements and the balance of his troops were withdrawing down the valley in the direction of Pretoria. We told ourselves that we had done our part in the day's work, and that they could do the rest.

My brother [Arnt] brought my roan and his own two riding-horses down the ravine, and we took two more horses from the English lines, where many stood picketed.

Searching out saddles and wallets[227] to match, we loaded our caravan with spoil in the shape of tea, coffee, salt, sugar, food, clothing, books and other luxuries of which we had long been deprived. Then we followed the other men who, having taken what they wanted, were riding along the foot of the mountain to the spot below the Old Wagon Pass from which we had started the evening before.

[225] He is referring to expanding, soft nosed or hollow point bullets rather than full metal jacketed ones.

[226] Wilkins Micawber is a clerk in Charles Dickens's 1850 novel David Copperfield. He is traditionally identified with the optimistic belief that "something will turn up." Wilkins Micawber - Wikipedia

[227] A bag or pouch. In this case saddlebags.

Thus ended our share in the fight.[228] [229]

The A.C.C. had lost five men killed and five wounded. Among the latter was our Commandant Krause with a bullet in his foot, and my Corporal Jan Nagel,[230] with his right shoulder-blade badly shattered. A French gentleman adventurer, Georges de Gourville,[231] who belonged to us, was also badly wounded, but the worst was Jan Joubert, who was carried down to the English camp with the other serious cases, and left there until the British could be asked to send them surgical aid, for we had neither drugs nor doctors.

Boer Auxiliaries: Corps of Frenchmen under the command of Lieutenant Gallopaud.[315]

My brother and I had a glorious feast, and then, having gone without rest for forty-eight hours, we slept the clock round.

Next day, such of the dead as had been carried down the mountain by friends or relations were buried in a single large grave that I helped to dig. General de la Rey was present and he addressed us in eloquent words that moved many to tears, for besides being a fighter he had a fine gift of simple speech.

Last honors to fallen adversaries.[232]

Here we remained for several days, during which time my brother and I enjoyed high living, after the straight diet of meat and maize on which we had subsisted for so long. We were refitted from head to heel, we carried a Lee-Metford rifle apiece, in lieu of our discarded Mausers,[233] and above all we were well found in horseflesh.

[228] In the Battle of Nooitgedacht on 13 December 1900, Boer commandos led by Generals Koos de la Rey and Christiaan Beyers combined to deal a defeat to a British brigade under the command of Major General R. A. P. Clements during the Second Boer War. Battle of Nooitgedacht - Wikipedia

[229] Reverse to General Clements De la Rey, J.C. Smuts and C.F. Beyers overran a British camp under the control of General Clements near the Magaliesberg.

[230] This is the only return at ABW.com: Nagel Jan Frederick Adrian Prisoner number: 27243 Captured: Veekraalkop 19 Dec 1901 Sent to: St. Helena Age: 13 Address: Rhenosterdrift. Google Earth returns: Granietkop for Veekraalkip and this is near the battle.

[231] Possibly: Gourville Henri (Henry) Volunteer from France serving with the Boer forces. Prisoner number: 18675. Captured: Leeuwfontein 08 May 1901 Age: 25 Address: Dinan

[232] A similar British service. "Army chaplain Gedge reading the Burial Service over dead Boer(s)" Navy & Army Illustrated Vol 10

"Boer Commandero, South Africa."[316]

My gentle loyal old roan was as flourishing as ever, and I had a fine little chestnut pony, which I had chosen in preference to the larger but less reliable chargers in the English camp. I gave the other horses away in order to reduce our stable to manageable proportions; my brother had the two horses which he had brought with him from the north. One was a toll-free chestnut[234] and the other was the strangest horse I have ever known. My father had purchased him in the Lydenburg district from a home going burgher, who omitted to tell us that he was possessed of the devil.

He indulged in such extraordinary antics that the police at the Government laager had declared him insane, and christened him 'Malpert' (the mad horse). Sometimes he would allow a single man to walk up and catch him without trouble, but at other times we had to turn out the whole Government from the Vice-President [Schalk Burger] downward to form a cordon around him. He would pretend to be quietly grazing, but as soon as he was completely hemmed in, he would look up in assumed surprise and start to back against the ring, kicking and lashing so furiously that we had to give way, when he would go capering off, heels in air, to crop the grass near by. If another cordon were made he would repeat his performance, until he left the men helpless between cursing and laughing.

The only persons for whom he had any respect were my brother Arnt and myself. He was afraid of me because once at the Lydenburg Camp, after he had twice kicked his way through I leaped at him from a distance of several feet and flung my arms about his neck. He reared and bucked and tried to bite and roll, but I locked my legs around his, so that he could not shake me off, and in the end I bested him. With my brother he was tractable, because he had doctored him for a badly ulcered back, and the 'Malpert' showed his gratitude by obeying him. His reputation had followed him down south, and he was quite an institution among the commandos. Often we would hear a warning cry 'Look out, here comes the "Malpert,"' and the burghers would scatter beyond reach of his heels. Nevertheless, his tricks and pranks were taken good-humouredly, for he had magnificent staying-powers and the men looked upon him with the admiration that born horsemen have for a good animal, and as for Arnt and myself, we had a very soft spot in our hearts for this queer outlaw.

Well mounted as we were, my brother and I felt that we could ride anywhere and be ready for anything, and we looked forward-with interest to the next move. General de la Rey, restless as usual, had gone off, leaving us behind with General Beyers. On Dingaans Day,[235] December 16th, he and the reverend Mr. Kriel held a religious gathering on a neighbouring hill. They invited all to join in piling a cairn of stones like that raised at

[233] Ammunition for these weapons will be more easily obtained from captured supplies as opposed to the imported Mauser bullets.

[234] Original Footnote: Among the Boers a chestnut horse with white face and four white stockings is called toll-free, there being a tradition that in the old days horses thus marked went through the toll-gates free of charge.

[235] The Day of the Vow (Afrikaans: Geloftedag) is a religious public holiday in South Africa. It is an important day for Afrikaners, originating from the Battle of Blood River on 16 December 1838, before which about 400 Voortrekkers made a promise to God that if he rescued them out of the hands of the approximately 20,000 Zulu warriors they were facing, they would honour that day as a sabbath day in remembrance of what God did for them. Day of the Vow - Wikipedia

Paarrde-Kraal in 1880, during the first English war.[236] My brother and I thought the proceedings somewhat theatrical and kept away, but so far as I know the beacon is still standing in testimony of vain hopes.

On the following day General French,[237] the English cavalry leader, was reported to be moving [West] up the Hekpoort valley from Pretoria, and General Beyers marched out to meet him.

Near Hekpoort the A.C.C. was ordered to take post in a range of hills skirting the valley, to guard our main body against surprise. We spend an uneasy night on a rocky crest overlooking the broad 'Moot' as it is called, and at day-break next morning we saw an English force of three or four thousand horsemen approaching. We had strict orders from Beyers to keep out of sight, so Krause led us into a gorge to hide our horses, after which we climbed up and peered over the rocks to view the enemy advance. The valley here lies four or five miles broad, and the British scouts were strung across it from side to side, their nearest horsemen passing so close beneath us that we could easily have shot them down. We could see General Beyers's men riding to occupy a line of hills farther up the valley, and before long they were hotly engaged. We had a fine view of the fight, but our interest in the spectacle was dampened by the fact that by now the British were between ourselves and our parent body, so that if General Beyers fell back we should be left isolated in the rear.

And as the gun-fire increased and the troops began to mass, we saw our men running for their horses and galloping away. This was not unexpected, considering the heavy odds, but nevertheless there went the commando, pursued hotfoot by the English, and here were we stranded far behind.

Krause decided to recover contact by riding round the left flank of the troops, but we fell foul of so many of their patrols, and were so often shelled by Pompoms and field guns, that we were forced to take refuge in the parallel hills. From here we could make out our men about six miles off, moving west from the valley, white puffs of shrapnel breaking over them as they went, and horsemen hard on their heels.

Krause now led us into the broken country north-west of Johannesburg (Skurwe-bergen), where a small force like ours could lie in hiding for a while, and by dark we were well within the region of its tumbled hills.

Although the rainy season was overdue, we had thus far experienced only sunny days, but now the weather broke, and for the next six days it rained without ceasing. We sought out a lonely farmhouse, and here we lay over, waiting for the deluge to lift. The farm was deserted, but sheep were straying near, and dry fuel was stacked in the barns, so we fared moderately well.

Georges de Gourville, the Frenchman, had insisted on accompanying us in spite of his wound, and he nearly died here, my brother and I nursing him through his fever. On the evening of the sixth day the clouds thinned, and we decided to set out that night in quest of General Beyers.

We had no idea where he was, but we counted on finding him sooner or later, so we started off immediately after dark. We rode all night, save for short halts, and towards morning came to a farm where a woman told us that General French was camped close by with his columns. She said that the English had given up the chase after Beyers and were on their way back to Pretoria, but had got weather-bound by the rains and were waiting for dry roads before going on.

On hearing this, Commandant Krause in his impetuous way rode off with only two other men to look around, and that was the last we ever saw of them. He and his companions must have ridden straight into the arms of the English, because soon after they left we heard a few distant shots and then silence. We waited until daylight, after which we decided to go on without him.

As a heavy mist hung over the veld, we made our way carefully, and it was as well that we did so, for when my brother and I with another man turned aside to water our horses at a dam lying off the road, four English troopers came riding out of the fog and let their horses drink about a hundred yards away. We gazed at each other suspiciously in the uncertain light, and then one of the soldiers shouted: 'Look out, those men are Boers,' and pulling round they galloped away. We dismounted and fired before the mist swallowed them, bringing two to the ground. Riding up, we found one dead and the other rolling in agony, but both their horses had bolted after the others. While we were trying to help the wounded man the mist lifted somewhat, and a large English

[236] The Paardekraal People's Assembly was a rally of between 6 000 and 8 000 citizens of the ZAR held on M.W. Pretorius' farm Paardekraal from 8 to 14 December 1880. The original stone pile was erected on 13 December 1880 at the suggestion of Paul Kruger; to serve as proof of their intention to fight for their freedom and the restoration of the republican state. Paardekraal People's Assembly - Wikipedia

[237] John French, 1st Earl of Ypres - Wikipedia

camp was outlined close by. We could hear our men splashing over the muddy ground in haste to get away, for they too had caught sight of the camp.

Bugles started calling and we caught glimpses of soldiers scurrying for their horses, so we galloped away, guided by the tracks lying clear across the sodden veld.

After a long ride we caught them up, and, when the sun broke through a little later and the fog dissolved, we could see some five hundred English horsemen coming towards us, but had no difficulty in outpacing them, for in spite of our long night's journey their heavy troop horses were no match for our hardier and lighter mounts.

Krause having been captured, we were now leaderless, but with the Boers each man is practically his own commander, so the loss did not weigh heavily upon us and, having shaken off our pursuers, we rode along westward at our leisure, until in two days' time we came on General Beyers and his men camped around the source of the Mooi River[238] in the district of Potchefstroom.[239]

Beyers appointed Corporal Jan Nagel to be our new Commandant, a popular choice, for, although a rough and illiterate man, he was well liked. He was still suffering from the wound received on the berg, but had remained in the saddle nevertheless.

We passed Christmas Day [1900] undisturbed, but next morning an English column came from the direction of Potchefstroom, and as there was no object in fighting except on ground of our own choosing, General Beyers gave them the satisfaction of thinking that we were running away, and at dark we drew off to spend the night near the village of Ventersdorp.[240] From here we moved about at random, seeking for a chance to strike, but no favourable opening presenting itself, we saw the year 1901 in without further incident.

[238] Mooi River (Vaal) - Wikipedia

[239] Potchefstroom is on the Mooi Rivier (Afrikaans for "pretty river"), roughly 120 km (75 mi) west-southwest of Johannesburg and 45 km (28 mi) east-northeast of Klerksdorp. Potchefstroom - Wikipedia

[240] Ventersdorp - Wikipedia

XVI. From West to East

We now came into the Lichtenburg[241] district, within a day's ride of Mafeking[242] and the Bechuanaland border. This was General de la Rey's military area, but he was farther south at the time, and we only met an occasional roving patrol of his.

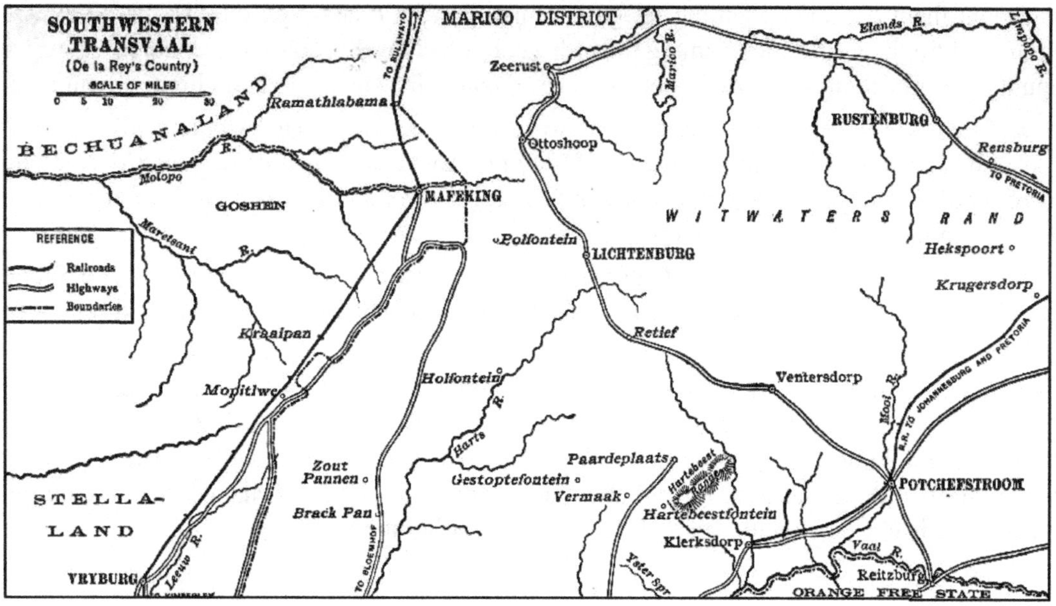

Southwestern Transvaal.[317]

Beyers had a mind to raid Mafeking, but before he could do so a messenger arrived from the east with orders from General Botha to bring all his men to the Emlelo district. This meant transferring our force right across the breadth of the Transvaal, a matter of three hundred miles. Fortunately our horses were in good condition, thanks to the abundant rains and consequent good grazing, so we prepared to start at once. We now began our long march, riding eastward until some days later we reached back to the Magaliesberg Valley once more, close to where we had taken General Clements's camp. Since that event the English had apparently decided to garrison this fertile stretch, for we found a strong force a few miles down, busily constructing a fortified camp as though they intended to hold the valley permanently.

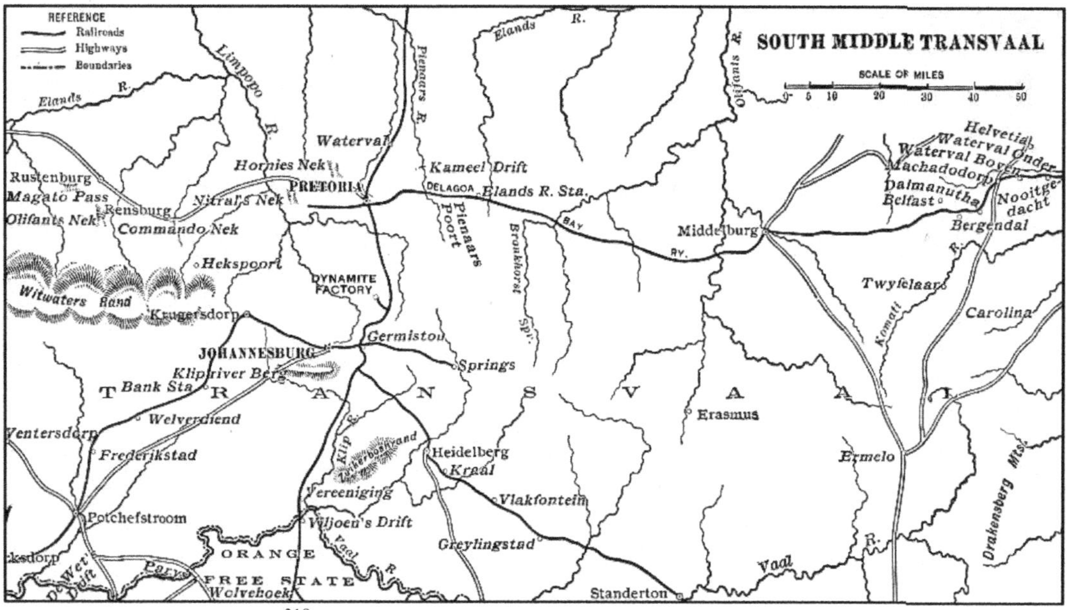

South Middle Transvaal.[318] Ermlo is lower right, above the Vaal River.

[241] Lichtenburg, North West - Wikipedia

[242] By now, 1901, the siege of Mafeking has been relieved. The Siege of Mafeking lasted 217 days from October 1899 to May 1900, and turned Robert Baden-Powell into a national hero. Mafikeng - Wikipedia

We left the camp alone, but for a few hours we lay on a ridge overlooking the troops who sent some shells at us, one of which killed a Waterberg man within a few yards of me. While we were halted here, I rode across the valley to the farmhouse near General Clements's camp to see my friend Jan Joubert, who had been left there with our other seriously wounded, after the fight. Although not yet out of danger, he was on the mend, and he told me that the British were treating them very well. A surgeon came over nearly every day, and medical orderlies were on duty to tend to their wants. They had brought his old mother from Pretoria to be with him, and officers from the camp down the valley often brought them fruit and other luxuries. I did not stay long for fear of a stray English patrol, and rode back after a hurried greeting. I did not hear what happened to him afterwards, but I believe that he made a complete recovery.

We now resumed our march, making for the railway line between Johannesburg and Pretoria. This had to be carefully approached as the track was patrolled night and day by armoured trains.

Armoured Train.[319]

Armoured Train manned by HMS Terrible's Marines.[320]

H.M.S. Philomel men with a Hotchkiss 3-pdr.

And close up.[321]

The A.C.C. was sent ahead to reconnoitre, and after an absence of twelve hours we reported back that the crossing was practicable. On our return journey to rejoin the main commando at the hill called **Swart Kop**, we found a suspicious-looking native watching our movements, and arrested him for further inquiry. We interrogated him closely, without getting any evidence that he was a spy, but for safety's sake we kept him by us during the night, intending to release him next morning. Just before daybreak, however, while I was lying asleep by my saddle, I heard shouts, and looking up, saw the prisoner running as fast as he could go towards Swart Kop Hill. Several of our men were already astir, and as he would not stop when called, they dropped the unfortunate savage dead in his tracks, for which I was sorry, but in the circumstances it could not be helped.

That evening we started off on an all-night march for the Johannesburg railway, which we struck soon after sunrise. The A.C.C. were scouting well in advance, so we were the first to cross the metals. As we were riding over, an English trooper come cantering up to see who we were, and he was considerably taken aback when he found out. We relieved him of his horse, rifle, and equipment, and told him to go off, a command which he obeyed with alacrity, and when we saw him last he was marching steadily for Johannesburg.

On the far side of the line stood a small homestead, out of which several more soldiers came tumbling. They all surrendered except one man, who made a bid for liberty by running into an orchard, but we shot him dead before he had gone far. At the same moment a trolley came down the line [North] from Kaalfontein Station, and as the crew tried to turn back, we fired on them, killing one and wounding a native. The dead man turned out to be a railway-ganger, as were the other four white men with him. They were on their morning round of inspection, so we let the survivors go after telling them that it was their own fault for not halting when challenged.

Our main commando now coming up, some of the burghers foolishly rode along the track towards the station, with the result that they left three men killed and several wounded in the hands of the garrison there. Our whole force now being across the line, we rode to a large farm over the rise to off-saddle.

This crossing of the railway proved a relatively simple matter, but later on in the war, when the English had completed their blockhouse system, it became increasingly difficult; and from all accounts it required something like a pitched battle to negotiate a passage over the Transvaal and Free State lines, though in the Cape Colony I saw none of this.

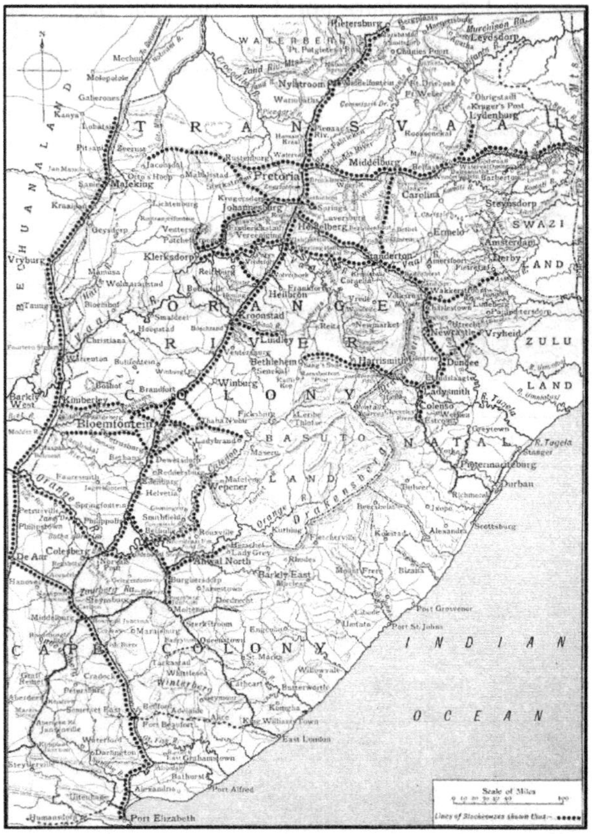

The Nearly Completed Blockhouse System.[243] [244] .[245] Blockhouse Near Wolseley, Western Cape.[322]

While we were resting at the farm, patrols were sent out, and in less than an hour one of these came galloping back to say that a strong force of English was approaching from Johannesburg. Scarcely had they returned, when we heard a succession of loud bangs and the roar of shells tearing overhead.

Our horses were out grazing, so that there was a good deal of confusion before we had the frightened animals collected, but we managed to saddle up and get away without loss, our whole force moving in a north-easterly direction with the English following. We could hardly have fought them on such open ground, armed as they were with field-guns, but in any case our object was to effect junction with the Commandant-General, so we made no pretence of standing. The troops tried to head us off, and an armoured train from Pretoria made a belated appearance, firing heavy Lyddite shells, but with our greater mobility we left the soldiers far behind, and by midday were able to come to rest after our long ride.

[243] The distance between blockhouses was calculated to be able to extent the reach as far as possibly but at the same time to prevent the Boers from passing unnoticed between them. The average distance apart was about 1,000 yards (910 m). Line of sight was vital and this required a greater density of blockhouses in some areas. Anglo Boer War - Blockhouses

[244] By the end of March 1902, the British had enclosed in barbed wire more than 23,000 square kilometres (~ 9,000 sq. miles) of the Transvaal and 27,000 square kilometres (~ 10,000 sq. miles)of the Orange Free State. More than 50,000 troops patrolled the wire and manned the 8,000 blockhouses studding the line. Over 80,000 mounted troops were prepared to meet the Boers in battle. Adams, Sharon (2022). "Canada and the Battle of Harts River". Legion Magazine. And: https://en.wikipedia.org/wiki/Battle_of_Hart%27s_River

[245] Times History, Vol 5 page 600.

A gun carrying carriage named "Cock of the North" of No. 3 Artillery Train at the railway station at Bronkhorstspruit, Gauteng Province.[323] Naval 6-Inch Gun on a specially-designed truck.[324]

Guns sent to Kimberley after Relief. Mr. Rhodes in front of gun in civilian clothes, wearing white pants.[325]

Next day we trekked to Olifantsfontein,[246] where the seven A.C.C. men had been killed the year before. At this place I suffered a most serious loss, for a young fellow claimed the pony which I had brought from the English camp after the Magaliesberg [Nooitgedacht] fight. He brought me and the horse before General Beyers, to whom he proved that it had belonged to his father, who was among those killed that morning during de la Rey's attack. I could not but admit the justice of his case, and handed over the pony, but this left me with only my old roan, a narrow margin of safety in these days when to be horseless meant almost certain capture.

At Olifantsfontein, General Botha came up with a small escort. He looked thinner than when I had last seen him in the Lydenburg Mountains four months ago,[247] but he was full of energy and confidence. He told us that Lord Roberts had decided to bring the Boers to their knees by a series of drives, in which vast numbers of troops were to sweep across the country like a drag-net. To that end, all through this month of January (1901) they were assembling fifty thousand men along the Johannesburg-Natal railway line, ready to move over the high veld on a front of sixty or seventy miles, with the intention of clearing the Eastern Transvaal, after which the process was to be repeated elsewhere, until every one was dead or taken.

The Boers were as yet mercifully unaware that this new system was to include the burning of farms, the destruction of crops and herds, and the carrying into concentration camps of their wives and children, but they were soon enough to learn that the British had taken the dread decision of laying waste the two republics, regardless of the suffering of the non-combatant population, and ignorant of the fact that these methods, so far from subduing the Boers would merely serve to stiffen their resistance.

At first we could hardly credit this, but when one wild-eyed woman after another galloped by, it was borne in on us that a more terrible chapter of the war was opening.

[246] Olifantsfontein, also known as Clayville, is a small town on the East Rand in the Gauteng Province. Olifantsfontein - Wikipedia

[247] In September of 1900, it is now January, 1901.

The intention was to undermine the morale of the fighting men, but the effect was exactly the opposite, from what I saw. Instead of weakening, they became only the more resolved to hold out, and this policy instead of shortening the war, prolonged it by a year or more.[248]

Two days after General Botha had ridden away, the storm broke upon us. As the sun rose, the skyline from west to east was dotted with English horsemen riding in a line that stretched as far as the eye could see, and behind this screen every road was black with columns, guns, and wagons slowly moving forward on the first great drive of the war.[249]

General Beyers, when he grasped the situation, divided his force in two, and rode away with one half to find the left flank of the enemy, while the rest of us were told to do what we could in front of the advance.

Far away, to our left, parties of General Botha's men were visible from rise to rise, scattered specks before the great host.

All that day we fell back, delaying the enemy horsemen by rifle-fire as far as possible, and breaking away when the gun-fire grew too hot. This went on till sunset without heavy losses on our side, despite the many batteries brought into play from every knoll and kopje. Once I saw my brother disappear from sight as a shrapnel shell burst on him, but he rode out laughing, he and his horse uninjured.[250]

During the course of the morning, pillars of smoke began to rise behind the English advance, and to our astonishment we saw that they were burning the farmhouses as they came. Towards noon word spread that, not only were they destroying all before them, but were actually capturing and sending away the women and children.

Towards dark the chase slowed down. It rained steadily all night, and we spent a miserable time lying in mud and water on the bare hillsides. At daybreak we were all on the move again, but, owing to the rain and the heavy going, the English could only crawl in our wake, and we had little difficulty in keeping our distance. By now, the news had spread that the English were clearing the country, with the result that the entire civil population from the farms was moving

The plain was alive with wagons, carts, and vehicles of all descriptions, laden with women and children, while great numbers of horses, cattle, and sheep were being hurried onward by native herd boys, homes and ricks[251] going up in flames behind them.

Family of Boer refugees passing by a flock of sheep.[326]

[248] Original Footnote: The Times History of the War relates:
The policy of burning down farmhouses and destroying crops as a measure of intimidation had nothing to recommend it and no other measure aroused such deep and lasting resentment. The Dutch race is not one that can be easily beguiled by threats, and farm-burning as a policy of intimidation totally failed, as anyone acquainted with the Dutch race and the Dutch history could have fore seen. Applying this system against a white race defending their homes with a bravery and resource which has rightly won the admiration of the world was the least happy of Lord Roberts' inspirations and must plainly be set down as a serious error of judgment....'

[249] 1/28/1901 General French marched eastward, clearing the valley of the Wilge River.

[250] Winston Churchill also recounts an instance where he emerged unscathed after a shell bust above him. Others in his company were not so fortunate. See, London to Ladysmith.

[251] Straw, hay etc. stored in a stack for winter fodder, commonly protected with thatch.

British Occupation

Burning of a farmhouse near Dewetsdorp in May, 1900.[327] “No Help For It. Pack up…..”.[328]

Worth Carrying Away. Confiscated wool which afterwards fetched £120 for Government.[329] A useless haul. A captured wagon for which no oxen are available is burnt.[330]

Wagon Loading (PF: Confiscation?).[331] A Domiciliary Visit Firing the outhouse [outbuilding] of a Boer who has been shooting natives.[332]

“Rough Justice in Wartime.”[333] “The Penalty of Foul Play. South African Light Horse Burning a Farm which has fired on Our Troops.”

A Boer family looks on at their house that was set alight by the British forces during the South African War.[334]

Family of Boer refugees.[335]

A Boer refugee family peering out from their covered wagon.[336]

Boer Women Visiting Their Imprisoned Husbands at Standerton.[337]

Camps

Over 100,000 women and children were held captive in 45 camps. Disease was rampant and about 22,057 children under the age of 16 years died. A total of 1,421 aged, (No sex, or age value provided in article). 4,200 women older than 16 years, as well as 210 boys older than 16 years died.[338]

Camps in South Africa

Transvaal	Orange Free St.	Natal	Cape
Barberton	Bloemfontein	Ladysmith	Aliwal-Noord
Balmoral	Brandfort	Colenso	Kimberley
Belfast	Bethulie	Howick	Norvalspont
Heidelberg	Heilbron	Pietermaritzburg	Doornbult
Irene	Harrismith	Wentworth	Kromelmboog
Johannesburg	Kroonstad	Merebank	Port Elizabeth
Klerksdorp	Ladybrand	Eshowe	Oos-Londen
Krugersdorp	Springfontein	Jacobs	Vryburg
Meintjieskop	Vredefortweg	Mooirivier	Uitenhage
Middelburg	Winburg		
Mafeking			
Nylstroom			
Pietersburg			
Potchefstroom			
Standerton			
Vereeniging			
Volksrust			

Table source.[339]

Inside one of the British concentration camps.[340]

Concentration Camp.[341]

Children in Camp.[342]

Boer women and children in a British concentration camp during the Boer war, circa 1901.[343]

Norvals Pont Camp.[344]

One of the Boer concentration camps, probably Bloemfontein judging from the image source.[345]

A Boer Encampment Waggons Massed on the Veldt.[346]

Boer family with a wagon on a farm in South Africa circa 1900. Titled: Father and son going to the front by Boer War Colourised Photographs.[347]

General Botha directed all non-combatants, wagons, and live-stock to make for Swaziland, and he ordered us to give way before the troops, and let them expend their blow in thin air.

Owing to these measures the drive went to pieces during the next few days. The British could not maintain a continuous front over the increased distances, and the troops were left groping about after the elusive Boer forces, which easily evaded the lumbering columns plodding through the mud far in the rear.[252]

The drive caused an immense amount of material damage to farmhouses and crops, and much live-stock was taken, but so far as its effect on the burghers went it was a complete failure, for it left them more determined than ever to continue fighting.

As soon as the main pressure of the drive had eased, we of the A.C.C. were deputed to patrol the areas around Bethal and Ermelo, and although we did much hard riding on the flanks of various columns that were trekking along, we had no actual fighting.

I now suffered the loss of my dear old roan horse. One morning he came staggering and swaying up to me from the grazing-ground, and I saw at once from his caving flanks and glassy eyes that he was stricken with the dreaded horse-sickness, from which scarcely one animal in a hundred recovers.[253] Nosing against me he seemed to appeal for help, but he was beyond hope, and in less than an hour, with a final plunge, he fell dead at my feet. This was a great sorrow, for a close bond had grown up between us in the long months since the war started, during which he had carried me so well.

I had to leave him, for as he fell, English scouts came swarming over the rise. I threw my saddle on a borrowed mount, and galloped away with the rest, and as my brother was absent on patrol with both his horses, I was forced to rely upon a succession of loaned animals, but there was no help for it.

On his return a week later he unselfishly made over the 'Malpert' to me, and we now spent some time scouting over the plains in small parties to see what the English were doing, for, although the drive had petered out, large columns were still aimlessly wandering over the country, and our orders were to keep count of them. We had several sharp brushes with their outposts, and one evening were caught by a burst of shrapnel which blew a Carolina man to pieces within an arm's-length of me, and wounded his horse so severely that I had to shoot it.

After thoroughly scouring the countryside as far as the Swaziland border, our various A.C.C. patrols reassembled at a place called Klipstapel, said to be the highest and coldest spot in the Transvaal.[254]

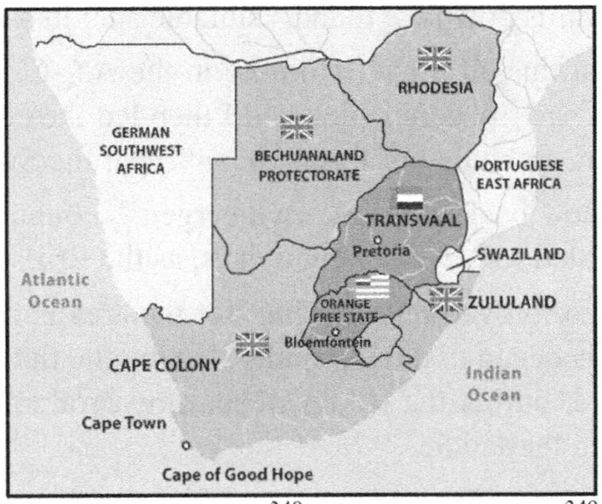

Dordrecht, 1902. Engine buried in snow for 8 days before dug out by 100 boys.[348] Note Swaziland.[349]

That same day instructions came from General Botha that the A.C.C. were to be with him by midnight, as he was going to attack an English force camped near Lake Chrissie.[255]

[252] 2/1/1901 General French continued to operate against Botha in the Eastern Transvaal.

[253] African horse sickness (AHS) is a highly infectious and deadly viral disease, spread by insect vectors. Horses are the most susceptible host with close to 90% mortality of those affected. African horse sickness - Wikipedia

[254] Breyten is situated at the foot of Klipstapel, the highest point on the watershed between the westward-flowing Vaal River system and the eastward-flowing Olifants and Komati River systems. Breyten - Wikipedia

[255] Lake Chrissie - Wikipedia

The attack took place, but we were held in reserve. We heard the sound of heavy firing shortly before dawn, and a rumour passed that the camp was taken. This proved only partly correct, for by sunrise came the trampling of many rider less horses, followed later by galloping riders, who told us that, although they had succeeded in overrunning the camp, the English troop horses had stampeded, creating such confusion in the dark, that our men gave way when success was practically in their grasp.[256] [257]

As it grew light, we could see the English camp lying intact below, and, as the troops turned several guns on us, and on such others as were in sight, we went scurrying away for the shelter of a neighbouring valley, where we found General Botha and most of his men. There was dejection in the air, for our loss overnight had been heavy, about forty killed and many wounded, with nothing to show for it.

But General Botha was not discouraged, for he addressed us, saying there were bound to be ups and downs, and he talked the men into a better frame of mind without any trouble.

After this, he split up his commandos. We had to live by foraging for sheep that were roaming about the veld, and by gathering maize from the unharvested fields, so he hived us off into smaller bodies, for easier provisioning.

As the drive had gone to pieces[258] and there was no immediate need for our further services in the Eastern Transvaal, he ordered Beyers to take his men north to the new storm-centre in the Waterbergen,[259] which the English were now penetrating and from which urgent calls for help had come.

Most of Beyers's men were Waterbergers, so it was only natural to send them up to the threatened parts, but we of the A.C.C. were a mixed community, a sort of Foreign Legion, not hailing from any particular area. As, moreover, we were not over-fond of General Beyers, we decided instead to return west and rejoin General de la Rey, especially as messages had come that there, too, another great drive was in progress. Beyers accordingly started [North] on his long march without us, and we took our leave of General Botha next day.

After some days' riding [West] over the plains we got back to the country between Pretoria and Johannesburg, in order to recross the railway line.

This time the track was better guarded, for we found numerous little camps on either side, and the line was closely patrolled by mounted men.

Jan Nagel, our commandant, sent me with another man after dark to find a convenient crossing. We crawled right up to Irene Station, ten miles [South] from Pretoria, passing within a few yards of the tents there, and a thousand yards down we found a suitable spot in a hollow.

We were at this all night, returning to the A.C.C. at a neighbouring farm by daybreak. Here we rested quietly until dark, and my companion and I then led the commando over the railway line without any trouble. We rode on all night, reaching our former haunts near the Swart Kop by morning.

We were now back in the Skurwe-bergen[260] country near Johannesburg, where we had taken refuge before, and here we rested for seven or eight days, as the British troops rarely visited these broken hills.

This mistake proved our undoing, for the area is notoriously unhealthy for horses during the rainy season, and our animals began to die so rapidly that, by the time we made for higher ground, more than half the A.C.C. were dismounted, but as the 'Malpert' and my brother's toll-free chestnut had escaped the contagion thus far, we were still in the saddle.

Jan Nagel decided that we must reach General de la Rey's commandos in the west, in the hope of getting fresh horses, so we started off, a miserable band, most of the men on foot, carrying their saddles and equipment on their backs, and the rest of us not knowing when we should be doing the same.

[256] 2/6/1901 General French, after encountering little resistance, entered Ermelo. General Smith-Dorrien repulsed 2,000 of the enemy at the Battle of Chrissiesmeer. Chrissiesmeer - Wikipedia. His losses were 23 killed and 52 wounded. Horace Smith-Dorrien - Wikipedia

[257] On 6 February 1901, Smith-Dorrien's troops were attacked in the Battle of Chrissiesmeer. Chrissiesmeer - Wikipedia
And: Horace Smith-Dorrien - Wikipedia

[258] 2/9/1901 Eastern movement continued in deluges of rain, but invasion of Natal by Botha eventually frustrated.

[259] Waterberg is one of the 5 districts of Limpopo province of South Africa Waterberg District Municipality - Wikipedia

[260] Skurweberg means 'rough mountain' in Afrikaans.

XVII. I Start for the Cape Colony

We struggled along for two days, making west, and losing more horses as we went, until we found our way barred by a body of troops stationed on the height known as Ramagothla, from which they had a wide view over the surrounding country.

Hampered as we were with dismounted men, we could not venture into sight across the open, so we turned north to the friendly refuge of the Magaliesbergen that we could see once more on the far horizon.

By stealing down kloofs and valleys, we made the foot of the mountains unobserved, reaching the very spot where the 'Old Wagon Pass' comes down, the same road along which we had travelled from the Waterbergen, two months before.

On this march the 'Malpert' showed signs of distress, his staring coat and lagging steps telling only too plainly that his course was spent.

A column of English horse was near, so we had to hurry up the pass as it grew dark, and I was obliged to leave him in an orchard below, to give him this slight chance of recovery. Next morning, on looking over the cliffs, I saw him lying dead. He had been game to the end—'*puure paert*' (an 'all-horse')—as the men called him, and my brother and I climbed down to pay a last visit to his poor emaciated carcass.

On our return Nagel discussed the situation. With more horses dying every day, our prospect of safely crossing the open plains and reaching General de la Rey was a poor one, and he suggested that we should return to the hill country from which we had just come. He said that we could wait there until word was got through to General de la Rey, who would doubtless send sufficient fresh horses to enable us to join him.

About half the men agreed to this, but the others refused. They said de la Rey was short of horses himself, and, with [Southern Hemisphere] winter approaching once more, they preferred making north over the mountain-range for the warmer climate of the bush-veld. My brother and I decided to go north, too, although for a different reason. I was without horse, and he might be so at any moment, so we decided to find my father, as we counted on him to fit us out again, after which we would return south.

There was very little discussion and no ill feeling, for Nagel was a sensible man, and that same afternoon all was concluded. He, and those who were accompanying him, said good-bye, and filing down the pass again were soon lost to view. With him went de Gourville the Frenchman, the two sons of our late Commandant Malan, Fred Hancock, an old Bloemfontein school-fellow, and many other friends and companions, none of whom I ever saw again.

I think most were hunted down, or shot out, before they could obtain remounts, for, as I was to discover, their hopes of getting horses from General de la Rey were practically nil as he had none to give.

The men who had chosen for the north now started over the mountains, some on horseback, the majority on foot, and thus ended the A.C.C.

My brother and I were in no hurry, so we remained resting among the boulders that night.

My father was somewhere in the Lydenburg country, and this was three hundred miles away, [East] no easy journey with a single horse between us, but we reckoned that by riding and walking in turn we should find him in the end, and we hoped that as a member of the Government he would be able to recondition us.

Next day we loaded our gear on the chestnut, and descended the north side of the Magaliesbergen into the great valley which we had crossed with Beyers the night that we found the burning convoy.

View of Magaliesbergen valley and reservoir of Hartbeespoort Dam.[350]

We reached the bottom in the afternoon, by which time the A.C.C. men who were travelling the same road had long since disappeared, leaving us seemingly alone in the broad rift. Towards evening, however, we came on an ox-wagon, outspanned in a patch of scrub, and owned by a stout-hearted old Boer lady who told us that she was on her way to the bush country. Her husband was fighting with de la Rey, and as the English were harrying and burning in the west she had thrown her belongings on a wagon, and, leaving the farm to look after itself, she had crossed the Magaliesbergen with her children and a native herd boy. She said she was going north at once, and offered us a lift.

Crossing A South African Drift.[351]

As this would save us much walking, we helped to fetch and yoke the oxen, and started off overjoyed at this timely relief. Before we had gone many yards, however, I remembered that by some oversight I had forgotten my saddlebags where we had halted that morning on our way down the mountain: a slight matter that altered the whole subsequent course of the war for me.

Saddlebags were scarce and valuable, and moreover mine contained a supply of salt that I had been lucky enough to discover at a deserted farm the week before, so it was agreed that my brother [Arnt??] would go on with the wagon, while I returned up the pass on his horse to retrieve the missing articles. We thought that I should easily be able to catch up next day, so, leaving the wagon to jolt along, I trotted the chestnut back along the road towards the mountain.

By the time I reached the spot where I had left my wallets it was dark, so I built a fire and spent the night there. At daybreak when I untied the hobbles of the horse he savaged my arm, a sure sign that he was not himself, for ordinarily he was gentle, but I saddled up none the less and rode him down the mountain and halfway across the floor of the valley, by which time the tell-tale flecks of foam at his nostrils showed that the horse-sickness was on him. Seeing an empty farmhouse not far away, I led him there and kept him in the shade (which is said to be the only chance), but in less than an hour he was dead.

I could see up the northern road for ten miles or more, with no sign of the wagon, and, as the sun was blazing down, I decided to rest in the house until it grew cooler, before walking on in search of my brother.

The farmhouse had apparently been used as a place of call for English supply columns, coming and going between Rustenburg and Pretoria, for there was a litter of empty tins and other debris outside and the floors within were strewn with cigarette-ends, matches, and other marks of recent occupation. In going through the rooms looking for a cool spot, I found an unopened packet of newspapers on the floor, and as I had not seen a paper for over nine months I lost no time in reading them. I learned for the first time that Queen Victoria was dead; that there was a war in China; that Lord Roberts had been superseded by Lord Kitchener;[261] and I read of a great many other events that had been passing in the outside world. What interested me most was an account of a Boer commando that was raiding far down into the Cape Colony, and I resolved at once to go south in search of them.

[261] 11/29/1900: Kitchener took over the command in South Africa.

It was not so clear how I was to set about this, for the Cape lay many hundreds of miles away, [South] and I was alone and on foot. But my mind was made up and, abandoning all thought of overtaking my brother, I threw my saddle across my shoulders, and carrying my rifle in one hand, and my cooking-tin in the other, I started back on a journey that was to take me very far indeed.

As I trudged along the road I had come, clouds of dust rising from the direction of Pretoria warned me that an English column or convoy was approaching, so I took to the bush and spent the night there. Starting long before daylight next morning, I reached the foot of the Magaliesbergen once more, at the Old Wagon Pass, and here I found about fifty Rustenburg men[262] who had come in overnight They were horseless like myself, owing to the ravages of horse-sickness, and, having observed the approach of the English troops, they, too, had made for the shelter of the Berg.

By climbing a little distance we could follow the movements of the soldiers, whom we made out to be about three thousand strong, with guns and many wagons, all making in our direction. We fled up the heights to a gorge near the top, where we lay secure, but the weather took a turn for the worse and it rained for eight days and nights without stopping, during which time we had no shelter but over-hanging rocks. It was impossible to find dry fuel, so we existed on biltong without a mouthful of anything warm.

We could not descend the mountain, because parties that went down reported that both north and south there was not a dwelling or a barn within reach that was not occupied by weather-bound English troops, who had been driven in from all directions to take cover against the unprecedented downpour. The knowledge that our opponents were comfortably housed merely added to our misery, and every time that the clouds and mist swayed aside we could see smoke cheerfully ascending from every building and shack below.

When the rain ceased, after a dreadful week, we impatiently watched the troops getting under way, and then made haste down the north side of the mountain, to re-occupy some of the farms which they had vacated.

My boots meanwhile had rotted and I had to climb down the sharp slopes on my bare feet, which became so swollen and blistered that I lay for a fortnight in a tobacco shed to which I was carried. The men were kind to me, one old 'takhaar' actually walking twenty miles to fetch a piece of leather of which he knew, to make me a pair of rawhide sandals which served me for many a month to come.

"An old Takhaar."[352]

These burghers were Doppers, a religious sect said to be somewhat like the Quakers. They held strange views on many things, but for all their primitive ways they were brave, unspoilt men.

Among them were some who had been serving with General de la Rey before they lost their horses, and they now began to discuss the possibility of returning to him.

This meant crossing over the Magaliesbergen to the south once more and going thence on foot over the open plains of the high veld, at the mercy of the first mounted enemy patrol that caught sight of us, so that it would not be at all plain sailing.

[262] Rustenburg Commando 572 records: https://www.angloboerwar.com/?option=com_content&view=article&id=3062

However, after discussing the pros and cons of the venture, thirteen of us decided to take the risk, in spite of the warnings of those remaining behind.

The prospect of a two-hundred-mile march on foot, burdened with our saddles and other belongings, was not an attractive one, but we began our preparations. While out after game one morning I chanced to pass the derelict English wagon-convoy that had been burnt last year, and looking over the remains I had an inspiration. There was not a single whole wagon, but it seemed to me that by taking an undamaged wheel here and there, and axles and planks from other parts, we might piece together a composite vehicle.

Returning to my companions, I put forward this suggestion, which was so well received that they went off in a body to inspect the skeleton convoy, and in two days' time had built up a strange-looking but quite trustworthy wagon.

Broken Wagon.[353]

Drakensberg. Men fixing overturned ox wagon.[354]

The question of transport animals was just as easily solved. In a deep kloof some miles away ran a large herd of cattle, kept hidden there by General de la Rey's orders, as a reserve supply depot for his commandos in the west. We visited the cattle-guards, and got a dozen good trek oxen from them.

Native African workers tending cattle. They probably are employed by the British Army to look after livestock.[355]

My fellow-travellers were born stockmen, who could tell at a glance which animals were trained to the yoke, and after a trial run they could even tell the exact place at which each ox bad previously pulled in the span, so they quickly selected an excellent team, with which we returned to complete our plans by rough-hewing sufficient yokes and curing the necessary straps and riems.[263] In a few days' time all was ready, and loading up our goods we said good-bye to the rest of the Rustenburg men, and with our improvised wagon jolted up the Old Wagon Pass on our journey of discovery.

We climbed over the mountain range in one day, reaching the orchard where poor 'Malpert' had died, in time to see the English still crowning the rise at Ramagothla. The night was black with heavy rain under cover of which we crawled on to the open plain, hurrying the oxen past the camp until by daylight we were well out of sight in the folds of the hills, cold, drenched, and weary, but pleased with the good progress that we had made. We

[263] A belt, strap, thong, or collar.

found ourselves on the scene of recent fighting, for the ground was littered with rotting carcasses of horses and mules, and there were newly made graves.

I heard later that Mr. Smuts, the State Attorney, had here besieged a force of Australians, who defended themselves so bravely that they beat off our people with loss.[264]

We now travelled onward for several days through devastated country. The counterpart of the great drive we had witnessed in the east had since rolled over this area, leaving behind it only blackened ruins and trampled fields, so that our course lay through a silent unpeopled waste, across which we navigated our wagon like a lonely ship at sea.

Ruined town-appears to be Frankfort, northern Orange Free State.[356]

My companions were big heavily bearded men of the old school, who looked on me as something of an alien, for I was town-bred, and they did not always understand my ways, but they were simple kindly souls and we got on well together. The Boers had their full share of laggards, but they had a full share, too, of steadfast yeomen such as these; men whose farms were lying in ruins, whose wives and families were scattered they knew not where, but who, unpaid and unbidden, returned to risk their lives in the fighting that swayed continually backward and forward over the western plains, and I got a truer insight into the fine courage and high qualities of their fighting-men during this journey than at any other time of the war.

Rustenburg lies near the Magaliesburg Mountains.[357]

On our fourth or fifth day out, a woman came walking towards us with her two small children and a native servant girl. She told us that she had been sheltering for the past ten days in a wooded kloof where her wagon was hidden, but the oxen having strayed into the open had been captured by passing soldiers, so that she was stranded. She said that General de la Rey s wagon-laager was camped at a place called Rietpan, near Tafel Kop,[265] so, ascertaining that she was in no immediate want, we turned south-east, coming in view of the shimmering waters of the pan by noon next day. There was a considerable number of wagons on the shore of the lakes with many horses and cattle out at graze, and we were just congratulating ourselves on having at last found de la Rey when we saw a sudden stir—men began to run for their horses, and oxen were rapidly driven in, and, as we realized that something was amiss, we halted our wagon to await events.

Soon horsemen came riding furiously past, followed after an interval by the wagons and carts, also urged forward in a panic. We managed to stop one rider, who only stayed long enough to say that the English were

264 Unable to identify this engagement. It occurred sometime after November of 1900.

265 Rietpan is a village in Ngaka Modiri Molema District Municipality in North West Province. Reed Pan - Wikipedia

upon us. We looked in vain for any sign of the enemy, but assuming that they must be near, we turned our tired oxen, and rattled along behind the retreat, which did not slow down until we had covered about six miles. Then horsemen from the rear overtook us with word that the alarm had been a false one.

Oxen on Veldt.[358]

All concerned had the grace to look ashamed of the stampede, but there was at any rate some excuse for them, as de la Rey had been severely handled early that morning in a surprise attack, in which he had lost over a hundred men. Some of his wounded, galloping up to the wagons, had precipitated a rout among the drivers, cattle-guards, and camp-followers, who formed the bulk of the laager.

In the afternoon General de la Rey himself rode in with his men, and he used withering language.[266]

We now went into laager at Tafel-Kop,[267] where de la Rey's commando and the wagons remained for the next few days. There were about a thousand mounted men and perhaps two hundred wagons. A few of these were used to bring maize from the Magaliesberg foothills, but the majority belonged to non-combatant refugees. De la Rey looked with an unfavourable eye on these and their vehicles, and the men said that the old man prayed to the Almighty night and morning for the enemy to relieve him of this incubus.

I saw a good deal of him at this time, as he held a daily levee beside his cart, where all were free to hear his views.

Attached to his person was a prophet, van Rensburg,[359] a strange character, with long flowing beard and wild fanatical eyes, who dreamed dreams, and pretended to be possessed of occult powers. I personally witnessed one of the lucky hits to which he owed his reputation, for one morning while we were congregated around the General's cart, van Rensburg was expounding his latest vision to-a hushed audience. It ran of a black bull and a red bull fighting and goring each other, until at length the red bull sank defeated to its knees, which he interpreted to mean that the British would soon be in like case. As he stood before us, his arms outstretched and his eyes ablaze, he suddenly called Out: 'See, who comes?;' and, looking up, we made out a distant horseman spurring towards us front the cast. We waited in silence for the rider. When he came up, travel-stained and weary, he produced a letter from General Botha, hundreds of miles away.

Nicolaas Janse van Rensburg.[360]

[266] Possibly, 12/19/1900 Boers under Delarey routed. Boer raid into Cape Colony.

[267] Tafelkop is a town in the Elias Motsoaledi Municipality in Sekhukhune District of the Limpopo Province. https://af.wikipedia.org/wiki/Tafelkop

When General de la Rey opened and read it, his face lighted up, and in a voice ringing with emotion he said: 'Men, believe me, the proud enemy is humbled' (Die trotse vijand se nek is gebuig). He went on to tell us that the letter contained news that the English had proposed a peace conference. Coming immediately upon the prophecy, it was a dramatic moment and I was impressed, even although I suspected that van Rensburg had stage-managed the scene. Of General de la Rey's sincerity there could be no doubt, for he was not a man to stoop to subterfuge, and I knew that he firmly believed in the seer's predictions.

These tidings created a great deal of stir and excitement during the next few days, and many of the men thought that the war was as good as over.

A peace conference did in fact take place a little later between General Botha and Lord Kitchener, but it proved abortive, and the only immediate result was to establish van Rensburg's reputation more firmly than ever.

Middleburg Conference

In early February of 1901, General Kitchener, with the consent of the Government decided to make peace overtures to General Botha well before the February 14th debate in the House of Commons. The actual conference took place from February 28th, until its unsuccessful conclusion on March 16th 1901.

On February 13,[268] a week after the failure of his night attack at Lake Chrissie,[269 270] Louis Botha received a message through his wife, a verbal message, from Kitchener.

If General Botha desired Kitchener, it, so the message ran, a meeting could be held with the British Commander to arrange terms of peace; the one condition being that the question of the independence of the two republics was not to be discussed in any way.

The invitation arrived at an opportune moment; for Botha, though he had effectively countered the converging movement on Ermelo, had nevertheless seen French's columns sweeping unchecked across the high veld, and was fully aware of the searchings of heart which this great display of force had caused among his burghers.

The reply, despatched the same day, but not received by Kitchener until the 22nd, was an acceptance. In the same letter Botha suggested Middelburg as a meeting-place, and requested a safe conduct to and from that place. Kitchener agreed, and February 28 was fixed for the conference.

Middleburg stands on a tributary of Olifant's River and is 94 miles from Pretoria.[361]

Kitchener had no illusions as to the character of the struggle in which he was engaged. Even without the eloquent policy examples of history, events in South Africa since his accession to the command had shown the enormous capacity for resistance possessed by a pastoral race skilled in arms. He was conscious, too, that his mounted army was passing through a period of transition and for a time, at any rate, must inevitably decline in usefulness. The efforts of the Burgher Peace Committee[271 272] having totally failed, there were two courses open:

[268] The Times' History of the War in South Africa 1899–1902 by Amery, L. S. (Leopold Stennett), Vol 5 Ch. VII pages 183–193.

[269] In an attempt to invade Natal in force and later his unsuccessful attempt to overwhelm a British Column. He eventually did break out back to his base area which the British operation had been designed to deprive him of access to.

[270] The Battle of Chrissiesmeer (Lake Chrissie) during Anglo-Boer War 2 started with an attack by General Louis Botha on the British forces under General Smith-Dorrien. Botha's main aim was to cripple the advance of Smith-Dorrien into the eastern Transvaal. According to oral tradition Smith-Dorrien's march had been kept under observation by San (Bushmen) who also acted as messengers. The Boers capitalised on the San's knowledge of the terrain and attacked at 02:50 in pitch darkness.
The burgers achieved initial success, but could not gain a foothold on the slopes and were not able to open direct fire on the main camp situated on a plateau. At 04:30, covered by thick morning mist, Botha ordered them to retreat. The Boer commando suffered about eighty casualties and the British seventy-five. About 300 horses of the British force were killed or stampeded. Despite heavy losses on the Boer side the British advance into the eastern Transvaal was delayed by the loss of their horses, which gave the Boer commandos time to re-group. https://sahistory.org.za/dated-event/san-involvement-battle-chrissiesmeer See Also: Lake Chrissie is a lake in Msukaligwa Local Municipality, Mpumalanga, South Africa. It is a shallow and large lake, with a maximum depth of about 6 metres (20 ft.), 9 kilometres (5.6 mi) long, 3 kilometres (1.9 mi) wide. https://en.wikipedia.org/wiki/Lake_Chrissie

[271] Peace Committee Times History Vol 5. P. 91–93 In December of 1900 Kitchener placed himself in communication with influential men among the surrendered Boers, and it was partly on their advice that the proclamation of December 20 was framed. Nor did he stop here. At a meeting held early in December, under the presidency of Mr. Meyer de Kock, of Belfast, a Peace Committee was formed, with de Kock as secretary, for the purpose of urging upon the commandos the hopelessness of resistance and the need for a general surrender.
A meeting of surrendered Boers convened by this Committee was held in Pretoria on December 21, and was addressed sympathetically by Kitchener. He promised the Committee his advice and support, spoke of the military conditions which exacted the formation of concentration camps, of his determination to give the enemy every possible chance of submitting voluntarily and of the just and progressive government which at the end of the war it was proposed to establish. The speech was printed in Dutch and English, and steps were promptly taken to give it and the proclamation of December 20 wide circulation. To this end emissaries were appointed to visit the commandos and preach the gospel of peace.

to maintain an attitude of unbending reserve, and accept with philosophy the prospect of an incalculably long and costly struggle; or, should a favourable opportunity arise, to persuade the enemy, on the promise of certain terms from which independence was rigidly excluded, to abandon a ruinous and futile resistance.

In the middle of February such an opportunity appeared to have arrived. De Wet and Steyn, the two most stubborn foes of British supremacy, were riding for their lives in Cape Colony. In the Transvaal French's great movement was throwing into panic the whole of a very important district. Kitchener, … . resolved to seize the opportunity; and, while de Wet and Steyn were incapable of interference, to effect a settlement with Botha. ……………..On February 28 Kitchener and Botha met at Middelburg. By a stroke of dramatic irony this was the very day on which de Wet and Steyn, after a period of desperate peril, crossed the Orange and won their freedom. It might have been read as a bad omen for peace.

The Middleburg[273] Conference. Group photograph of the participants in the Middelburg peace conference. (February-March 1901)[362 363] Standing (left to right); Colonel G. F. R. Henderson, Dirk Overgauwe ("Dick") van Velden,[274] Major Watson, H. Fraser, Captain Maxwell, H. de Jager. Sitting (left to right); Christiaan de Wet, (Sic. Possibly Wynand Viljoen) General Louis Botha, General Lord Kitchener, Colonel Ian S. M. Hamilton.

Note the black armbands worn by all five British officers to commemorate Queen Victoria's death a month earlier. (January 22nd, 1901).

Annie Botha, the wife of general Louis Botha.[364 275] "Mrs. Botha is descended from Robert Emmet,[276] the famous "United Irishman" who was executed for high treason, September 20, 1803."

Like the proclamation itself the peace propaganda produced the contrary effect to that hoped for. It exasperated the fighting burghers, who viewed its promoters, sincerely patriotic as they were, as cowards and traitors, and, when they obtained the opportunity, treated them as such. Though most of the emissaries escaped in the course of time to the British lines, many were tried by court-martial and sentenced to fines and imprisonment. Meyer de Kock himself, having ridden into Viljoen's lines at the end of January, was tried, sentenced to death for high treason, and executed. In one notorious case there is, at least, a grave suspicion that an irregular and violent course was taken.

On the whole, however, the treatment accorded to the peace emissaries was regular and not, under the circumstances, unduly harsh. In taking the risk of attempting to seduce the men on commando, they acted with their eyes open according to their own sense of patriotism. The fighting burghers, holding a different standard of patriotism, were within their rights in applying even extreme penalties.

[272] Meyer de Kock (5 October 1849 – 12 February 1901) After the British had captured the capital Pretoria and annexed the Transvaal, he surrendered on 10 December 1900. In January 1901, he went to the eastern Transvaal to persuade burghers still in the field to surrender. He was arrested, court-martialed, and later, on 12 February 1901, executed by firing squad. https://en.wikipedia.org/wiki/Meyer_de_Kock See also Viljoen Chapter 30.

[273] It was initially established as a halfway station between Lydenburg and Pretoria. The British built a large concentration camp in Middelburg during the Second Boer War. A Memorial Museum was built at the site next to the 1,381 graves of women and children who died in the concentration camp. Middelburg, Mpumalanga - Wikipedia

[274] Van Velden Dirk Overgauwe Kaptein Commando: Staff General Botha. Awarded the DTD.

[275] Annie Botha (née Emmett; 3 July 1864 – 20 May 1937) was a South African civic leader and political hostess. She was the wife of Louis Botha, who served as the first Prime Minister of South Africa. She established an orphanage in South Africa and, with Georgiana Solomon, co-founded and chaired the South African Women's Federation. Annie Botha - Wikipedia

[276] Robert Emmet (4 March 1778 – 20 September 1803) was an Irish Republican, orator and rebel leader. Emmet's Proclamation of the Provisional Government to the People of Ireland, his Speech from the Dock, and his "sacrificial" end on the gallows inspired later generations of Irish republicans. Robert Emmet - Wikipedia

Newspaper Coverage.

There were several fragmented reports that a meeting was taking place published at the time.

Several of the reports were of this brief nature.

Highland News - Saturday 02 March 1901—The Fighting in South Africa—Reported Surrender of Botha.

> Various reports were current on Thursday [February 28th] that General Louis Botha had formally surrendered to Lord Kitchener at Middleburg, but they remain totally unconfirmed. The War Secretary, replying to a question on the subject in the House of Commons, said no official report of the surrender had come to his Department.

The most accurate summary appeared in the Weekly Irish Times on Saturday 09 March 1901.[365]

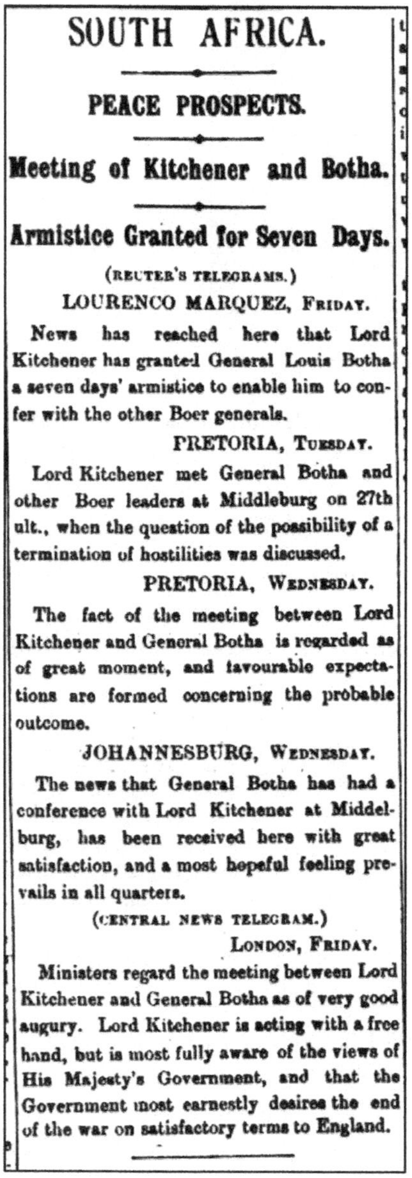

SOUTH AFRICA.

PEACE PROSPECTS.

Meeting of Kitchener and Botha.

Armistice Granted for Seven Days.

(REUTER'S TELEGRAMS.)

LOURENCO MARQUEZ, FRIDAY.

News has reached here that Lord Kitchener has granted General Louis Botha a seven days' armistice to enable him to confer with the other Boer generals.

PRETORIA, TUESDAY.

Lord Kitchener met General Botha and other Boer leaders at Middleburg on 27th ult., when the question of the possibility of a termination of hostilities was discussed.

PRETORIA, WEDNESDAY.

The fact of the meeting between Lord Kitchener and General Botha is regarded as of great moment, and favourable expectations are formed concerning the probable outcome.

JOHANNESBURG, WEDNESDAY.

The news that General Botha has had a conference with Lord Kitchener at Middelburg, has been received here with great satisfaction, and a most hopeful feeling prevails in all quarters.

(CENTRAL NEWS TELEGRAM.)

LONDON, FRIDAY.

Ministers regard the meeting between Lord Kitchener and General Botha as of very good augury. Lord Kitchener is acting with a free hand, but is most fully aware of the views of His Majesty's Government, and that the Government most earnestly desires the end of the war on satisfactory terms to England.

Peace Prospects. Meeting Kitchener and Botha. Armistice Granted for Seven Days. (Reuter's telegrams.) Lourenco Marquez, Friday.

New has reached here that Lord Kitchener has granted General Louis Botha seven days' armistice to enable him to confer with the other Boer generals.

PRETORIA, Tuesday.

Lord Kitchener met General Botha and other Boer leaders Middleburg on 27th inst., when the question the possibility of termination of hostilities was discussed.

Pretoria, Wednesday.

The fact of the meeting between Lord Kitchener and General Botha is regarded as of great moment, and favourable expectations are formed concerning the probable outcome.

Johannesburg, Wednesday

The news that General Botha has had a conference with Lord Kitchener Middelburg, has been received here with great satisfaction, and a most hopeful feeling prevails in all quarters.

(Central News Telegram.) London, Friday. Ministers regard the meeting between Lord Kitchener and General Botha as of very good augury. Lord Kitchener is acting with free hand, but is most fully aware the views of His Majesty's Government, and that the Government most earnestly desires the end of the war on satisfactory terms to England.

In spite of the express condition laid down by Kitchener draft terms. and implicitly accepted in the letter of the Boer general, Botha began by asking that the Republics should retain some sort of independence. Kitchener having refused to discuss the point, a conference began upon the details of a possible settlement. At the end of the discussion, which was conducted in a very friendly and reasonable spirit, Botha took his departure, and the terms of a proposed settlement were thrown into the shape of a draft letter from the British to the Boer general.

This letter was submitted to Milner, who met Kitchener at Bloemfontein, and was sent by him for approval to the Home Government.

The terms subsequently published in England for general readership.

THE BOER WAR.

THE RECENT PEACE NEGOTIATIONS.

THE TERMS REJECTED.

The official papers relating to the recent negotiations between Lord Kitchener and General Botha were issued this (Friday) morning. After detailing the events leading up to the interview, the following terms are specified by Lord Kitchener:—That, in the event of a general and complete cessation of hostilities and the surrender of all rifles, ammunition, cannon, and other munitions of war in the hands of burghers or in Government depots or elsewhere, His Majesty's Government is prepared to adopt the following measures:—His Majesty's Government will at once grant an amnesty in the Transvaal and Orange River Colonies for all bona-fide acts of war committed during the recent hostilities. British subjects belonging to Natal and Cape Colony, while they will not be compelled to return to those colonies, will, if they do so, be liable to be dealt with by the law of those colonies, specially passed to meet the circumstances arising out of the present war. As you are doubtless aware, the special law in the Cape Colony has greatly mitigated the ordinary penalties for high treason in the present cases.

All prisoners of war not in St. Helena, Ceylon, or elsewhere, will, on the completion of the surrender, be brought back to their country as quickly as arrangements can be made for their transport.

At the earliest practicable date military administration will cease and will be replaced by civil administration in the form of Crown colony government. There will, therefore, be in the first instance in each of the new colonies a Governor and an executive council, consisting of a certain number of official members to whom a nominated unofficial element will be added. But it is the desire of His Majesty's Government as soon as circumstances permit to introduce a representative element, and ultimately to concede to the new colonies the privilege of self-government. Moreover, on the cessation of hostilities a high court will be established in each of the new colonies to administer the law of the land, and this court will be independent of the executive.

Church property, public trusts, and orphans' funds will be respected.

Both the Dutch and English language will be used and taught in the public schools where parents of the children desire it, and allowed in courts of law.

As regards the debts of the late Republican Governments, His Majesty's Government cannot undertake any liability. It is, however, prepared as an act of grace to set aside a sum not exceeding £1,000,000 to repay the inhabitants of the Transvaal and Orange River Colonies for goods requisitioned from them by the late Republican Government or subsequent to annexation by commandants in the field being in a position to enforce such requisitions. But such claims will have to be established to the satisfaction of a judge or Judicial Commission, appointed by the Government, to investigate and assess them, and, if exceeding in the aggregate £1,000,000, they will be liable to reduction pro rata. I also beg to inform your honour that the new Government will take into immediate consideration the possibility of assisting by loan the occupants of farms, who will take the oath of allegiance to repair any injury sustained by destruction of buildings or loss of stock during the war, and that no special war tax will be imposed on farmers to defray the expense of the war.

As regards the extension of the franchise to Kaffirs in the Transvaal and Orange River Colony, it is not the intention of His Majesty's Government to give such franchise before representative government is granted to those colonies, and, if then given, it will be so limited as to secure the just predominance of the white races. The legal position of the coloured person will, however, be that which they hold in Cape Colony. In conclusion, I must inform your honour that if the terms now offered are not accepted, after a reasonable delay for consideration, they must be regarded as cancelled.

The terms were published in England on March 23rd.[366]

Previously the Government made changes to Kitchener's proposed language and this was sent to Botha on the 7th. (March) In its final form it ran as follows :-

THE BOER WAR.

THE RECENT PEACE NEGOTIATIONS.

THE TERMS REJECTED.

The terms were published in England on March 23rd.[367] Previously the Government made changes to Kitchener's proposed language and this was sent to Botha on the 7th. (March) In its final form it ran as follows:—

To his Honour Commandant-General Louis Botha. Pretoria, March 7, 1901.[277]

Your Honour,

With reference to our conversation at Middelburg on the 28th February, I have the honour to inform you that, in the event of a general and complete cessation of hostilities, and the surrender of all rifles, ammunition, cannon, and other munitions of war in the hands of the burghers, or in Government depots, or elsewhere, His Majesty's Government is prepared to adopt the following measures. His Majesty's Government will at once grant an amnesty in the Transvaal and Orange River Colony for all bona fide acts of war committed during the recent hostilities. British subjects belonging to Natal and Cape Colony, while they will not be compelled to return to those Colonies, will, if they do so, be liable to be dealt with by the laws of those Colonies specially passed to meet the circumstances arising out of the present war. As you are doubtless aware the special law in the Cape Colony has greatly mitigated the ordinary penalties for High Treason in the present case. All prisoners of war, now in St. Helena, Ceylon, or elsewhere, being burghers or Colonists, will, on the completion of the surrender, be brought back to their country as quickly as arrangements can be made for their transport.

At the earliest practicable date military administration will cease, and will be replaced by civil administration in the form of Crown Colony Government.

There will, therefore, be, in the first instance, in each of the new Colonies, a Governor and an Executive Council, composed of the principal officials, with a Legislative Council consisting of a certain number of official members to whom nominated unofficial element will be added.

But it is the desire of His Majesty's Government, as soon as circumstances permit, to introduce a representative element, and ultimately to concede to the new Colonies the privilege of self-government. Moreover, on the cessation of hostilities a High Court will be established in each of the new Colonies to administer the laws of the land, and this Court will be independent of the Executive.

Church property, public trusts, and orphan funds will be respected. Both the English and Dutch languages will be used and taught in public schools when the parents of the children desire it, and allowed in Courts of Law.

As regards the debts of the late Republican Governments, His Majesty's Government cannot undertake any liability. It is, however, prepared, as an act of grace, to set aside a sum not exceeding one million pounds sterling to repay inhabitants of the Transvaal and Orange River Colony for goods requisitioned from them by the late Republican Governments, or subsequent to annexation, by Commandants in the field being in a position to enforce such requisitions.

But such claims Will have to be established to the satisfaction of a Judge or Judicial Commission, appointed by the Government, to investigate and assess them, and, if exceeding in the aggregate one million pounds, they will be liable to reduction pro rata.

I also beg to inform your Honour that the new Government will take into immediate consideration the possibility of assisting by loan the occupants of farms, who will take the oath of allegiance, to repair any injuries sustained by destruction of buildings or loss of stock during the war, and that no special war tax will be imposed upon farms to defray the expense of the war.

[277] Part of appendix B from Three Years' War By: Christiaan Rudolf De Wet. Charles Scribner's Sons, (1902)

When burghers require the protection of fire-arms, such will be allowed to them by licence, and on due registration, provided they take the oath of allegiance. Licences will also be issued for sporting rifles, guns, &c., but military fire-arms will only be allowed for purposes of protection.

As regards the extension of the franchise to Kaffirs in the Transvaal and Orange River Colony, it is not the intention of His Majesty's Government to give such franchise before representative Government is granted to those Colonies, and if then given it will be so limited as to secure the just predominance of the white race. The legal position of coloured persons will, however, be similar to that which they hold in the Cape Colony.

In conclusion, I must inform your Honour that, if the terms now offered are not accepted after a reasonable delay for consideration they must be regarded as cancelled.

I have, &c.,

Kitchener, General,
Commander-in-Chief, British Forces, South Africa.

But in the meantime the prospects of peace had been declining. French, hampered by rain and scarcity, had for some time been at a complete standstill on the Swazi frontier, and the burghers behind him had had time to recover their spirits. Botha had consulted his Government and found it stubborn; he had corresponded with De la Rey and had met with no encouragement. Before the time came to reply, it was known that de Wet and Steyn had escaped capture, and their opinion of the negotiations scarcely needed asking.[278] Nor would they have held themselves bound by any decision of the Transvaal.

In a letter dated March 16 Botha, without Botha giving any reasons, declined to go any further with the negotiations.

Botha's Reply. These terms were refused by Botha and reported on March 23rd, in the following message:[368]

> I have the honour to acknowledge the receipt of your Excelleney's letter stating what steps your Excellency's Government is prepared to take in the event of a general and total cessation of hostilities. I have advised my Government of your Excellency's said letter, and, after. mutual exchanges of views at Middleburg on the 28th February, it will certainly not surprise your Excellence to know that I do not feel disposed to recommend that the terms of the said letter shall have the earnest consideration of my Government. I may add also that my Government and my chief officers here entirely agree to my views.

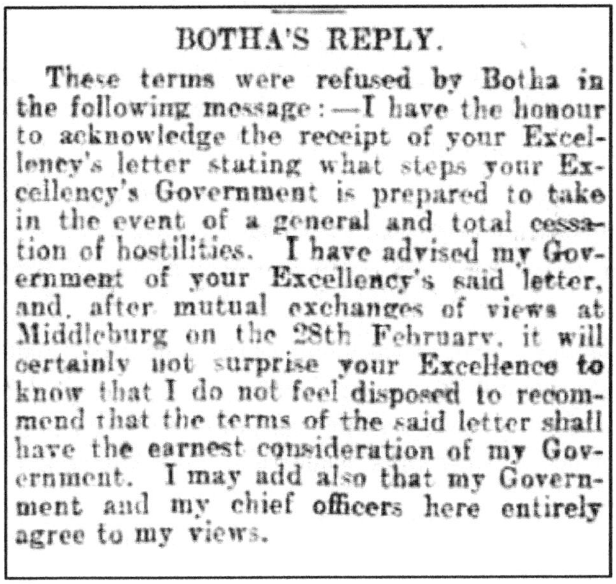

BOTHA'S REPLY.

These terms were refused by Botha in the following message:—I have the honour to acknowledge the receipt of your Excelleney's letter stating what steps your Excellency's Government is prepared to take in the event of a general and total cessation of hostilities. I have advised my Government of your Excellency's said letter, and, after mutual exchanges of views at Middleburg on the 28th February, it will certainly not surprise your Excellence to know that I do not feel disposed to recommend that the terms of the said letter shall have the earnest consideration of my Government. I may add also that my Government and my chief officers here entirely agree to my views.

[278] Original footnote: * Botha. wrote to de Wet and Steyn, asking them to come north and confer with him, but the letter arrived too late and the meeting did not take place till April. De Wet's secretary, however, went by train to Pretoria, only to find that negotiations had been broken off. He could have been of no use, for his master's temper, as shown by a circular to his burghers, was as unbending as ever. In this circular, which is dated April 1, he inveighs against various points in the proposed settlement, and asks finally, what is the use of examining it in detail, since nothing short of independence will induce the Boers to make peace.

Kitchener was always inclined to feel that with a little a little stronger support, especially on the question of amnesty, he might have obtained peace. However long the war lasted, it must, he held, eventually end in some such settlement as this. There was much truth in that; but the time had not yet come. His overtures were a bold but premature experiment. The Boer spirit-that new spirit born amidst the disasters of the previous year and matured in an offensive revival still quite recent, was far from being broken.

Although exhaustion and despondency existed in some quarters, when it was put to the Boers in black and white that their cause was lost irretrievably and that it only remained to settle subordinate issues, the national will revolted. The conditions which favoured this revolt are conditions which it can never be easy for a highly civilised industrial nation fully to understand.

The causes which bring such a nation to its knees were absent here. There were no vital industries to dislocate, no teeming population of peaceful workers to plunge into starvation. The people lived on the land, and though its fruits could be destroyed and the farms upon it gutted, the land could not be damaged. A year's crops were a relatively small matter, farm-buildings were of little value, and when a farm was once gutted it signified little whether it was rebuilt a year sooner or later.

Meanwhile, the male population, in other words the army, was inured by habit and inherited instinct to a life of peril and adventure under the stars. These men, moreover, were scattered over a vast expanse of territory under circumstances so varied as to render a common judgment unattainable. A few were suffering acutely; some were fighting with fair success; some were living in perfect safety and comfort. In short, there had never come to the Boers one of those moments of national agony which wring from a people a cry for peace.

As events fell out, Botha had effected a very useful reconnaissance in two directions. He had tested the sentiment of his burghers, made them familiar-and that was something-with the idea of negotiation; and he had found out on what lines an ultimate settlement was to be framed. ....

Commander-in-Chief by right of undoubted military merit, he nevertheless had not risen without effort. A keen ambitious purpose had helped him to gain the high position which he occupied. To crown its tenure by signing away independence as long as there remained the faintest hope of carrying his career to a triumphant military climax, was a hard step to take.

He knew very well that it was impossible to recover the lost territories with the forces at his disposal, but there always remained the chance of a general rising in Cape Colony. The chimera of foreign intervention still survived, as did the dim hope that pro-Boer sentiment in Great Britain would grow and triumph.

Meanwhile he could salve his conscience with the knowledge that the British policy of concentration camps freed him from responsibility for the women and children, whose misery was the most cogent argument for submission. There was little to lose; there was much to gain. For Botha had the shrewd perception that, whatever the ultimate issue, the Boers could fare no worse than they would fare under the proposed settlement.

His mental attitude is revealed clearly in a letter addressed to his burghers on March 15.

"Virtually, "he writes, "Lord Kitchener's letter contains nothing more, but rather less than what the British Government will be obliged to do should our cause go wrong." ....

"The cause is not yet lost," he writes, "and since nothing worse than this can befall us, it is well worth while to fight on." ....

Such were the abortive negotiations of 1901. Their failure caused much disappointment in England.

On public platforms and in Parliament the Government was bitterly attacked by a section of the Opposition for having thrown away the chances of peace by attempting to exact too much, and the differences of opinion, frankly set forth in the despatches, between Mr. Chamberlain, Lord Milner and Lord Kitchener, were used to further the argument. ....

The nation at large instinctively grasped the truth, and resigned itself with unflinching determination to a further struggle of indefinite length and severity.

Philip Botha

General Botha's brother, Philip,[279] was killed in action and his two nephews wounded on March 6th.[369] while the negotiations at Middleburg were proceeding. It was reported: "From Lord Kitchener to Secretary of State for War. Pretoria, Wednesday, 4.30 p.m.—General P. Botha, brother of the Commandant-General, was killed on the Doornberg. His two sons were wounded."

Philip Rudolph Botha.[370]

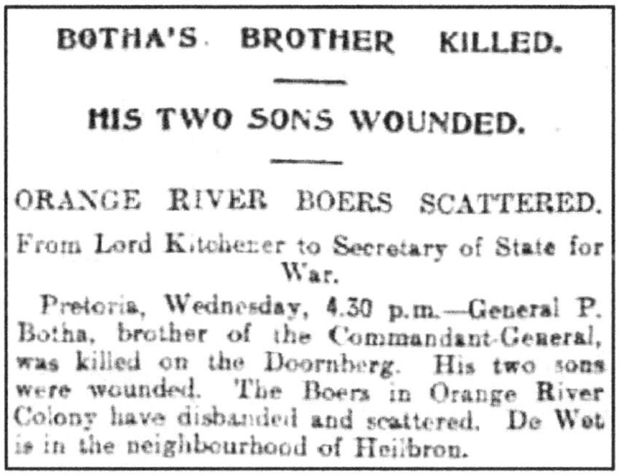

BOTHA'S BROTHER KILLED.

HIS TWO SONS WOUNDED.

ORANGE RIVER BOERS SCATTERED.

From Lord Kitchener to Secretary of State for War.

Pretoria, Wednesday, 4.30 p.m.—General P. Botha, brother of the Commandant-General, was killed on the Doornberg. His two sons were wounded. The Boers in Orange River Colony have disbanded and scattered. De Wet is in the neighbourhood of Heilbron.

Treaty of Vereeniging Terms

On May 31st this resolution was notified to Lord Kitchener, and at half-past ten of the same night the delegates arrived at Pretoria and set their names to the treaty of peace. ....

The following were in brief the terms of surrender:—

1. That the burghers lay down their arms and acknowledge themselves subjects of Edward VII.
2. That all prisoners taking the oath of allegiance be returned.
3. That their liberty and property be inviolate.
4. That an amnesty be granted—save in special cases.
5. That the Dutch language be allowed in schools and law-courts.
6. That rifles be allowed if registered.
7. That self-government be granted as soon as possible.
8. That no franchise be granted for natives until after self-government.
9. That no special land tax be levied.
10. That the people be helped to reoccupy the farms.
11. That 3,000,000 pounds be given to help the farmers.
12. That the rebels be disfranchised and their leaders tried, on condition that no death penalty be inflicted.

These terms were practically the same as those which had been refused by Botha in March 1901. [At the Middleburg Conference] Thirteen months of useless warfare had left the situation as it was."

Quote from The Great Boer War by Arthur Conan Doyle.[280] [281]

[279] Botha Philip Rudolf, was born in 1851. While in command of the Harrismith burghers, he made a breakthrough at Sprinkaan's Nek. He was killed in action while attacking the British position near Ventersburg on March 6th, 1901. See Also: Philip Botha - Wikipedia

[280] Doyle, Arthur Conan. The Great Boer War 17th Edition, 1902 with the two volumes as one document: The Great Boer War by Arthur Conan Doyle - Free Ebook (gutenberg.org) Chapter 39. Pages 232–234 of the Word document.

[281] General Article: Treaty of Vereeniging - Wikipedia

Resume

I asked General de la Rey for a horse, but he said I should have to wait until parties he had sent into the Free State had returned, and, indeed, I could see for myself, from the number of dismounted men with the laager, that he was unable to assist me, so in the meanwhile my companions and I clung to our wagon and team, taking turns at herding the oxen and foraging for maize. Like everyone else, we had to find ourselves in food, excepting meat, which was supplied from a communal drove.

By this time my clothes had fallen from my body, owing to the rains, and my entire wardrobe consisted of a blanket and a pair of sandals, so that, as it was towards the end of March (1901) by now, with winter coming on, I felt the cold pretty severely. General de la Rey had noticed my scanty attire, and one morning he walked over to our wagon with a pair of breeches and a coat, a gift I much appreciated, for he could have been none too well supplied himself, but it was of a piece with his natural kindliness and consideration towards all.

General Jacob Hercules De La Rey.[371] General de La Rey and Colonel Guorko, Russian military *attaché*.[372]

From Tafel Kop, after some days, he rode off with the mounted men, and we poor-infantrymen and other camp-followers were ordered to a place ten miles away, where there was better grazing for the trek-oxen. Here we remained, my prospects of reaching the Cape Colony looking pretty uncertain, although I was as determined as ever to get through.

On the 3rd of April, or the day we calculated to be that date, the wagon crew and myself were busy preparing a dinner to celebrate my birthday with an ox-tongue and a few odds and ends that we had managed to collect. Suddenly there came the crash of a field-gun, followed by the roar of a shell exploding close by, and almost immediately some of de la Rey's horsemen came riding out of the heavy mist that lay over all the country. They were riding hard to warn us that the English were coming. In a moment there was a wild rush to fetch the oxen, and fortunately the fog was so thick that none of the other shells fell among us and the entire wagon-laager was able to get under way without being seen.[282]

To the sound of small-arm fire where our horsemen were engaged in the mist, the convoy spread out for better speed, each wagon striking out a course of its own, without regard to the others. The fog held, and heavy rain setting in still further decreased the visibility, so that my companions and I got clean away, although at times the English horsemen splashed past so close to us that we could hear them shouting to each other for their bearings.

The bulk of de la Rey's fighting men were not in the vicinity at all, but only some twenty or thirty of them had been approaching, the wagon-laager, for a few days' rest, when they unexpectedly stumbled on an English force that was stalking the convoy under cover of the mist. But for this we should have been taken by surprise, and I doubt if a single man would have escaped. As it was, the warning we received enabled us to escape wholesale

[282] This may be near to the events described: 3/26/1901: Victory by General Babington over Delarey at Ventersdorp. Nine Boer guns captured.

capture, but it was touch and go, for the boom of the guns, the splutter of rides-through the fog, and the hallooing and galloping sounded very near. As a matter of fact a number of wagons were taken, although we did not know it at the time. As for our party, we ran beside the team, keeping the oxen at a trot, and as our wagon was light, carrying only our saddles and cooking-tins, we made good progress. The other wagons and carts were almost immediately lost sight of in the- mist, but we forged ahead by ourselves, and towards evening reached broken ground to the north, where we were safe enough.[373]

We built a large fire on the banks of a stream and resurrected the remains of my birthday-dinner, which had come along in a leather bag. Afterwards my companions stood in a circle round the blaze, and solemnly sang hymns of thanksgiving for our escape. Then we camped for the night.

Next morning as it grew light, other wagons came crawling over the plain, also making for the safety of the hills, so we yoked our oxen and joined them, heading for a valley in the heart of the rugged tract known as the 'Swart Ruggens,' where we found the remnant of our laager assembled.

We could not yet tell how many wagons had fallen into the hands of the enemy. Nearly half were missing, but most of these drifted in during the next day or two, and in the end it was found that less than a dozen had come to grief.

While we were camped in the valley, General de la Rey came up with some of his men to see how the convoy had fared, and, before riding back to the fighting area, he left orders for the wagons to proceed to a spot some thirty miles away, where there were two hundred horses from the Free State, which were to be given out by ballot to the dismounted men of the laager.

Many English columns were moving west at this time in continuance of a drive they were making, but we were able to trek along in comparative safety behind the line of their advance, and in due course reached our destination Here we found a patrol that had come up from the Free State in charge of the remounts. The animals were mostly unbroken mares, and, as there were over three hundred horseless men, the drawing of lots was followed with keen interest.

I drew a blank, but no less than nine of our original wagon party got a horse apiece. I was cast down at the result, although we were comforted by an assurance that further horses were expected, and those of us who were unsuccessful in the drawings had at least the fun of seeing the winners break in their mounts, a diverting spectacle.

Nearly two hundred horses were bucking and squealing at the same time, their riders biting the dust in all directions, while we sat on the wagon-rails cheering them on. It was wonderful how quickly the men mastered their unruly steeds, so that by the third day practically every animal was broken to the saddle, and the newly mounted force could ride off to the war.

With them went the nine men from our wagon, leaving the rest of us to wonder disconsolately when we should sit astride a saddle again, and leaving me to wonder as well whether, at this rate, I should ever see the Cape.

The wagon-laager was now instructed to return to the Swart Ruggen Hills from which we had come, so we lumbered slowly back, halting after some days at the head of a beautiful stream that gushed from the rocks. Here we went into permanent camp, and the place soon became a sort of base depot for de la Rey's fighting commandos. The sick and wounded were brought here, and there was a steady influx of horseless men drifting in from the plains, and of men who had left the firing-line for a rest.

Our numbers varied from two hundred to over four hundred at times, and so securely were we hidden that not once did the British troops come near us, although their columns were ranging far and near over the adjoining open country. Two days after our arrival at the new camp I tried to smash a log for fuel by bringing down a heavy stone on top of it. The stone came back at me like a shot from a catapult, breaking my right tibia halfway between knee and ankle, and for three weeks I lay in great pain and discomfort with splinters of bone working their way out from a suppurating[283] wound.

Luckily there were some Germans here, marooned like the rest of us for want of horses, and one of them had a working knowledge of surgery. Thanks to him and to a healthy constitution, I was at length able to hobble about, so I was doubly fortunate in having his assistance, and in being left in peace by the British while I was incapacitated. By now all but one of our original wagon complement had received horses and ridden away, and

[283] To form or discharge pus.

all the Germans except two were gone, and it began to look as if I was to become a professional camp-follower, for winter was upon us, and the arrival of more horses from the Free State was very unlikely.

A chance turn, however, brought me relief. One morning, soon after I was able to limp around, a small party of Germans rode into camp under command of a little hunch-backed Field-Cornet named Mayer.[284] At this time there was still in the field a German contingent about thirty strong, with Mayer at its head. In a sense they were the direct descendants of Baron von Goldeck's force, with whom I had served the year before; for although most of the Baron's men had melted away during the retreat to the Portuguese border, some of them had remained, and had collected themselves into a self-contained little commando forming part of General de la Rey's army.

Mayer had heard that there were a few of his compatriots stranded at the wagon-laager, so he had brought some spare horses to remount them. This stood me in good stead, for there were only two horseless Germans left in camp, the others having drawn animals in the lottery and ridden off, and Mayer agreed to let me have a little grey mare on condition that I joined him. The wound in my leg was still far from healed, but as I could not afford to miss the chance I made over my half-share in our wagon and team to the only remaining member of our original company; and with my leg in a splint accompanied the Germans when they set out on their return to de la Rey's forces.

It was a harsh journey. My leg throbbed and ached at every stride, and the winter had set in with great severity. I had not felt the cold so much in the seclusion of the valley where we had been camped up to now, but out in the open it was a different matter. By day clouds of dust and biting winds drove across the bleak plains, and at night we could hear the crackle of ice forming on the pools, as we lay shivering beneath our threadbare blankets. From now lay onward, indeed, until the end of the cold season, five long months ahead, we endured great hardship and suffering, for never, even in the memory of the oldest man, had there been so prolonged a spell of bitter weather all over South Africa.

For three days we rode on without meeting anyone. Another great drive was in progress with de la Rey's men hanging on to its flanks, the country behind being left wasted and ruined. At length we came on some of his commandos in the ridges near the village of Hartebeestpoort,[285] where they lay in attendance on a large concentration of English troops, said to number twelve thousand. Here we found the balance of Mayer's Germans, with whom we took post in the line that had been established in a half-moon before the enemy columns. For the moment the troops were resting, so we lay watching their camps, guns and convoys, waiting for them to make the first move.

They sent shells over us at times, but were obviously holding their hand until they were ready, an event to which we looked forward with considerable misgiving, as there were-only about six hundred burghers present, the last of de la Rey's men being elsewhere engaged.

But the General himself was with us and, happening to ride past one afternoon while my leg was being dressed by one of the Germans, he sent me off to a kind of field hospital that he had established some miles back. I found it in a ruined farmhouse, with a young Hollander doctor doing what he could for the sick and wounded under his care. It was a cheerless place, with only dried grass to lie on, and in the absence of medicines or bandages there was little enough comfort for the patients. They were mostly lighter cases, as the serious casualties were left for the British ambulances to pick up. Amid all the cruelty of farm-burning and the hunting down of the civilian population, there was one redeeming feature, in that the English soldiers, both officers and men, were unfailingly humane.

This was so well known that there was never any hesitation in abandoning a wounded man to the mercy of the troops, in the sure knowledge that he would be taken away and carefully nursed, a certainty which went far to soften the asperities[286] of the war.

[284] PF: Two possibilities listed: Mayer Wilhelm Volunteer from Germany serving with the Boer forces; and Klussmayer Volunteer from Germany serving with the Boer forces

[285] Hartbeespoort - Wikipedia

[286] The quality of being difficult or unpleasant to experience.

Kroonstad Refugee Camp Hospital.[374] Red Cross Nurses and their hospital wagon.[375]

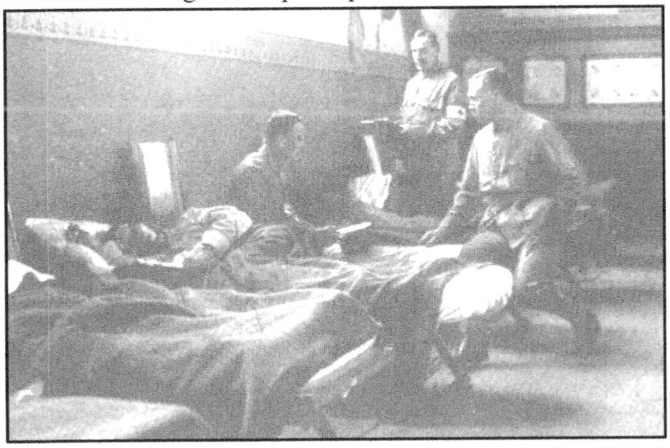

Dr. Conan Doyle, attending sick Canadians in the Club-house at Bloemfontein.[376] Dr. A. Conan Doyle in his tent at Bloemfontein, South Africa.[377]

Field Hospital at Orange River.[378] At the Orange River Hospital; a group of sisters and surgeons.[379] An ambulance wagon converted to a hospital tent.[380]

A few days after I joined the hospital we heard the sound of heavy gunfire about nine in the morning, and realized that the expected attack had come. I had kept my horse by me all the time, so I saddled and rode back to where I had left the Germans in the firing-line, while the doctor loaded his charges on a mule-wagon and made off to safer quarters. When I reached our men I saw the English attacking away to our left. Under cover of a severe bombardment their cavalry rode cheering towards a spot in our position a mile away, and broke through almost at once.

We of the German contingent were merely spectators, but we were close enough to see the men at the threatened point run for their horses and take to flight. Here and there a man went down, but our casualties were not heavy, as there had been scarcely any resistance in face of the shell-fire to which we could not reply, and when the burghers on either side of the breach saw what was happening, they too fetched their horses and beat a retreat. Naturally the Germans and I went as fast as the rest, to join the stream that flowed to the rear. We fell back until two in the afternoon, when the English gave up the chase and we were able to come to a halt in a bush-covered hollow, where General de la Rey addressed us in his half-humorous half-serious manner, and soon he had the men laughing and making light of their misfortunes.[287 288]

We rested our horses here until dark, and then rode west for several hours, as there was word of further columns converging upon us.

There had been a magnificent double-tailed meteor in the sky of late, the two streamers of which looked like the letter 'V,' and van Rensburg, the General's Prophet, had been giving out that this stood for 'Vrede' (peace), but on this night as we rode along, I heard a boyish voice from the darkness ahead call out: 'Mijnheer van Rensburg, that letter V up there does not mean Vrede, it means Vlug (retreat).' There was wry laughter in the ranks at this sally, which the discomfited oracle bore in meek silence, although it did not diminish his output of prophecies, which continued right up to the peace.

Great Comet of 1901.[289]

We spent a cold night beside a pan, and at daybreak could make out a large body of English troops in the distance, but we could not tell whether they were our pursuers of yesterday or a fresh column. In any case they were too far away to cause us any immediate anxiety, so we shot some oxen (a number from the larger communal herd having accompanied us) and began to prepare breakfast.

After this General de la Rey divided us into two parties. With half the men he rode south, and before sunset we heard the boom of guns thirty miles away, where the indomitable old man was once more attacking an enemy column.

He left Commandant Jan Kemp[290] in command of the rest of us, with orders to carry out a raid into the territory of Bechuanaland,[291] with a view to capturing a supply train on the Rhodesian railway line. Accordingly, for the next two days, we rode on through barren country, until we reached a point on the Harts River,[292] from which we were to make a night march to our objective. Mayer asked leave to scout in advance with his men, a request that was granted, so we started two hours ahead of our main body, riding through the Cunana Native reserve[293] until, at four in the morning, we crossed the British border.

[287] Probably this engagement: 6/1/1901 Severe engagement between General Dixon and Delarey at Vlakfontein, in the Magaliesberg. Enemy repulsed with heavy loss. Our casualties also heavy.

[288] Vlakfontein is a small settlement in the Mpumalanga Province of South Africa. During the Second Boer War it was the site of a guerrilla action against the British forces [Later on July 1st, 1901] where a Victoria Cross was awarded to William John English of the Scottish Horse for conspicuous gallantry. Vlakfontein - Wikipedia

[289] See: Great Comet of 1901 - Wikipedia Discovered April 12th, 1901. Reached Maximum on May 5th. The comet was readily visible to the naked eye until about May 20. And: File:Great comet of 1901.jpg - Wikimedia Commons

[290] Jan Christoffel Greyling Kemp (10 June 1872 – 31 December 1946) was a South African Boer officer, rebel general, and politician. Jan Kemp (general) - Wikipedia

[291] Bechuanaland Protectorate - Wikipedia Botswana - Wikipedia

[292] The Harts River is a northern tributary of the Vaal River, which is the largest tributary of the Orange River: Harts River - Wikipedia

[293] Probably Griqualand West - Wikipedia or Bechuanaland Protectorate - Wikipedia

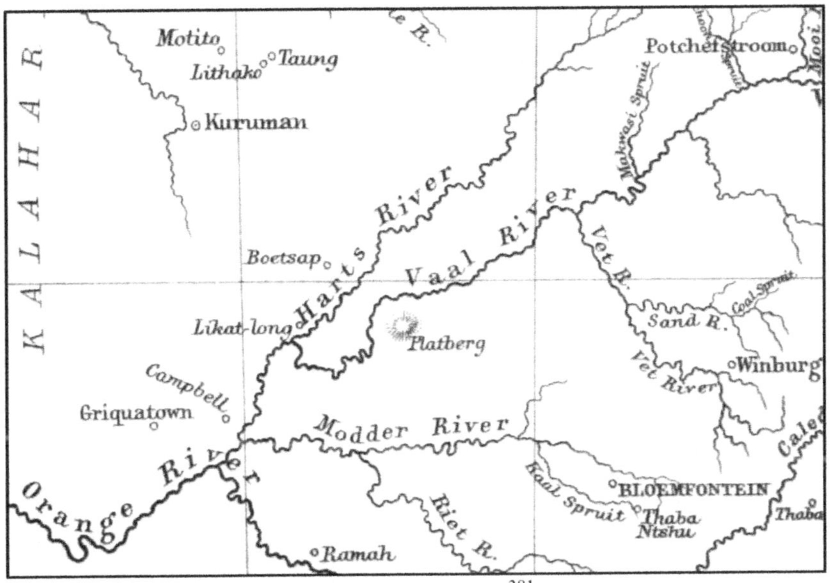

Harts River (Center) Northwest of Val river.[381]

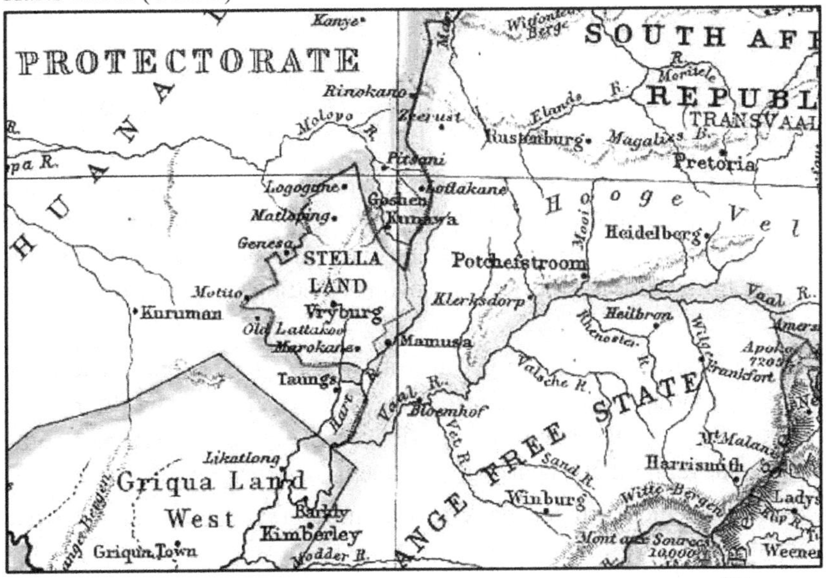

Map of Stella Land (center) and portion of Grinqua Land West.[382]

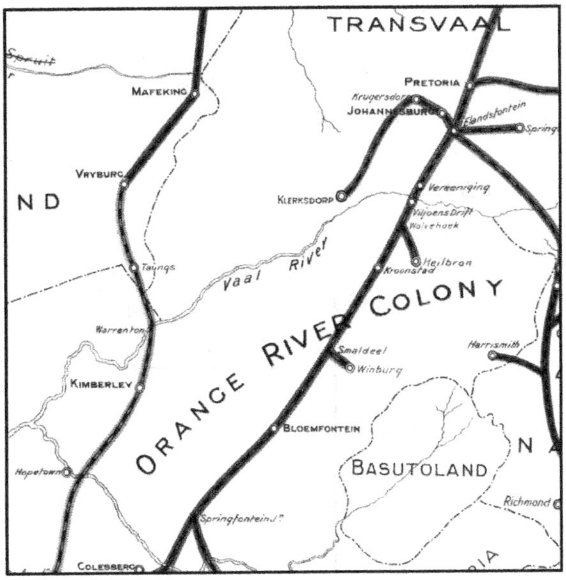

African Railway System 1901. Bechuanaland (on the left is cropped. The railroad through Kimberley is the Rhodesian line.[383]

The railway ran only a mile or so beyond, and we soon reached the metals. We started tearing up the track, with poles from a neighbouring fence for our only tools. Presently Mayer handed me a pair of pliers, and told me to climb up one of the telegraph standards and cut the wires, no easy task with my sore leg. As I was swarming up, there came a sudden volley from a culvert about fifty yards away. I slid down and in a moment we were on our horses, and away, for now that we were discovered there was no hope of surprising a train.

We rode back to a hill half a mile in the rear, to wait for daylight. When we got there we found a little native village nestling behind it, and went there to get information as to the number of soldiers guarding the line below. It was growing light as we rode through the gateway of the thorn enclosure surrounding the 'stad,' and when we entered, two khaki-clad white men rushed out of a hut, rifle in hand, followed by a native also armed.

Showing a circle of Zulu homes placed around a Kraal or cattle pen.[294] Building a Zulu Homestead.[384]

Mayer was off his horse at once, firing almost before he touched the ground. He hit the native through the chest, and the other two men put up their hands. This single shot caused a panic in the village. The natives grabbed their karosses,[295] mats, babies, and whatever else they could lay hands on, and fled down the slope in the direction of the railway line to seek protection from the British troops, the women setting up a long-drawn wail as they ran.

Mayer ordered me to head off the fugitives, so I galloped round to get in front of them. Unfortunately three or four of the Germans had lagged behind on the way up from the railway line, and, when they saw the rush of natives coming at them in the uncertain light, they opened fire, thinking themselves attacked. Before I could stop the shooting, they killed four and wounded several others, nearly getting me too.

We did what we could for the wounded and had them carried back to the huts, greatly upset at what had happened, for there were two women amongst the dead.

The white men captured in the kraal turned out to be renegades (National Scouts)[296] from Potchefstroom. They were handed over later on to General de la Rey, who had them both executed, I believe.

It had by now grown fully light, and we could see Kemp's men approaching through the bush. They were riding in small parties, making for the railway line, so we sent off a man to warn them that the English were there, but before the messenger got halfway, shells came screeching through the air from a hill beside the line. So far from expecting to be shelled, we had believed this part of the railway to be unguarded, but we could now make out quite a considerable number of tents among the trees and, as the gun kept dropping shells near us, Kemp ordered a withdrawal, and we trekked back into the Transvaal after a profitless venture.

We rode thirty miles that day, reaching Leeuwpan[297] by nightfall, where we found General de la Rey awaiting us. He said there were several English columns halted in a semicircle ahead, and he instructed Kemp to get out

[294] The kraal is made of the smaller branches of trees trimmed and placed in the ground, then bound together near the top, forming a strong hedge. It is covered with clay and grass mixed together and then thatched over with long grass. South Africa; its History, Mackenzie 1899

[295] Kaross: A blanket of softened skins, used both as a cloak and (now more usually) as a covering for a bed or floor.

[296] The National Scouts were mainly recruited from Boers who had surrendered to the British or chosen to fight for the imperial forces. As 'Handsuppers' [Hands Uppers] and 'Joiners' they were despised and ostracised by the 'Bitterinders' and their womenfolk. So much animosity abounded that a separate Afrikaan church, the 'Kruiskerk' was established shortly after the war to accommodate them 1595 records https://www.angloboerwar.com/?option=com_content&view=article&id=434

[297] Near the headwaters of the Hart's River. And: Leeuwpan coal mine - Wikipedia

of their way that night. Having told us what to do, he rode away with his handful of retainers, and I did not see the doughty old warrior again, for from that night our roads lay far apart.

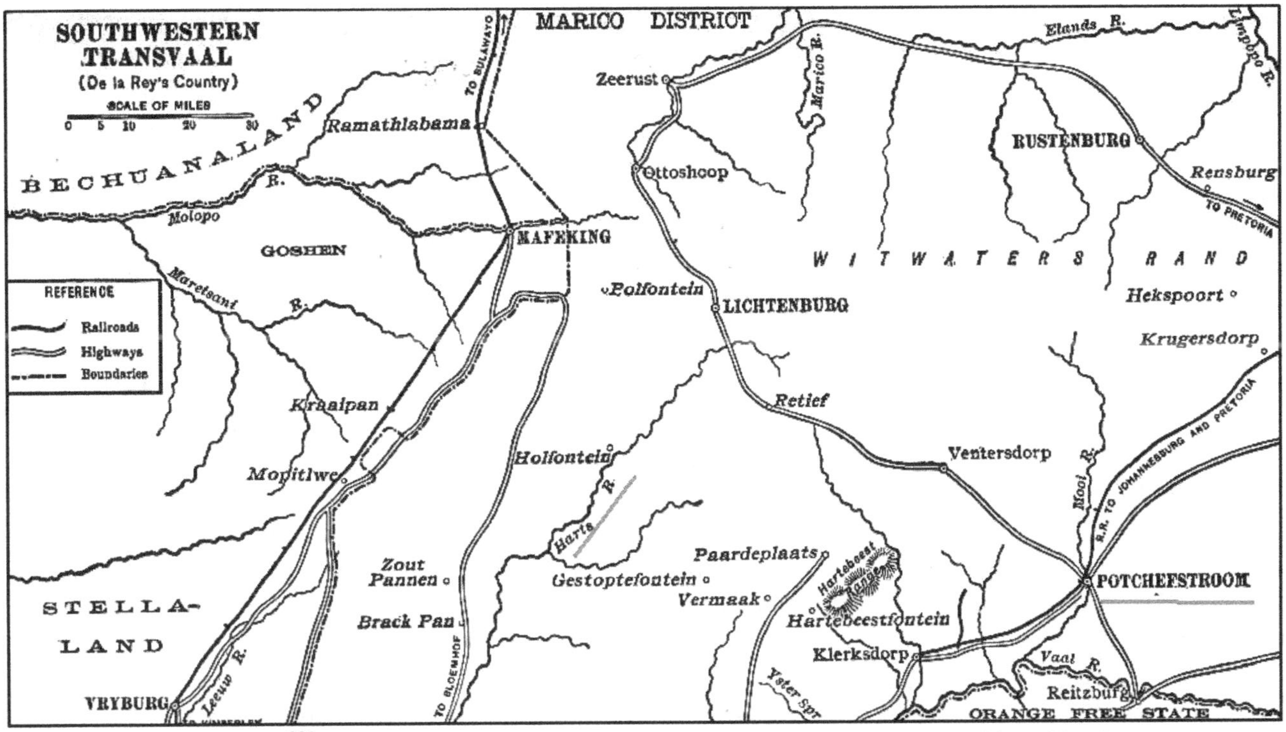

Southwestern Transvaal.[385] They are somewhere in this area. Pontchefstroom, Harts River Headwaters.

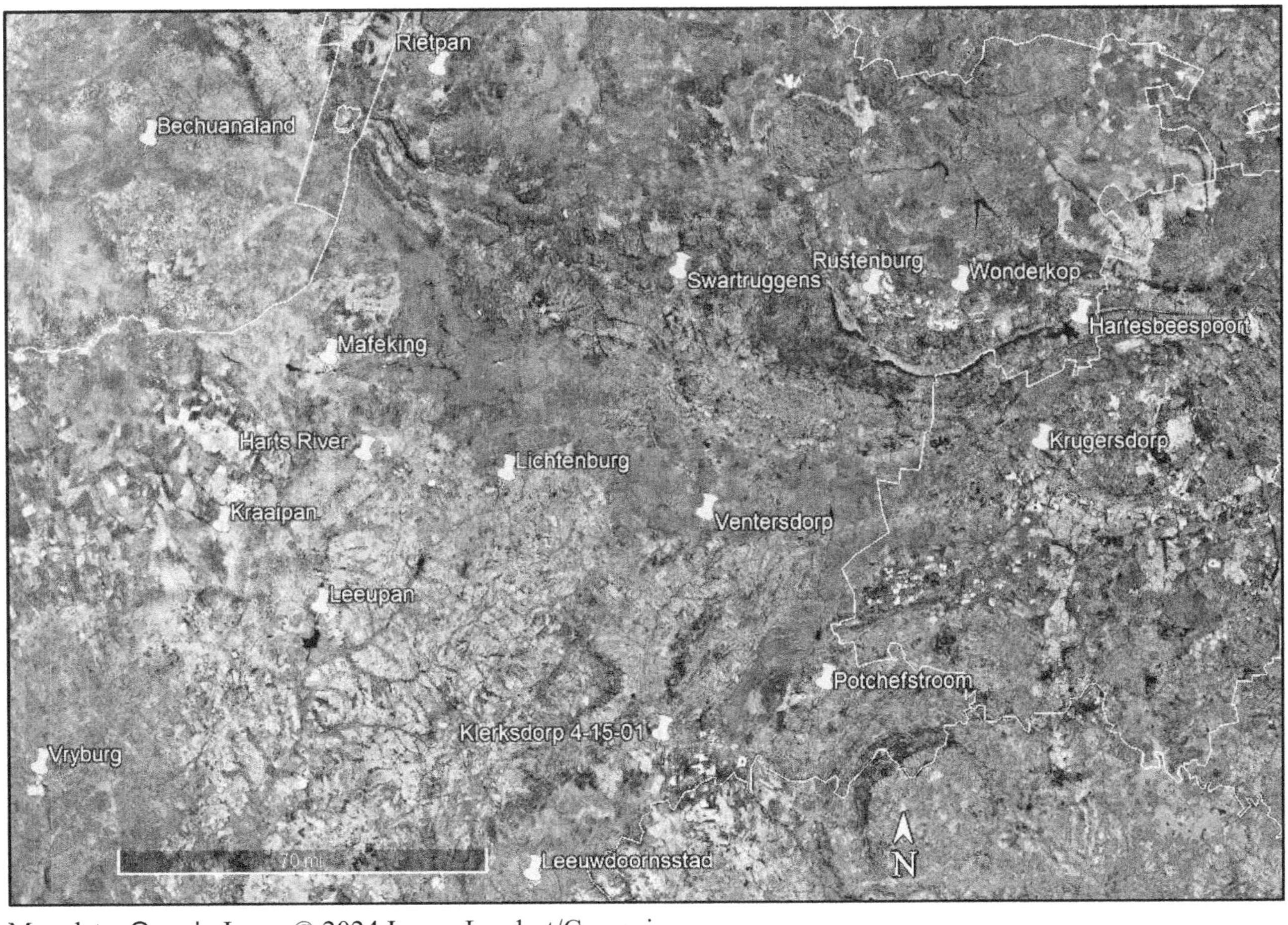

Map data: Google Image© 2024 Image Landsat/Copernicus.

XVIII. The Next Stage

As soon as it was dark enough, Kemp led us off in the direction of the English camps, between which he intended passing. We were skilfully guided, for we slipped through a gap between two camps, passing so close that we could hear the murmur of voices, and could see the forms of soldiers outlined against their fires. No alarm was raised until we were through, when there was some shouting and firing which did no damage. Once we had put the camps behind us the men were ordered to march on foot to spare their horses, for we had been on the move for over thirty hours, with no rest worth mentioning.

As I still had a limp, I gradually fell behind; and to make matters worse, my poor little mare was delivered of a still-born foal. With this travail coming upon her she had borne the long treks so unfalteringly, that I had not even known that there was anything wrong with her, but now her strength was gone. After a while she staggered to her feet, and as I could not risk remaining in too close proximity to the English camps, whose fires were still visible in the distance, I led her slowly forward.

By this time the rest of our men had long since vanished in the darkness, and I had to plod on alone for an hour or two, dragging my horse behind me, until she could go no farther, when I decided to halt till morning. It was bitterly cold—so cold that earlier in the evening I had heard the men say that it was the coldest night they had ever known. As I could find no fuel for a fire, I wrapped my blanket round my shoulders and sat with chattering teeth until sunrise. When it grew light I found myself on a cheerless expanse, with a view that extended for many miles, but there was no sign of the commando.

By a distant thorn tree, however, I found four of my German friends, huddled together against the cold. They said that they had missed me during the night, and, knowing that I was crippled, they had generously remained behind to wait for me.

After collecting what fuel we could in such barren country, we built a small fire to fry some meat, and then set out on the spoor[298] of the commando, making slow progress, for my companions' horses were not in very much better condition than my own. By nine or ten o'clock in the morning, ominous pillars of dust rising in the rear warned us that the English columns of last night were turning in our direction.

Troops of one of the British mounted regiments on the move.[386]

The troops were not as yet in sight; but considering the state of our animals, we stood a poor chance of keeping ahead of them once their scouts topped the skyline, so we hurried on as best we could. Just as we were beginning to see an occasional horseman far behind us, we providentially came on the Harts River. It was more of an earth crack across the plain than a river. No trees stood on its edge, and fifty yards off the banks were invisible, but it was our salvation, for hardly were we and our horses out of sight in the dry bed below, when the troops came swarming towards the river. All went well, however. The English, when they reached the bank, set to digging gradients for their guns and wagons, and, although it was hours before they got their transport through, during which time we anxiously peered over the top in fear of discovery, we had the satisfaction before dark, of seeing the tail of their convoy vanish over the horizon.

[298] The track, trail, droppings or scent, of an animal.

This was good as far as it went, but the question was how to catch up again with our own people, though the fact of being temporarily cut off was not of vital importance having regard to the fluid nature of guerrilla warfare, and we were not greatly troubled on that score, our worst anxiety being the weak state of our horses.

For the next three or four days we toiled on behind the enemy, who in turn were following our men. We kept some distance to the rear, only moving when the dust-clouds ahead of us showed that the English were advancing too, and in this manner we crawled along on foot, leading our horses by the reins.

One evening a burgher rode up, the first we had seen since the night we dropped behind. He told us that General de la Rey had dispersed his commandos into smaller bands, owing to the pressure from converging British columns. These bands, he said, were by now scattered all over the Western Transvaal, waiting for the end of the drive.

After taking a look at our animals, he said he thought we stood a poor chance of finding Mayer and the other Germans, then he rode off on business of his own, leaving my companions and myself to digest the information received. They were for attempting to work north to the milder climate of the bush-veld, in order to escape the infernal cold of these open plains. They said, as far as Mayer was concerned, we could always find him again once the winter was over, and in any case it mattered little if we didn't.

Seeing that they attached no particular moral value to the necessity of rejoining our unit, I now sprang on them my long-cherished scheme of making for the Cape Colony. They stared at me in surprise when I first broached the subject, but after I had explained my views, and had pictured the Cape to them as a land of beer for the taking at every wayside inn, they became eager converts, and we agreed to start without delay.

The four Germans were a mixture The eldest, Herman Haase,[299] was a man of about forty-five, in looks the typical sausage-eater of the English comic papers, but, as I found out, a kindly, good-natured gentleman, a Johannesburg merchant, who had been in the field from the beginning. He was the last man one would have suspected of a liking for war, as his talk was all of his wife and family, and the joys of home life.

Next came W. Cluver, a clever, cynical Berlin student, who told me many interesting things of life in the old world; then there was Pollatchek, also a Berlin student, who had come out to fight for the Boers, as on a crusade. He told me that his initial ardour had long since evaporated, but he liked the life of adventure and so had remained, a pleasant, cheerful fellow whom I grew to like very much.

Lastly, there was a farmhand named [Heinrich] Wiese, a clumsy, slow-witted rustic, but brave enough. With these four men my lot was now cast. Wiese and Cluver did not get very far, but with Haase and Pollatchek I was long associated, although they turned back in the end.

Our preparations for going to the Cape were quickly made. We slaughtered a stray sheep, and cut the meat into strips for drying in the wind (as we had no salt), and we ground a quantity of maize into meal in a small coffee-mill that Haase carried on his saddle-tree, and next morning we started.

On the evening of our first day out, we had an exciting interlude. Another of de la Rey's men who, like ourselves, had been left stranded in the rear of the British drive rode up to see who we were. He said he had been lurking around here for some days, seeking an opportunity to regain the commandos. While stripping mealie cobs in a field that afternoon, he had seen a native ride from a 'stad' close by to meet an English patrol. The native had conferred with the officer in command for some minutes, after which the patrol went off. As the native was obviously a spy in British pay, the newcomer suggested that we should try to capture him after dark. Accordingly, about nine o'clock that night, we started on foot for the kraal, going quietly so as not to frighten our quarry. When we reached the village, however, we found the headman and all his followers on the 'qui vive.'[300] They denied all knowledge of the spy, so our guide cut matters short by seizing the Induna[301] by the throat, threatening to shoot him. On being thus roughly handled, he wrested himself free and shouted loudly, 'Help, help, the Boers are killing me! In a moment we were surrounded by thirty or forty braves, most of them brandishing assegais[302] and knobkerries in our faces.

They began leaping and dancing about us, uttering fierce yells and menaces, while some tore thatch from the huts, lit it at a fire, and held it aloft instead of torches. From beyond the circle of angry savages came the cries of

[299] Possibly: Haase Adolf, Volunteer from Germany serving with the Boer forces
[300] A state of heightened vigilance, on the alert.
[301] A South African tribal counselor or headman (under the king); someone in authority, a boss.
[302] A slim hardwood spear or javelin with an iron tip, especially those used by Bantu peoples of Southern Africa.

the women, urging them to kill the white men, and things began to look unpleasant. We could have opened fire, but the space inside the stockade was cramped and they were far superior in numbers. Moreover, all through the war the Boers had observed an unwritten law that it was a white man's quarrel, and that the native tribes were to be left alone. For these reasons we did not fire, but closed up, and slowly backed out through the gateway by which we had entered. Once free of the enclosure we had no trouble in safely reaching our camping-place, much annoyed at the poor figure we had cut. To add to our humiliation, at dawn next morning we saw a native boldly riding from the 'stad' and when we sent a few shots after him, he waved a defiant arm and disappeared over the rise.

Having thoroughly bungled the affair, de la Rey's man took his leave of us, and we continued on our way.

For several days we were unable to travel in a direct line, for we found the countryside alive with British troops moving in all directions, and we calculated that we saw twenty-five thousand of them before we got clear. It was plain from the way in which they swept forward on an enormous front that they were conducting another of their drives, but as we did not see a single burgher, or the vestige of a commando during all this time, they must have had little to show for their activity. Clearly General de la Rey was resorting to his usual tactics of avoiding these huge concentrations of troops by scattering his men until the blow was spent.

My knowledge of veld-craft brought our party safely [southeast] through to the Vaal River, for my early experience was of value, and we threaded and twisted successfully between the enemy columns, never having occasion to fire a shot. Once we were held up for half a day while a body of English troops camped within hail of where we lay hidden in a patch of thorn. Another time Cluver and I tried to ambush two officers, but he showed himself too soon and they got away. In the course of these operations we had to jettison Heinrich Wiese. His horse gave in and he himself had blistered feet, so we abandoned him near an English column, where he was sure to be picked up and cared for. My leg was on the mend, but we suffered a great deal from the cold at nights. Otherwise we almost grew to enjoy the excitement of dodging the enemy forces and patrols, and the Germans said that it was the best time they had had in the war.

At length, on the fifth or sixth day, we breasted the long rise at Leeuwdoorn,[303] from which the country slopes down to the Vaal River, and we saw the wide plains of the Free Slate stretching beyond. We slept a night in an unburnt farmhouse on the Transvaal side, and next morning, as we were riding off, we saw a body of English approaching, so we climbed a kopje to see what their plans were.

The soldiers made for the farm we had just vacated, and soon smoke and flames were issuing from door and windows. As we looked on, two old fellows rode up from the direction of the Vaal River, and joined us on the hill. They reminded me of my former commander, General Maroola, and his brother, for they both wore rusty bell-toppers, and the tails of their ancient claw-hammer coats flapped in the breeze as they came. With a curt greeting they dismounted and sat down on the rocks, silently watching the work of destruction below. for a long time neither of them spoke, and it was only when the roof fell in amid a shower of sparks, that the elder of the two sighed and turning to the other said: 'Brother John, there go those teak-wood beams I brought from Pretoria after the Jameson Raid.' This was his sole comment on the loss of his home, then the couple remounted their horses to ride back to the river.

As we were also going there, we fetched our animals and overtook them, learning on the way that they were making for a women's laager halted beside the water.

We found that the laager consisted of about thirty wagons, with perhaps three or four times that number of women and children, all under the care of our two worthies. Now that the British were capturing the civil population it had become the practice for the women on the farms, when hostile forces approached, to load what they could on their wagons, and join hands with others similarly situated, in order to form a joint laager. Once the immediate danger was past they returned to what was left of their homes, to subsist as best they could, rather than be taken for internment to the British concentration camps.

[303] Leeudoringstad (Afrikaans for Lion thorn city) is a small farming town situated on the main Cape Town - Johannesburg railway line in North West Province of South Africa. Leeudoringstad - Wikipedia

"Outside the women's laager at Mafeking."[387]

We rested at the laager for a few hours, while one of the old church-wardens rode out to see what the English were doing. He came galloping back to say that about fifteen hundred English horsemen were on the river bank eight miles higher up. In view of this it was decided that the laager must cross through to the Free State shore, and immediately all was in a bustle.

There was a ford of sorts close by, over which we helped the women to get the wagons, but it was pitiful to see them standing waist deep in the icy water, tugging at the wheels, and urging on the oxen in their anxiety to put the river between themselves and the column. After working hard for nearly two hours, we had the whole laager safely over.

Ox wagon stuck in the Crocodile River.[388]

We then waded back to the Transvaal bank to fetch our horses, and on our return found the wagons under way, making across the unlimited open country of the Free State that stretches southward for hundreds of miles. The Germans and I remained resting under the trees to dry our clothes, and slept the night a few miles farther on.

In the morning the Englishmen came moving down the river, and by nine o clock they were fording the stream into the Free State, so we made off, and for the next three days rode leisurely along, following the south bank of the river, and warning the women on the farms, as we passed, that the troops were behind.

A big horse train crossing the Wilge river. A wagon is overturned in the bed of the river.[389]

I took this route as I had decided to make for the neighbourhood of Hoopstad,[304] a village and district that I knew well, for my brothers and I had camped and hunted here as boys.

I counted on finding the herds of semi-wild horses that used to frequent the river, that we might remount ourselves, for there was small chance of the horses we rode being able to last out the bitter winter that was upon us. We found the homesteads along the river bank intact, but as the bulk of the male population of these parts had been captured with old General Cronje at Paardeberg more than a year ago, the farms were for the most part tenanted by women and children. They told us that an occasional British column had marched through during the past six or eight months, but thus far the policy of farm-burning and the removal of the civilian population had not been put into operation here.

This happy condition was now coming to an end, for pillars of smoke were rising far behind us, and at night the sky was reddened with the glare of burning homesteads, to tell the unfortunate inhabitants that their long immunity-was over.

Farms set alight.[390]

[304] The town of Hoopstad is situated at the intersection of the R34, R59 & R700 in the Free State; an area of maize-producing. The town is located near the Bloemhof Dam which is situated on the Vaal River. Hoopstad - Wikipedia

The women took the matter bravely, although there were tears and weeping at times, but each family, as soon as they realized the danger, fetched the oxen, inspanned their wagon, and trekked away south across the plains, out of harm's way.

At length we came to Hoopstad town.[305] The place was deserted, the English garrison, for some unexplained reason, having set fire to their stores and marched away two days before. That same afternoon a solitary German trader rode in from the south-east, where he was serving with a small Boer commando. He proved a useful ally, as he showed us a quantity of maize concealed in an underground receptacle, on which we fed our horses for some days, enabling the wretched animals to pick up condition a little.

Our new acquaintance said he must now return to his commando, and he suggested that we should accompany him. He said that the horses running on the veld were so shy that, unless we got assistance, we should never succeed in catching any of them, as all fences were down throughout the district. We agreed to go with him and, travelling over endless rolling plains, within a few days reached the force to which he was attached. This consisted of twenty men under an officer of the now defunct O.F.S. Artillery.[306]

Corps of the State Artillery, Orange Free State. Telegraphists, heliographers and artillerymen commanded by Lieut. Baay. Commando of genl. Olivier outside Colesberg.[391]

They made us welcome, but shook their heads when we mentioned horses, and indeed we found next day that we might as well have tried to catch antelopes. There were lots of horses roaming about, but when the men spread out, the mustangs went racing away, manes and tails in the wind, and, in spite of our endeavours, we failed to capture a single one of them. It appeared that the British troops had been firing on them with machine-guns and rifles, ever since they discovered that General de la Rey was getting remounts from here, which accounted for their bedevilment. In any case they were too fleet for us, and, as there were no wires to stop them, we gave up the business for fear of foundering the horses we had.

We spent a week with our friends. They were a ragged crew carrying a queer assortment of weapons, and what little ammunition they had they reserved for shooting game, which abounded, and they seemed quite happy so long as they could keep out of the hands of the British.

They gave us unfavourable news of the grazing to the south, so I determined to make for the mountain country farther east, where, I was told, the grass would be better, and where we might procure fresh horses, as de la Rey himself had latterly been sending thither in search of remounts.

[305] The town of Hoopstad is situated at the intersection of the R34, R59 & R700 in the Free State Province. This area is considered to be the richest maize-producing district in South Africa. The town is located near the Bloemhof Dam which is situated on the Vaal River. Hoopstad - Wikipedia

[306] Orange Free State Artillery.

So, when we had shot enough springbok and blesbok[307] for biltong, the three Germans and I set out once more. On the night after we started, a bitter cold wind drove across the plains. We were halted beside a water-hole without shelter for man or beast, with the result that, towards morning, my little grey mare broke loose, maddened by the pelting earth and pebbles. She fled down the storm and I never saw her again.

One of our late friends had given Pollatchek a horse, which he made over to me, so the loss of my own, serious as it was, did not prevent my going on. Having ridden out the blizzard without further damage, we pushed on going roughly east by south. We passed Kopje-Alleen,[308] where my brothers had been ridden down with the 'A.C.C.' last year.

It is a curious isolated hill visible for sixty miles around. As a boy I had sat on its summit watching the game-covered plains below, and now, as then, great herds of antelope and troops of wildebeest were grazing at its foot.

I climbed to the top, partly for old times' sake and partly to see whether the land was clear, but there was really no need for anxiety, as we were in empty country.

The farmhouses stood abandoned, the fields lay unploughed, and we saw neither human beings nor domestic animals, even the natives having fled.

We rode through this unpeopled waste until, some days later, we fell in with a party of Freestaters near the Sand River railway bridge.[309] There were nine of them under a Field-Cornet named Botha.[310] They had been trying to derail a train, but the English were building block-houses along here and their attempt had failed, so Botha and his men were returning to their own haunts in the mountain ranges to the east. As we were bound for that region ourselves, we joined them.

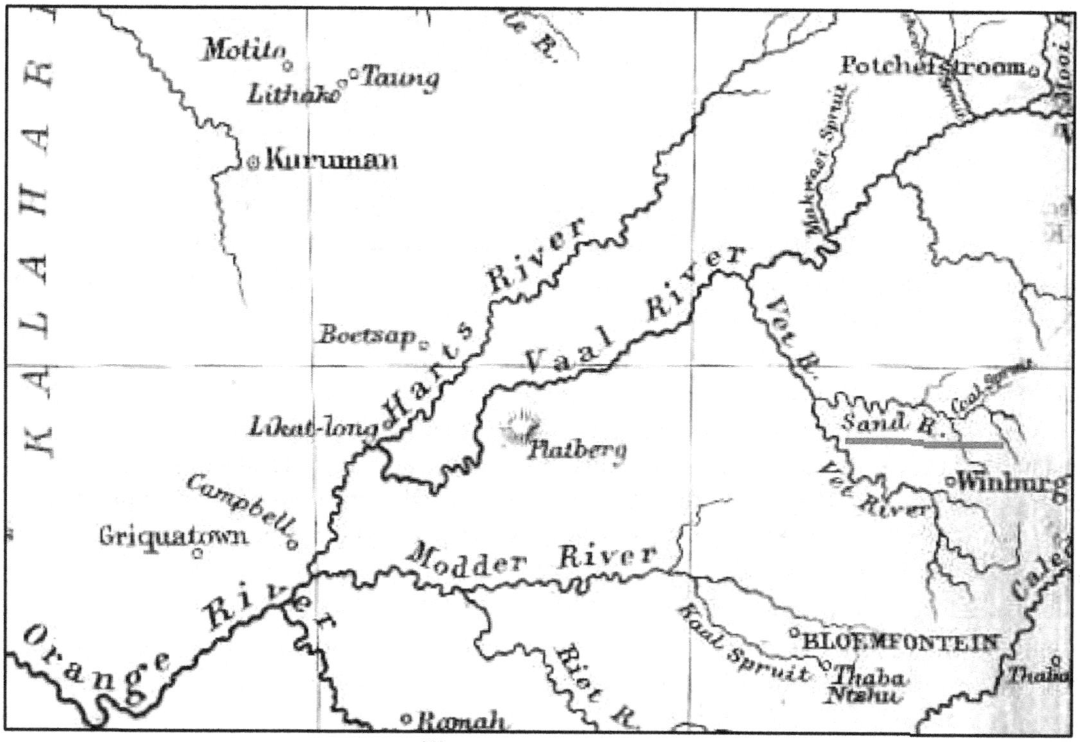

The Sand River is north of Winburg.[392]

[307] The blesbok or blesbuck is a subspecies of the bontebok antelope endemic to South Africa. It has a distinctive white face which inspired the name, because bles is the Afrikaans word for a blaze such as one might see on the forehead of a horse. Blesbok - Wikipedia

[308] Riebeeckstad Riebeeckstad - Wikipedia

[309] Near Virginia, South Africa: Virginia, Free State - Wikipedia

[310] Likely to be: David Hercules Botha Veldkornet Kroonstad Commando and General Smuts Commando.

Map data: Google Image© 2024 Image Landsat/Copernicus.

At eleven o'clock that night we set out to cross the railway line (the main line coming up from Bloemfontein to Johannesburg). The cold was intense and the darkness so thick that those in front had constantly to shout and whistle to give those behind their bearings. After we had been going for some hours we found that Cluver was missing. We called and whistled and fired shots, but got no reply, so Pollatchek and I rode back to look for him. We retraced our steps for more than a mile, and then, as the directing shouts of the others were growing fainter, we reluctantly rejoined our companions. The trouble was that Cluver was subject to epileptic fits. Three or four times since being with us he had fallen into convulsions, and we supposed one of these attacks had come upon him while he was lagging behind. The night was so freezing that to lie unconscious on the ground meant certain death, but Field-Cornet Botha said it was essential to cross the block-house line before it drew light, and, as we were still a long way from the point at which he intended to slip over, he insisted upon our pushing on, so we continued our journey, somewhat heartlessly I admit, but we could not afford to lose the chance of getting through under the guidance of men who knew the lie of the block-houses.

Towards four in the morning we reached a hollow between two fortified posts, and crossed the line without any trouble. We never heard of Cluver again, and I have no doubt that he was dead by morning, for he was not strong enough to have survived that winter night in the open.

Native village taken over by the British Army unit as a picquet post.[393] Blockhouse showing wire fence.[394]

By sunrise we were well into the foothills beyond the railway, and now for two days we rode east into increasingly mountainous country, until we reached a lonely farm lying high up amid the crags near Wonder Kop.[311] At this place Botha and his men had their lair. From here they made periodic raids down to the plains, and they must have given a good account of themselves, judging by what they told us, and by their horses, weapons, and equipment, all of which were of British origin. Small private bands like this were common enough-in the Free State nowadays, remnants of larger forces that had dwindled away under the misfortunes of war.

Botha told me that he had at one time commanded over three hundred men. Many of them had been killed or captured, others had surrendered, and others again had left him to join larger commandos in the north. He clung to his title of Field-Cornet, and as two of his men called themselves Corporals, the rank and file of his army consisted of but six men. We had reached a very snug haven. The owners of the farm were gone, but there was plenty of good clean straw in the barns to lie on, and in a cave in the cliff over-hanging the-homestead was a store of wheat, a welcome change from our eternal diet of maize, while in the valleys below half-savage pigs were to be had for the shooting.

Moreover, there was a huge copper cistern, and, as water and fuel were plentiful, we could prepare a hot bath as often as we liked, a rare luxury in this freezing weather.

For ten days the two Germans and I revelled in the unaccustomed joy of good food, cleanliness, and comfortable sleeping quarters, but then I became restless once more. I succeeded in converting Field-Cornet Botha and his followers to my scheme of raiding into the Cape Colony They were at first disinclined to move so far from their beloved mountains, but eventually I swung them round, and about the end of June (we were vague as to dates and time) we started down the mountains and headed due west, intending to recross the railway line to the plains beyond, and then make a wide detour round Bloemfontein to strike the Orange River somewhere near Ladysmith.[312] On the morning after our departure from the farm, as we were descending the pass near Breslersflat, a young fellow named Jacobus Bosman[313] came riding up. When we told him that we were going to the Cape he said he would come too. As he was one of the Cape rebels who had joined the Boers during their temporary occupation of Colesberg in the beginning of the war, I advised him to stay where he was, for if he were captured on British territory, it would go hard with him. He said he would take the risk, so we enlisted him, but my warning was justified, for he was taken and hanged, as will be seen later on.

[311] Not this place. Too far away. Wonderkop is a town in Bojanala District Municipality in the North West province of South Africa. Wonderkop - Wikipedia

[312] Unable to locate this 'Ladysmith' near the Orange river.

[313] Possibly: Bosman Jacobus Johannes Prisoner number: 29507 Captured: Straithnairn 27 Feb 1902 Sent to: India Age: 21 Address: Vechtkop Source: Anglo Boer War Museum 2016

"Where General French showed his mettle: View of Colesburg with Coleskop in distance (re-occupied by General Clements), Feb. 27th, 1900."[395] Coleskop.[396]

Suffolk Hill near Colesberg. On January 6th, 1900 four companies of the 1st Suffolk Regiment advanced by night against a low hill commanding the road to Colesberg Bridge. In addition to the many killed and wounded, seventy were forced to surrender, including seven officers.[397] Graves of men killed in the Suffolk Disaster,[398] Coleskop in the background.

After three days of steady progress we were back on the open plains, within sight of the Bloemfontein-Johannesburg railway line, and we scouted round for a suitable crossing. This was becoming more and more difficult to find, now that the English were perfecting their block-house system.

As I spent the rest of the war roving the Cape Colony, I did not experience the full effect of this network, but I have heard that it caused the Free State and Transvaal Commandos great trouble, and in the end contributed largely to the break-up of the Boer resistance.

There were working parties of soldiers dotted along the railway track, engaged in putting up these block-houses, but we had no difficulty in galloping across the metals, despite a fairly heavy rifle-fire, and, having safely negotiated the line, we rode on, passing not far north of Brandfort village. Towards dark we came on a field of maize into which we turned our weary horses, and here we spent the night.

Next morning, just before sunrise, we heard the crack of distant rifle-shots, and shortly after two burghers rode by. They pulled up to say that an English column was coming our way, and they advised us to go along with them. They were making for a women's laager not far away, to warn them of the enemy's approach so we caught our horses and rode off. one of these men, Piet Marais, was an old acquaintance whom I had known as a compositor in a newspaper office at Bloemfontein. He had since inherited money and had bought a farm nearby, which he and his companion had been about to visit when they came on the advance-guard of the British force.

As the sun rose we were able to see the column crawling along the road some miles behind. We judged them to be fifteen hundred horsemen, with a string of wagons, and several guns. The women's laager consisted of some fifty wagons and carts, with over two hundred women and children, the collective non-combatant population of the district.

"Inside the women's laager at Mafeking."[399]

On hearing our news, they began to pack up; the poultry were run to earth, children and household effects were rapidly stowed, and in a very short time the laager was moving off. Women trotted alongside the teams with whips and quirts,[314] while the children peered out anxiously from beneath the hoods at the dust-clouds in the distance, which betokened the approaching enemy. Fortunately the going was easy on these undulating plains, and the wagons made good headway in the direction of the Vet River, lying ten miles north. [Of Brandfort]

By now other men were riding up in twos and threes from various points, until we were about forty strong, and we began to practise our old tactics of galloping across the front of the English advance to fire a few shots, and then falling back to repeat the process farther on, as soon as the shell-fire grew too hot.

This served its purpose sufficiently well to give the laager time to get away, and after a few hours we had no fears for their safety.

Towards three in the afternoon the English halted. We could see them turning out their animals to graze, so we rode up to a kopje near by, where we lay resting under the trees while our horses ready-saddled cropped the grass close by. Before long we heard a clatter, and jumping to our feet, saw a troop of mounted soldiers coming at a gallop.

Our look-out had not troubled to keep awake, and the horsemen were within six hundred yards of us before we heard them. As they had two pom-poms with them, we loosed only a single volley, and then ran for our horses.

Maxim Machine-Gun (the "Pom Pom").[400 315]

Pom Pom Maxim[316]

[314] A rawhide whip plaited with two thongs of buffalo hide.

[315] During the Second Boer War, the British used Maxim machine guns, and the Boers used a modified Maxim, a belt-fed, water-cooled machine gun that fired explosive 1-pound (0.45 kg) rounds (smokeless ammunition) at 450 rounds per minute, which became known as the "pom-pom." Maxim Nordenfelt Guns and Ammunition Company - Wikipedia And: See also this 15 minute video on the Forgotten Weapons channel describing the weapon. Maxim Pom-Pom 37mm Machine Gun - YouTube See Also: QF 1-pounder pom-pom - Wikipedia

[316] Identified is 210 Private John McCracken (left) and 320 Private Lucian Franklin Hayward. The centre soldier is unidentified, but believed to be Private William James Neild. All three were involved in the capture of a Maxim gun; for this action Hayward was awarded the Distinguished Conduct Medal (DCM) and McCracken and Neild were both mentioned in despatches. Hayward's name is wrongly entered in the records as Lessian." File:AustralianTroopsWithQF1pounderBoerWar.jpg - Wikimedia Commons

Rifle bullets were soon spitting about our ears and shells were bursting farther on, but as we were riding wide no one was hit, although once, when we were bunched together at a gap in a fence, a shell pitched right among us and killed a led-horse, whose blood was spattered in my face.

We spread out once more, and sprinted for the shelter of the Vet River banks, two or three miles beyond, and here we gave our winded animals time to recover. The British made no further attempt to advance that day, realizing perhaps that we had the heels of them. After satisfying ourselves of their intentions, we crossed over the north bank by a bridle-path, to build fires and cook a meal, for we had eaten nothing that day, and after dark we trekked down-stream for a few miles, to be in safety. Burghers, who had ridden away earlier to see how the women's laager was getting on, returned to say that all the wagons had escaped, so we spent an easy night under the thorn trees beside the water.

At sunrise yesterday's English approached as far as the river, where they came to rest. Our little party from the mountains thought that we had come far enough out of our course, so we decided to double back behind the column in order to resume our journey southwards. We took our leave of the Freestaters after hearing from them of an insurmountable barrier that we should meet at the Modder River,[317] where, they said, the English had a line of fortified posts that would stop our progress to the south.

Pontoon Bridge on Modder River 1899.[401]

Recrossing the Vet River, we rode back along the way we had come the day before, and the soldiers halted along the stream did not even send a shell at us when we skirted round their camp.

We were getting short of ammunition, so during the next two days we followed the road by which the English force had travelled, to pick up Lee-Metford cartridges. The English soldiers were notoriously careless with their ammunition. If a round or two dropped from their bandoliers they would never trouble to dismount, as they knew they could get more, and at their halting-places one could almost always find cartridges lying spilt in the grass. So much was this the case that latterly it had become a regular practice to trail the columns, sometimes for a week on end, to glean these crumbs from the rich man's table, and I doubt if the British ever realized to what an extent the Boers were dependent upon this source of replenishment.

We followed the road back until we came once more in sight of Brandfort village, by which time we had picked up nearly a hundred rounds. Satisfied with this we now turned west, and then south-west, towards the Modder River.

[317] The Modder River is a river in South Africa. It is a tributary of the Riet River that forms part of the border between the Northern Cape and the Free State provinces. Modder River - Wikipedia

XIX. Farther South

We rode on steadily for three days, making a bend so as to strike the river midway between Bloemfontein and Kimberley. Nothing of interest happened, save for a brush with an English patrol that suddenly bore down on us one morning out of the mist.

The light was so uncertain that both sides drew off, after firing a few shots, for neither party could make out how strong the other was.

We passed the 'dorp'[318] of Bultfolltein[319] on our way and looked in for what we could find, but the place was gutted. For the rest we rode over interminable plains devoid of human beings. We did not see a single homestead that was not in ruins, and at some places lay hundreds of sheep clubbed to death or bayoneted by the English troops, in pursuance of their scheme of denuding the country of live-stock to starve out the Boers.

Boer family photographed outside their house.[402]

Destroyed Farmhouse.[403]

In spite of this there was generally an odd sheep or two straying about, and there were plenty of springbok besides, so we did not lack meat. Nevertheless we were leading a pretty hard life. Bread and salt, soap, tobacco and books were things of the past. We had almost forgotten the taste of tea, coffee, sugar and vegetables. If a man was not lucky enough to possess a tinder-box,[320] he had to expend a valuable rifle-cartridge every time he wished to light a fire.[321] It was midwinter, with ice on every pool, and we went in tattered clothing and slept under threadbare blankets at night.

We crossed the Modder River at midnight above Abrahamskraal, and daybreak found us facing the line of military posts we had been told of, that the English had established all the way from Bloemfontein to the Diamond Mine at Koffiefontein,[322] a distance of sixty miles. This cordon was formed to guard thousands of sheep and cattle taken from the farms. Each post consisted of ten or twelve tents, at intervals of perhaps two thousand yards, and as our horses were in too poor a condition to make a dash for it, we turned westward, seeking a convenient spot at which to slink by without attracting attention. We could see patrols of cattle-guards riding about, but they probably took us for their own men, and for a while we went in peace.

Blanket Tents.[404]

[318] Town or Village.
[319] Bultfontein - Wikipedia
[320] A tinderbox, is a container containing flint, firesteel, and tinder (typically charcloth, but possibly a small quantity of dry, finely divided fibrous matter such as hemp), used to help kindle a fire. Tinderbox - Wikipedia
[321] By using the gunpowder inside and the percussion cap for an ignition source.
[322] Koffiefontein is a small farming town in the Free State. The name means coffee fountain in Afrikaans. Koffiefontein - Wikipedia

The cattle guards by Clive M. Dixon.[405] Remount Camp at Stellenbosch. "The heavy mortality amongst the horses and mules during the campaign put the British forces to considerable inconvenience, and large numbers were purchased in North and South America, Spain and Italy."[406]

Then a heliograph started winking from a kopje, and before long there was activity; soldiers running for their horses, and others galloping in from the plain.

Heliograph in use.[407]

Tinder Box with flint and sparking (carbon) firesteel.[408]

Seeing this, we boldly headed between two of the camps for a chain of hills beyond, firing from the saddle and getting well peppered in return, but we gained the cover of the hills before a real pursuit could be organized. A considerable number of horsemen came towards us, and when we opened long range fire they contented themselves with spattering some volleys against the rocks about us, and then returned to their business of minding the herds. We were now safely across the much-talked-of Modder River barrier, and, continuing our journey, reached the mineral spring near Lotshoek by next day, where we fell in with a small commando of fifty or sixty men under a Field-Cornet named Blignault.[323]

[323] Probably 'Blignaut' but too many possibilities in ABW database to identify further.

Abraham's Kraal & Driefontein.

This is the vicinity they are traveling through.

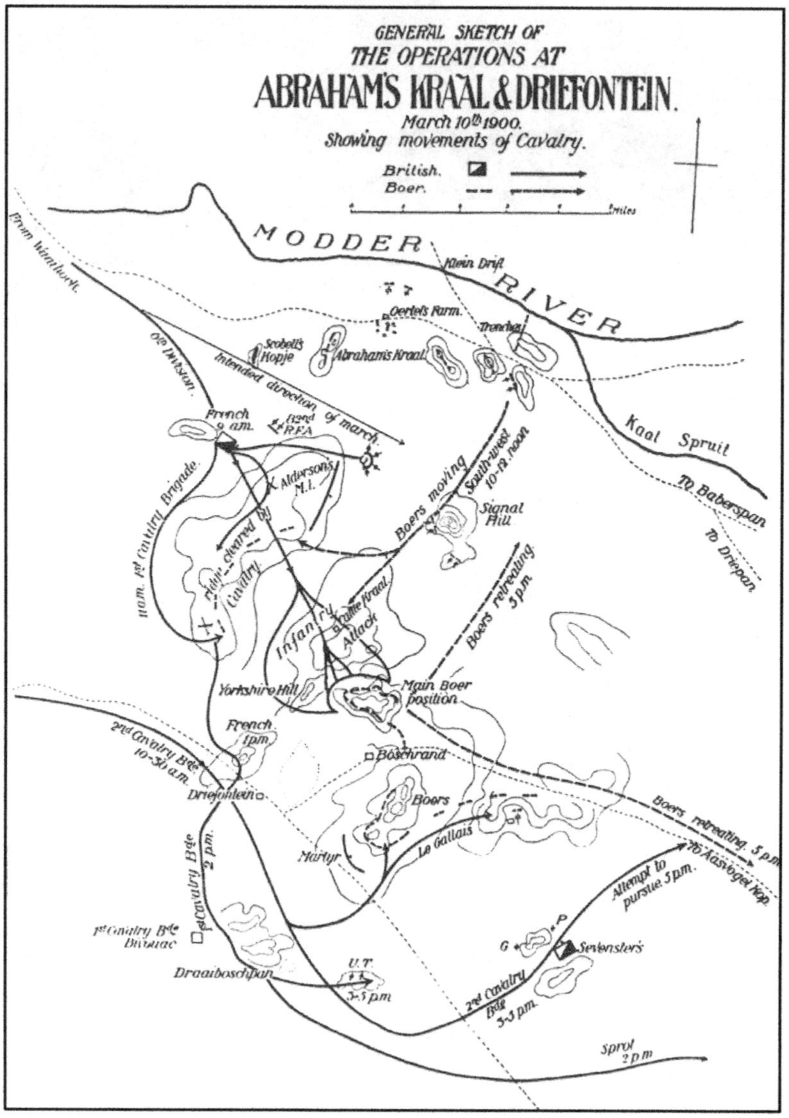

Driefontein (Abraham's Kraal) March 10th.[409] Abraham's Kraal[324] Boers: 124 men killed and captured, while the British lost 82 killed and 342 wounded.

[324] The Battle of Driefontein on 10 March 1900 followed on the Battle of Poplar Grove in the Second Boer War between the British Empire and the Boer republics, in what is now South Africa. In the first half of 1900, the British made an offensive towards the two Boer republic capitals of Bloemfontein and Pretoria. The Boer forces under the command of Christiaan de Wet were holding a 7-mile (11 km) line covering the approach to Bloemfontein. Roberts subsequently ordered a division under Lieutenant General Thomas Kelly-Kenny to attack the position from the front, while Lieutenant General Charles Tucker's division moved against its left flank. The Boers were subsequently forced to withdraw losing 124 men killed and captured, while the British lost 82 killed and 342 wounded. Battle of Driefontein - Wikipedia

See Also: Battle of Driefontein, 10 March 1900 (historyofwar.org) The battle of Driefontein (Boer War) was the last attempt to prevent the British under Roberts from occupying Bloemfontein, the capital of the Orange Free State. The Boer position in the Orange Free State had collapsed very quickly. Only four weeks earlier their main army had been camped at Magersfontein, blocking the direct route to Kimberley. On 11 February General Roberts had begun his great flank march, which had seen Kimberley liberated on February 15th, and the army from Magersfontein captured at Paardeberg on 27 February. A first attempt to stop the British advance had failed with hardly any fighting at Poplar Grove (7 March). In the aftermath of the rout at Poplar Grove, President Kruger had made valiant attempts to rally the fleeing burghers, but only 1,500 of the 6,000 men from Poplar Grove could be convinced to stay. Under the command of Christiaan De Wet, with assistance from De La Rey, the Boers organised a new defensive line at Dreifontein. This time they fought, and fought well. Roberts organised his men into three columns and attacked along the entire Boer line. Despite being heavily outnumbered, the Boers managed to hold on all day, but as darkness fell the British had seized the northern end of the Boer line, and were threatening the southern flanks. Once again the burghers fled – De Wet described it as a panic. This time there was no stopping them. Three days later the British captured Bloemfontein without facing any more significant resistance. British losses at Driefontein had been 82 dead and 342 wounded. Boer losses were at least 102 dead and 22 captured. To the British it began to look as if the war would soon be won.

We were now in the Fauresmith district,[325] not more than fifty miles from the Orange River, beyond which lay the Cape Colony, but I was yet to discover that it was farther off than it seemed.

When we told Blignault or our plan of invasion from here, he strongly advised us against the attempt. He said that if we crossed the Orange River here, we should find ourselves in open and arid country, where our horses would surely starve, and where we would be ridden down by the first British force that saw us. On the other hand, if we went east towards the headwaters of the Orange River, a hundred and fifty miles away, and then turned south into the Cape, we should find mountains and good grazing for our horses, and there so small a band as ours would stand some chance of survival. Blignault and his men had much to say of the hardships they had endured when they accompanied General de Wet into the Colony some months previously.

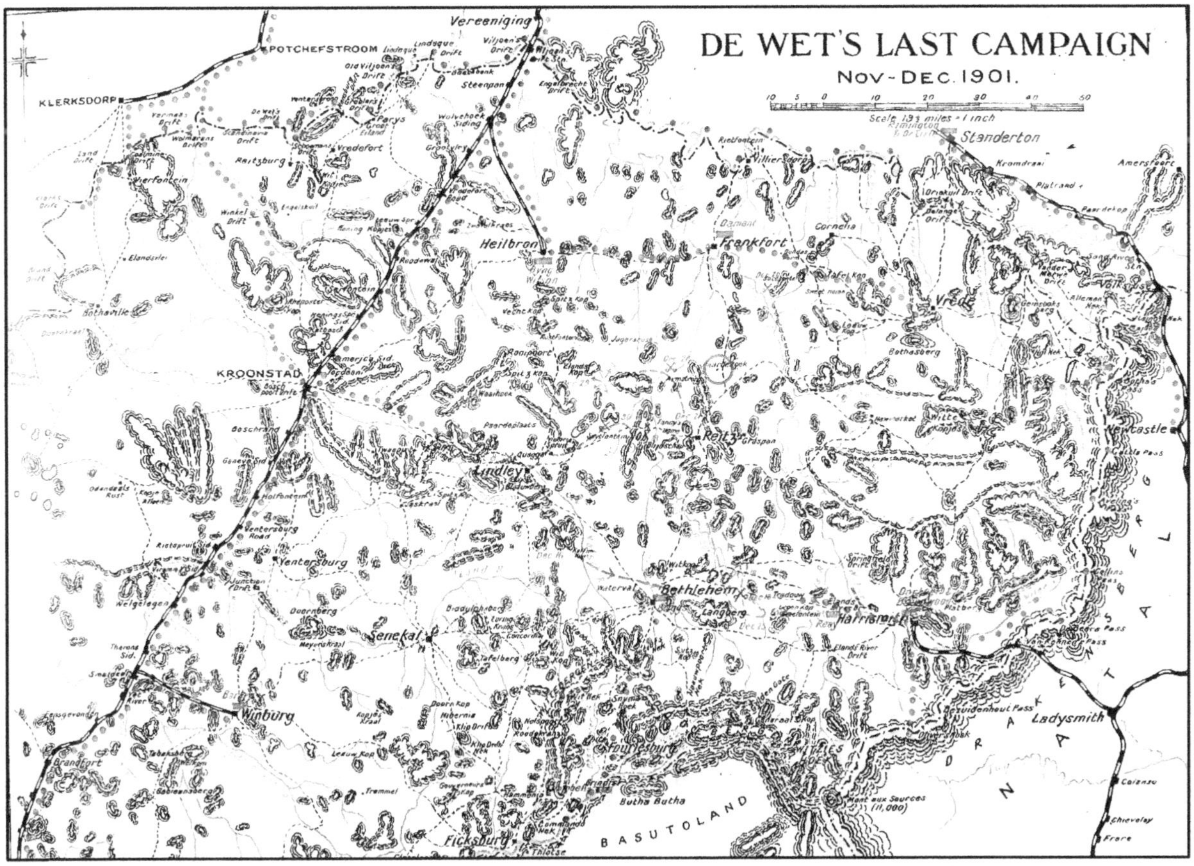

Blockhouse system in the Cape Colony in late 1901.[410] .

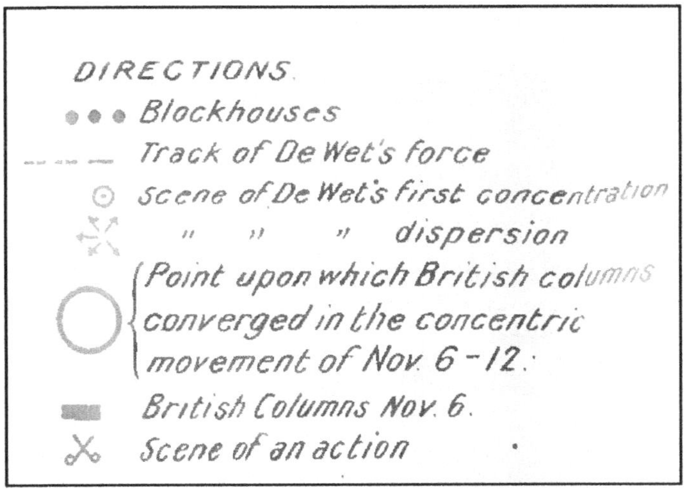

Judge (now Commandant) Brand Hertzog.[411] [326]

[325] Fauresmith is located 130 km south west of Bloemfontein, South Africa. Fauresmith - Wikipedia

Fifteen hundred men had crossed the river, but they were so beset by enemy columns that they had to return to the Free State with the loss of many men and horses. In fact everyone we met in these parts spoke against further raids into British territory.

We were reminded, too, of the disaster that overtook General Hertzog's[327] venture to the south, and of the fate of almost every other party that had crossed the Orange River. It seemed that, like the cave in the fable, many tracks led over the river, but few came back.

This was not encouraging, and Field-Cornet Botha and the rest of my companions were so impressed that they began to waver, but Jacobus Bosman and I talked them into a better frame of mind, and finally persuaded them that, by accepting the advice we had received and going east, we could get into the Cape Colony and hold our own once we were there.

Next day we said good-bye to Blignault's commando and rode away east, coming within sight of Edenburg[328] village the following afternoon. This place lies about forty miles south of Bloemfontein on the main railway, and, as we found the line strongly block-housed and saw a big English camp on the outskirts of the village, we lay over for the night, to spy out the land next day.

The night after that we saddled at dark and rode towards the railway track, hoping to get over unobserved at a spot which we had investigated during the day. Just before we started, a sturdy little Shetland pony came wandering up from the English camp, and I took him along with me, which was just as well, as it turned out. After an hour's ride we reached the railway line at what we took to be the point which we had selected, but the night was so dark that we had gone astray, and we ran into a block-house instead. We were met with the usual 'Halt! Who comes there?' Followed by rifle-shots, so we bore away to look for a more suitable crossing, and some five hundred yards farther down we made another attempt.

As is the case with most railways in South Africa, a fence ran on either side of the line, composed of thick strands of wire which had to be cut before the horses could be led through. The only implement we possessed was a large file, and with this a young fellow named Verster and I tried to saw the wires, while the others waited a hundred yards back. The file grating across the taut wires made a tremendous noise, and before we had cut even one strand, we were again challenged and fired at by a sentry, who sounded not twenty yards away. We hurriedly mounted to rejoin our companions, but our horses began to plunge and flounder over obstructions staked along the ground. In approaching the railway we had somehow or other missed these entanglements, but now we were in the thick of them, and-the tins always attached to them were clanging and jangling, and increasing the terror of our animals. To this din was added a blaze of musketry from a block-house standing only a few yards away, which in the darkness we had mistaken for a mound of rocks.

[326] General James Barry Munnik Hertzog (3 April 1866 – 21 November 1942), was a South African politician and soldier. He was a Boer general during the Second Boer War who served as the third prime minister of the Union of South Africa from 1924 to 1939. Hertzog advocated for the development of Afrikaner culture and was determined to prevent Afrikaners from being excessively influenced by British culture. He founded the National Party in 1914.

In 1941, Hertzog, who had resigned after South Africa - against his efforts - entered World War II on the side of the Allies, issued a statement in which he openly praised Nazism and said it needed to be adapted to South African needs under a fascist dictatorship. https://en.wikipedia.org/wiki/J._B._M._Hertzog

[327] 1/12/1901 Boers driven eastward from Witwatersberg by General French. Activities in Cape Colony to frustrate Hertzog's advance. General James Barry Munnik Hertzog KC (3 April 1866 – 21 November 1942), was a South African politician and soldier. He was a Boer general during the Second Boer War who served as the third prime minister of the Union of South Africa from 1924 to 1939. Throughout his life he encouraged the development of Afrikaner culture, determined to prevent Afrikaners from being influenced by British culture. J. B. M. Hertzog - Wikipedia

[328] See: Edenburg - Wikipedia

British Army piquet post with wire fence.[412]

Masonry Rock blockhouse.[413]

Rifle-fire at point-blank range is unpleasant at the best of times, but when one is on a maddened horse-staggering amid wire loops it is infinitely more so, and had there been even a glimmer of light to guide the soldiers we should both have been shot. It was so dark, however, that they were firing at the sounds, and not at us, and Verster managed to wrench his horse free, but mine was shot and I was nearly pinned underneath him. I undid the buckles of the girth, and dragged my saddle from under the prostrate animal and, stumbling over the rest of the obstructions, we got clear away to where the others stood whistling and shouting to us and anxiously watching, not daring to shoot for fear of killing us. I had left the Shetland pony with them when I went forward to the fence, so I now put my saddle on him, and we galloped off, leaving the soldiers firing blindly into the night.

We made a half-moon, until we again reached the railway line, intending to have another try, but as soon as I began to use the file, we heard the sound of men running along the track towards us, so we lost no time in decamping, and abandoned all thought of crossing that night. We spent a cold night behind a kopje, and, when it grew light, we were nearer the English camp at Edenburg than we imagined. We had barely time to get our horses saddled before a hundred or more troopers came racing at us, but, riding fast, we got safely into the hills at Boomplaats, [East of Edenburg] my Shetland pony going surprisingly well.

As the English horsemen turned back after a while, we halted by the little cemetery where those English soldiers lie buried who were killed at the battle here between Sir Harry Smith[329] and the Boers, in 1848.[330] This graveyard was of some personal interest to me, because it had almost caused my father the loss of his position as President of the Free State when I was a boy. The British Government many years before had erected headstones over the fallen soldiers inscribed with the words: 'Killed in action against the Rebel Boers.' Many of these stones, in course of time, fell to pieces, so my father ordered replicas with the original inscription faithfully copied.

This gave rise to much ill feeling, for there were indignant patriots who considered the epithet 'Rebel' an insult to the Boers, and my father very nearly lost the next presidential election in consequence I have another reason for remembering the Boomplaats cemetery, for while we were resting here, Field-Cornet Botha came to me and said that he and the other men had for some days been reconsidering the matter of continuing to the Cape Colony, and he now asked me to give up the plan. Apparently last night's affair and the chase of that morning had put the finishing touch to their indecision, and, with the exception of Bosman and myself, they all declared their intention of returning to the Winburg Mountains.[331] [125 miles Northeast]

[329] Sir Harry Smith, 1st Baronet - Wikipedia

[330] The Battle of Boomplaats (also referred to as the Battle of Boomplaas) was fought near Jagersfontein at on 29 August 1848 between the British and the Voortrekkers. The British were led by Sir Harry Smith, while the Boers were led by Andries Pretorius. The British were victorious after one Boer opened fire too early and betrayed their position. Battle of Boomplaats - Wikipedia

[331] Winburg is the oldest proclaimed town (1837) in the Orange Free State, South Africa and thus along with Griquastad, one of the oldest settlements in South Africa located north of the Orange River. Winburg - Wikipedia

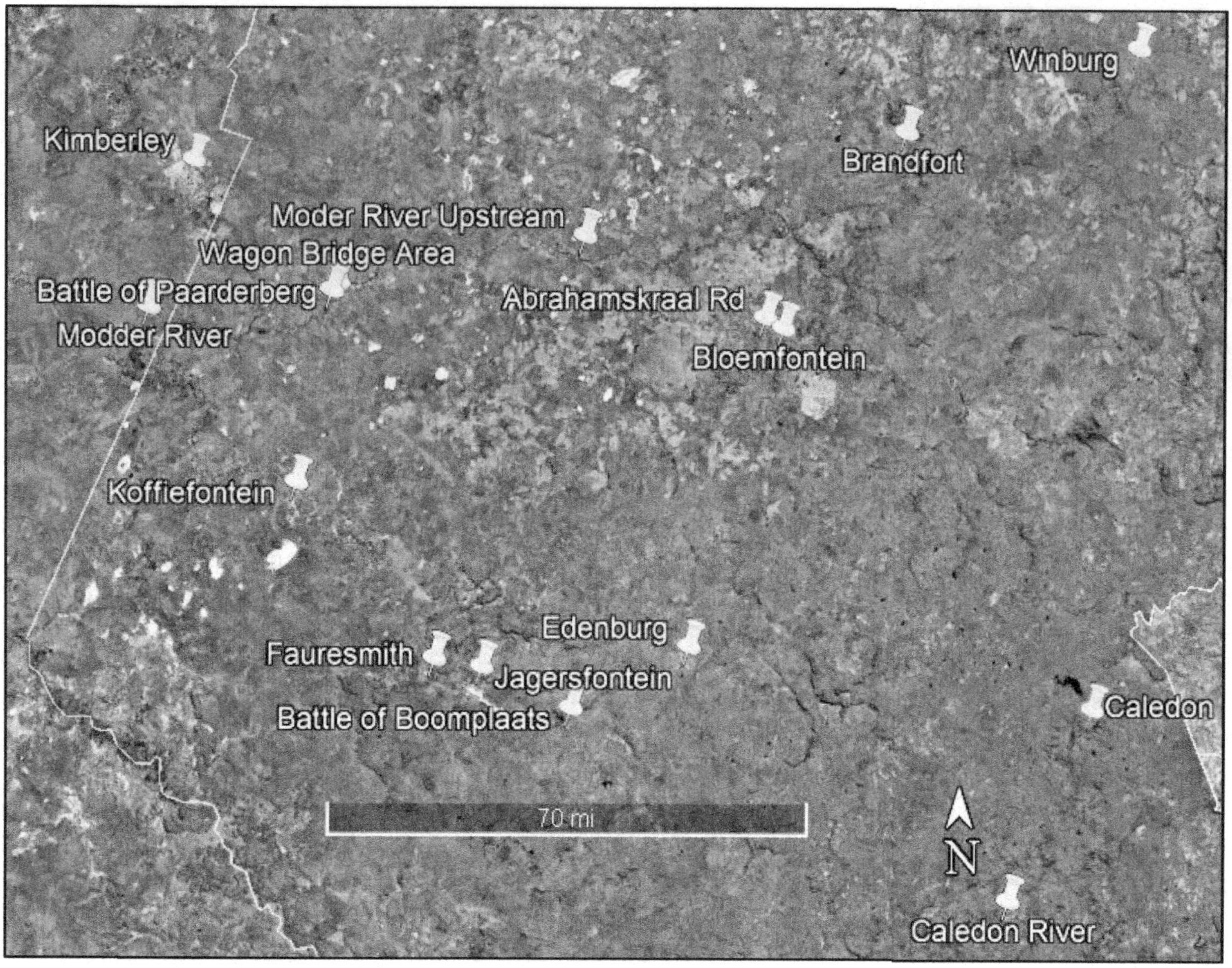

Map data: Google Image© 2024 Image Landsat/Copernicus.

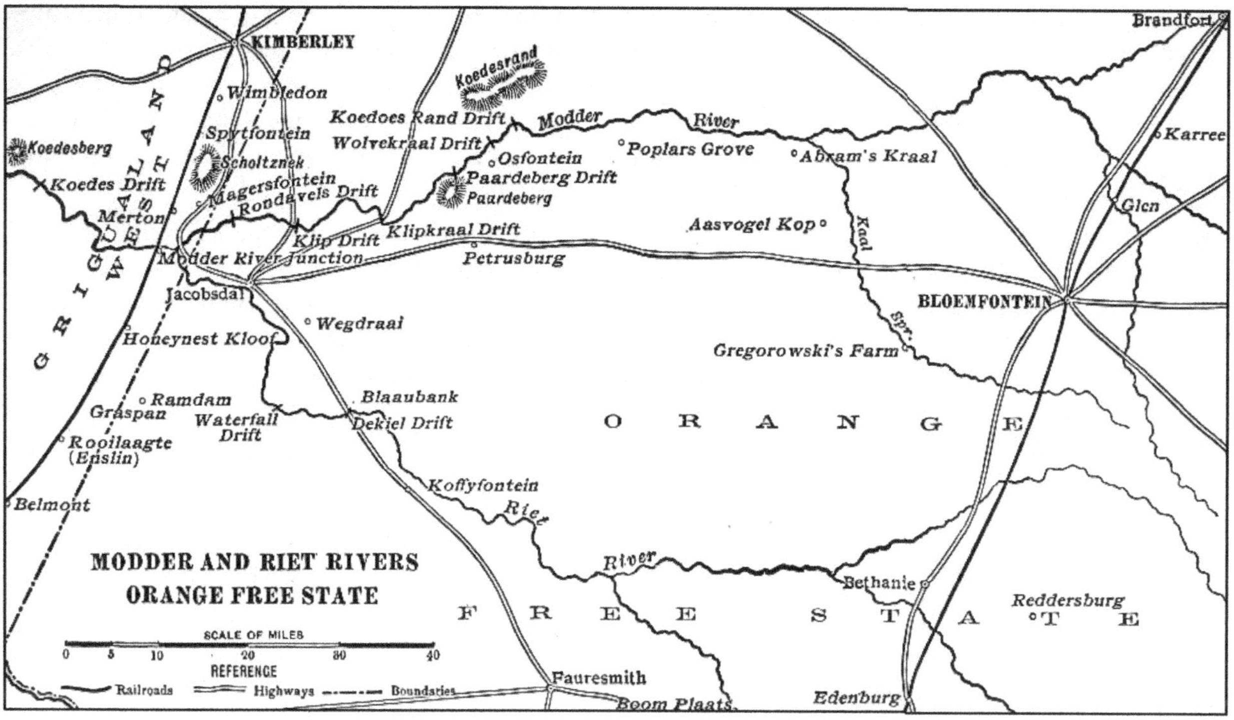

Modder and Reit Rivers.[414]

We two said that we had not ridden thus far to turn back now, and we told them that we were going to the Cape, even if we had to go alone. We argued and entreated for hours, but in the end Botha and his men, with my two German friends Haase and Pollatchek, saddled their horses and rode off, leaving us behind.

Bosman and I spent the night beside the graveyard, feeling too depressed to light a fire, and next morning we found our horses strayed out of sight. It took us the better part of five hours to trail their spoor, and when at last we got back to where we had left our saddles and other belongings, there was nothing there. This was a serious blow, for saddles were indispensable, and blankets and cooking-tins even more so We could see from the hoof-marks that there were two mounted thieves, and although we could not say whether they were Boer or British, we decided to follow; them. After careful going, on our bare-backed horses, for ten miles or more, we saw two men seated beside a fire below a kopje, and, stalking them on foot came close enough to cover them with our rifles before they could stir. They proved to be two Boer boys from a neighbouring outpost, and they had all our missing gear beside them. They said that they thought the things belonged to English scouts, a palpable lie, and we told them plainly what we thought of them, but we did learn from them of a commando under General Hertzog lying in the hills beyond the Riet River forty or fifty miles away, [West] and we decided to go there.

For two days we rode on, passing through the deserted mining town of Jagersfontein,[332] and reaching the next village, Fauresmith,[333] by sunset. This place had also been abandoned by its original inhabitants, but we found it occupied by men who were little better than bandits.

As we went up the main street a number of unkempt individuals rushed at us, rifle in hand, and ordered us to halt. They crowded round threateningly, with shouts of 'Maak dood die verdomde spioene' ('Kill the cursed spies'); although they must have known that we were nothing of the sort, and only wanted a pretext to rob us. We sat on our horses, uncertain what to do with such ugly customers, and it looked as if we were in for serious trouble, for already greedy hands were clutching at our wallets, and trying to pull us from our saddles.

Just then two gaunt, famished-looking women rushed from a neighbouring house, and shrilly ordered them off. They had so much influence over the rabble, that they stood aside growlingly and allowed us to accompany our forbidding guardian angels to their home. They were the daughters of a once-wealthy farmer, ruined and killed during the war, and they had taken refuge in an empty house in the village, where they were living in poverty. They told us that the people who had molested us were riff-raff ejected from the fighting commandos, existing on what they could rob and loot. We now heard that General Hertzog's commando had moved west, so we took our leave next morning, and rode for two days in search of them, until we were right on the Griqualand[334] border, where we met a solitary burgher, from whom we learned that we were a long way out of our course and that the commando was back on the [upper portions of the] Riet River.

We turned north-east, and by dark of the following evening we crossed the river by the wagon-bridge which, strange to say, was still intact.

An example of a wagon bridge. 'Passage of Infantry over a bridge of Wagons.'[415]

[332] About 9.6 million carats (1,900 kg) of jewel-quality diamonds were extracted during the mine's century of operation. Since it was first established in 1870, two of the ten biggest diamonds ever discovered, the Excelsior and the Reitz (now called the Jubilee), were mined from Jagersfontein. Jagersfontein Mine - Wikipedia

[333]The town, named after Rev Phillip Faure and Sir Harry Smith, is the second oldest town in the Free State. Fauresmith - Wikipedia

[334] Griqualand West is an area of central South Africa that now forms part of the Northern Cape Province. It was inhabited by the Griqua people – a semi-nomadic, Afrikaans-speaking nation of mixed-race origin, who established several states outside the expanding frontier of the Cape Colony. Griqualand West - Wikipedia

On the far bank was a cattle-post in charge of an old man and his two sons. They were looking after the meat supply of General Hertzog's men, lying, we were glad to hear, only ten miles upstream. As it was late, we spent the night with our hosts, an unsavoury trio, who did not belie their looks, for we found next morning that our saddlebags had been rifled. So I stood them in a line while Bosman went through their pockets and recovered our knives, tinder-boxes, and other property, after I had grazed the eldest ruffian's arm with a bullet to teach him better manners.

We then rode on, not particularly impressed by the specimens of Fauresmith men whom we had met thus far, for we had been the victims of three attempted robberies in less than a week. These men, however, were only off-scourings, and our subsequent association with the real fighting-men of the district was much happier.

By midday we came on General Hertzog's force, camped on the banks of the river at a place called Bethal.[335] I knew General Hertzog from the old days, a thin high cheeked man with angry eyes, though his speech was pleasant and I saw that his men held him ill great respect. He had been a Judge of the Supreme Court at Bloemfontein before the war, but was now in command of the south-western districts of the Free State. He had with him about three hundred men, the largest commando that I saw on my passage through, and he had other smaller bodies scattered about.

A large commando on the move.[416]

We remained with the commando for ten days, hoping to gain recruits for the Cape, but not a man would come. They said once was enough, and every time we broached the subject we were met with emphatic refusals, and with tales of the privations and losses which they had sustained during the previous expeditions, so it began to look as if Bosman and I would have to go by ourselves after all. During the time spent with the commando we earned our keep by taking part in a sharp little affair in the hills above the Riet River wagon-bridge.

An English column, a thousand strong, crossed [Northwards] to our side of the river one night, on their way to Kimberley, and we held them up for two days, until at length they fell back over the bridge and returned by the way they had come.

General Hertzog lost only two killed and a few wounded, although the English rifle-fire was very heavy at times, and they had guns as well. On the afternoon of the second day, while two of us were riding down to a dam to water our horses, an English trooper came galloping towards us from round a kopje. He was plucky enough, for when he realized his mistake he fired from the saddle, from so close that before he could do so again I was alongside with my rifle in his ribs. He said that he had been scouting and had lost his way, and as my companion was barefooted, we relieved him of his boots, as well as his horse, and we left him gingerly picking his way over the sharp surface towards his own side, a thing I did not envy him, for he had to make a semicircle of several miles to get around our flank, before he could rejoin his companions.

When we found next morning that the English had withdrawn overnight, General Hertzog ordered us to return to our former halting-ground higher up the river, and here Bosman and I had good luck, for, off-saddled close by, was a small party of newcomers, ten in number, amongst whom I recognized several old acquaintances from the

[335] Unable to identify this location near Reit River.

Transvaal, and to our great joy they told us that they were on their way into the Cape Colony. This was good news indeed, and we told them of our own plans at once.

As from now onward I was to be intimately associated with this little band, and as so many of them died tragically, I shall give their names:

The leader was Jack Borrius,[336] a short thick-set man of twenty-eight, from Potchefstroom, whom I had met while serving with Captain Theron's scouts and in the 'A.C.C.' Next came Benjamin Coetzee,[337] from Pretoria, whose reckless bravery had earned him a reputation as long ago as the Natal days, when I had known him as a member of the Pretoria commando. Then Nicolas Swart and Cornelius Vermaas,[338] both sons of wealthy Transvaal farmers. Vermaas had been wounded and captured by the British six months before, but he had jumped from the train in the mountains near Cape town while being taken for shipment to Ceylon, and after many trials he had got back on foot to the Transvaal.

Ben Coetzee. [417]

Next were Percy Wyndall and Edgar Duncker,[339] two English-speaking boys from Johannesburg, and Frits Balogh, a young Austrian from Pretoria.

Then Jan van Zijl,[340] illiterate but witty, and Piet de Ruyt, a Hollander, and lastly young [Henry] Rittenberg, also from Potchefstroom.

They were all Transvaalers and, with the exception of Borrius, not one was yet twenty years old; and like myself, they had all been taken with the idea of free-lancing it into the Cape Colony.

By way of ironic comment on their tattered clothing and ragged appearance they called themselves the 'Rijk Section' (or 'Dandy Fifth,' as Duncker translated it). of this small band four were to meet their death by execution, and six were wounded or captured, a result that fully justified the many warnings which we had received against entering the Colony.

While we were discussing whether to try the Orange River in this vicinity or to take the advice we had received, and go east towards the headwaters, there arrived another party of horsemen, who decided us for the latter course.

These were Commandant George Brand[341] and a score of followers from the south-eastern districts, come to confer with General Hertzog on military affairs.

He was a son of Sir John Brand,[342] who had preceded my father as President of the Free State. He told me that as soon as he had finished with General Hertzog, he was returning east across the railway line to the area

[336] Cited supra.

[337] Coetzee Benjamin Pieter Veldkornet Commando: General Smuts.

[338] Vermaas Cornelius Lourens Prisoner number: 32509 Captured: Doornspruit 07 May 1902 Sent to: Unknown Age: 35

[339] Perhaps his father: Duncker Claudius Frederick Prisoner number: 26901 Captured: Johannesburg Sent to: Unknown Age: 48 Address: Johannesburg. This is the only Duncker on ABW.

[340] Seven possibilities under this spelling on ABW.

[341] Brand George Alfred Asst Head Commandant Commando: OVS. Awarded the DTD and received the Wound Ribbon.

[342] Previously cited: Sir Johannes Henricus Brand, commonly known as Jan Brand) was the fourth State President of the Orange Free State, 1864 until his death in 1888. He had 11 children, 8 sons and 3 daughters. Johannes Brand - Wikipedia

between the Caledon River[343] and Basutoland,[344] and when he heard or our intention to enter the Cape Colony he advised us to accompany him as far as the Caledon River, where the passage would be safer than here.

As we had already received similar advice, and as he assured us that we could get as many horses as we liked in the mountains, we agreed to travel with him.

The very next day, saying good-bye to Hertzog and his men, we set out and made such good progress that by the following afternoon we were back in the same hill overlooking Edenburg village from which we had started with our previous companions on our unsuccessful attempt to cross the railway line.

Now things went better, for we were with men who knew the exact position of every block-house and every sentry along the track, and by midnight we were over without a single casualty, although there was a good deal of firing from block-houses on either side of us as we went through.

The railway is regarded as the dividing line between the Western and Eastern Free State. To the west lie the great plains, while to the east the country grows more and more mountainous as one approaches the Basutoland border.

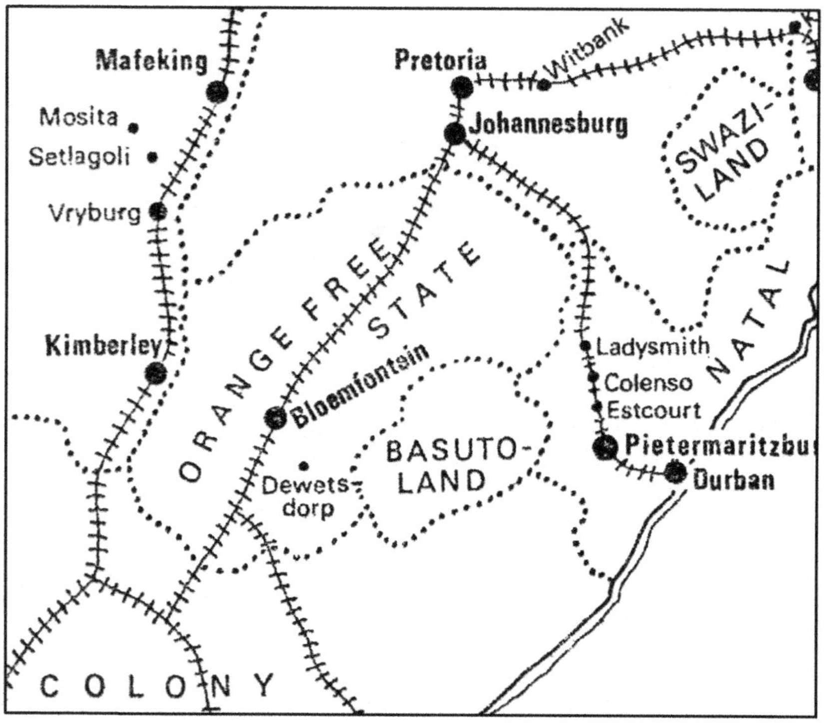

I knew this region quite well, having hunted here as a boy, so after sunrise we took our leave of Brand and his men, whose course lay in another direction, and we rode on for the next few days in the direction of the upper reaches of the Caledon River. Only once did we see an English column, to which we passed so close that we heard the bugles sound the alarm and saw the soldiers running for their horses; but after the exchange of a few shots, we went on our way.

Most of the live-stock lay clubbed to death around the burnt farmhouses, and there was no sign of the civilian population, for those who had escaped capture were hiding in caves and gorges.

We did not see many fighting-men either, for they were dispersed in small guerrilla bands among the mountains.

Troops of wild horses were fairly common between Sikonyella's Peak and the Elandsberg, so we searched for a local outpost, and with their help were able to corral sufficient horses to give each of the 'Rijk Section' two fresh mounts, my own share being two spirited mares, a brown and a roan. The horses were entirely untrained, but in a few days we had them broken to bit and rein and our prospects were vastly improved, for the animals

343 The Caledon River (Sotho: Mohokare) is a major river located in central South Africa. Its total length is 642 km (399 mi), rising in the Drakensberg Mountains on the Lesotho border, flowing southwestward and then westward before joining the Orange River near Bethulie in the southern Free State. Caledon River - Wikipedia

344 Lesotho

we had been riding thus far were in wretched condition. To these we gave their freedom, knowing that when they had picked up flesh they could be recaptured and used by others, and it was pleasant to see them trotting up the mountain-side to a well-earned rest, my little Shetland pony kicking up his heels and whinnying joyously at their head.

Commando on campaign.[418]

It was near the end of August (1901) by now, and the rainy season might be upon us at any moment, when a single freshet might render the Orange River impassable, so, as we were come to within fifteen miles of it, we made ready for our final effort.

Upon the very morning, however, on which we hoped to cross, a large body of horsemen appeared over the shoulder of a distant hill. By their formation and manner of riding we knew them to be Boers, but as there was no commando of that size round here, we waited for them with considerable interest. After an hour the column came abreast of us and we were astonished to see that at its head rode Mr. Smuts, the Transvaal State Attorney, now a General, who, after hearty greetings, told us that he was on his way to the Cape Colony with three hundred men.

This was another stroke of luck! Falling in with the 'Rijk Section' had been the first, but to meet with a whole commando making for the river was the greatest fortune of all, far greater than we realized at the time, for our subsequent experiences went to show that a small force like ours would not have lasted a week in the troublesome country beyond.

It was the finest commando with which I ever served. The rank and file were mostly keen young farmers from the Western Transvaal, the pick of de la Rey's fighting men, and in command of them was perhaps the one man in South Africa who could have led us through the perilous days to come.

General Smuts halted his men and ordered them to off-saddle. This gave me time to go among them, and I had another pleasant surprise, for I came on my Hollander uncle, Jan Mulder, whom I had not seen since that time in Natal, nearly two years before. I also found the General's brother-in-law, Krige, one of the few survivors of Isaac Malherbe's corporalship, who had been so badly wounded.

I learned from them of the great hardships and dangers which they had encountered on their way through the Free State. The English had wind of their intention to invade the Cape Colony and strenuous efforts were made to head them off. Large forces were hurried up from all sides, giving them no rest either night or day, and it was only by hard riding and hard fighting that they had escaped, with the loss of many men and horses.

From what I could gather, General Smuts proposed a flying raid into the central districts of the Cape, to test a large-scale invasion later on, in order to relieve the increasing pressure in the north.

Whether this was really his object I do not know, for he was an uncommunicative man, but at any rate here he was, and we were only too eager to go with him, whatever his plans might be. When we went to tell him that the 'Rijk Section' proposed joining his force he said he was very pleased, and he appointed us to be his scouts, a distinction which we accepted gladly, but for which we paid dearly an the long run.

Smuts' Commando

Smuts' Commando in April of 1902.[345]

Annotated with the individuals identified from the' Our Military Officers Album.' [346]

[345] Jan Smuts is seated. Viewer left standing is General Jaap van Deventer wearing a Mauser C96 and standing viewer right with his leg up is General Manie Maritz. General Smuts Vol 1 by Millin Sarah Gertrude (1936) p.159 File:Smuts Boereoorlog.jpg - Wikimedia Commons Also at higher resolution: **https://af.wikipedia.org/wiki/L%C3%AAer:Smuts_Boereoorlog.jpg**

[346] Our Military Officers. Album of portraits with life sketches. Original Publication: Onze krijgs-officieren: album van portretten met levens-schetsen der. by Gustav Schoeman Preller, Frans Vredenrijk Engelenburg 1904 Publisher "Volksstem" kantore. In Dutch with a preface by Louis Botha. Transcribed and published by: Mr. Pieter van der Merwe see infra. Original Publication : Onze krijgs-officieren: album van portretten met levens-schetsen der. by Gustav Schoeman Preller, Frans Vredenrijk Engelenburg 1904 Publisher "Volksstem" kantore Catalogue Record: https://archive.org/search?query=title%3A%28Onze%20Krijgs-Officieren%29
Direct Link: https://archive.org/details/onzekrijgsoffic00engegoog

1. Gen. Jan Smuts	10. Field Cornet Willie Kotzé
2. Burger B.D. Bower S. "Oom Barend"	11. Corporal Hennie Vermaas
3. Gen. Jacob Louis van Deventer	12. Field Cornet Jan Van Brummelen
4. Field Cornet A. G. Boshoff	13. Commandant Stoffel Schoeman
5. Commandant Ben D. Bouwer	14. Jacob Brink, Military Commissioner
6. Field Cornet Cornelis Brink	15. Cornet A. Standers
7. Field Cornet Jan Van Der Berg	16. Field Cornet Ben Coetzee
8. Cornet Piet F. Visser	17. Marquis De Kersauson "Roberts"
9. Gen. S. G. Maritz	18. Carl T. Möller

The Smuts Commando. The group of officers posted here was taken near Okiep, then General Smuts, who sat in the middle of the group, had taken possession of this place. A page was devoted elsewhere in this Album to General Smuts himself. About the other officers are reminded of the following: Starting with those on the right side of General Smuts in a standing position are located, the spectator encounters a bearded warrior, namely:

Burger B.D. Bower Sr., Known conveniently as "Oom Barend," 62 years old, he fought with the Free Staters 1865 against the Basoetoes, when he was seriously wounded.

Next to this one: General Van Deventer. to whom a separate sketch is dedicated elsewhere in this Album.

On his right side stands Field Cornet A. G. Boshoff, who fought under Commandant Bouwer and on 25 February 1902 in Windhoek, distr. Vanrhijnsdorp, was seriously injured during the capture of the Police camp.

In addition to these is located Commandant B. D. Bouwer, already mentioned elsewhere in this album. On the corner on this side of the group is:

Field Cornet Cornelis Brink who served under the aforementioned Commandant and came from Haaskraal, distr. Potchefstroom.

Rest at his feet Field Cornet Jan Van Der Berg, also from Potchefstroom, who in July 1901 during the storming of a railway fort in Brandfort was seriously injured, which did not prevent him from entering the Cape Colony insisted. There he was seriously injured for the second time in January 1902 of which he will remain disabled for the rest of his life. Rests on his left side

Field Cornet Piet F. Visser, a former Transvaal mine official, who became an officer in the K. Colony appointed. The group to the left of General Smuts is composed as follows: the heavy-set officer is next to him

General S. G. Maritz, years old, for a time a member of Theron's Reconnaissance Corps and present at the death of this beloved captain. He passed the Cape with General Wijnand Malan colony and, after fighting for a while, was sent to Calvinia district with three citizens where he managed to organize a commando, whose Commandant he was is appointed.

After some beautiful battles, "ou Manie" grew and became a general Maritz was called by his comrades—armed forces every day, so that on 7 August, 1901, took possession of Niewoudtsville and invaded Clanwilliam, where he settled with Captain Jan Theron and Captain Smit, and together into the district Piketberg, Moorreesburg, Darling and Malmesbury were extended. From there it went trip to Tulbach district, where the Town Guard of twenty-four rivers Wellington is disarmed. In November of this year General Smuts was met the Van Rhijnsdorp district; In January 1902 Maritz was appointed General. When the peace had become a fait

Reproduction: of Our Military Officers. Album of portraits with life sketches of the Transvaal Generals and Commandants), Volksstem, Pretoria 1904. Transcribed and published by: Mr. Pieter van der Merwe..
File:Onze Krijgs-Officieren. Album van portretten met levens-schetsen, Pretoria, Volksstem, 1904.pdf
Download Access: https://en.wikipedia.org/wiki/File:Onze_Krijgs-Officieren._Album_van_portretten_met_levens-schetsen,_Pretoria,_Volksstem,_1904.pdf
Direct Link to File: https://upload.wikimedia.org/wikipedia/commons/5/52/Onze_Krijgs-Officieren._Album_van_portretten_met_levens-schetsen%2C_Pretoria%2C_Volksstem%2C_1904.pdf
Source Page: https://en.wikipedia.org/wiki/Annie_Botha#cite_note-4

accompli, “old Manie” left the Colony in dismay and visited German S.W. Africa and Europe. He was captured at Tontelboskolk, in Galvinia Dec. 1901 seriously injured. Standing behind him, next to Genl. Smuts,

Field Cornet Willie Kotzé, from Calvinia district, a known good marksman, and to the left of this

Corporal Hennie Vermaas, living in the Klerksdorp district, one of the brave four that died in January 1902 from Van Rhijnsdorp district on foot and armed the Bovenland, Paarl and Stellenbos spies for fourteen days with excellent results.

On his left side is Field Cornet Jan Van Brummelen, Dutch by birth, a determined and widely loved soldier.

In addition to these stands Commandant Stoffel Schoeman, a Free Stater, who succeeded General Maritz as Commandant, was previously a member of Theron’s corps, and one of the three that entered Calvinia district with Maritz. He followed General Maritz to German South West Africa. Next to this one

Jacob Brink, Military Commissioner of Commandant Bouwer, from Johannesburg, competent and helpful. Finally, on this corner the group stands

Field Cornet A. Standers belonging to the command of Commandant Schoeman, and originating from the district Ventersburg, Orange Free State. He also used to belong to the late Danie’s corps Theron and then followed Maritz to the old Colony.

At his feet sits the fearless Field Cornet Ben Coetzee, before the war an official at the Pretoria post office. He originally belonged to it Kommando van Bouwer, but later transferred to Schoeman.

On his right is seated Marquis De Kersauson, known to his fighters under the name of “Roberts,” a Frenchman from nationality, who moved to Transvaal during the war out of sympathy for the Afrikaners had come. His bravery is especially evident as a depeche[347] driver to separate. Twice he traveled through D. S. W. Africa to Europe and back with reports and has sacrificed a lot for the Boers.

At the latest: Carl T. Möller, from Graaff Reinet, at the outbreak of the war, student of medicine in Edinburgh, mot Sievwright’s ambulance came to Transvaal, under Dr. Neethling in Lijdenburg employed and then head of General Smut’s ambulance. He nursed with success of General van Deventer, when he was seriously injured; as well as Commandant Bouwer.

[347] depeche (third-person singular simple present depeches, present participle depeching, simple past and past participle depeched)

XX. We Go into the Cape Colony and Meet with a Warm Reception

"The adventures of this handful of resolute men led by General Smuts forms one of the most interesting episodes in the whole course of the guerrilla war."—The Times History, Vol 5 p.302.

The place of our meeting with General Smuts and his commando was in sight of the little village of Sastron,[348] about fifteen miles from the Orange River, and his intention was to march nearer that day, and cross during the night. By noon a start was made, and towards five in the evening we could see a dark line in front of us marking the gorge, at the bottom of which runs the river between high mountain walls.

Unfortunately this was not all that we saw. Our side of the canyon was held for miles in each direction by a cordon of British troops, stationed there to bar our way. Whenever a footpath led down the cliffs, there stood a tented camp, and the intervening ground was patrolled by strong bodies of mounted men who clearly knew of our coming.

On seeing this, General Smuts led us back into a range of hills, where we waited until next day, whilst men were sent in search of some neighbouring outpost to act as guides. At dusk a young officer named Louis Wessels arrived with fifty men, a hard-bitten crew, with whom he had been operating for over a year.

Louis Wessels with his Mauser C96 and his Secretary, Mr. MacDonald.[349] Jan Smuts wearing a C96 Mauser under his left arm.[350] PF: This is a different picture, taken in the same sitting from that reproduced above.

He reported enemy columns closing in on us from the rear, and said that unless we were able to effect a crossing that night, we should be trapped. He said, moreover, that the river was everywhere difficult, owing to the depth of the gorge and the perpendicular cliffs, but he had brought with him a veteran of the Basuto wars who knew of a path which might be practicable.

General Smuts decided to start at once, and in the falling dark our force rode out, accompanied by Wessels and his men, who agreed to enter the Colony with us we travelled on, hour after hour in the dark, over rough ground, and then, towards three in the morning, we caught a glint of white far below, where the Orange River boiled and eddied in its narrow channel. It was yet night when we commenced the final descent, but after toiling down the precipitous path to which our guide had brought us, and along which assuredly no other mounted troops had

[348] Zastron is a town in the Free State province, some 30 km from the border of Lesotho. Zastron - Wikipedia

[349] Wessels Louis Frederik Prisoner number: 32074 Captured: Moabsvelden 26 Apr 1902 Sent to: South Africa, Umbilo Age: 26 Address: Nooitgedacht. Image courtesy of Neville Constantine. Notice Mauser C96 with stock attached.

[350] Jan Smuts wearing a Mauser C96.

ever passed, we reached the edge of the water. In single file we began to cross the river, a strong and turbulent mountain torrent, not broad, but so swift that our horses could scarcely maintain their footing, and as dawn lit the cliffs above, the hindmost man was through, and I stood in the Cape Colony at last.[351]

After a short halt we took a path that led to the top of the cliffs opposite, by a deep cleft, up which we tugged[352] our leg-weary animals, until, far above, we emerged on a wide grass-covered tableland, pleasantly dotted with native villages and herds of cattle at pasture.

We were now actually on British territory, but the country here lies in an angle between the Free State, Basutoland, and Cape boundaries, and the region seemed exclusively occupied by Basutos, for there were no European habitations in sight.

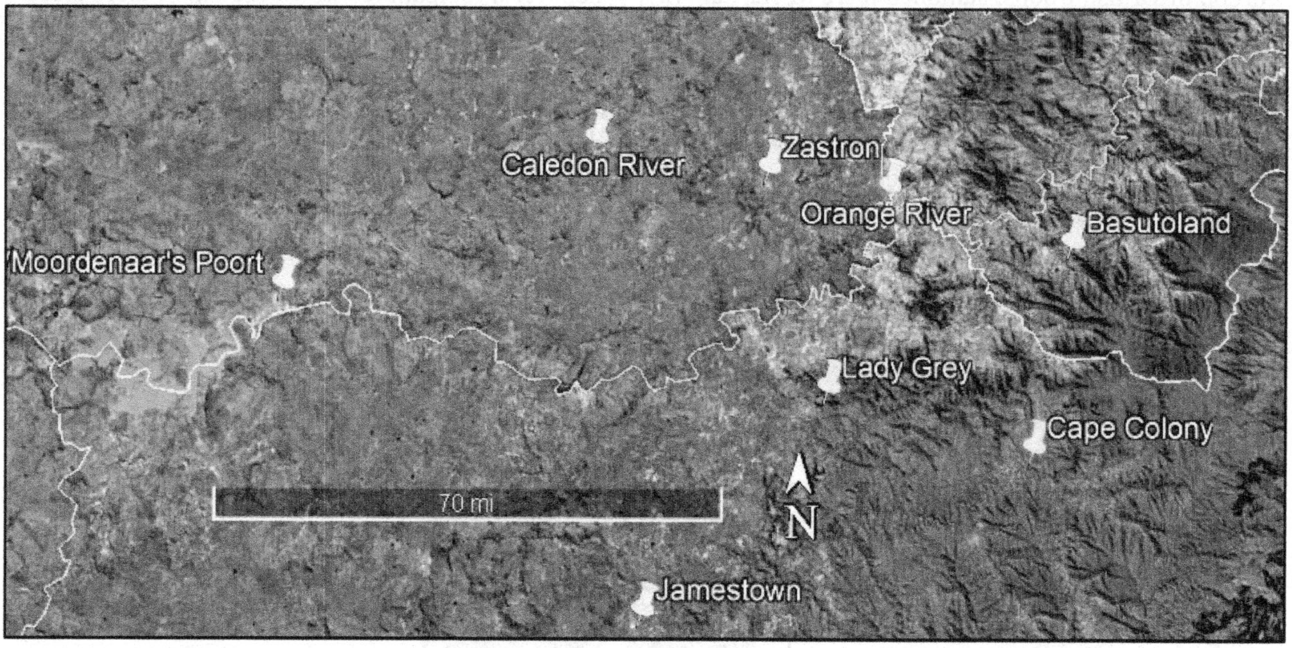

Map data: Google Image© 2024 Image Landsat/Copernicus.

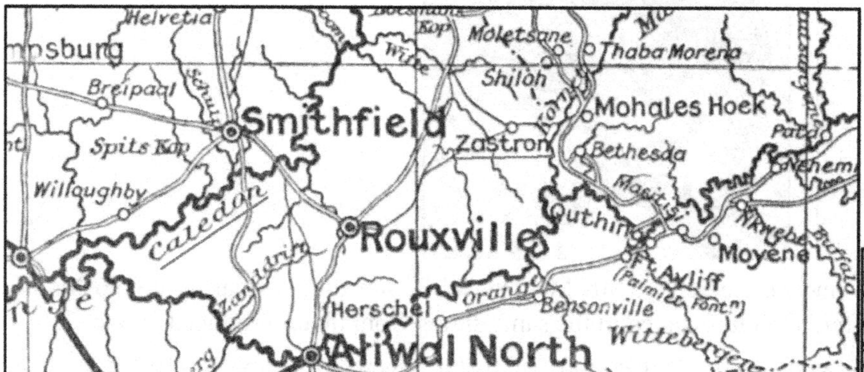

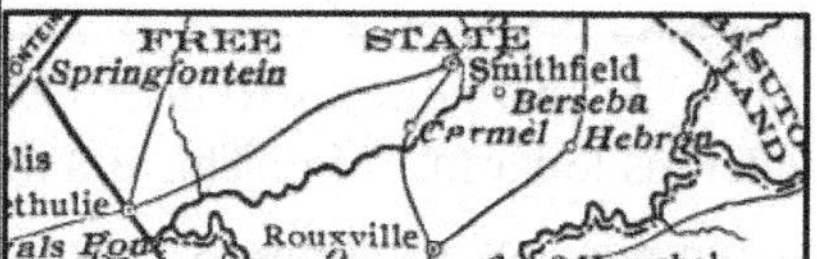

Part of the same area with different features noted. Lady Grey Noted.

As soon as we gained the top, we scattered into small parties, riding from one native village to another, in quest of tobacco and fodder for our horses. While we were thus ranging, a body of mounted Basutos, about three hundred strong, came moving swiftly towards us. Some were armed with rifles, others carried battle-axes, assegais and knobkerries, which they brandished in the air as they approached. We did not know what to make of this, but we thought that they could not be contemplating an unprovoked attack on a white force equal to their

351 September 3rd 1901 Crossing the Orange River at Kiba Drift.

352 To pull or drag with great effort.

own. General Smuts, therefore, contented himself with sending word for the various foraging parties to close in, and the commando continued on its way without paying much attention to the horsemen, who at a shouted word from their leader came to a halt on a knoll close by, where they sat their horses in silence, watching the Boers pass.

At this stage, my uncle and I, with five others whose names I did not know, lagged behind to feed our horses from the grain-baskets to be found in every native village, and as we were not frightened by the Basuto parade, which we put down to curiosity, we allowed the commando to get a considerable way ahead of us. At length, seeing that we were being left too far behind, we mounted and followed our men, the last of whom were just vanishing over the edge of the tableland by a road leading to the plain below.

By the time we could look down, the bulk of the commando was already at the bottom. They were riding along a road flanked on the left by a ledge of overhanging natural rock, part of the footwall of the tableland which they had just quitted, and on the right by a Mission Church and a long rubble fence, separating the road from fields and gardens. A force moving down this enclosed alleyway could be easily ambushed, and we were alarmed to see that the Basutos had left their horses above, and were scrambling down the final shelf of rock overhanging the road, crawling forward to the edge, to look straight down on our men riding unconsciously below. We expected to see them open fire on the crowded ranks at any moment.

Indecision, however, came over the natives; they began nudging one another as if each wanted someone else to start shooting, and by the time that they had made up their minds the opportunity was gone, for the commando was already debouching from the confined space of the road into the open plain. My fellow-stragglers and I were worse off, for although the Basutos had hesitated to attack the larger force, their intentions were clearly hostile, and we wondered how they would deal with our little band left isolated in the rear.

After hurried consultation we decided to follow on, and attempt to catch up with the commando, so we began to descend the slope. We reached the bottom unmolested, but as we passed the church beside the road we caught sight of many dark faces pressed against the window-panes, and white eyeballs peering at us from within. Then came a deafening crash, as a volley was fired at us point-blank from the building, sending showers of splintered glass about our heads. Fortunately the native is a notoriously bad marksman, for he generally closes his eyes when he pulls the trigger, so not one of us was hit, although the range was under ten yards.

When the Basutos lying on the rocky shelf overhanging the road heard the volley, they took courage and also opened fire. The five men with us did the only reasonable thing under the circumstances. They dug their spurs in and rode off as fast as they could, but my uncle with his usual impetuosity loosed his pack animal, swung his horse in behind a massive boulder that had calved from the ledge above, and jumped to the ground. I had to follow suit, relinquishing my own led-horse, and riding in behind the rock, a huge cube that leaned against the parent crag in such a manner as to give us cover against the Basutos overhead as well as from those firing through the shattered windows of the church across the way. We opened fire in turn at the church, but we saw at once that our position was untenable. Immediately above, the natives were excitedly shouting as they fired at our retreating companions, and at such of the rearmost commando men as were still within range.

Already we could hear the voices of several of them craning over to get at us from above. With some of the enemy standing on the roof, as it were, and others shooting from the church not fifteen yards off, we realized that to remain here could have only one ending, and we prepared to mount once more, although our chances of escape seemed desperate.

Looking down the road, we could now see only two of our men, riding for their lives across the fields, for they had succeeded in leaping the dividing wall. The other three men were nowhere to be seen, but two dead horses lay in the road, and a third was galloping rider less in the distance. This looked bad, but no other course was open to us, so we leapt into the saddle and rode out from the sanctuary of the fallen rock. The moment we did so, the natives in the church saw us, and redoubled their fire, while those on the bank above raised blood-curdling yells and also fired.

As we sped past, more natives rose from behind the fence lining the road. Fortunately these last were armed only with assegais and knobkerries, which came whirring about our ears. In this pandemonium we took every moment to be our last, but we ran the gauntlet safely for perhaps sixty yards, when the road fell suddenly into a deep spruit which neither of us had noticed in our excitement. It meant salvation although at first it looked as if it only meant more danger, for, riding down, we saw some fifteen or twenty natives squatted in a circle, intent upon something that lay on the ground between them. Before they could do more than spring to their feet and strike blindly at us, we were through. Instead of riding up the opposite bank of the spruit, where we should come under fire again, we galloped along the bed under cover of the high banks, until we were able to emerge out of range.

Of the five men who had been with us, the two whom we had seen making across country got clean away, but the other three were either killed on the road, and then dragged into the spruit, or else they were destroyed on reaching there, for we learned long after that their bodies were found on the causeway, dreadfully mutilated by the natives for medicine, in accordance with their barbarous custom. I have little doubt that when my uncle and I rode down amongst them, they were busy at their grisly task of dissection,[353] which we ourselves so narrowly missed.

We were out of danger now, but the prospect was not reassuring. True, we had got off without a scratch, but our pack animals were gone with most of our gear, and looking over our saddle-horses we were dismayed to find that they were both badly wounded. My brown mare had been hit by a jagged missile that had smashed her lower jaw, and my uncle's horse had a bullet through the crupper[354] and another through the hinds leg. I put my poor animal out of her misery at once, but my uncle thought that his might recover (which it did). I shouldered my saddle, and, leading the injured horse, we advanced on foot, dolefully speculating upon our future, for we had only a wounded horse and a handful of cartridges between us in this inhospitable country.

Horse Killing.[355]

Far away we could see the commando posted on a ridge, to cover the retreat of those who were still in danger, for some of the men had been within range when the firing started on us at the church, and there were several wounded painfully making their way across the plain.

When at last we reached the commando and had time to look around, my uncle and I had a pleasant surprise, for there stood our two led-horses safe and sound, with our blankets and cooking-tins intact. They must have fled straight on when we loosed them on the road, and they had overtaken the commando while we were fighting below the shelf. Now they were contentedly ranged in line with the rest of the horses behind the hill, as if nothing out of the common had happened, so at any rate we still had a riding-horse apiece.

While we were halted here, we saw that another party of ten or twelve men was in difficulties. They, too, had got separated from our main body, while foraging on the table-land earlier in the morning, but no one had missed them until now when we heard the sound of distant firing and saw them riding into view two miles away, hotly pursued by numbers of mounted Basutos.

353 Doctoring for War.

354 The rump. A crupper is a piece of tack used on horses to keep a saddle, harness or other equipment from sliding forward.

355 To the Bitter End: a photographic history of the Boer War, 1899–1902, by Lee, Emanoel C. G Harmondsworth: Penguin, 1986. https://archive.org/details/tobitterend00eman_0

African scouts and warriors.[419]

Another view of native scouts and warriors.[420]

African soldiers prepared to repel raiders into Basutoland/Lesotho.[421]

Native African British Scouts.[422]

They were in grave danger, for between them and ourselves ran a deep ravine, towards which they were being shepherded, and as the ravine seemed impassable, it looked for a time as if we should have to stand by helpless and see them killed. In order to do what we could, the whole commando mounted and rode to the edge of the chasm, and here; fortunately, we found a piece of high ground from which we over-looked the scene on the other side, and were able to drop our bullets among the advancing natives with such good effect that they reined in. This gave the cornered men time to search the cliff for a way down, which they succeeded in finding under cover of our fire, and ultimately they rejoined us without casualties.[356]

Boer's shooting down into a valley.[423]

This final episode reduced my ammunition to four rounds, and indeed many of the rest were no better off, for the long chase to which they had been subjected during their dash through the Free State had depleted their

[356] This action took place on September 4th, 1901 Near Wittenberg Mission station.

bandoliers to such an extent that the question was becoming a very serious one for a column such as ours, starting to invade hostile territory.

After this we halted for an hour to give the wounded men a rest, and to enable those whose horses had been killed to get remounted. There were seven men hit, and, as we had no lint, bandages, or medical supplies, there was little that we could do for them.

After a while the injured men were placed in their saddles, and we trekked away, with several bands of natives hovering in our rear, as if they contemplated a further attack, but in the end they retired. After a wearisome ride we got beyond the area of the Native Reserve, and towards afternoon we came across the first European farmhouse, where we left our wounded to be fetched in by the British.

We rested our exhausted animals till dusk, and then, saying good-bye to the wounded men in the house, we rode on for five or six miles before camping.

For months past we had experienced an unbroken spell of fine weather, bitterly cold at night, but cloudless sunshine by day. Now, however, there was a change, and it came on to rain heavily, so that we spent the long hours of darkness dismally lying in mud and water. This weather, coming on top of the crowded events of the last twenty-four hours, gave us our first taste of what was awaiting us in the Cape Colony, and thus early we began to appreciate the fact that our road was likely to be a thorny one.

Next morning the sky cleared somewhat, although a penetrating drizzle continued for most of the day, through which we rode shivering, our thin clothing being but little protection. My own wardrobe was typical; a ragged coat and worn trousers full of holes, with no shirt or underwear of any kind. On my naked feet were dilapidated rawhide sandals, patched and repatched during eight months of wear, and I had only one frayed blanket to sleep under at night. Few of the men were better off, and we looked with apprehension on the change of weather, for it meant that the rainy season was upon us, with its attendant hardships, the full extent of which we were yet to learn.

Our course during this day took us through more settled parts, and for the first time we looked at farms and homesteads untouched by the hand of war. There were men peacefully working in the fields, and women and children standing unafraid before their doors as we passed, a very different picture from that to which we were accustomed in the devastated republics.

The people were almost exclusively of Dutch origin so they gave us unselfish hospitality. In the matter of clothing they were hardly able to assist us, on account of the military embargo which prevented them from buying more than certain quantities, but gifts of coffee, sugar, salt and tobacco were ungrudgingly made, and the first slice of bread and butter and the first sip of coffee I had tasted for a year almost made the long journey worth while.

In spite of the bad weather, our first day among a friendly population was a pleasant experience, which put the men in good spirits, and I dare say we posed a little before the womenfolk, laughing and whistling as we rode along.

Our course, later in the afternoon, took us up a mountain-pass, and when we reached the top towards evening, we could see in the distance the comfortable hamlet of Lady Grey[357] nestling to the left, while in a glen below was the old familiar sight of a British column, crawling down the valley. This gave us no anxiety, for, being without wheeled transport of any kind, we turned across the heath and easily left the soldiers far behind.

That night it rained again, and a cold wind drove against us from the south. Our commando presented a strange appearance as we wound along; we had no raincoats, so we used our blankets as cloaks against the downpour, and the long line of draped horsemen looked like a tribe of Red Indians on the warpath.

Long after dark we came to a halt, spending a wretched wet night, and at dawn, cold and miserable, we trekked over bleak country, the biting wind in our faces, until at four in the afternoon we came to rest near a place ominously called Moordenaar's Poort (The Murderer's Way).[358] The rain how ceased, and, a passing herdboy having told us that English troops were camped a few miles off, General Smuts decided to go and see them for himself. He took with him two young Freestaters who had joined him on the way down, and another man named Neethling from Pretoria an old friend of mine. With these he left, saying that he would be back by dark. At

[357] Lady Grey is situated in the foothills of the Witteberg mountains, near the border with Lesotho. Lady Grey, South Africa - Wikipedia
[358] Bethulie is a small sheep and cattle farming town in the Free State province of South Africa. Bethulie - Wikipedia

sunset he had not returned, and for hours we anxiously waited for him until, shortly before midnight, he walked in among us on foot and alone.

He had been ambushed by a British patrol, who had killed all three of his escort and all the horses, he alone escaping down a nullah. Had he been killed I believe that our expedition into the Cape would have come to a speedy end, for there was no one else who could have kept us together.

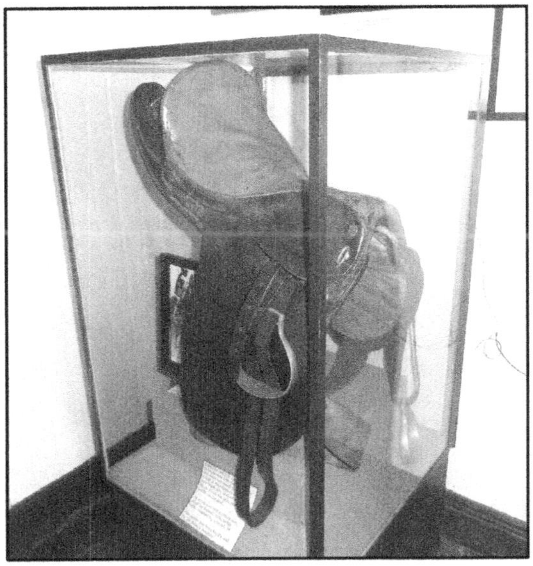

Jan Smuts' riding saddle in display case at the Smuts House Museum.[359] Commandant Smuts, with a similar saddle, if not the same one.[424] Commandant Barend Daniel Bouwer.[425]

The commando was divided into two portions commanded by Jacobus van Deventer[360] (Later Sir Jacobus van Deventer in the Great War.) and Ben Bouwer[361] [362] respectively, both good fighting-men but neither of them

[359] General Smuts' saddle: In September 1901 General Smuts' men were ambushed by a patrol at Moordenaars near Dordrecht in the Cape [Colony]. Smuts' horse was shot and [the] saddle confiscated. The saddle has been kindly lent by the Howick Museum. File:Jan Smuts se rysaal, Smutshuis, a.jpg - Wikimedia Commons Author JMK.

[360] Lieutenant-General Sir Jacob Louis van Deventer KCB CMG DTD (18 July 1874–17 August 1922) was a South African military commander. A serious wound at the end of the Anglo Boer War affected his ability to speak. In the German East Africa Campaign (1916–1918), he commanded a South African Overseas Expeditionary Force mounted brigade, then a division, and finally (1917–1918) all the British imperial forces in the region (though, allegedly, he could hardly speak English). He was knighted twice for his services. Jacob van Deventer (general) - Wikipedia

[361] Bouwer Barend Daniel General: He was born in 1875 and grew up in Angola. A good linguist, he served as an interpreter at the law courts in Pretoria. He was involved in the Jameson Raid as a despatch rider. During the Boer War her served in the Ermelo Commando and rose to the rank of vetch general. And: Commandant Commando: General Smuts. Awarded the DTD and received the Wound Ribbon. His commando surrendered on 19 Jun 02. He participated in the Peace Conference and afterwards worked as a prospector. He dies in Cape Town in 1938. See: https://af.wikipedia.org/wiki/Ben_Bouwer Permalink: https://af.wikipedia.org/w/index.php?title=Ben_Bouwer&oldid=2700841
See also: Ben Bouwer - Wikipedia

[362] Commandant Barend Daniel Bouwer. Literal translation from Dutch from Our [Transvaal] Military Officer Album..
About twenty-five years ago a number of Boer families left the Transvaal in order to settle in Damaraland and it was here, during this "Dorstland-Trek," that Barend Daniel Bouwer, on January 31. 1875, born beheld. Returning to Transvaal, he received his education in Pretoria and entered the profession subsequently in state service, initially at the Commandant General's office, then on the Department of the State Attorney. When the war broke out, he was a public servant Prosecutor in Ermelo. Then, during the last days of 1895, it became known Jameson had crossed the border, Bouwer informed him, by order of the Government, to return immediately before any fighting had taken place. As a witness in the known He visited the Parliamentary Inquiry that was subsequently instituted against Jameson next year England. At the outbreak of the latest war he joined it Ermelo commanded, contributed to the main battles in Natal and later joined the Transvaal squadron of Pres. Steyn, under Commandant J. Boshoff, to the west districts. At the end of 1900 he was appointed Commandant of General de la Rey one of the commandos who, under General Jan C. Smuts, went to the Cape Colony. Here about 150 Colonials joined him and he took then, until the end of the war, an active share in military operations there. He was wounded four times: once in Rustenburg, then in Windhoek, Cape Colony, then in the Van Rhijnsdorp district, Cape Colony, and finally, at the siege of O'okiep, Namakwaland, on April 9, 1902. Literal translation from Dutch from: Our [Transvaal] Military Officers: Dutch: Onze krijgs-officieren: album van portretten met levens-schetsen der. by Gustav Schoeman Preller, Frans Vredenrijk Engelenburg 1904 Catalogue Record: https://archive.org/search?query=title%3A%28Onze%20Krijgs-Officieren%29

possessing the personality or the influence over men that General Smuts had, to save us from going to pieces during the difficult period upon which we were now entering.

We spent the night where we lay, and there was more trouble before daylight, for a porcupine came grunting through our lines, with the result that the horses stampeded in a body. They thundered off in the dark, crashing through fences and undergrowth in blind terror, and at sunrise there was not an animal to be seen.

Commandant Barend Daniel 'Ben' Bouwer.[426]

With an English force in the vicinity this was a serious predicament, for they would make short work of us if we were dismounted, so all hands turned out to hunt for the horses. Luckily a few of the men had hobbled their mounts, which prevented them from going as far as the rest, and as these were run to earth in a hollow not far away, they were used to track the others, and after three or four uneasy hours, expecting to see the English appear at any moment, we brought back all the missing horses safe and sound.

We now travelled on for the next three days across windy barrens, heading south-west. The weather grew more and more tempestuous as we went, and we suffered severely from the cold, and from the intermittent rains that accompanied us. Both horses and men began to show signs of distress. The animals looked thin and gaunt, and the man sat on their saddles pinched, shivering, and despondent, for South Africans are peculiarly susceptible to the depressing effects of bad weather. They can stand cold and other hardships as well as anyone, but continued lack of sunshine soon makes them miserable, and for the time being we were a dispirited band, wishing we had never come.

By day we were wet and cold, and the nights were evil dreams. Dry fuel was almost unprocurable, and after a weary day we had to spend the hours of darkness cowering together to snatch a little sleep on some muddy mountainside, or in an equally sodden valley.

Soon we were losing horses freely, and not a trek was made without some wretched animals being left behind with tuckered flanks and drooping heads, waiting for the end. We had three days of this, but our real troubles were only beginning.

Towards sunset one evening we came in sight of the village of Jamestown,[363] and saw a strong English column to our right, so General Smuts moved us on. It grew pitch-dark, and a driving rain smote straight in our faces. The night was so black that it was impossible to see even the man immediately before one, and the cold so bitter that we became stiff and numbed, and it was only with difficulty that we could drag our horses along, for we were ordered to go on foot to husband their strength.

[363] Jamestown, Eastern Cape - Wikipedia

The British Army column, both infantry and mounted regiments, crossing a river.[427] Unit of British cavalry or mounted yeomanry crossing a river.[428]

When I was crossing a spruit, my sandals stuck in the heavy pot-clay and came to pieces when I tried to withdraw them, and it was only by cutting corners from my blanket and wrapping one about each foot that I was able to go on at all. Our guide, a young man from a local farm, had lost his bearings, so we had to grope our way through icy rain for five hours, until we could continue no longer, and stood huddled together ankle-deep in mud and water, praying for sunrise.

When it grew light, over thirty horses lay dead from exposure, besides others abandoned overnight, and our spirits, low before, were at zero now.

The rain continued pitilessly until midday, when the sky cleared and the blessed sun shone upon us once more. We moved forward, and not far away saw a large farmhouse and outbuildings containing plenty of fuel. Soon we were warming our numbed bodies, and cooking our first hot meal for days.

The housewife at the farm gave me a pair of old-fashioned elastic-sided boots, and I unearthed an empty grain-bag in which I cut a hole for my head, and one at each corner for my arms, thus providing myself with a serviceable great-coat. My appearance caused much laughter, but I noticed that during the next few days, whenever we passed a barn, grain-bags were in great demand, and soon many of the men were wearing them.

As the people here told us that there was an English force in the neighbourhood, we moved on later in the afternoon, first saying good-bye to Louis Wessels, the young Free State officer and his men, who turned back from here as they had only come thus far to see our force well launched into the Colony. I believe that they reached their own country again in safety.

We continued for an hour, and then halted in a valley. Whilst we were idly resting in the grass two field-guns banged at us from a hill, and shells came tearing overhead. More followed and, taken by surprise, we leaped into the saddle and made for the cover of a line of hills to the rear.

The artillery was poor, and neither man nor horse was hit.

Brabant's Horse Artillery.[429]

Once in safety, we put our animals out of harm's way, and climbed up to see what the English meant to do. We could now see a column of horse coming down towards us, around a spur that had previously hidden them from view.

There were about six hundred of them, and they had with them two fifteen-pounder Armstrong guns, and several Pompoms which unlimbered and opened fire, while their horsemen cautiously approached us. After a time they quickened pace as though to attack, but coming under our fire, they took cover behind some farmhouses and kraals.

15 pounder Armstrong Gun.

British Pom Pom.[430]

The Royal Artillery crew in action. Note the Navy Bluejacket in the background carrying the next shell.[431]

In spite of the shelling, and a lively exchange of rifle-fire, there were apparently no casualties on either side, and the affair terminated after dark. I did not fire a shot on account of the state of my cartridge belts, and the others fired no more than was necessary to stave off the enemy, because, as I said before, the ammunition question was an exceedingly serious one.

When the light went out we withdrew to a farm close by, hoping for a real rest this time as we had not enjoyed a full night's sleep since we crossed the Orange River, more than a week before.

We did not get that night's rest, for at three o'clock next morning we were ordered up in the dark, and started in a cold drizzle of rain on a record march. Our men were weakened by long privations, our ammunition had dwindled to vanishing-point, and our horses were in the last stages of exhaustion, yet during the next six days, beset on all sides, we marched and fought, and in the end successfully got through.

XXI. Horses and Men

When the sun rose and the rain ceased, we found ourselves crossing a high shoulder of land with a wide expanse of mountains to the south of us, and there, in every valley and on every road, stood the white tents of English camps, to bar our progress.

Scots Greys' Camp at Orange River.[432]

General Smuts surveyed the blocking forces for a while, and then led us due east across the front of the enemy posts. Our road took us through rough country and he ordered every man to go on foot to spare the horses. The English made no attempt to come after us, their orders apparently being to hold the roads and exits, so we trekked all day, seeking to turn their extreme right flank. Throughout the expedition into the Cape we had no difficulty in getting local sympathizers to act as guides, and on this occasion a young farm-hand volunteered to lead us. He picked his way so unerringly that towards nightfall we had not only succeeded in finding the end of the British line, but had even got round behind them, and could see the town of Dordrecht[364] in the distance.

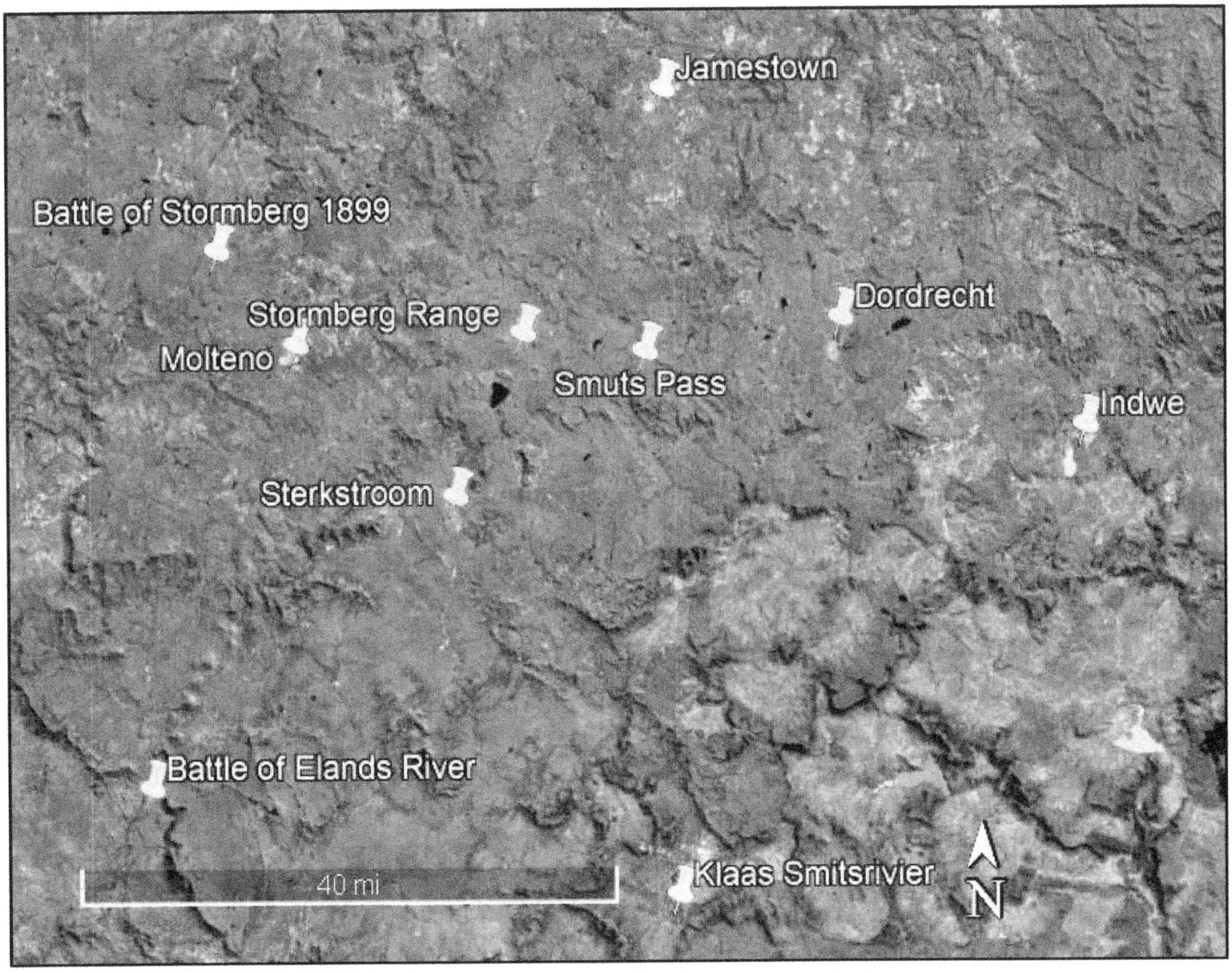

Map data: Google Image© 2024 Image Landsat/Copernicus.

[364] Dordrecht, South Africa - Wikipedia

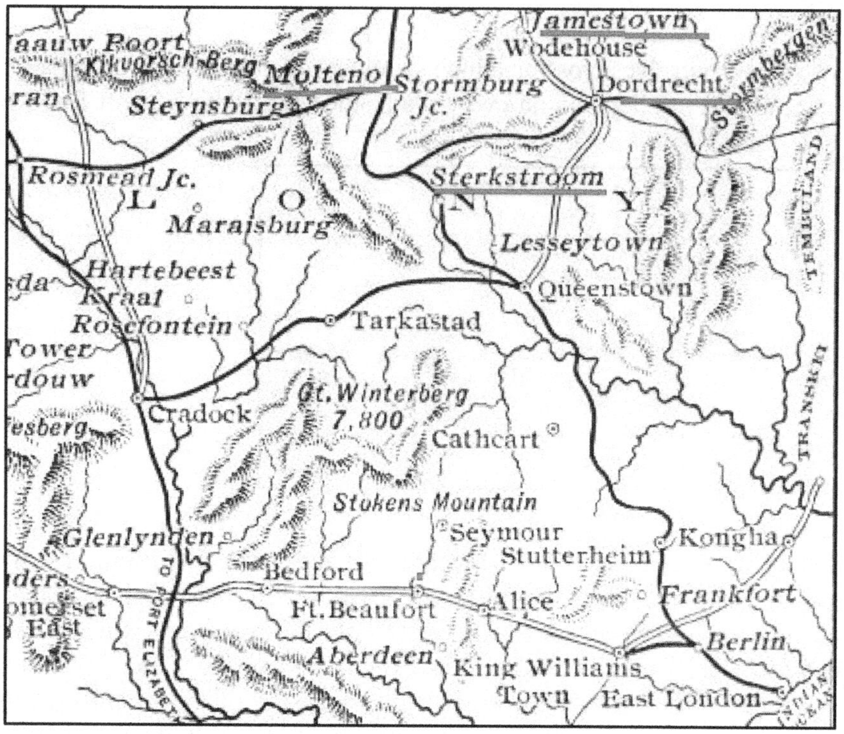

Jamestown, Molteno/Stormberg and Sterkstroom Features Noted.

We must have covered nearly thirty miles since getting out that morning, but as we were not yet out of danger of being headed back, we continued after dark, and, hungry and weary but in good cheer, we trudged all through the night, with only an occasional halt, mostly along steep mountain paths, wet and slippery from the rains. When daybreak came our young guide had done his work with such skill that we were well beyond the cordon, and there now lay before us the long mountain chain of the Stormbergen,[365] stretching east to west as far as the eye could see.[366]

Stormberg Mountains.[433]

He told us that we could cross almost anywhere, so he was allowed to return home, and we made for a large farm lying at the foot of the range, where we turned our horses into the fields, and set about preparing a meal, once again hoping to spend the rest of the forenoon in sleep, for we had been on the move for twenty-four hours. But we had scarcely slaughtered a few sheep and broken our fast, when the well-known cry of 'Opsaal! Opsaal!'[367] sent us scurrying to fetch our unfortunate animals; for coming down the slopes was a long column of English horse making our way. Near by ran a pass up the mountain, and as it seemed clear of troops, we made

[365] The Stormberg is a range of mountains situated in the Eastern Cape province of South Africa, are an easterly extension of the Bamboesberge and form an outlier of the greater Drakensberg mountain range. Stormberg Mountains - Wikipedia

[366] PF Likely to be: Smuts Pass. A mountain pass situated in the Eastern Cape province of South Africa, on the road between Dordrecht, Eastern Cape and Molteno. Smuts Pass - Wikipedia

[367] 'Saddle up'

for it, and in an hour stood on the top of the Stormbergen, with the enemy force slewed round and following us. The summit was a grassy tableland about three miles wide, sloping gently to where the southern face of the mountain fell abruptly down to the plains below.

There was no sign of troops up here, but as those in the rear were coming after us, General Smuts disposed his men to hold them back while we of the 'Rijk Section' were ordered to ride forward to the far edge of the plateau, to see whether the way in that direction was clear. We set out in couples, making for different points from which to look down on the Karroo.

Anysberg mountains overlooking the Karoo.[434]

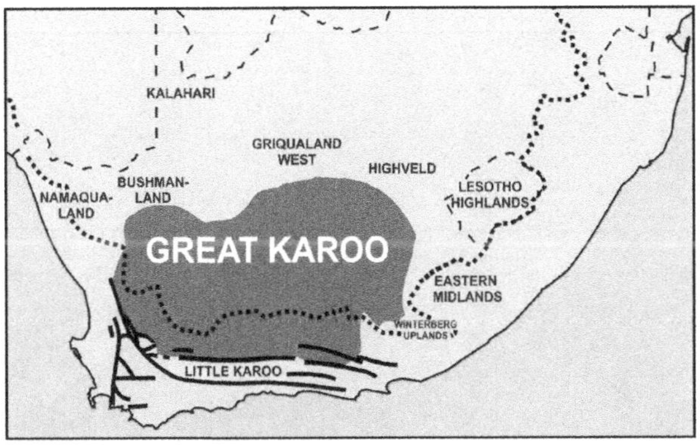

Great Karro Area.[368]

My companion was Henry Rittenberg, and when we reached the rim of the plateau we saw the narrow ribbon of a railway track winding across the plain at our feet, with train after train steaming up to a small village station and disgorging large numbers of soldiers.

The British, having failed to stop us at their first barrier, were now hurrying troops round by rail to establish a second, and already several mounted columns had detrained, and were beginning to climb up the mountain.

One body was so far advanced that their scouts were appearing on the tableland itself, so Rittenberg and I rode nearer to examine their strength. We had not gone half a mile when some two dozen troopers rode at us from behind a roll in the ground, firing from the saddle as they came. We whipped round and galloped away, but had not the balance of the 'Rijk Section' come to our aid we should have been shot or captured, for there was no cover in which to make a stand, and our horses were in no fit state to compete with the well-fed English chargers.

We now returned to General Smuts, to report what we had seen, and he looked grave enough, for, with the original column of that morning closing in behind and all those fresh troops coming up in front, we were almost invested.

A strong north wind, which had sprung up earlier in the day, had steadily increased to a violent gale, and most of the men were crouching with their backs to the storm to escape the flying grit that stung like buckshot.

General Smuts, with Commandants van Deventer and Bouwer, was, however, on the look-out, standing well forward near the head of the pass up which the commando had come. While we were explaining the situation, about three hundred soldiers appeared on foot, having left their horses below. They did not seem to be expecting us here, or, for, when our men sprang up at a shout from van Deventer, they turned and ran back.

They were out of sight almost at once, and when we reached the edge we saw them scrambling down, but the wind in our faces made accurate shooting impossible, and I do not think any were hit. The soldiers only fired a few shots in reply, and a young man standing near me, named de la Rey, a nephew of General de la Rey, threw up his arms and dropped dead with a bullet through his brain. We left him where he fell, for we had no spade or other implement, nor had we time to bury him, for looking back we could now see more and more English horsemen emerging on the tableland, until we were practically encircled, although they were not yet strong enough to prevent us from moving freely inside the wide ring that they had formed about us.

[368] Extent of the Karoo (olive-green) and Little Karoo (bright green) The thick interrupted line indicates the course of the Great Escarpment which delimits the Central South African Plateau. To the immediate south and south-west the solid lines trace the parallel ranges of the Cape Fold Belt. File:Map of Karoo.png - Wikimedia Commons Author: Oggmus.

Nevertheless we could not break out, for soon there were machine-guns at every point of vantage, so commanding the terrain that a burst-through during daylight would have cost us more men than we could afford, and our only course was to try to stave off the pressure until after dark. To that end, with the gale roaring about our ears, General Smuts led us hither and thither all the afternoon, now pushing back one portion of the enemy line, and then another, avoiding the machine-gun fire by using dead ground, and generally preventing them from hustling us too closely. With our tired horses and men the strain was great. Ammunition was at such a low ebb that some had not a round left, and when, towards evening, many more troops had come up our case seemed hopeless.

Pumping Lead.[435]

We and our horses had marched for forty hours on end, and we were all but finished for lack of sleep and rest, while the noose around us had slowly tightened, until by dusk we were at bay around a small farmhouse and kraal, lying somewhat in a hollow, where for the moment we had comparative shelter, but where our speedy capture seemed certain.

When the English troops saw us preparing to make a stand they stayed their advance, in the belief, no doubt, that having cornered us, they could afford to wait for our surrender in the morning.

General Smuts stood before the homestead in whispered consultation with his two lieutenants, while the rest of us leaned on our rifles, too weary to care very much what happened. Then out of the house came a hunchbacked cripple,[369] who said that he would lead us through the English troops to the edge of the table and, by a way which was unlikely to be watched, for it ran through boggy soil. His offer was eagerly accepted and orders were given to mount at once.

Six or seven men had been wounded during the day, two of them so badly that they had to be left behind, but the others chose to accompany us, and in a few minutes we were silently filing off into the darkness, the cripple crouching insecurely on a horse at our head. He took us along a squelching path, that twisted for a mile or two so close to the investing troops that we could hear voices and the champing of bits, but at the end of an anxious hour he had brought us undiscovered to the escarpment. From here the mountain-side fell sharply away into black depths below, how steeply we could not tell, but our guide warned us that it was very steep indeed. Dropping from his horse, he plodded off into the night on his crutches, carrying with him our heartfelt thanks, for he had risked his life and goods on our behalf.

We now began to descend what was probably the nearest approach to the vertical attempted by any mounted force during the war. I doubt whether we could have accomplished it by day, but horses are more tractable and surer-footed in the dark, so we pulled them over the edge and went slithering down. At times whole batches of men and horses came glissading past, knocking against all in their course, but luckily the surface was free of rock, and covered with a thick matting of grass which served to break the impact, and after a terrible scramble we got down without serious damage. For the time being we had shaken free of the enemy once more. Our most insistent need now was sleep, but this was still denied us.

Somewhere on the plain before us ran the railway line on which we had looked down that morning, and many miles beyond that lay still another track, both of which had to be crossed before sunrise, if we did not wish to

[369] Probably Hans Kleynhans. See Battle of Elands River article infra.

have the troop-trains hurrying up more men. So General Smuts implacably ordered us on, and, leading our horses, we tramped obediently but wearily forward, little dreaming that another twenty hours of unbroken marching lay before us, and several days of even greater trials to come.

It was about ten o'clock by now, and the storm that had been raging throughout the day was subsiding, though the aftermath still blew cold; a blessing perhaps, for it served to keep us awake, and it made us step out to keep warm. After an hour we reached the first railway, a branch line from the Indwe Coal Mines.[370] As we approached we saw the lights of a train, but General Smuts would not allow us to pile boulders on the metals nor to fire as the engine thundered by, for fear of killing civilians,[371] so we stood aside, catching a glimpse of officers and others seated in the dining-car, smoking and taking wine, all unaware of the men looking at them from the darkness. General French, the English cavalry leader, told us long after that he was on board that train with his staff, hurrying round by rail to control operations on the berg where he imagined us still to be, so unknowingly we missed a great opportunity.

Boer sharpshooters sighting a British patrol.[436]

Stones on the track intended to derail train.[372]

Engineers repairing a Railway destroyed by the Boers near Kroonstadt.[437]

Destruction of a train of supplies.[438]

After crossing the rails, we went on mile after mile, dazed for want of rest. Whenever there was delay at a fence or a ditch, whole rows of men would fall asleep on their hands and knees before their horses like Mohammedans at prayer, and it was necessary to go round shaking them to their feet to prevent them being left behind. Save for occasional halts we continued thus all night, for it was imperative to cross the remaining railway. As we had no guide we travelled by the stars, and the sun was rising before we struck it at a small siding about five miles east of Sterkstroom[373] village, where, from the activity at the station, it was obvious that our escape from the

[370] Indwe is a small town in Chris Hani District Municipality near Dordrecht in the Eastern Cape province. Indwe - Wikipedia

[371] Commandant Lotter will be executed for train wrecking.

[372] Note Guard Rail installed to keep train on track around the curve. Credit: Neville Constantine. at Angloboerwar.com.

[373] Sterkstroom is a settlement in Enoch Mgijima Municipality of Chris Hani District in the Eastern Cape. Sterkstroom - Wikipedia

mountain was known. Several trains were unloading troops, and there was no time to be lost. So mounting, we galloped as fast as we could across the rails for fear an armoured train might cut us off.

Pattern 1 Gun Truck and Train.[439]

No. 3 Artillery Train at the railway station at Bronkhorstspruit.[440]

Pattern 2 train.[441]

An Armoured Train.[374]

Commandant van Deventer and a few of us remained behind to search the railway buildings for anything that might come in useful, and while we were busy at this a long goods-train came clanking up, and we brought it to a standstill by switching the points. It was an empty coal train in charge of a driver, stoker, and brakesman, whose faces were a picture when they saw what we were doing. As the train consisted of nothing but

[374] Frontiers and Wars four early books, 1962. Credit, Mr. Helmut Gernsheim.

incombustible steel trucks, we let them proceed, after extracting a mail-bag from the guard's van. The letters were all private ones, seemingly written under censorship, for not one of them made any reference to the war, but the newspapers were less reticent, and in one of them was unflattering mention of ourselves, for it said that General Smuts had invaded the Cape Colony 'with the riffraff of the Boer Armies,' which caused much merriment when later on I was able to read it to the men.

There was another surprising feature in the mail-bag in the shape of a Proclamation by Lord Kitchener, wherein every burgher under arms after the 15th of September[1901] was sentenced to perpetual banishment from South Africa. This was news to us, and seeing that it was the 13th of that month by now, we were left with a bare two days in which to comply. This announcement was received with equal derision, when made known to the commando, and from what I have heard since, it had equally little effect up north in the Republics where the 'paper bomb,' as it was called, got treated with the scorn it deserved.[375]

After speeding the goods-train on its way, we overtook the commando at the Klaas Smits River,[376] where we halted for about an hour, to give our poor horses a chance to pluck a few mouthfuls of grass, and to prepare a hasty meal for ourselves. Longer than that we were not given, for a column of troops with guns came up, and kept us on the move all day, slowly retiring from hill to hill, half dead with fatigue, but keeping them at arm's length until sunset, when they turned back and left us free to camp at last at a large farm, where we lay like dead men until morning after sixty hours of continuous marching.

This full night's rest was a great relief, but the strain was by no means over, and the worst was yet to come.

Towards nine o'clock next day, an English column appeared from the direction of Sterkstroom, so we saddled up and rode away, skirting the base of some hills running south. The column contented itself with following us slowly, apparently having been sent to keep us under observation, and they dogged our steps until sunset, when it came on to rain and we saw them go into camp. We halted in a patch of thorn trees and, as it poured heavily till dawn, we had another of those wet and miserable nights which had been so frequent since our entry into the Colony.

Daylight saw the troops once more coming after us, and, owing to the shortage of ammunition and the condition of our horses, we had no option but to give way. The going was heavy, and at times the swollen spruits and dongas held us up, but we were in no great danger of being overhauled, for the column had wagons and guns, which impeded them so much that we had miles to spare.

In the afternoon the English camped again, and we halted for the rest of the day at a small farmhouse standing out on the plain. The rain had ceased since morning, but it was cold and threatening; black clouds hung low in the sky, and there was every promise of more dirty weather to come.

We could see smoke curling from the English camp four miles behind, where whole streets of comfortable tents had sprung up, at which we gazed wistfully, for there were warmth and rest, whilst we stood shivering in the biting wind, wondering how it was all to end.

[375] Original Footnote: The Times History says of this Proclamation, Vol. V, 321:
Lord Kitchener made his first and last attempt to end the war by a minatory proclamation.
It began by giving in a solemn legal preamble an account of the military situation, which must have seemed strangely unconvincing to the Boers, and which, it must be confessed, was replete with unconscious humour. The fourth paragraph in particular where the Boers were informed that they were incapable of carrying on regular warfare was a strange tissue of perverted logic.... The operative art of the Proclamation demanded the surrender of all Boers before September 15th [1901] under pain of stringent penalties. The results were not encouraging: Botha, Steyn and de Wet sent defiant replies, and among the burghers at large there was a sullen silence.

[376] The Klaas Smits River is a river part of the Great Kei River system in the Eastern Cape, South Africa. It originates south of Molteno and flows through Sterkstroom, first southwards and then southeastwards before joining up with the Black Kei River Klaas Smits River - Wikipedia

Large British Camp.[442]

Camp Barber (Hairdresser).[443]

Camp Kitchen.[444]

Afternoon Tea.[445]

Shaving in Camp.[377]

Arrival of the English Mail.[446]

Camp butchery a pleasant change from Bully Beef.[447] Smiling soldier of one of the British Army units at the table at his camp.[448]

377 Anglo-Boer War Album, Cape Town, D. Edwards & co. [1902?]

The English numbered about a thousand, and it was useless to attempt an attack in our present condition, for we were wet, cold, and in low spirits, and our ammunition was almost finished. So, when darkness fell, General Smuts gave orders for us to saddle up, intending to make for a larger farm where there was said to be ample shelter.

As we started, hard rain came down once more, and the darkness was so intense that we could not see a yard ahead. We had not gone three hundred paces before we heard horsemen splashing through the mud in front, and ran into the tail of an English patrol or column, we could not tell which, evidently making for the same farm.

Neither side was prepared to risk a fight in the rain and dark. The troopers galloped away, and we sheered off, too, but with this difference, that they were able to continue on to the shelter of the farm, whilst we were adrift on the open veld.

The night that followed was the most terrible of all. Our guide lost his way; we went floundering ankle-deep in mud and water, our poor weakened horses stumbling and slipping at every turn; the rain beat down on us, and the cold was awful. Towards midnight it began to sleet. The grain-bag which I wore froze solid on my body, like a coat of mail, and I believe that if we had not kept moving every one of us would have died. We had known two years of war, but we came nearer to despair that night than I care to remember. Hour after hour we groped our way, with men groaning who had never before uttered a word of complaint, as the cold searched their ill-protected bodies. We lost fourteen men that night, and I do not know whether they survived, but we never again had word of them.

We also lost a large number of horses, and I remember stumbling at intervals over their carcasses. We went on till daybreak, dragging ourselves along, and then, providentially, came on a deserted homestead and staggered into shelter, standing huddled together in rooms, stables, and barns until dawn, still shivering, but gradually recovering from the dreadful ordeal. When it grew light, some fifty or sixty horses lay dead outside. My little roan mare was still alive, but both my uncle's horses died here, and he, with thirty or forty more, was now a foot-soldier.[378] This night's 'Big Rain,' as we named it, left such a mark on all of us that later we used to call ourselves 'The "Big Rain" Men' (Die Groot Reent Kerels) to distinguish us from those who had not experienced it, and for my part I passed through no greater test during the war.[379]

The day was cold and wild, but the rain stopped. We broke up the floors and windows, tables and chairs, and everything else that would burn, and made great fires to dry our clothes and blankets, and to warm our chilled limbs. Towards noon, General Smuts ordered us on to another large farm, eight or nine miles away, which had, a native told him, plenty of fodder for the horses.

No attempt was made to send back for the missing men, because we were too exhausted, and they had to be abandoned.

We plodded over the waterlogged country, a quarter of our number on foot, and the rest soon likely to be, for there was not a fit horse in the commando.

We found this farm also deserted, but there was protection for all, and a good store of oat-sheaves, as well as sheep for slaughter, so that, although the rain came down again, we at last spent a comfortable night.

Although we had managed to avoid the different cordons thrown in our way and had eluded the columns sent in pursuit, we were not yet out of danger, for local natives now told us that southward every road, valley, and outlet was blocked by English troops. This meant that they were once more trying to head us back out of the Cape; but, with so many enemies in our rear, our only alternative was to go forward. Next morning, we set out on what was to be an eventful day (September 17th, 1901).

[378] Original Footnote: As practically every man had crossed the Orange River with two horses, the number of dismounted men did not necessarily correspond to the number of horses that were lost.

[379] September 15th, 1901

Kitchener's Proclamation

LORD KITCHENER'S PROCLAMATION.

SURRENDER OR BANISHMENT.

GOVERNMENT DETERMINED TO END THE WAR.

A Parliamentary paper, issued to-day, contains the proclamation issued by Lord Kitchener on August 6 with the concurrence of the British Government. It orders surrender before September 15, or permanent banishment for all commandants, field cornets, leaders of armed bands, and all members of the Boer Governments. The full text of the proclamation is as follows :—

Whereas the late Orange Free State and the late South African Republic have been annexed to his Majesty's dominions, and whereas his Majesty's forces are and have been for some considerable time in complete possession of the seats of Government of both of the aforesaid territories with their public offices and the whole machinery of administration as well as of all the principal towns and the whole of the railway lines, and whereas the great majority of the burghers of the two late Republics to the number of 35,000 exclusive of those who have fallen in the war are now either prisoners or have submitted to his Majesty's Government and are living peaceably in towns or camps under the control of his Majesty's forces, and whereas the burghers of the late Republics still in arms against his Majesty are not only few in numbers but have also lost almost all their guns and munitions of war and are devoid of regular military organization and are therefore unable to carry on regular warfare or to offer any organized resistance to his Majesty's forces in any part of the country, and whereas those burghers who are still in arms though

UNABLE TO CARRY ON REGULAR WARFARE

continue to make isolated attacks upon small posts and detachments of his Majesty's forces to plunder or destroy property and to damage the railway and telegraph lines both in the Orange River Colony and Transvaal and in other portions of his Majesty's South African dominions, and whereas the country is thus kept in a state of disturbance checking the resumption of agricultural and industrial pursuits, and whereas his Majesty's Government is determined to put an end to a state of things which is aimlessly prolonging bloodshed and destruction and inflicting ruin upon the great majority of the inhabitants who are anxious to live in peace and to earn livelihood for themselves and their families, and whereas it is just to proceed against those still resisting and especially against those persons who being in a position of authority are responsible for the continuance of the present state of lawlessness and are instigating their fellow burghers to continue their hopeless resistance to his Majesty's Government: Now therefore I, Lord Kitchener, &c., under instructions from his Majesty's Government proclaim and make known as follows :—

"All Commandants, Field-Cornets, and leaders of armed bands, being burghers of the late Republics, still engaged in resisting His Majesty's forces, whether in the Orange River Colony and the Transvaal or in any other portion of His Majesty's South African Dominions, and all members of the Governments of the late Orange Free State and the late South African Republic, shall, unless they surrender before September 15 next, be permanently banished from South Africa; the cost of the maintenance of the families of all burghers in the field who shall not have surrendered by September 15 shall be recoverable from such burghers and shall be a charge upon their property moveable and immovable in the two Colonies."

LORD KITCHENER'S REPLY

Lord Kitchener replied on August 2nd:

"Ministers, Cape Town, entirely concur in proclamation. Have not yet heard from Natal." Natal."

On the following day Lord Kitchener telegraphed:—"The Government of Natal express approval of the proclamation."

The correspondence further shows that on July 24th, Sir H. E. McCallum, Governor of Natal, telegraphed to Mr. Chamberlain as follows:—

"I sent you by last mail long minute submitted by Ministers, containing the following suggestions:—(Begins). Protracted continuance of hostilities, vitally affecting interests of Natal, is viewed with grave concern. Raids into the Colony frequently render it impossible for loyalists to return to farms and avocations. Feeling of unrest among natives created by raids. Revenue suffering, trade paralysed, railways monopolised by military, towns overcrowded with refugees, and persons awaiting return to Transvaal; stock being infected with disease due to introduction of captured stock from the new Colonies; famine prices prevailing; Colony still subject to censorship and martial law. Under these circumstances Ministers advocate sterner measures to crush present guerilla warfare. They point out that Boers still fighting have little to lose, that their women and children are protected and well treated, and that their farms are safe from confiscation. Therefore, Boers free from anxiety and encouraged to continue in the field; growing accustomed to life of pillaging and looting, and communicate frequently with refugee camps, which thus are source of danger. Those who are not rebels know that if captured they will be treated as prisoners of war and released at conclusion of hostilities. Ministers believe that excellent effect would be produced if it were made generally known that if burghers now in field do not surrender before given date say within one month, cost of maintenance of all women and children will be chargeable against immovable property of burghers in field; also that Boer generals and leaders in field should be informed that unless they and their commandos surrender by date specified, they will be banished from South Africa for life when captured.

"In making these observations and recommendations Ministers disclaim intention of appearing to reflect on military operations, of which they realise the immense difficulties. They will continue to render Imperial Government every assistance to secure settlement and pacification." (Minute ends.)

Proclamation Text.[380] Kitchener's Reply to a request to obtain consent from the Colonial Governors.[381]

Lord Kitchener's Proclamation.
Surrender Or Banishment.
Government Determined To End The War.[382 383]

A Parliamentary paper, issued to-day, contains the proclamation issued by Lord Kitchener on August 6 with the concurrence of the British Government. It orders surrender before September 15, or permanent banishment for all commandants, field cornets, leaders of armed bands, and all members the Boer Governments. The full text of the proclamation is follows:

Whereas the late Orange Free State and the late South African Republic have been annexed to his Majesty's dominions, and

Whereas his Majesty's forces are and have been tor some considerable time in complete possession of the seats of Government of both of the aforesaid territories with their public offices and the whole machinery of administration as well as of all the principal towns and the whole of the railway lines, and

Whereas the great majority of the burghers of the two late Republics to the number of 35,000 exclusive of those who have fallen in the war are now either prisoners or have submitted to his Majesty's Government and are living peaceably in towns or camps under the control of his Majesty s forces, and

[380] Kitchener's Proclamation: Pall Mall Gazette August 9, 1901 p. 7 col. 1 BNA File BL_0000098_19010809_042_0007.pdf.

[381] Kitchener's Proclamation: The Northern Guardian, (Hartlepool) Friday August 9th, 1901 p. 3 col 3. BNA File: BL_0003039_19010809_040_0003.pdf.

[382] Kitchener's Proclamation: Pall Mall Gazette August 9, 1901 p. 7 col. 1 BNA File BL_0000098_19010809_042_0007.pdf.

[383] Fife Free Press - Saturday 10 August 1901 https://www.britishnewspaperarchive.co.uk/viewer/bl/0001062/19010810/032/0002.

Whereas the burghers of the late Republics still in arms against his Majesty are not only few in numbers but have alto lost almost all their guns and munitions of war and devoid of regular military organization and are therefore unable to carry on regular warfare or to offer any organized resistance to his Majesty's forces in any part of the country, and

Whereas those burghers who are still in arms though unable to carry or regular warfare continue to make isolated attacks upon small posts and detachments of his Majesty's forces to plunder or destroy property and to damage the railway and telegraph lines both in the Orange River Colony and Transvaal and in other portions of his Majesty's South African dominions, and

Whereas the country is thus kept in a state of disturbance checking the resumption of agricultural and industrial pursuits, and

Whereas his Majesty's Government is determined to put an end to a state of things which is aimlessly prolonging bloodshed and destruction and inflicting ruin upon the great majority of the inhabitants who are anxious to live in peace and to earn livelihood (or themselves and their families), and

Whereas it is just to proceed against those still resisting and especially against those persons who being in position of authority are responsible for the continuance of the present state of lawlessness and are instigating their fellow burghers to continue their hopeless resistance to his Majesty's Government:

Now therefore I, Lord Kitchener, &c., under instructions from his Majesty's Government proclaim and make known as follows "All Commandants, Field-Cornets, and leaders of armed bands, being burghers of the late Republics, still engaged in resisting His Majesty's forces, whether the Orange River Colony and the Transvaal or in any other portion of His Majesty's South African Dominions, and all members of the Governments of the late Orange Free State and the late South African Republic, shall, unless they surrender before September 15 next, be permanently banished from South Africa; the cost of the maintenance of the families of all burghers in the field who shall not have surrendered September 15 shall be recoverable from such burghers and shall Ire charge upon their property moveable and immovable in the two Colonies."

Times History.[384]

> Mr. Chamberlain's Message Mr. Chamberlain, in telegraphing the text of the proclamation to Lord Kitchener, on July 30th, said His Majesty's Government proposed to instruct Lord Kitchener to issue the proclamation with the least possible delay. "But," added the Colonial Secretary, "they desire you before doing so to communicate its terms to the Governors of Cape Colony and Natal, and to ascertain whether their Governments agree to them. You should state that in the opinion of His Majesty's Government its terms seem to be fully warranted by the existing situation, and calculated to have a good effect in bringing a more rapid termination of hostilities."

[384] Times History Summary of Proclamation of August 7th, 1901:[In mid 1901].............Kitchener, at the instance originally, of the Natal Government and with the approval of Milner and the Home Government, made his first and last attempt to end the war by a minatory proclamation.

Dated August 7, and issued while the great Free State drive was still in progress, it began by giving, in the form of a solemn legal preamble, an account of the military situation which must have seemed strangely unconvincing to the Boers, and which, it must be confessed, was replete with unconscious humour.

The fourth paragraph in particular, where the Boers were informed that they were not only few in number but had lost nearly all their guns and munitions of war and their proper organisation, and were therefore incapable of carrying on regular warfare, was a strange tissue of half-truths and perverted logic.

It might have been answered that the reduction of numbers had gone hand in hand with aggressive vigour; that the loss of guns, so far from being a weakness, had inspired more effective tactics; that .... the Boer organisation, if not "proper," was eminently suited to its purpose, and that whether the hostilities were called regular or irregular was not of much account.

The operative part of the proclamation demanded the surrender of all Boers before the date of September 15th under pain of certain stringent penalties.

These penalties were of two sorts. Government officials, commandants. Field cornets, and "leaders" were to be banished for ever from South Africa; burghers were to be rendered chargeable for the cost of their families' support in the British camps. The charge to be levied on their property and land. The immediate results of the proclamation were not encouraging.

Botha, Steyn and de Wet all sent defiant replies. [Received by Kitchener on August 27th, 1901]

Meanwhile the columns, worked by Kitchener at the highest possible pressure, endeavoured to supply the missing arguments for submission. Times History Vol 5 page 321–322.

Lord Kitchener's Reply

Lord Kitchener replied on August 2nd: "Ministers, Cape Town, entirely concur in proclamation. Have not yet heard from Natal." On the following day Lord Kitchener telegraphed:— "The Government of Natal express approval of the proclamation." The correspondence further shows that on July 24th, Sir H. E. McCallum, Governor of Natal, telegraphed to Mr. Chamberlain as follows: "I sent you by last mail long minute submitted by Ministers, containing the following, suggestions:—(Begins),

Protracted continuance of hostilities, vitally affecting interests of Natal, is viewed with grave concern. Raids into the Colony frequently render it impossible for loyalists to return to farms and avocations.

Feeling of unrest. among natives created by raids.

Revenue suffering, trade paralyzed, railways monopolised by military, towns overcrowded with refugees, and persons awaiting return to Transvaal; stock being infected with disease due to introduction of captured stock from the new Colonies; famine prices prevailing; Colony still subject to censorship and martial law. Under these circumstances Ministers advocate sterner measures to crush present guerilla warfare. They point out that Boers still fighting lave little to lose, that their women and children are protected and well treated, and that their farms are safe from confiscation.

Therefore, Boers free from anxiety and encouraged to continue in the field; growing accustomed to life of pillaging and looting, and communicate frequently. with refugee camps, which thus are source of danger.

Those who are not rebels know that if captured they will be treated as prisoners of war and released at conclusion of hostilities.

Ministers believe that excellent effect would be produced if it were made generally known that if burghers now in field do not surrender before given date say within one month, cost of maintenance of all women will be chargeable against immovable property of burghers in field; also that Boer generals and leaders in field should be informed that unless they and their commandos surrender by date specified, they will be banished from South Africa for life when captured.

In Making these observations and recommendations Ministers disclaim intention of appearing to reflect on military operations, of which they realise the immense difficulties. They will continue to render Imperial Government every assistance to secure settlement and pacification"

Kitchener's Despatch of October 8th 1901

From Lord Kitchener to the Secretary of State for War, London, S.W. [From] Pretoria 8th October, 1901.[385]

Sir, In continuation of my despatch, dated 8th September, 1901, I have the honour to report that during the past month there has been a general recrudescence[386] of activity on the part of the enemy, which has made itself felt more particularly in the extreme south-east portions of the Transvaal.

The immediate cause of this was, undoubtedly, the necessity felt by the Boer leaders to tide over the 15th September, the date fixed by the Proclamation of 7th August [1901] as the limit of time within which by voluntarily surrendering, the leaders might avoid certain penalties threatened by that Proclamation. In this they have succeeded. There has been no general surrender, but the device to which the Commandant-General resorted for turning the thoughts of his burghers in another direction has probably cost him and his cause more heavily than a simple pursuance of the usual evasive tactics would have entailed.

[385] Kitchener Despatch October 8th 1901 p. 587 https://babel.hathitrust.org/cgi/pt?id=hvd.32044089026009&seq=587&view=1up

[386] From Late Latin recrudescentia, ("to become raw again");("to grow harsh or violent; to become worse") (from crūdus ("bleeding, bloody, raw") The word is cognate with French recrudescence. https://en.wiktionary.org/wiki/recrudescence

Battle of Elands River 1901

Our road ran south down a long valley. The sky was clear, and the sun warm and bright for the first time for weeks, so that the men were cheerful again, although there was little other cause for optimism.

As a fighting force we were on our last legs. In front walked those who still had horses, dragging scarecrows behind them; then came a trail of footmen in twos and threes, their saddles slung across their shoulders, and in the rear rode the wounded in charge of their friends.

Soldier of one of the yeomanry regiments and his horses.[449] Fair Sample of how the Horses Looked in Ladysmith.[450]

However, the sun was shining after the wet and cold and we went hopefully along. After a few miles General Smuts ordered the 'Rijk Section' to scout ahead of the commando, so those of us who still had horses mounted and rode forward as fast as our weakened animals could carry us. When we got to where the valley widened into more open country, a Dutch farmer rushed from a cottage beside the road and, in a voice hoarse with excitement, told us that English cavalry were waiting for us lower down. He said that they had mountain-and machine-guns, and he estimated their strength at two hundred men, with over three hundred horses and mules, all of which proved substantially correct.

A British Ordnance BL 15 pounder gun at camp near Boshoff 1900.[451] Colt Machine Gun and Dundonald Galloping carriage with crew. To the left a Red Cross Wagon.[452]

Large British Camp.[453]

Edgar Duncker was sent back to report, and before long he returned with General Smuts, accompanied by Commandant van Deventer and a dozen men. General Smuts immediately decided to attack, and I heard him say that if we did not get those horses and a supply of ammunition we were done for.

He ordered van Deventer forward with the men who were with him and the 'Rijk Section,' to locate the British force, while he himself waited here to bring up the rest of the commando. We set off at once, and in a few minutes reached the banks of a small river which we crossed. As we were going through the fringe of thorn trees on the other side, we rode straight into fifteen or twenty troopers cantering towards us. Most of our men were still among the trees, but four or five of us were in advance, and when we leaped to the ground the soldiers were not more than ten yards away. Opening fire, we brought down several, and the rest turned and galloped back along the road. I fired my last two cartridges here, and my first thought was to run to a dead soldier and seize his rifle and bandolier, abandoning my own rusty weapon, then I rushed for my mare and joined in the chase.

The troop-horses were in much better condition than ours, but the soldiers were delayed by a gate, so we got close again, dropping two or three more from the saddle.

At the gate van Deventer himself and half a dozen men turned aside to a kopje for observation, but the rest of us, about twelve in number, followed the retiring patrol to a low stony ridge farther down the road.

They got there several lengths ahead of us and, abandoning their horses, took to the rocks. It was too late for us to retire back across the open plain behind, so we galloped on.

Before we reached their outcrop the soldiers opened fire almost point-blank, and worse still, a mountain-gun unexpectedly fired on us from a point to our left, not thirty yards off, and a machine-gun rattled into action close by.

Firing from amongst the rocks.[454]

With The 3rd "Queen's."[455]

So near was the mountain-gun that smoke from the discharge billowed over us although the shells went wide. It was astonishing that any of us escaped, but, owing no doubt to our sudden appearance behind the flying patrol, the firing was wild, and only three men and some horses went down before we reached the rocks in which the soldiers were. Here we, in turn, loosed our horses and ran up, to find ourselves within a few feet of our original quarry and a number of others, who had been posted here before.

Now that we could look over to the far side, we were surprised to see a large English camp less than a stone's throw away, buzzing like a disturbed ant-heap. Officers were shouting orders, and men tumbling out of their tents, some running towards us; others going to the right and left to take their stations.

This placed us in a remarkably tight corner, as we were so far ahead of our main body that they could not help us, for the English, having recovered from their first surprise, were sweeping the plain with gun-and rifle-fire.

Infantry behind a Sangar.[456]

Infantry Line Firing.[457]

The result was that our little party was stranded on the very edge of an armed encampment, and practically mixed with the English soldiers. Fortunately General Smuts had hurried the commando on, and in a few minutes they opened fire from a hill in the rear, thus preventing us from being overwhelmed, for our opponents were forced to take cover and could not surround us.

Those before us were in rough alignment along the bank of the ledge, so we were able to form a similar front, with a space of two or three yards separating us, while along the perimeter of the camp lay the rest of the troops in a half-moon. A young Transvaaler named Muller and I lay at the end where the rocks ran dead, and from here we could see the mountain-gun close by, busy shelling our commando. The gunners could not fire on us as they would have to hit their own men, and in any case they did not seem to realize that we were so near, for they were unconcernedly loading and firing at our men on the hill six hundred yards back.

Standing behind the gun was a tall man handing shells to the three at the breech. I fired at him, and he spun round and sank in a sitting position against the wheel, where I found him dead when the fight was over. The other three ran for the camp at their backs. I fired at one, and he pitched forward dead, while Muller brought down a third, but the last man got away among the tents. Having disposed of the gun-crew in a matter of seconds we turned to the other work on hand. The place we were fighting in was an outcrop of loose rocks, jutting up like a reef, nowhere much higher than a man, although the rear slope fell somewhat more steeply into the English camp.

In this narrow space, where we were facing each other almost at handshake, a grim duel began. As the soldiers raised their heads to fire we brought them down, for they were no match for us in short-range work of this kind, and we killed twelve or thirteen and wounded several more, at a distance of a few yards. We did not suffer a single casualty, except for the three men hit as we rode in.

Of these, one was Edgar Duncker with a bullet through his foot, and another a Jew named Cohen, with a smashed ankle. These two had been able to crawl forward to the firing-line and were taking part in the attack, but the third man, Raubenheimer[387] (a brother of Vera, Countess of Cathcart[388]), lay out in the open with his thigh broken, and his dead horse pinning him down.

Before he could reach his men, I hit a sergeant who came running up from the camp, a big heavily built man. He doubled up like a knife, and rolled about, shot in the stomach; then he died.

Nicolas Swart by my side shot two other soldiers in quick succession, as they tried to join those in the rocks. There was a young lieutenant a few feet from me. I found out afterwards that his name was Sheridan,[389] and they said he was a cousin of Winston Churchill. Twice he rose to fire at me and missed, at his second attempt I grazed his temple, and he dropped out of sight, but only dazed, for in a moment he was up again, swaying unsteadily on his feet, with his face streaming with blood, but still trying to level his rifle at me. While I was

[387] About 40 individuals with this name on ABW.com name search.

[388] Vera Estelle Cathcart, Countess Cathcart (14 May 1892 – April 1993) was a South African writer and actress. Vera Cathcart, Countess Cathcart - Wikipedia

[389] Sheridan Richard Brinsley Lieutenant He was killed in action at Modderfontein, near Tarkastad, on September 17th, 1901 aged 27. He was born in May 1874, and educated at Harrow. He entered the 17th Lancers from the 3rd Battalion Royal Scots (Lothian Regiment), in June 1896, being promoted Lieutenant April 1898. Lieutenant Sheridan accompanied his regiment to South Africa in February 1900, and served throughout the war up to his death. QSA (4) Source: Donner 17th (The Duke of Cambridge's Own) Lancers. Anglo Boer War - Name search

hesitating what to do, Jack Borrius shot him through the brain. Another soldier fired several hasty rounds at me, and I put a bullet into his heel, which was protruding from behind the rock near which he was lying. The sudden shock made him leap up, and again Jack Borrius, who was wonderfully quick, shot him dead as he rose.

In this manner the fight went on, until a mile beyond the camp we saw a small force of English troops approaching from the south. There were not many of them, but for all we knew they were the advance-guard of a relief force and, should sufficient reinforcements arrive to drive off our commando, those of us here in the rocks would be marooned, so we decided to clear the rocks by charging. After a whispered consultation from man to man, Jack Borrius gave the signal, and, rising together, we leaped in among the surviving soldiers. There were only ten or fifteen left, and so far as I can remember not a shot was fired on either side. Our sudden onslaught took them unprepared, and they surrendered at once.

Without troubling about our prisoners we ran down shouting and cheering into the camp, before the rest of the defenders knew what had happened.

When they saw us among the tents in their rear, something like a stampede set in. Soldiers went running in all directions, some making away into the thorn trees, others coming towards us and throwing down their arms. One man rushed to the horselines, and mounting barebacked, flourished a revolver and tried to ride off. I shouted to him to halt, but as he gave no heed I shot him dead. When the commando saw us enter the camp, they came galloping across, and the fight was over.

I took part in a final episode, for William Conradi and I, walking through a patch of trees to disarm some soldiers, came on a stone cattle-kraal, in which a dozen men were holding out. When we looked into the kraal, they were leaning on their rifles on the far wall and firing at some of our commando men moving in the distance. We called out 'Hands up! Hands up!' but they turned instead and blazed a volley into our faces. Only our eyes were showing or we should both have been shot.

Conradi killed one man and wounded another with a single bullet, and I wounded one, but even now they did not surrender, for, rushing across the kraal, they ranged themselves against the near wall, which alone separated us, and one of them thrust his rifle so near my face that his shot scorched my cheek and neck with cordite, fragments of which had to be picked out for days afterwards, with the point of a knife. When I seized the muzzle he gave an oath and jerked it back so forcibly that the sharp foresight gashed the ball of my thumb and the palm of my hand, and I had to let go.

Few in number but well sheltered.[458] 17th Lancers at Elands River, 1901.[459] Battle of Elands River 1901[390] AKA: Battle of Modderfontein (farm).[391]

The situation was fast becoming dangerous, when, to our relief, we heard the sound of voices through the trees, and a number of our men came running up to see what the firing was about. The soldiers now threw their rifles over the wall, but even this was not the end, for, as I hastened round to the entrance of the kraal to receive the prisoners, I collided with a soldier who came crouching along to get us in the flank. He did not know that the

390 Battle of Elands River (1901) - Wikipedia

391 Near Tarkastad. Tarkastad - Wikipedia

fight was over, and if I had not rammed him when I did, in another moment he would have been round the corner, shooting us down while we were engaged with the men inside. He said I was a 'surprise packet,' offered me a cigarette, and came with me to join his captured companions in the kraal with his hand amicably on my shoulder. The whole incident had not lasted five minutes, but it had been sharp enough, and Conradi and I reckoned ourselves well out of it as we hastened back to the camp to take part in the looting.

The commando was up and there was a great ransacking of tents and wagons. The small relieving force that had given us cause for alarm turned out to be only a patrol, and it had the doubtful satisfaction of watching us from afar as we turned the camp inside out. When we had done we were like giants refreshed. We had ridden into action that morning at our last gasp, and we emerged refitted from head to heel. We all had fresh horses, fresh rifles, clothing, saddlery, boots and more ammunition than we could carry away, as well as supplies for every man.

Moreover, we had renewed confidence in our leader and in ourselves, a factor of considerable importance to a body of men in a hostile country.

In the fight we lost only one man, who was killed when we rushed the camp, and six wounded, whereas the enemy had thirty killed, many wounded, and many taken prisoners.

I did not count the number of soldiers opposed to us, but there must have been about two hundred. They belonged to the 17th Lancers,[392] one or the crack regiments of the British Army. Among their wounded was their Commander, Captain [Victor] Sandeman,[393] and Lord Vivian,[394] whom I found among the rocks where we first rushed them. He it was who told me the fate of the three men killed and mutilated by the Basutos, the day that we crossed the Orange River. He pointed out his little bivouac tent, and said that it would be worth my while to have a look at it I was not slow to take the hint, with the result that having started that morning with a grain-bag for my chief garment, a foundered horse, an old rifle, and two cartridges I now appeared in a handsome cavalry tunic, riding- breeches, etc., with a sporting Lee-Metford, full bandoliers and a superb mount, a little grey Arab, which his coloured groom said had been the property of Lieutenant Sheridan.

I also selected a strong riding-mule in preference to another horse, for my experience during the past fortnight had taught me that a good mule for long marches and a light nimble pony for use in action were the ideal combination. After I had completed my equipment, commissariats and ammunition supply, I walked around the camp.

[392] The 17th Lancers (Duke of Cambridge's Own) was a cavalry regiment of the British Army, raised in 1759 and notable for its participation in the Charge of the Light Brigade during the Crimean War. 17th Lancers - Wikipedia

[393] Sandeman V S Captain Severely wounded. Near Tarkastad, 17 September 1901 Source: South African Field Force Casualty Roll. The words of Lord Kitchener's despatch of 8th October 1901, para 11, are as follows: "On the 17th September Smuts' commando arrived at Modderfontein, eighteen miles north-west of Tarkastad, where the Boers made a most determined attack upon a squadron of the 17th Lancers, under Major Sandeman, posted to close all egress to the south. The enemy being dressed in khaki were taken for our own troops and got to close quarters, with the advantage of ground, before the mistake was discovered. Thus taken at a great disadvantage, our men offered a most gallant resistance, and worthily maintained the traditions of their regiment. The losses of the squadron were very severe, 3 officers and 20 men being killed, and 2 officers and 30 men wounded. The Boers, who had evidently made the attack in order to elude the close pursuit to which they were subjected, also suffered heavily before the approach of another squadron of the 17th Lancers compelled them to break off the engagement." See: Battle of Elands River (1901) - Wikipedia

[394] George Crespigny Brabazon Vivian, 4th Baron Vivian (21 January 1878 – 28 December 1940) was a British soldier from the Vivian family who served with distinction. Recovering from wounds received in the battle, Vivian returned home December 1901. He later served in the First World War, being appointed to the Distinguished Service Order in 1918 George Vivian, 4th Baron Vivian - Wikipedia

"Tommy's pet: Dinner for two."[395]

We considered that the taking of it was chiefly the handiwork of our original storming party, for while we could not have done it without the protection and covering fire of the commando, yet by riding in on the heels of the English troopers and taking post on the very edge of the camp, we had served as the spearhead that made success possible. I also saw the dead gunners and other men whom I had shot, and I looked on them with mixed feelings, for although I have never hated the English, a fight is a fight, and though I was sorry for the men, I was proud of my share in the day's work.

Lastly, I went to see what had become of my roan mare. She was still patiently standing where I had left her at the ledge. On each side of her lay a dead horse, but she had escaped unharmed. The gallant little beast was, however, so exhausted that when I tried to lead her away she could scarcely put one foot before the other, so I unsaddled her, throwing the saddle aside, for it was old and worn with much use since I had taken it from general Clements's camp ten months before. Removing the bridle and halter, I turned her loose in the hope that some neighbouring farmer would look after her, for she too had shown the mettle of her Free State pasture, and the marvellous endurance of the South African horse.

General Smuts now ordered us to set the tents and wagons on fire, and to destroy the mountain-and machine-guns, as well as such surplus ammunition and other supplies as could not be removed. Then, leaving the prisoners, mule-drivers and native servants to shift for themselves, we rode off in triumph.

Commando on Horseback.[460]

[395] Black & White Weekly Vol 1–2 1899–1900 Photo by D. Barnett. Original Caption: "Tommy's pet: Dinner for two." PF: Mules vary widely in size, and may be of any color. They are more patient, hardier and longer-lived than horses, and are perceived as less obstinate and more intelligent than donkeys.

XXII. Moss-Trooping

Next morning we rode out of the mountain country into the open plains of the Karroo. In the face of great odds we had broken across the successive barriers placed in our way, and although we had still many troubles to meet, the English had failed to turn us back. We now slowly marauded southwards. At the village of Maraisburg,[396] a large number of troops was waiting for us, but General Smuts skilfully led the commando through at night without firing a shot, and we continued unmolested.

During this time the 'Rijk Section'' came into its own. Our share in the attack on the 17th Lancers had enhanced our reputation, and in this open country our services as scouts were in greater demand, so we ranged far ahead hospitably entertained by the Dutch-speaking population, and philosophically tolerated by the English farmers with whom we came in contact.

In our fine khaki tunics, and on our well-found horses, our appearance had undergone such a transformation that when asked at the English farmhouses who we were, our stock witticism was to say that we were 'English-killing Dragoons.' We thoroughly enjoyed it after all the hardships of the past.

The weather had improved, the long winter was over, and cloudless sunny days put still further heart into us. Then we had another stroke of luck, for we were joined by Field-Cornet Botha with twenty-five men, the remnant of a band of free-lances that had been roaming the midlands, until their numbers had so dwindled that they had been forced into hiding among the mountains. Hearing of our passage they had hastened to find us, and practically replaced our wastage since coming into the Cape.

But we had to abandon Raubenheimer, whose thigh had been smashed during the last fight, and a day or two later we had to leave Cohen[397] behind, as his wound became gangrenous.

Besides being a brave man, Cohen must have been a bit of a wag, for I subsequently read in an English newspaper that when he was captured and asked by a British officer why he, a Jew and a Uitlander, was fighting for the Boers, he replied that he was fighting for the Franchise.[398]

The next loss was heavier for me. My friend Jacobus Bosman,[399] who had so loyally stood by me when the others turned back in the Free State, was taken ill with some malignant fever. He gamely tried to keep up but we had to leave him delirious at a farm. I went off with a heavy heart, for I knew that he was doubly in danger. If the disease spared him, the English would be waiting with a charge of high treason, and my fears were only too well founded, for about three weeks later I read that he had been sentenced to be hanged as a rebel at Graaf Reinet.

He was the first of our 'Rijk Section' men to meet a humiliating death by execution, but not the last, for three more were destined to stand before a firing party, and also other members of the commandos

Two days after we had left Bosman behind, we reached the foot of a high mountain range, the name of which I have forgotten. There was a road running into a narrow defile called Lily Kloof into which the 'Rijk Section' was sent scouting. We rode up the gorge for some distance, until we saw an English foraging party going off, each man with sheaves of oats tied to his saddle.

Troops of one of the British mounted regiments collecting forage for their horses.[461]

[396] Now Roodenpoort and in Gauteng province, not the Cape.

[397] Cohen David Louis Prisoner number: 30612 Captured: Near Cradock 21 Sep 1901 Sent to: India Age: 26 Address: Johannesburg. See: Cradock, South Africa - Wikipedia Which is near the battle site.

[398] By serving in the military he would be entitled to full voting citizenship in the Republic.

[399] As cited previously this may be: Bosman Jacobus Johannes Prisoner number: 29507 Captured: **Straithnairn 27 Feb 1902** Sent to: India Age: 21 Address: Vechtkop Source: Anglo Boer War Museum 2016

Ben Coetzee killed one of them, a local farmer named Brown, who had joined the troops, and the rest of the patrol raced off. A woman ran out of a cottage to warn us that there were English strongly posted at a narrow point farther up, so we turned back to report to General Smuts.

Some miles west lies an equally deep ravine, through which we now tried to find a way, but here again we were warned that the route was held. General Smuts said that he was not going to squander men in forcing his way, when there were other means of crossing, so we retraced our steps, and that night, led by a local guide, we picked our way over the range by a bridle-path. It was a long march with no chance of sleep, but dawn found us on the far slopes, in the English-speaking, district of Bedford.[400] From here there was a glorious view, across deep mountain valleys and green uplands of one of the loveliest and most fertile parts of South Africa.[401] We had left all serious pursuit so far behind that for the next few days we rode leisurely on our way, while the men scattered about, visiting farmhouses and enjoying themselves.

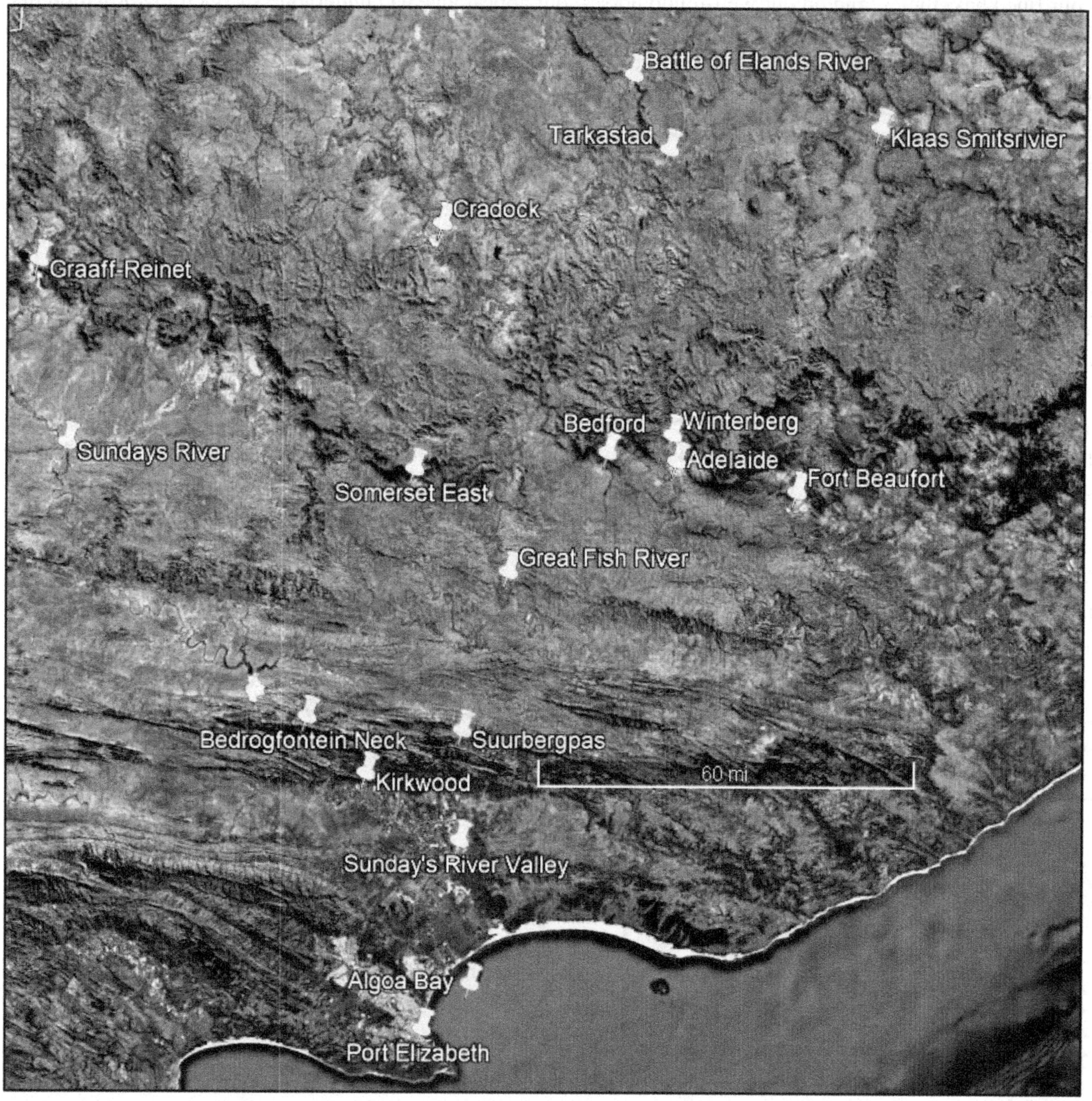

Map data: Google Image© 2024 Image Landsat/Copernicus.

400 Bedford or Nyarha is a rural town in the centre of the Eastern Cape province of South Africa. Bedford, South Africa - Wikipedia

401 The Great Winterberg is an extensive east–west mountain range lying immediately north of the small towns of Bedford, Adelaide and Fort Beaufort Fort Beaufort - Wikipedia in the Eastern Cape of South Africa. The Great Winterberg Peak north of Adelaide rises to 2371 metres, high enough to be occasionally covered in snow. Winterberg (Eastern Cape) - Wikipedia

The inhabitants took our coming in good part, and there was never any sign of ill feeling, although they hoped that we should get rounded up, and told us so.

We saw an occasional Defence Force patrol, which, however, gave us no anxiety, for they were local levies merely keeping their eye on us, and there was no trouble until one afternoon when a column of horse showed on a hill and opened long-range fire in our direction. Not wishing to be involved in fighting other than of our own making, we turned up a defile that wound into a mass of rugged mountains before us. These were the Great Winterbergen, and we halted at sunset amid gorgeous forest-covered steeps, where we built huge log-fires, and spent a comfortable night, greatly taken with the fine country we had reached.

Next day our path ran through even more beautiful scenery. Around us was primeval forest, and through an occasional tunnel in the trees we glimpsed green fields and white homesteads in the valleys far below.

The following morning, while halted in a picturesque glade, a woodsman in a log cabin told us of a tavern and trading-station at the foot of a pass leading down close by. The men were all for going to see what was to be had, and towards sunset we reached the bottom, where stood a substantial wayside inn flanked by well-stocked warehouses.

So little were we expected in this remote part, that no effort had been made to remove the goods to the protection of the nearest military post, which was generally done when we were approaching, and the owner suffered from our visit, for we were masters by now in the gentle art of commandeering. I would not have mentioned this excursion had it not cost us the life of another member of the 'Rijk Section.'

There was plenty of beer and spirits at the inn, and although few of the men had tasted liquor for a year or more, there was no drunkenness, but Piet de Ruyt, our Hollander companion, took too much, and when the commando moved away at dusk, he was left asleep unnoticed.

Weeks later we learned that he had been discovered in a room, and as, like most of us, he was dressed in a British uniform. The poor fellow was executed, in all probability before his fuddled brain had time to take in what was happening.

Neither then, nor for weeks later, did we know that the death penalty attached to the wearing of khaki, and although after a while rumours reached us through the country people that our men were being executed, these stories left us doubting and perplexed. We could not believe that the English were resorting to the shooting of prisoners, and it was only after many had been executed that we learned of Kitchener's proclamation ordering the death of all Boers caught in khaki. As far as I know no steps were ever taken by the military to acquaint us with its contents.[402]

From the foot of the Winterbergen we rode on for some hours in the dark, by a footpath winding across a bush-covered plain, and we camped in all open space for the night. At daybreak we made out the little village of Adelaide[403] in the distance, but, as it seemed to be strongly garrisoned, we left it alone, and continued all that day slowly making south through broken country.

Here again the inhabitants were chiefly English-speaking farmers, who submitted with good grace to our depredation, for we slaughtered what sheep we required, and helped ourselves freely from their larders and orchards. Towards evening a column came following us from Adelaide, halting when we halted, and moving when we moved, and when after dark we went into camp, their fires were soon winking at us five or six miles away. During the afternoon an armed Colonial had ridden up from a farm with a story that he wished to fight the English. He rode a spirited horse and spoke fluent Dutch, and as he seemed genuine he took us in, but now, while we were off-saddling by a pool, he suddenly pulled his horse round and galloped away. He was gone before anyone could give chase, and, as he was obviously a spy sent to learn our intentions, we re-saddled and rode on for an hour or two before resting for the night.

Next day we bore away in a somewhat more westerly direction, reaching the Great Fish River by sunset. We crossed at a ford, and a mile or two beyond, at Commadaga Station, we passed over the railway line that comes up from Port Elizabeth through the midlands. As there were no block-houses we had no difficulty in getting to

[402] PF: I have been unable to find this directive explicitly mentioned. In general, engaging in combat while dressed in an enemy uniform does warrant a sentence of execution. The law of war is more vague regarding simple ruse de guere (ruse of war) of infiltrating enemy lines while dressed as an enemy soldier. Generally, if a member of a combatant force is caught behind enemy lines dressed either as a civilian or as an opponent, execution is warranted.

[403] Adelaide, South Africa - Wikipedia

the other side after dark, and then, as our horses were tired, General Smuts ordered us to camp at a farmhouse five or six hundred yards beyond. We had scarcely turned our animals out to graze, when an armoured train came puffing up, the beams of its powerful searchlight sweeping the veld, but, as we lay in a fold of the ground, the crew could not see us although they must have suspected our presence, for they sent a number of shells howling into the night, only one of which came within measurable distance of us.

Great Fish River.[462]

Canadian Troops in Armoured Train with Signal Light on top.[463]

It burst harmlessly near by, but we thought that we had been discovered, and ran to find our horses, with the result that there was a good deal of bumping and confusion in the dark before we realized that the English were shooting at random. Then we had a good laugh at ourselves, and spreading our blankets slept in peace till daybreak.

From here we headed south-west, riding at our ease through the district of Somerset,[404] until by sunset we made the foot of the Zuurbergen,[405] the last great escarpment before the country drops away to the sea.

Suurberg Cycads.[464]

We were by now within fifty miles of Algoa Bay.[406] I do not know what General Smuts's intentions were at any stage of our expedition, for he was a silent man, but I think that at this particular juncture he was contemplating a sudden raid on Port Elizabeth,[407] for next morning, when we saw what looked like over three thousand troops coming after us, he kept us quietly resting at the foot of the mountain, instead of slipping away east or west as he could easily have done. These troops had been brought by rail, and had detrained at Commadaga Station,

[404] The forested, mountainous backdrop frames the town (which is within sight of 16 waterfalls) Somerset East - Wikipedia

[405] Suurberg Pass, (English: sour Mountain) Suurberg Pass - Wikipedia And: Suurberg - Wikipedia

[406] Algoa Bay is a maritime bay in the Eastern Cape,. It is located in the east coast, 683 kilometres (424 mi) east of the Cape of Good Hope. Algoa Bay - Wikipedia

[407] Gqeberha, formerly Port Elizabeth is the town located at Algoa Bay. Gqeberha - Wikipedia

where we had crossed two nights before, and from their converging front it looked as if they hoped to bring us to bay against the slopes of the range.

We watched them slowly approaching for most of the day, until their scouts were almost within rifle-range, and then General Smuts led the way straight up the steep slope behind us. We spent the night on the crest of the first of the great parallel hogbacks that constitute the Zuurbergen.

From the point where we reached the top, we looked on a world of more mountains, line upon line of high ranges, each separated from the next by deep wooded gorges, and the prospect of being driven into these fastnesses was not inviting. However, with troops closing in on us from the rear, there was no help for it, so we built fires and camped.

Port Elizabeth Area.

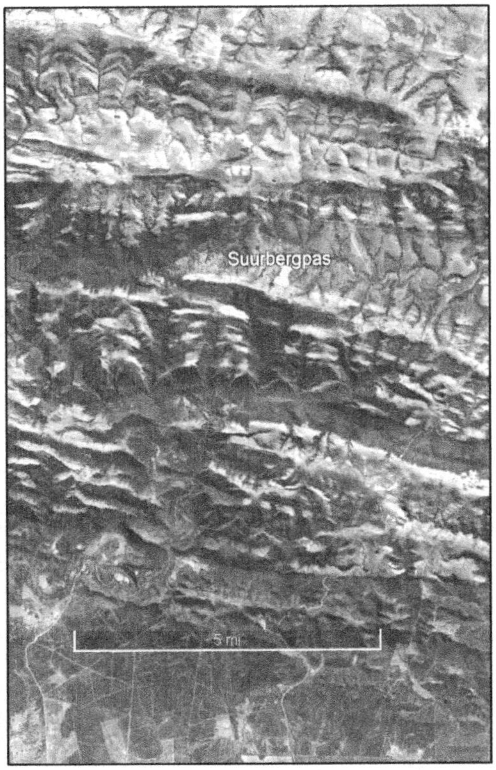

Suurbergpas.[408]

On the way up Jack Borrius and I had met a native herd-boy, who informed us that there was a big troop of horses in a neighbouring kloof.[409] On all our journey through the Cape we had not found a single riding-horse on any of the farms, as the English had cleared the country to prevent us from getting remounts, and the boy said over five hundred horses had been collected by the military within the last few days, on hearing of our approach.

Martial law in Cape Colony: handing in bicycles to the military authorities At Graaff Reinet.[465] Five hundred bicycles handed in under military orders at Graaff Reinet.[466]

[408] Map data: Google Image© 2024 Image Landsat/Copernicus.

[409] A deep glen or ravine.

Accordingly Jack and I left the commando before daybreak and followed the ridge for some miles, until we found a practicable descent down which to lead our horses into the kloof which the native had pointed out. At the bottom we came on a deserted homestead, with the ground trampled by many hoof marks, but a Red Kaffir,[410] who appeared out of the forest, told us that a patrol had come the day before, and driven all the animals away to the coast. We therefore gave up the idea of finding them and, turning our mounts into a paddock, stretched ourselves under a shady tree and fell asleep.

Some hours later we were waked by a couple of men who had ridden out in search of food. They roused us and told us to listen. Bumping to our feet, we heard the sound of distant rifle-fire coming from the quarter where we had lefts the commando that morning. It was clear that they were fighting, and before long we caught sight of our men on the slope of the next range, crawling like ants up a steep side. They were miles off, but we could see that all was not well with them, for they were strung out in a disorderly line and the firing was coming from an enemy force some where out of view.

We caught our horses and led them up the range which the commando was climbing, and on getting above we could see our men hurrying towards us along the top. It took more than an hour for them to come up with us, and then we learned that while they were encamped a number of English had unexpectedly opened fire on them.

After an interchange of shots the commando had fallen back deeper into the mountain, with only one man wounded and a few horses killed. The wounded man was badly injured about the face, but had come on. We were now on the second range. From here we saw many English horsemen riding about on the first crest. When came opposite us, with only the deep kloof, through which we had passed, lying between, and then opened fire upon us with several machine-guns and a field-piece. I do not know how they had succeeded in dragging those up the mountain, but there they were barring the way should we try to break back, while behind us lay mile upon mile of tumbled forest-clad mountains and gorges.

"All together boys" – the well-known image of a Naval 12-pdr QF 12-cwt gun on "Scott" improvised carriage, taken by society photographer Horace Nicholls.[467] Dragging a Maxim up.[468]

The immediate danger from the troops was not pressing as we had good cover for ourselves and our horses, but now a fresh-complication set in.

Up to now we had found so little difficulty in commandeering supplies from the farms we passed, that no one ever thought about the next day, with the result that when we unexpectedly found ourselves in a wild region without habitations, the men had little or no food with them, and were already beginning to feel hungry. Scattered about stood a strange growth known as 'Hottentot's bread' (Encephelartos Altensteinii), [411] a wild fruit not unlike a large pineapple.

It is edible only at certain seasons of the year, but coming from the north, we did not know this, and as one of the men sampled it and found it to his liking, many unfortunately followed suit. I had not eaten any, and returning to the firing-line, after going to tie up some horses that had broken loose, I was astonished to find more than half our men groaning and retching on the ground in agony, some apparently at their last gasp.

[410] Referring to the red ochre traditionally used by Xhosa peoples to smear the body and clothing.

[411] Encephalartos altensteinii is a palm-like cycad in the family Zamiaceae. It is commonly known as the breadtree. There are usually two to five greenish-yellow cones up to fifty centimeters long, the female scales covered with protuberances. The cones are poisonous to humans. Encephalartos altensteinii - Wikipedia

General Smuts was worse than the rest, so, with half our number out of action, we were also leaderless, for he was lying comatose.

Breadtree, Eastern Cape giant cycad.[469] Forest buffalo (S. c. nanus).[470]

The horsemen of the strong enemy force before us were even now descending the opposite slope to attack us, while behind were mountain wastes stretching as far as we could see. We had no food, and could not move without abandoning the sick, so our position was critical.

For the moment our most urgent concern was the soldiers advancing towards us. They had by now reached the bottom of the kloof, and some of them, leaving their horses behind, were already swarming up, firing as they came.

Commandant van Deventer was too ill to take charge, but Ben Bouwer, though bad, was able to order every man who could still handle a rifle to extend along the top. It was nearing sunset, and the light was uncertain, so I do not think we did much damage with our shooting, but it served to turn our assailants, for they went back to their horses, and then climbed up the other side, until their camp fires began to shine in the dark, which meant that they were settling down for the night.

Those of us who had been lining the forward crest now had time to look around, and what we saw was not comforting. The sick men were worse than ever. General Smuts was very bad indeed, and van Deventer, his second in command, not much better. From the groans and cries on all sides it was clear that the sufferers could not travel, and there was nothing to do but to wait, although it was urgently necessary to get away before daylight would enable the English to surround us. I shall not soon forget that night. It was dark, and a chill wind blew from seaward. We dared not light a fire, and those of us who were not ill from the poisoned fruit were starving.

We knew that if the men did not recover in time to avoid the pursuing column, our expedition into the Cape would come to a speedy end next morning, and we sat beside the sick men not knowing when we might be fallen upon. However, as the darkness slowly passed, one man after another recovered sufficiently to stagger to his feet and towards dawn there were not more than twenty unable to stand. General Smuts was still prostrate, but able to take in the position, and he gave orders that the men who could not help themselves were to be tied to their saddles, and that the commando was to march deeper into the mountains. He himself had to be held on his horse, and we started off in the dim light, following a game-track that led down into the next gorge. At the bottom we halted to rest the sick men, and then crawled up the far slope, which was almost as steep as the place we negotiated on the night of our retreat from the Stormbergen.

Bringing up the sick men was a difficult task, made more difficult by the fact that the English troops had actually dragged their gun to the top of the second range where we had spent the night, and opened fire on us as we were climbing. The distance was so great that only a few shells fell among us, their chief effect being to spur the men to greater activity and rouse the invalids, some of whom asked to be lifted to the ground, for they did not relish the idea of being trussed up in the circumstances.

As one of the last to gain the top, as I had led my horse and mule very slowly, a few yards at a time, to save their legs. But when I got above and chanced to look back, I was surprised to see that General Smuts was still lying below; with three or four men attending him, while down the path above them came a number or English scouts trailing the spoor which the commando had made. At the rate at which the scouts were descending they would soon come on General Smuts and his men, and I realized that there was no time to be lost. I tied my two animals to the nearest tree, and rushed and slid to the bottom unharmed by the bullets of the scouts. When the men with General Smuts heard the firing overhead they lifted him to his horse, and were already starting, a couple on each side to hold him, by the time I got down.

We did not follow the commando, as we should have presented too easy a mark, but kept away to the right where there was a gully up which we could make our way unseen. When we reached the top I fetched my two animals, and we followed on to find the commando waiting for us in a glade.

We were now in the very heart of the mountains, so far from any farms that buffalo[471] were seen, and their tracks and mud-wallows were frequent.

Not one of the sick men had died and most of them were better, perhaps because of the shaking and movement. General Smuts and a few other cases were, however, still in danger, and it was decided to spend the day here, as our scouts reported that the enemy had turned back, and it was thought necessary to give the sick a long rest before continuing. The trouble now was to get food. Parties were sent out in search of native kraals, for smoke had been seen rising from a distant part of the forest, and here, after struggling for hours through dense under-growth, we found a few poverty-stricken huts whose in-habitants had fled. They belonged to a destitute tribe of Red Kaffirs, but we unearthed a supply of millet, enough to give the commando some sort of a meal.

In the afternoon we moved deeper into the mountains, and, breasting a high grass-covered shoulder, caught a distant view, thirty-five miles away, of white sand-dunes, and of a grey haze, which was the Indian Ocean. We were elated, for we knew that we had now penetrated farther south than any other commando during the war, and that we were the first to come within sight of the coast.

After dark, camped on a height, we could see the lights of Port Elizabeth shining far off, and this strengthened our belief that General Smuts, in spite of his illness, still intended to go there. Next morning we went down into a beautiful valley filled with yellow wood trees, centuries old, and here we camped for the rest of the day, still subsisting on boiled millet.

One of our men recognized this part, having hunted buffalo and elephant here long before, and he said that he remembered a path running south, by which we could get out of the mountains into the valley of the Sunday River,[412] where we might take the small village of Bayville.[413]

On hearing this a small party, mostly 'Rijk Section,' was made up to raid the place. Unluckily both my animals had strayed into the forest, and I had to follow on long after the others had gone. I tracked them until I got clear of the mountains into the wide valley, down whose centre the Sunday River runs to the sea. But the floor of the valley was dense scrub, ten or twelve feet high, in which I got completely bushed, so I had to work my way back with difficulty, reaching the commando at dark, tired and disgusted.

Jack Borrius and his buccaneers had not yet returned, but that night we marched on, as we had exhausted our grain supply. We led our horses in the dark between high ranges; along a well-beaten path that twisted down a valley, until towards daybreak we reached the Sunday River, at a point still closed in by the mountains, but where, in a clearing, lay a well-stocked farm.

[412] The source of the 250 kilometres (160 mi) long Sundays River is in the Sneeuberge (the highest mountain range in the former Cape Province) near Nieu-Bethesda. Sundays River - Wikipedia

[413] Kirkwood is a town in the Eastern Cape Province of South Africa. It is situated on the banks of the Sundays River in the eponymously named Sundays River Valley. Bayville is nearby.

During the Boer war, Jan Smuts' Commando of 250 men passed through the valley on their epic campaign to the northwest. Indeed, the three Boers who died the furthest south during the war were killed on Cecil John Rhodes's farm Brakkefontein.

Shortly after these casualties were incurred a major battle took place on the Bedrogsfontein mountain pass. Kirkwood, South Africa - Wikipedia

General Smuts remained pale and weak, but he sent for me and thanked me for having come down to warn him two days before.

Some time after sunrise Cornelius Vemlaas, Henry Rittenberg and I were ordered to scout up the course of the river, and we set out on what proved to be my companions' last ride. After following the stream and making inquiries from occasional natives, we learned that there were some white men off-saddled close by, and going thither came on Jack Borrius and his detachment, halted under the trees. Jack himself was suffering from terrible wounds. They had entered Bayville unopposed, but on their way back they fell foul of an English patrol, and, in the ensuing encounter, a bullet had entirely blown away his left eye, leaving nothing but a cavity filled with dried blood. In addition his right hand was smashed to pulp, but he had refused to be left behind, and we found him lying in great pain, but determined to remain with the commando. While we were busy with him, General Smuts rode up. He told Ben Coetzee to take over command of the 'Rijk Section,' and ordered him to reconnoitre down a lateral valley, to see whether it would bring us out of the mountains, as he intended to enter the lower Sunday River Valley.

Ben Coetzee took with him Rittenberg and myself together with Vermaas and one of van Deventer's men, named van Onselen, who volunteered to come. We five went ahead carefully picking our way, and discovered late in the afternoon that the valley we were in ran to a dead end, against what we took to be the final range of the Zuurbergen.

Coetzee kept van Onselen with him, and told the three of us from the 'Rijk Section' to climb the height ahead, and see what lay on the far side. As we began to lead our horses up the slope, we heard a shout, and van Onselen overtook us, with a message to say that I was to return, as Coetzee thought my horse looked poorly, and that van Onselen was to go in my stead. The fact that my grey pony had a staring coat saved my life, for the three men, on reaching the top, walked straight into the arms of an English force lying in ambush there, and, as they were dressed in khaki uniforms, were executed out of hand. They lie buried where they fell, their graves being the most southerly of any of our republican dead.

We only knew of their actual fate long afterwards. For the moment all we heard was a burst of firing, and looking up we saw large numbers of soldiers on the skyline, so we rode back along the valley full of anxiety.

At dusk we met the commando coming in our direction, and when we told General Smuts that our progress was barred, he halted us for the night.

At daybreak next morning, while we were saddling our horses, the troops opened fire from above our heads. No one was hit, but we were forced to retire back into the deeper part of the mountains, instead of getting out into the open country to the south. During the retreat we lost three more men. They were looking for their horses when the firing began and, as no one noticed their absence, they were left behind. According to an English newspaper which we saw afterwards, one of them was hanged as a rebel British subject, the sixth man of our force to be executed, not counting the three that had been murdered by the Basutos.

As I have said before, we had not heard of Lord Kitchener's proclamation against the wearing of British uniforms, and I went about wearing Lord Vivian's khaki tunic, with regimental badge and buttons, and the 17th Lancers skull and cross-bones in my hat, not a little proud of my well-earned trophies, and never dreaming that I was under sentence of death.

Cap badge of the 17th Lancers.[472]

We made a long trek through the kloofs of this wild region until, after midday, we found a disused pass, made, I was told, by Sir Harry Smith in the 'fifties' during the native wars.

This pass ran up a dark ravine, flanked with dense timber, and heavily overgrown with brushwood, but otherwise in good preservation. Its chief drawback was that it led northwards over the mountains back to the plains of the Karroo, from which we had come, whereas our endeavour had been to break seawards; but with the English column of that morning pressing steadily in our rear, we had to go wherever there was an opening. We started to ascend the pass, not knowing if another force was waiting for us above. If so we should have been trapped, but luckily this particular loophole had not been closed, and we found the top clear. Edgar Duncker and I, who were scouting in advance, saw a human skeleton beside the road, a relic or some past tragedy, and we placed the grinning skull on a log as a warning to our pursuers.

Not far from the top was a space of level ground with the ruins of an old building, and the remains of walled gardens and orchards, and as we could see the English troops halted in a glade far below, General Smuts stopped too for he and others were still very ill. The sick and wounded, with poor Jack Borrius, withdrew to a distance, while the rest of us turned our horses out to graze and lolled about at our ease, admiring the grand forest scenery and enjoying the luxury of our beautiful surroundings.

Then some half-savage cattle came out of the woods and we shot several, and had the added satisfaction of eating our fill around big fires. Afterwards, most of the men stretched themselves under the trees, at peace in this pleasant place.

But towards five in the afternoon we began to bestir ourselves. Ben Coetzee, Nicolas Swart and I sat basking in the sunshine on a wall not far from where the pass came out. Suddenly, while we were talking, we saw the foremost ranks of a body of horsemen appear at the head of the gorge, not a hundred yards away. Shouting an alarm, we seized our rifles and ran down, followed by the men who had been lying close by. The English must have been under the impression that we had gone straight on over the berg, for they were riding in a compact body of thirty or more with no advance-guards, and evidently our presence here was as great a surprise to them as their sudden appearance was to us, but we were the first to recover ourselves, and started to fire as we ran.

The English could not deploy on the narrow road, so they pulled round, and made back as fast as they could, for the ground above and below was so steep that they had to keep to the causeway, down which they poured in disorder.

They seemed to be boring and pushing each other frantically under our fire, horses and men toppling over the edge of the road, and crashing into the timber beneath. The road became obstructed with dead and wounded horses, for we were firing into the brown, and we could hear angry shouts, as those behind tried to pass. Only a few men reached the bottom, where I caught a glimpse of them lashing their horses, as they rode through a clearing below. The rest abandoned their animals, and took cover in the forest, directing at us so hot a fire that we dared not climb down to get at the ammunition of the dead and wounded, nor at the holsters or wallets of the fallen horses. The main English column, that had been halted at the foot of the pass, now moved nearer, and when their fire was added we withdrew to the top to avoid casualties, and escaped without any.

After dark we went down the far side of the mountains, still following Sir Harry Smith's road, and daylight found us back in the Somerset district. We had now left the mountains on the same side on which we had entered them five days before, and, although we had failed to break out to the south, we were all heartily pleased to be clear at last of this appropriately named range.

As soon as it was light we halted at a farm for a few hours, and here General Smuts called us together. He said that we had reached a turning-point in the expedition, and he told us that from now onward he was going to make for the Atlantic seaboard, and the old-established districts of the South-Western Cape. After thanking the men for the way in which they had borne themselves, he told us that he was dividing our force, partly to mislead the English columns, and partly for easier provisioning, as the inhabitants of the districts through which we had passed had complained that so large a force as ours was too severe a tax upon them. Accordingly he proposed to send Commandant van Deventer off with approximately half the commando, while he himself would take command of the rest. Both units were to advance independently of each other, and ultimately reunite in the far west.

He still looked pale and ill, but his spirit was undaunted and at midday he ordered those of us who were remaining with him to saddle our horses, and we rode away amid cheers and farewells from van Deventer's men, Who were to start later. We went on until sundown, then halted for the night in a thorn-covered hollow.

Next morning we struck across a wide plain, with alternating patches of bush and open country, over which we continued till noon, when we came to rest in a wide tree-covered bottom.

Percy Wyndall and Frits Balogh, of the 'Rijk Section,' were sent to a neighbouring rise on outpost duty, and from where we lay we could see them beneath a tree evidently enjoying a quiet chat, their horses cropping the grass behind. Before long a dozen English troopers rode out of the thicket in the rear and surrounded them before our eyes.

Horses grazing in the field during a break in a march.[473]

The distance between us was perhaps half a mile, and when we realized what had happened, we of the 'Rijk Section' (now reduced to six) rushed for our horses and, followed by a few others who happened to have their horses at hand, galloped up the slope. The soldiers were so intent upon their prisoners that they did not see us until we were close by, when they loosed a ragged volley, and leaping into their saddles, abandoned the two captives, and made off.

Shouting to the rescued men to follow, we rode straight after the patrol, and got to within thirty or forty yards of them, bringing three to the ground. Two more, whose horses were hit, surrendered, but the rest scattered in the bush.

A friend of mine, Jack Baxter, one of Bouwer's men was riding next to me, and he and I singled out one of the flying soldiers. We got near enough to order him to halt, but he rode on, not heeding our shouts. We fired at him but missed, for shooting from the saddle is trick work and he might have escaped had he not been brought up by a wire fence across his path. He jumped to the ground, and leaving his horse, climbed through the fence and disappeared into a patch of scrub. Almost at once his bullets were singing about our ears. Fortunately for us his marksmanship was poorer than his courage, and Baxter and I had time to dash our horses out of sight into a thicket.

We dismounted, and, tying our animals, started to stalk him down. Crawling through the fence, we wormed our way from tree to tree, until we located him. As we could take no risks with such a resourceful opponent, we emptied our magazines at the spot where we had seen movement. Silence followed, and after a few more shots to make sure, we went up to find him lying face down, riddled with bullets, but still clutching his rifle. We smashed his weapon, and shared the contents of his ammunition-belts, after which we returned to fetch our horses and his, and rode back, feeling almost regretful at the way in which we had hunted down so brave a man.

The rest of our storming party were collected around the soldiers who had been killed. The prisoners and wounded were there, too, and after depriving the living and the dead of their boots (an unpleasant but necessary task, for there was continual shortage of footwear amongst our men), we rode back to rejoin the commando.

In the course of our return journey Ben Coetzee and Edgar Duncker branched away on their own, and soon after, hearing several shots, we galloped in their direction. When we came up we found them sitting their horses in considerable agitation, while on the ground lay an officer and a trooper, both dead. It appeared that shortly after leaving us, as they rounded a piece of thorn bush, they ran into a small English patrol. So unexpected was the encounter that they were alongside before they could think, and Duncker, on the spur of the moment, called out, 'Don't fire, we are the 17th Lancers' The officer in charge, a Captain Watson, said, 'I don't believe you; all Smuts's men are dressed in khaki. Put up your Hands.' Then Coetzee and Duncker, both of whom carried Webley revolvers fired simultaneously, killing Captain Watson and one of his men, and seriously wounding another, who, however, got away with the rest.

This was a very unlucky incident, for the wearing of British uniforms had without doubt been the proximate cause of the death of these two men, and although we knew nothing as yet of Lord Kitchener's proclamation, General Smuts pulled a long face when he was told of the business. Indeed, long afterwards, when we met Lord Kitchener himself, he cited this very case in defence of the execution of so many of our men for wearing khaki.

Well the harm was done, and I can only say that none of us ever wore captured uniforms with the deliberate intention of decoying the enemy, but only out of sheer necessity.

We met no more soldiers that day, and towards afternoon the commando moved to a camping-place ten miles farther on. Next morning, while scouting ahead, I met a British ambulance-wagon with a doctor and several stretcher-bearers, on their way to fetch in the wounded of the day before. The Medical Officer already knew the manner of Captain Watson's death, for he spoke heatedly of murder, and abuse of military uniforms, although he made no mention of the proclamation, perhaps thinking that we knew about it. After this we went steadily west for some days without anything happening of importance, except that one evening we came in sight of a tiny hamlet called Hobsonville, where there was a small garrison of a dozen men, with whom the 'Rijk Section' and two or three more had a sharp brush, one of Bouwer's men being shot through the thigh. To avoid further damage we galloped in among the houses, in time to see the defenders rush for their horses, and make good their escape, as they had seen the commando coming on behind.

There were two well-stocked shops, and a quantity of military stores, so we did quite well out of the place, and spent the night feasting on tinned food and other luxuries.

In this manner we journeyed slowly on, until at length we reached the Port-Elizabeth-Graaf Reinet[414] railway line which we crossed at night, an armoured train sending a few shells to speed us on our way. Next day we saw the town of Aberdeen,[415] lying seven or eight miles distant, and as there was a large camp on the outskirts, we did not need the local farmers to tell us that there was trouble ahead.

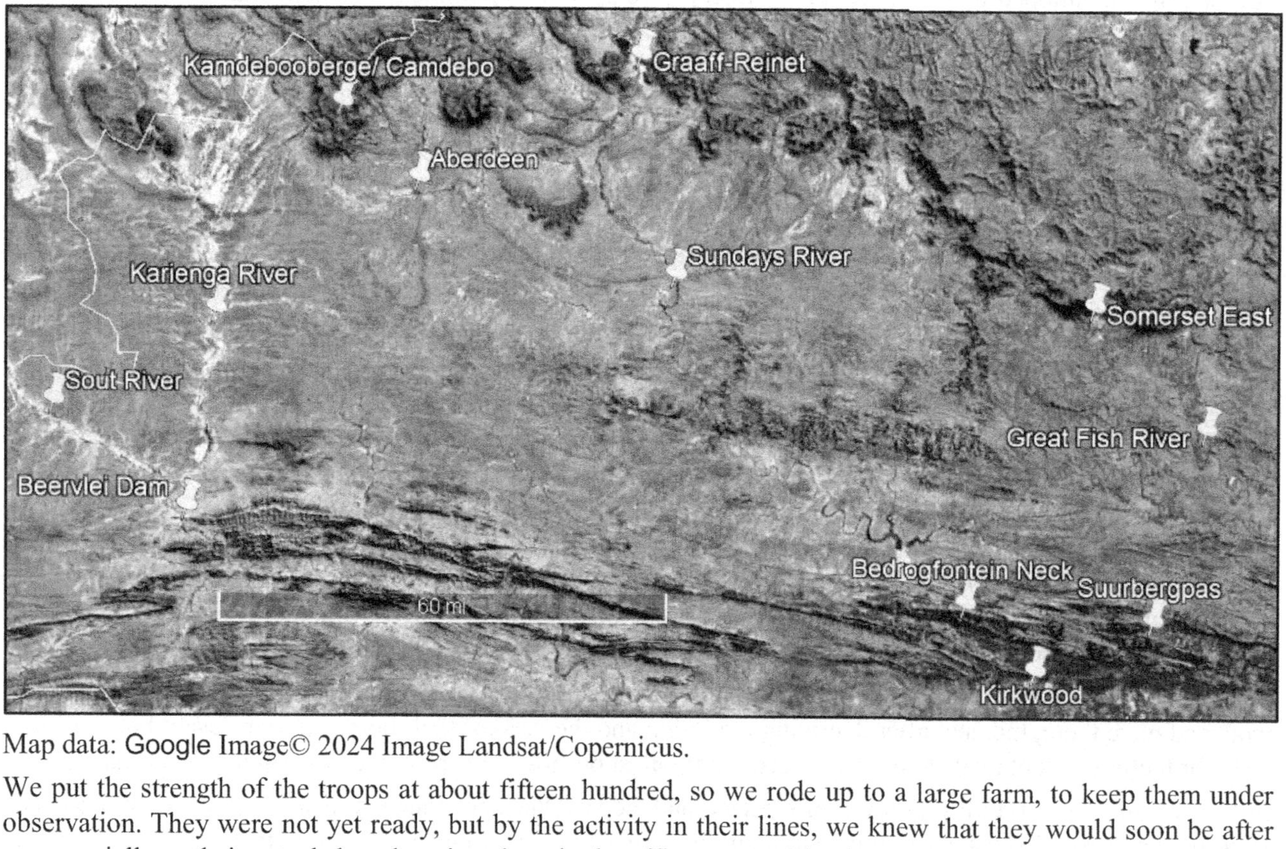

Map data: Google Image© 2024 Image Landsat/Copernicus.

We put the strength of the troops at about fifteen hundred, so we rode up to a large farm, to keep them under observation. They were not yet ready, but by the activity in their lines, we knew that they would soon be after us, especially as their patrols kept hanging about in the offing most of the day.

[414] Graaff-Reinet is a town in the Eastern Cape Province of South Africa. It is the oldest town in the province Graaff-Reinet - Wikipedia

[415] Aberdeen is a small town in the Sarah Baartman District Municipality. Aberdeen, South Africa - Wikipedia

That afternoon we moved on, bearing slightly north for the Camdebo Mountains,[416] ten or fifteen miles away. We reached the foot next day, by which time a long English column was marching on our tracks, while various smaller bodies were skirting round the base of the range, in an encircling movement. General Smuts did not wish to get involved in a fight against heavy odds, so we started up the mountain by a gorge, and at sunset reached a high saddle, over which we passed.

Unfortunately the fine weather which we had enjoyed of late now changed. It turned bitterly cold, and a biting rain set in, and when darkness overtook us we had to halt on the rear slope for fear of falling over precipices. The icy water came down in torrents all night and there was no hope of a fire, so we sat before our horses until dawn, cold and drenched. Commandant Bouwer and two others were so benumbed, that when it grew light we had to carry them down in blankets to a valley, where we at last succeeded in getting fires alight, while General Smuts, Jack Borrius, and the rest of the sick and wounded must have endured agony that night.

It was still raining when we marched over a bleak up land, apparently uninhabited. To the north lay a world of barren-looking peaks and heights draped in heavy clouds, a sight that made our hearts sink, for, with an English column in our rear, it seemed as if we were in for another spell of cold and hungry mountaineering.

By midday the rain had stopped, and the sun showed through, and after plodding for hours over sodden turf, we came upon a farm with fuel to dry our clothes, and a flock of sheep for supplies.

A picket had been left behind at the neck which we had crossed the night before, and these men rode in after a time to say that the English column was nearing the top.

General Smuts sent the 'Rijk Section' and other patrols to look for a way down the south face of the mountain, for he liked the idea of being forced higher up as little as we did.

We rode to what seemed a likely kloof, but when we got to the edge and peered down, an enemy force lay waiting for us below. The other parties brought similar reports. All the exits from the tableland were guarded and our position unpleasantly resembled that on the Stormbergen. However, things shaped better this time, for here again the owner of the farm where we had halted (and whose sheep we had freely slaughtered) volunteered to lead us to a bridle-path, and after five or six miles brought us out on another high saddle, over which a faint goat-track led to the bottom. As soon as it was dark, we took our horses along the easy gradient, and got down in an hour or two without trouble. In order to shake off the columns more effectively, and to get clear of the terrible cactus-belt that girdles the foot of the mountains, we trekked on till day-break.

This cactus (prickly pear)[417] was brought from Central America about fifty years ago, and found the Karroo such congenial soil that there are now vast tracts rendered valueless by it. Our way now ran through a veritable forest of this vile growth, standing twenty feet high in places.[418]

Prickly pear with fruits.[474]

Soon after sunrise we reached the Kareega River,[419] which rises in the mountains which we had just left, and runs due south across the plains.

[416] See: Kamdebooberge, AKA: Sneeuberge - Wikipedia

[417] Opuntia - Wikipedia

[418] See: Prickly pears in South Africa - Wikipedia

We camped for the day on its wooded banks, seeing no sign of the English. In the course of the morning the 'Rijk Section,' with the exception of my uncle and myself (and Jack Borrius), rode off to forage. I could not go, because General Smuts ordered me to ride through the river to a rise beyond, to watch the country towards Aberdeen. Before leaving I asked my uncle [Jan Mulder] to look after my riding mule during my absence, and that was the last I ever saw of either of them. It was well that I went off on my Arab pony, for his fleet legs were to be my salvation that day. I kept watch on the hill for many hours without seeing anything of note, but when General Smuts sent a messenger up in the afternoon to say that I was to remain until half an hour before sunset, I sent back a warning that I could not tell what might be going on at the river, as the heavily timbered banks made it impossible to see along its course.

When I judged from the height of the sun that my time was over, I rode down to the river, on my way back to where the commando was camped on the opposite side. Near the ford in a clearing stood a farmhouse, and, being thirsty after my long spell in the hot sun, I touched there to ask for a cup of coffee. The old lady inside was eager to grant my wish, but, as the kettle was not boiling, she asked me to sit down while she piled more wood on the fire. I knew that the commando was moving at sunset, so I decided not to wait, and, bidding her good-bye, rode off.

I had not gone thirty yards when I heard the trample of many horses, and, glancing over my shoulder, saw a swarm of English troopers gallop into the glade and surround the house.

Had I still been seated there, I should have been caught in full khaki, and that would have been the end of me, but out here in the open my English tunic saved me, for the soldiers took me for one of their own men, and let me ride away. Seeing this, I went slowly until I was out of sight among the trees, and then rode all out to rouse the commando.

I crossed the ford, and, as I breasted the other bank, looked back to see that the English had discovered their mistake and were streaming behind me like hounds on a hot scent. My grey pony was equal to the occasion and I was able to keep well ahead of my pursuers, firing alarm shots as I went, for I knew that if all these troops came on our commando unprepared and off-saddled among the trees, with their horses out at graze, there might be a serious disaster.

I had still about six hundred yards to go, and when I rode in among our men I was relieved to find that they were rushing about the thorn trees bringing in their horses.

There was not a moment to lose, for, at the rate at which the English were approaching, they would be there long before the horses could be brought in and saddled. About fifteen or sixteen men, however, were mounted, having had their animals by them, and General Smuts called to them to ride forward and delay the oncoming enemy until the rest were ready.

With Commandant Bouwer at our head, we galloped back, but, before we had gone far, a hundred or more English horsemen came charging at us. We jumped to the ground and fired, upon which the troopers opened out and also dismounted, giving rapid fire and obliging us to fall back behind the wall of a small dam, that very opportunely stood dose by.

From here we were in a more favourable position to rake them, which we did so effectively that the men withdrew into denser bush, from which they fired heavily, but we kept our heads well down and had no casualties.

Thus far we had satisfactorily carried out our orders to hold up the enemy until the commando was collected, but we could see parties of English horsemen filtering round us through the bush in constantly increasing numbers, and, to make matters worse, they were unlimbering a field-gun on rising ground to our left. In a few moments shells were bursting overhead, against which the wall offered little or no protection, so Bouwer, boldly riding on to the crest of the dam, scanned the country to the rear, until he saw the commando move from the clearing where we had left them which meant that our task was accomplished. When he called out that the commando was making fast down the river to the south, we retreated at once. As soon as the English saw us go, they came in hot pursuit.

[419] The Groot River originates at a point where the Kariega River and the Sout River meet, although they enter the Beervlei Dam as the Sout River. Beyond the dam, the river becomes the Groot River, running southeast. Near the town of Steytlerville it turns slightly southwards before heading eastward again, lastly bending southwards before it joins the Kouga River to form the Gamtoos at the confluence. Groot River (Eastern Cape) - Wikipedia

The sun was setting and the short twilight giving way to darkness, otherwise it would have gone harder with us than it did. The troops were on our heels, yelling and firing as they pounded behind, and, had there been more daylight, I think we should all have been captured.

My khaki uniform saved me for the second time that day, for a batch of troopers rode by in the dusk, and, mistaking me for one of their men, shouted that I was to hurry, but passed on without taking further notice.

Owing to the bush, both English and Boers got separated into smaller groups, and single horsemen and parties of twos and threes were galloping about. It was soon quite dark, and, to avoid being further accosted, as there were soldiers in front and behind me, I halted in a copse[420] to let them through. I waited for a long time, until the hue and cry had died down and I could see camp fires springing up ahead, which showed that the English were halting for the night.

A copse or coppiced woodland.[475]

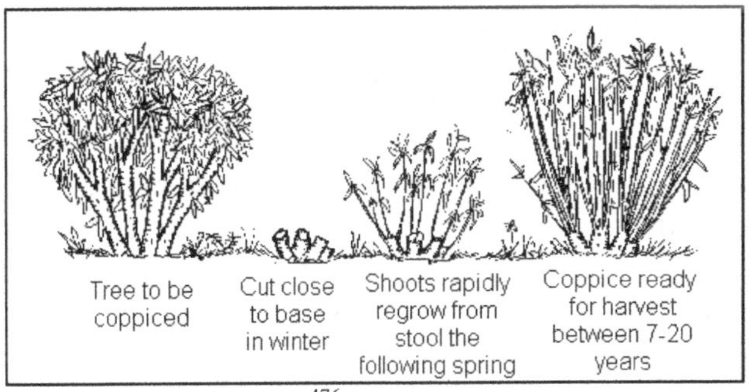

Method of Coppicing.[476]

I could not tell how my companions had fared, but my own position was sufficiently difficult, for I was alone with the enemy across my path and the commando gone.

When I thought that the soldiers had settled down, I cautiously rode on, picking my way with difficulty through the bush, for it was a dark night. After an hour or two I had skirted round their camp, and was following the course of the river once more.

At length I saw the gleam of a lighted window, and, stealing up, peered in to see five of our rearguard party of that afternoon standing in conversation with the inmates of the house.

I joined them at once, eager to hear news of the rest of our band, but they knew nothing, nor did they know what had become of the commando. As it was useless trying to pick up the spoor on so dark a night, and as our horses were tired, we got some food, and went a little distance to sleep amongst the trees, confidently expecting to overhaul our men next day. Shortly before sunrise we saddled up and crossed the river towards a conical hill lying on the other side, from which we should be able to see over the plains.

As we rode on, we found two more of our men, who had escaped from last night's affair, and had also taken to the thickets on the banks. There were now eight of us, and, when we got to the foot of the hill, we left our horses and climbed up. As we reached the top the sun was rising, and, like sailors adrift in a boat, we anxiously scanned the horizon for a sign of the commando, but the country to the south lay open before us without a horseman in view.

General Smuts must have ridden all night, for although we could see for half a day's journey, he and his men had vanished, and difficult weeks were to elapse before we found them again.

[420] Coppicing is a traditional method of woodland management which exploits the capacity of many species of trees to put out new shoots from their stump or roots if cut down. A coppice: an area of woodland managed by coppicing (periodic cutting near stump level)

XXIII. A Long Trail

For a long time we remained on the top, still hoping to see our men, but at last yesterday's English column came down the opposite bank of the river, so we made haste to get down to our horses and reach safer quarters. Searching out a less conspicuous hill, we left our animals to graze in a neighbouring hollow, while we lay amongst the boulders watching the progress of the enemy.

Horses grazing in the field.[477]

Boer Outpost.[478]

Instead of trekking on, as we expected them to do, they went into camp at a farm three miles away. This was a misfortune, as our chief desire was to make a start towards picking up the trail, but, with the English halted by the river and their patrols scouring the plain in all directions, we dared not venture out into the open, and were obliged to let the valuable hours go by while our commando was getting farther and farther away.

They kept us until nearly sunset before moving off, and then went south, from which we inferred that they were starting on a night march after General Smuts. So we only waited until the last horseman was riding away before we went down to the farm to get information.

It was dark before we reached there, and the owner, a well-to-do Dutch farmer, named le Roux, quickly told us such news as he had gleaned from the officers and soldiers during the day.

In the first place he said that three or four of our delaying party had been captured, and that one of them, my friend Jack Baxter, had been executed that morning at an adjoining farm for wearing khaki.

We were thunderstruck. The inhabitants of the districts through which we had passed could not have known of the death penalty or they would surely have mentioned it to us, and it was only when le Roux produced a recent newspaper, containing Lord Kitchener's proclamation, that we understood the position. We learned, too, for the first time, that other men of ours had been shot for the same reason, although it was only later, as more newspapers came into our hands that we found out their names.

From what I could make out, the executions had been kept quiet, but now, for some reason or other, perhaps the killing of Captain Watson, the military authorities were giving them publicity. From a farm labourer who came in, we had details of Baxter's shooting, which brought home to me how narrowly I had on several occasions missed a similar fate, so I lost no time in changing the tunic I wore for a coat which I borrowed from our host, who also supplied such of my companions as were in khaki with whatever he had in the way of civilian dress.

As to the commando, le Roux said that General Smuts was believed to be heading for the Swartbergen,[421] [422] a great range whose peaks we had seen during the day, looming fifty or sixty miles to the south. We said good-bye and rode on all night, and for the next three days made our way across the plains that lie towards the mountains.

[421] The Swartberg mountains (black mountain in Afrikaans). Swartberg - Wikipedia

[422] The pass runs between Oudtshoorn in the south and Prince Albert in the north. Swartberg Pass - Wikipedia

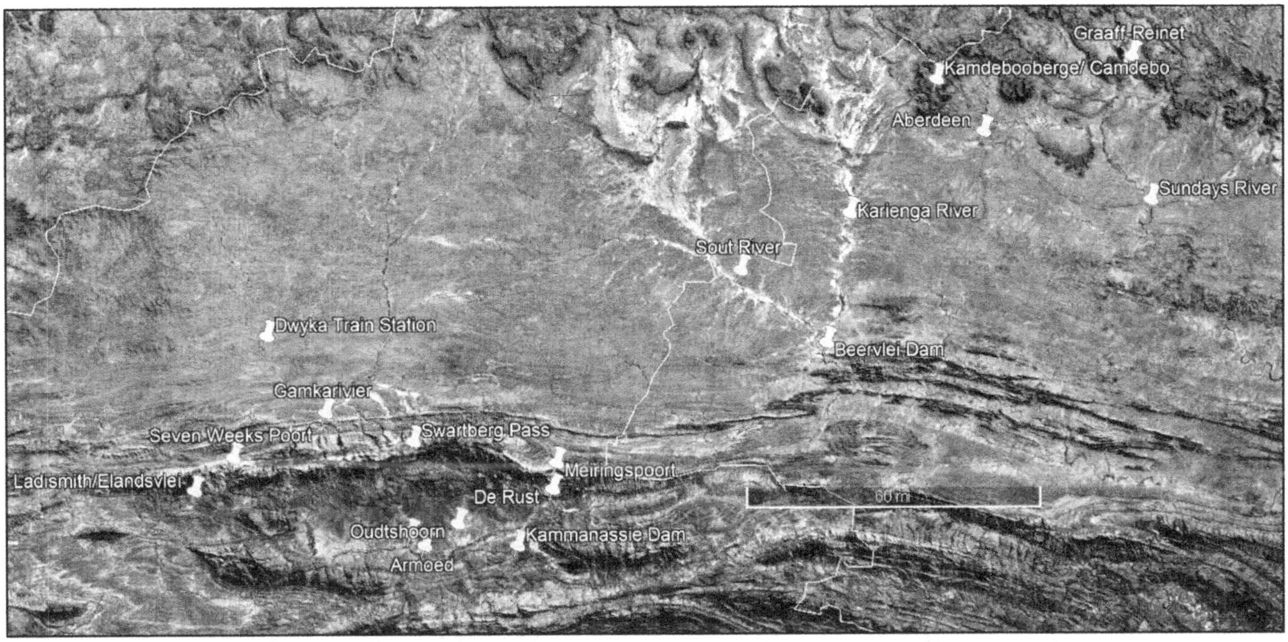

Map data: Google Image© 2024 Image Landsat/Copernicus.

Swartberg Range from the Laingsburg area.[479]

Flowering Karoo near Laingsburg.[480]

Map data: Google Image© 2024 Image Landsat/Copernicus

The local inhabitants gave us word that General Smuts had passed by, but there were several English columns moving between, and their patrols were so active that we had to go warily. Once at dawn we were hotly chased for many miles, so our progress was slow, and our chance of speedily overtaking the commando grew perceptibly less.

William Conradi, who was with us at the Kraal during the 17th Lancers fight, as the oldest and most experienced of our party, took charge. The others were: Albert von Rooyen, Albert Pienaar,[423] Cornelius Brink,[424] W. Pypers, W. van der Merwe and a boy named Michael du Preez, all Transvaalers, except Conradi, who came from the Western Cape, and all good, brave fellows. On the afternoon of the third day I was ahead to watch the doings of a small English force, when I saw a horseman detach himself from them and come riding up the road in my direction. I lay in wait for him behind some trees, and, as he passed, I leaped out and knocked him from the saddle with the butt of my rifle.

He turned out on closer inspection to be a Hottentot[425] soldier, such as the English employed as scouts and dispatch-bearers. He was more terrified than hurt, and, when on the off chance I ordered him to hand over the message which I thought he might be carrying, he did actually produce one from his boot.

Native African scout, employed by the British army, at Belfast, Transvaal Province.[481] Two men employed by the British Army seated beside a tree. Their brass arm bands denote military service.[482]

Natal Native Volunteer Corps.[483]

When my companions came up we carefully studied this document, which was addressed to a Colonel Scobell,[426] informing him that General Smuts had crossed the Swartbergen into the district of Oudtshoorn the night before. It added that he had been reinforced by nearly a hundred men, which was a mystery to us, but we found afterwards that a roving band of fifty men had joined him the day before. They were the remnant of a commando that had long been operating in these parts under Commandant Scheepers,[427] who had recently been executed for train-wrecking. All was therefore well with the commando, and we were still on the right track, so

[423] Pienaar Albertus Pieter Roux, Veldkornet. Commando: General Smuts.

[424] Brink Cornelius Veldkornet, Commando: General Smuts. Awarded the DTD.

[425] A member of the Khoekhoe group of peoples. (archaic, now offensive) A member of the broader Khoisan group of peoples.

[426] A major in the Queen's Own Worcestershire Hussars and the regiment's Honorary Lieutenant Colonel in 1896 He is the father of Major General Sir Sanford John Palairet Scobell, DSO (26 September 1879 – 2 March 1955), a senior officer in the British Army. John Scobell - Wikipedia

[427] Most likely: Scheepers Daniel Du Plessls, Veldkornet Commando: General Kritzinger. Awarded the DTD.

we divested our prisoner of his horse, rifle and ammunition, and told him to clear off, a command he obeyed with a cheerful 'Dag, mij haasies,'[428] as he trotted up the road.

We were by now within fifteen or twenty miles of the Swartbergen, but a ragged tract of foothills had to be crossed before we could reach the bottom of the range itself, and it took us the whole of the ensuing night to get there, for the going was dreadfully rough. Towards morning we came to a beaten highway, which, from information previously received, we knew to be the approach to Meiring's Poort, a pass leading over the mountains near here. The pass was not for us, because it was held by a garrison, but we decided to make use of the road for a while, as preferable to the boulder-strewn country across which we had been toiling.

Meiringspoort Pass Road.[484]

Great Swartberg Pass soon after being cut.[485]

This landed us in a mess, for we ran into a body of English horsemen. It was too dark to make out their strength, and we were so mixed up with them that no one could shoot. For a few seconds we were milling about, neither side quite certain whether we were dealing with friend or foe, and no one uttered a word for fear of precipitating trouble. Then we heard Williani Conradi shout to us in English to break away, so we disengaged ourselves and turned back into the rough, while the English clattered away along the road without a shot having been fired.

After this we went more carefully, and sunrise found us leading our horses up the street of a tiny village standing at the bottom of the pass.[429] Dogs began to bark, and windows to open, and we saw soldiers running to a large building, so we mounted and rode hastily out.

Before us rose the Swartbergen, steep as a house, but we climbed it all day, dragging our leg-weary horses, until we reached the top at dusk. From here we could look south over more mountains and deep valleys, and far beyond lay a grey haze, which we took to be the sea.

Greater Swartberg Range near Meiringspoort.[486]

Our ascent during the day had not been unreasonably difficult for the north face, up which we had come, though steep, was grass-covered and devoid of krantzes,[430] but the slope down which we had now to go was fringed with high crags, so, with darkness coming on, we were obliged to halt for the night.

We had eaten nothing for twenty hours, as the presence of the troops the day before had prevented us from foraging, so we sat cold and hungry, looking down into the black depths below.

[428] Afrikaans: 'Goodbye my bunnies.'

[429] Probably De Rust - Wikipedia Dutch for "There Rest" about halfway down the pass.

[430] An encircling or overhanging wall of rock.

After a while there was a faint twinkle of a light, evidently a farmhouse in some valley, and, as Michael du Preez and I were the youngest and hungriest of the party, we prevailed on the others to let us go down the mountain. It was a bad climb, for we had to feel our way in the dark by cracks and crevices, to the bottom of the cliffs, and it took us the rest of the night to do the remainder. So it was not till well after sunrise that we reached the farm whose light we had seen in the evening. The owner, an Englishman named Holm, gave us a generous meal, including an omelette made from ostrich egg, to which we did full justice.

Ostrich Farm, South Africa.[487]

Ostrich Farm, South Africa.[488]

After getting from him a further supply of food for our companions, and eliciting the information that General Smuts had passed down the valley the day before with enemy troops hanging on behind, Michael and I started up the mountain once more.

We were desperately tired, having had no sleep for two nights running, so it took us seven hours to drag ourselves to where the others were waiting. They had, in the meanwhile. succeeded in finding a practicable way down the cliff for the horses, and we found them considerably below the spot at which we had left them the night before. They were so famished that they had begun to look over the horses with a view to shooting the worst of them for food, but our arrival, each with a bag containing an ostrich egg, meat, and bread, made this unnecessary, and they fell to cooking a huge meal instead. When all had finished, Conradi ordered an immediate start, for we had told him that General Smuts was in the valley, and he was for not losing a moment in going after him, so Michael [du Preez] and I had to go along once more.

Climbing down was difficult, and in places we had to roll boulders into the torrent that rushed down the gorge we were descending, to form a bridge for the horses. Halfway down, we came on a cattle-path, which made progress easier, and we reached the foot of the mountain by ten or eleven that night. We halted in an orchard, and I was asleep almost as soon as I had the saddle off my horse.

We were now in the great valley that runs parallel to the Swartbergen towards the town of Oudtshoorn, [431] and I well remember how oppressed we felt in this region of mountains, for we were accustomed to the open country and wide horizons of the north.

As we made our way down the valley all that day, we were comforted with news of General Smuts at every farm-house, and we were hopeful of coming up with him at any moment, for the commando tracks lay plain before us on the road. By nightfall we were so hot on the trail that we passed a homestead where our men had halted that afternoon, and a little farther on we were told that Edgar Duncker and Nicolas Swart had gone by on foot only an hour before. Their horses had been killed during the retreat at the Kareega River, and, like ourselves, they had been following the commando ever since.

We hurried on, intending to catch up before halting, but after dark we lost the tracks owing to stony soil, and could find no farmhouse at which to make inquiries, so we camped in a gully for the night, feeling sure that we should see the commando the first thing in the morning.

We were up at sunrise, eagerly gazing down the valley. Sure enough there rose a cloud of dust, and we could make out horsemen riding among the trees, and we saddled in haste, congratulating ourselves that our search was over.

[431] Oudtshoorn the "ostrich capital of the world." Oudtshoorn - Wikipedia

But disappointment was in store, for, as we hurried down the road, a woman ran from a field with outstretched arms, to warn us that those men were English troops who had come into the valley overnight, and indeed, before long, so many of them came riding from farm to farm in our direction that we had to go up one of the smaller lateral valleys to escape their attentions. We did not know what had become of our men, but from what the woman told us, General Smuts had evidently got wind of the English movements and had escaped under cover of darkness, but in which direction we were unable to discover—a sad blow, after having so nearly rejoined the commando.

We continued up the smaller valley, then climbed over a height and descended into another of the broad valleys that abound in these parts.

I had never been here before, but our family clan is a large one, some of whose branches have spread far from the older settlements around Table Bay, so I was not surprised that afternoon to come on a connection of mine named Rex, a lineal descendant of George Rex,[432] the morganatic son of King George III by Hannah Lightfoot,[433] the Quakeress. George Rex had been sent out to South Africa in 1775, and given a large tract of land at the Knysna,[434] on condition that he did not again trouble his august parent. His descendants still live there, and one of them had married my mother's brother.

Rex and I spent an hour discussing family ties, and before I left he insisted upon giving me a pair of new boots, as mine were considerably the worse for wear. For this he was fined and imprisoned by the military, and I read in a newspaper that he was convicted of 'comforting the King's enemies,' which amused me greatly, although I was sorry that I had landed him in trouble.

Soon after this the ubiquitous English patrols were once more in evidence, on a house-to-house visitation, so a local farmer accompanied us to where a path led into a narrow kloof,[435] and, having put us on our way, slipped home again. This path ran between high crags, that sometimes almost met overhead, until at length it reached the side of a mountain up which it ran. We followed to the top, and got there long after sunset. We could not see the country beyond, but through the darkness shone a light from a farmhouse, and, as we were anxious to pick up the lost trail of the commando, W. Pypers and I went down to make inquiries.

The slope was steep, but clear of rocks, and by midnight we were hammering at the door of a large homestead. It be-longed to an Englishman named Guest, who, when he opened the door and saw who we were, exclaimed, My God! First come the Boers this morning and slaughter my sheep; then come the British, who kill more sheep instead of catching the Boers, and now I am hauled out of bed at this time of night by more Boers!

We spoke to him pleasantly, and the old fellow cooled down enough to explain that General Smuts had camped on his farm at eleven that morning, and was followed by a pursuing English force, that had also halted here and made free of his livestock, so he not unnaturally looked upon our arrival as the last straw in a distressful day. Becoming more affable, he roused the servants and gave us a good supper, during which he told us that Duncker and Nicolas Swart had caught up the Commando here. Having eaten well, and obtained as much information as we could, we persuaded our host to give us enough food for our friends, and started back.

As we went off he doubtless heaved a sigh of relief at having got rid of us, but, had he known it, his troubles were only beginning.

Pypers and I reached the others shortly before daybreak, and as Conradi was not one to let the grass grow under his feet, he started us off at once. At the first dawn of what turned out to be a lively day, we began leading our horses down the mountain, and towards eight o'clock were nearing Guest's house, when there swung into view,

[432] George Rex (29 August 1765 – 3 April 1839) was a British-born entrepreneur who spent most of his adult life in the Cape Colony. He founded the town of Knysna, Cape Colony and played a key role in its development. Rex filled a number of positions in the Cape Colony – including Marshal of the Vice-admiralty Court, notary public to the Governor and advocate for the Crown – before settling on the farm Melkhoutkraal, in the Knysna district. George Rex was the alleged first son of King George III. George Rex - Wikipedia

[433] Hannah Lightfoot (12 October 1730 – before December 1759), known as "The Fair Quaker," was a Quaker in Westminster. She married Isaac Axford in December 1753 but, before the end of the following year, had disappeared. Later gossip, originally in amusement and ridicule, first noted in print in 1770, but much embroidered in the late eighteenth and early nineteenth centuries linked her name, although some eight years his senior, with the extremely shy fifteen-year-old Prince George. Prince George became King George III in 1760 and was known to admire the simplicity of the Quakers. After George's death, rumours circulated that he had engineered her abduction, married and had children by her. However, no contemporary source connecting the Prince and Hannah has ever been found. Hannah Lightfoot - Wikipedia

[434] Knysna - Wikipedia

[435] A deep glen or ravine.

round a bend of the valley, some two hundred English horsemen, riding hard for the farm. Our presence there during the night must have been reported to them, for they were riding like men with a set purpose, and on nearing the homestead they divided to right and left, to surround the buildings and orchards. Luckily we were able to hustle ourselves and our animals into a gully, without being seen, and from our hiding-place we watched the activities of the soldiers with considerable interest. When they drew a blank at the farm, they deployed along the foot of the slope on which we were, and, splitting into parties, began a systematic search. Soon on every knoll and hillock men stood scanning the mountain-side, as if they knew that we were somewhere about, but, although some of the troopers came within a hundred yards of us, we were not discovered. We did not fire for we knew that if we did it would be all over with us, so we lay hidden, meaning only to shoot when there was no other alternative. A man on a white-faced Argentine [horse][436] came trotting up to within twenty yards of us.

Two Criollo stallions in Brazil.[489]

Argentine Criollo mare.[490]

He dismounted to examine the path for hoof marks, and was so close to us, that had one of our horses jingled a bit, he must have heard it. We held our breath until he got into the saddle again and rode away, little thinking on what a thread his life had hung, for we had him covered, and, had he seen us, he would have been a dead man, but we had no wish to precipitate a general battle against impossible odds, and we let him go.

After what seemed an age, the hunt died down and the soldiers gradually drifted back to the farmhouse, where they camped for two hours, during which we could not move. We saw the men flinging oat-sheaves from a loft, and chasing poultry, and I could not help feeling sorry for Mr. Guest, who was once more being put under the harrow,[437] and not for the last time either, as it proved.

When the troops at length saddled and rode away, we waited barely long enough for the last man to be off the premises before we hurried to the house.

When Mr. Guest saw us appear, with the soldiers only just going through the garden beyond, he looked as if he had seen an apparition, and when we laid him under further contribution he seemed on the verge of a fit. However, he complied with our demands, grumbling and complaining at first, and then laughing at his ill luck. Having satisfied our requirements we rode up the valley a little distance, to where there was a pleasant orchard and a large cultivated field hedged round with high branches of thorn, in the manner customary in this area. This was the one error of judgment we made throughout our trip, for instead of making for the wider country lower down, we had entered a cul-de-sac.

We thought that the English were finally gone, and prepared a meal, after which I made a second mistake, for, while the others kept their horses by them, I turned my little Arab, all saddled as he was, into the field, and, thoroughly weary from having been up all night, sought out a shady spot in the lee of the thorn fence and; without telling the others where I was, fell sound asleep.

[436] The Criollo (in Spanish), or Crioulo (in Portuguese), is the native horse of the Pampas (a natural region between Uruguay, Argentina, Brazil, and Paraguay, in South America) with a reputation for long-distance endurance linked to a low basal metabolism. The breed, known for its hardiness and stamina, is popular in its home countries. Criollo horse - Wikipedia

[437] In agriculture, a harrow is a farm implement used for surface tillage. It is used after ploughing for breaking up and smoothing out the surface of the soil.

I was awakened, I do not know how long after, by the clash of rifle-fire near by, and, starting to my feet half dazed, saw a number of English soldiers standing before their horses and blazing away at my seven companions, who were riding down the valley for their lives. I had only myself to blame for being left behind, as they did not know where I was, and were in any case unable to wait. My chief hope of escape-was my horse, but he was standing inside the field in full view of the firing soldiers.

They had not yet noticed me, as I was screened by the fence, so I parted some of the branches to see what chance there was of getting at my pony. By great good luck he was standing on the other side, within a few yards of me. The firing had alarmed him, for he was restlessly tossing his head and sniffing the air, and I could see that in another moment he would bolt, so I called to quiet him, and, worming my way through a weak spot in the fence, ran up to where he stood quivering with excitement. Jumping into the saddle, I rode for a small gateway in the far corner, which was the only outlet. The soldiers saw me at once, and turned their fire upon me, in spite of which I managed to get through the opening, but, just as I was gathering speed beyond, a bullet brought my poor horse headlong to the ground and flung me yards over his head.

Picking up my rifle I ran towards the homestead, thinking that my party might be making a stand there. The soldiers beyond the field kept firing at me as I appeared and disappeared amongst the trees, but I got within hail of the house unharmed. At the corner of a barn stood six or seven men, whom in my haste I took to be my friends, and I made straight for them. But as I came within thirty yards of them, one stepped forward, and, levelling his rifle, called on me to halt.

They were English soldiers, and not the only ones, for more came rushing round from the stables and out of the dwelling-house. Escape seemed impossible, but I made a bid for it. To my right was a small grove of poplars, and, swerving aside, I dashed for this cover before they could send more than a bullet or two after me.

Volleys came crashing through the trees as I ran, but I emerged safely on the other side into hummocky ground, where I twisted and turned to such good effect that, although the men came hurrying round to cut off my retreat, I got into a broken stretch with no more serious damage than a gash from a bullet, which ripped up the sole of my boot and made running difficult.

Breasting a knoll, I glanced back. The soldiers near the field had mounted their horses, and were coming after me. Of those around the homestead, some were running in my direction, and others were in the yard throwing saddles on their animals, and I had a final glimpse of Mr. Guest in his shirt-sleeves on the stoep wildly gesticulating, but whether he was urging on the men to my capture or protesting against the crowning disaster of a battle on his doorstep, there was no time to consider, for I was in a very tight corner.

There was no sign of my companions. The sharp ground was cutting my foot, the horsemen were close behind me, and already I could hear the men yelling at me to stop, and I was just deciding that I had better do so, when I came on a deep nullah running down the mountain-side. Here it flashed on me that if my pursuers saw me disappear over the bank, they would naturally think that I was making down its bed to the centre of the valley, or up towards the mountain. Looking aslant my shoulder, to make sure that they saw what I was doing, I went over the bank, but instead of trying to escape up or down the water-course, as they would expect, I found a spot on the opposite side, where the rains had washed out a shallow runnel,[438] and, crawling up this, went flat on my face into the bushes beyond, which stood just high enough to conceal a prostrate man. Having left the nullah unperceived, I worked myself forward another fifty yards to a slightly denser patch, and stopped there.

The soldiers, seeing me jump into the spruit, did exactly what I anticipated. On reaching the spot where they had, seen me vanish, they separated into two parties, one of which galloped up the mountain-side, and the other down towards the valley. I had a clear view of the search from where I lay, and after a while I could see, from the undecided way in which they were riding about, that, they were completely nonplussed.

In the end they must have concluded that I had got away on the upperside, for they spread out along the mountain slope like beaters at a shoot, moving farther and farther from my hiding-place. I knew now that I was comparatively safe, for the sun was setting, and before long I heard them clattering back to the farm, where presently their camp fires shone out, indicating that Mr. Guest was once more to be an unwilling host.

[438] A channel.

I felt proud of my successful ruse, but there was little else pleasant to contemplate. I lay in the bracken like a hunted rabbit; my foot throbbed painfully; my companions were gone, and so was the commando; my horse was dead and my saddle and belongings were in the hands of the enemy.

As thinking did not mend matters, I rose at length, and limped off in the dark.

After about an hour, I heard the sound of a hymn and the wheeze of a harmonium,[439] such as stands in almost every Dutch farmhouse, and knew that I was nearing friends. When I knocked at the door there was a hush at first, for in these disturbed times a visit late at night meant military requisition, but then I heard a shuffle of feet and the door opened.

Mason & Hamlin Organ.[491]

A whole family was peering from within. When I told them who I was, they almost dragged me into the house, so eager were they to help. I must have looked very dishevelled, for the women wept with pity while removing the boot from my sore foot, and during the more painful process of extracting a thorn, nearly an inch long, that had run into the palm of my hand when-I was thrown from my horse that afternoon. They fetched hot water and tore up clean linen for bandages; a meal was laid, with coffee, and the kindly people almost quarrelled for the right to serve me, so keen was their sympathy, although they knew that it might mean for them fines and imprisonment.

Having attended to my wants, they took further counsel. It was agreed that I could not remain here, for even if the continuous patrols did not ferret me out for themselves, my presence was certain to be reported by the coloured farm labourers, who all over the Cape sided with the British. As I assured them that I was well able to walk, it was decided that I must continue westward on the off chance of coming up with General Smuts, who might be held up somewhere. It seemed a forlorn hope, but as there was the risk of an enemy detachment coming by at any moment, I made ready to start as soon as my boot had been sufficiently repaired.

The head of the family, a patriarch of seventy, insisted on acting as my guide during the first stage of the journey and firmly refused to waive the right in favour of his sons, who offered themselves. A grain-bag was packed with food, and after an affecting leave-taking, the old man and I set out. We trudged along, hour after hour, until his strength gave out and I made him turn back, his voice shaking with emotion as he wished me God-speed.

My foot scarcely hindered me, and now that I was alone I made good speed on the well-marked wagon-road upon which he had set me, until, towards three in the morning, it dipped down into a ravine. By the time I reached the bottom the moon was clear, and by its light I saw several fresh hoof marks on the ground. On examining these, I recognized the slightly malformed marks of Michael du Preez's pony, and closer investigation showed me the footprints of men which I knew at once as those of some, if not all, of my seven missing companions, who had crossed the road here on their way down the ravine. This providential discovery cheered me immensely, for I had known all along that my hopes of overtaking the commando by myself were slim, but it was pretty certain that I could catch up with the men who had passed here so recently, and I lost no time in following their spoor.

[439] A pump organ. Pump organ-Wikipedia

After some miles they had branched off into a smaller kloof, and I followed along this without difficulty, for the tracks lay clear in the dusty cattle-path. At last, as day was breaking, I heard the whicker of a horse and, going forward carefully, found all seven men fast asleep beneath the trees. They were astonished to see me, as they had made certain that I was either dead or taken. They themselves had been hard put to it to make their escape from the farm, and although not one of them had received a scratch, yet out of the nine horses we had possessed between us, no less than six had been killed, and, what was almost as serious, We had lost the bulk of our saddles, cooking-tins, and blankets.

We agreed that the first thing to do was to replenish our equipment, so we continued along the gully, until a long march brought us into the thickly populated valley of the Caminassi River,[440] where we heard many rumours about our commando, but no certain news, beyond the fact that General Smuts was making west with a strong force of cavalry closely pressing him. The people willingly supplied our wants, and, for the next three days, we slowly felt our way down the broad valley.[441] Far ahead were tall pillars of dust, made, they told us, by General French with thousands of horsemen, engaged in a great drive behind our Commando, and occasionally we heard the distant boom of guns. As five of us were on foot, we had to proceed with great caution, and repeatedly we had to hide for hours at a time, to avoid bodies of horsemen passing from the rear to join their advance-columns ahead.

The more we saw of the valley, crowded with columns on the march, the less we liked it. William Conradi, after watching for some time, said that he had a better plan. Instead of following the commando any longer, he proposed to turn north across the Swartbergen, back into the Karroo country from which we had come. He said that General Smuts was almost certainly heading for the Western Cape, and, if we got that mountain range between ourselves and the troops, we could travel unmolested and perhaps join him when the chase had died down. We agreed, and set off at once. For two days we worked our way through the intervening country, towards where the Swartbergen stood like a wall on the northern horizon. At a village called Armoed[442] we had trouble, for a party of soldiers rode at us as we were leaving it, and only the falling dusk and our brisk reply to their firing enabled us to escape through the river, into the bush on the other side.

My foot gave me no pain, and we got along safely, thanks to the local farmers, who kept us well informed.

Map data: Google Image© 2024 Image Landsat/Copernicus.

On the morning of the third day we reached the foot of the mountain chain, just east of the Seven Weeks Gorge.[443] Into this poort[444] ran the main road to the open Karroo, but the pass was garrisoned, and our only course lay up the flank of the range. As there were English patrols riding about we began the ascent without

[440] See: Kammanassie Dam - Wikipedia

[441] Of the Olifants River: Olifants River (Southern Cape) - Wikipedia

[442] Volmoed is a small town in the Klein Karoo, about 15km from Oudtshoorn. It was formerly known as Armoed. Volmoed - Wikipedia

[443] Seven Weeks Defile Peak is the highest mountain in the Cape Fold Belt and the highest point in the Western Cape province. Seweweekspoortpiek - Wikipedia

[444] Middle Dutch, a gate.

delay, toiling upward steadily, until we made the summit by evening. These were the same mountains which we had crossed coming in the opposite direction days before. At that crossing the Swartbergen had consisted of a single clear cut barrier, but here it forked out into numerous sierras,[445] that looked like giving us much trouble.

As it was getting dark, and heavy rain began to fall, we dropped some distance over the crest to seek shelter for the night. It was too cold to sleep and too damp to light a fire, so we sat shivering until dawn, when we started to grope our way down the mountain-side enveloped in a dense mist. Towards four in the afternoon we were below the clouds, and could see a long narrow canyon lying at our feet, its sides closed in by perpendicular cliffs. On the floor of the chasm, a thousand feet below, we made out a cluster of huts, and, thinking to find natives there to guide us, we went down in a body to investigate, leaving the horses in a ravine to look after themselves. We climbed through a fissure in the crags, and reached the bottom soon after sunset. As we approached the huts, a shaggy giant in goat-skins appeared and spoke to us in strange outlandish Dutch. He was a white man named Cordier, who lived here with his wife and a brood of half-wild children, in complete isolation from the outside world. He knew all about us, for one of his sons had been up the mountain that morning, and, hearing the sound of men and horses in the mist, had stalked us and carefully noted our number and the language we spoke, after which he had vanished over the edge of the cliffs to warn his father.

We were received with uncouth but sincere hospitality, and we applied ourselves gratefully to the goats' meat, milk, and wild honey that were placed before us. Cordier told us that no British troops had ever penetrated this fastness and that we were the first Boers to do so. He had heard vaguely of the war, but his knowledge of the events of the last two years was scanty.

We spent that night and the next day with this curious Swiss Family Robinson, and in the evening toiled up the cliffs again, accompanied by our host and some of his colts,[446] who stayed with us around our camp fires, and led us the following morning across rugged mountains, until by dark we looked down at last upon the northern plains. Our intention now was to descend to the open country and, keeping the mountains well on our left, to strike west towards the districts of the Cape lying along the Atlantic seaboard two hundred miles away, where we hoped ultimately to get news of General Smuts.

We spent another night on the heights, and, parting from our guides at daybreak, climbed down the slopes to level ground and headed across the plains. We were now in the 'Gough' Karroo, as the Hottentots[447] call it, an arid, waterless action, sparsely occupied by wandering herdsmen. By next day we crossed the railway line that runs from Cape town to the north. There were no block-houses, as in the Transvaal and Free State, so we had no difficulty, although we saw the double-tiered watch-towers at either end of the bridge over the Dwyka River.[448]

Karoo National Park.[492]

[445] A rugged range of mountains.

[446] In this sense, his children.

[447] The indigenous peoples of Southern Africa who do not speak one of the Bantu languages, comprising the Khoekhoen and the Sān or Sākhoen Khoisan - Wikipedia

[448] It flows from the North-west, joining the Gamka River as a tributary at the Gamka Dam. Dwyka River - Wikipedia

On the far side of the line lay country even less inhabited, and we suffered severely from thirst and hunger, for water was to be had only by digging in the dry gravel-courses with our bare hands, and for food we had to subsist on what we had brought with us.

About a week after passing the railway line, always going due west, a patrol of English soldiers appeared on the baking plain. They opened fire at us, and when we replied they made off, doubtless to report to some larger force in the neighbourhood.

The day after that we reached a prosperous-looking farmhouse, the first we had seen since crossing the Swartbergen. At this place we had a miniature battle, for while we were talking to the owner and his wife, twelve or fifteen troopers suddenly rode on to a ridge and fired on us. We told our friends to go indoors, and ran down into the spruit, from which we worked forward, thinking to get near enough to our opponents to dislodge them and capture their horses, but by the time we got within a hundred yards of the troopers they were pinning us down with such accurate shooting that we had to hug the earth, and were only too glad to crawl back out of harm's way after dark, and retrieve our three horses and belongings at the farm.

For the next two days we travelled slowly on, gradually approaching the more thickly settled district that lies towards Calvinia.[449]

At every house at which we touched we made inquiries for General Smuts, but no word of him was known. However, a pleasant surprise was awaiting us. Early one morning, as we sat by our fire, we saw a diminutive cart come over the rise, drawn by two donkeys.

Family of Boer refugees in the horse cart approaching a British soldier on patrol.[493]

On the seat was a grey-bearded old Dutch farmer of the poorer class and beside him a smart English sergeant. When we stopped this queerly assorted couple, we were astonished to learn that they had quitted General Smuts and his commando only an hour or two before. The soldier said that he had been captured the previous day while scouting, and after spending the night with our men, had been released that morning.

We were so delighted with this unexpected good news that we insisted on shaking hands with our informant, who could not at first understand our elation. When we explained, he said that for his part he had less cause for congratulation because, having been deprived of his horse, and having no mind to walk ninety-odd miles to the nearest military post, he had ordered his fellow-traveller in the King's name to provide him with transport.

As the only available conveyance was the one we saw, he was not looking forward to the journey, particularly as relations were strained with the driver. The farmer was glum and angry at having to drive a verdomde rooinek[450] (as he called him) on a journey that would take him from home for the better part of a fortnight, and would subject him to the jeers of all his neighbours on his return, while the servant was no better pleased with his companion, who, he said, did not understand a word of English and only grunted when spoken to. We did not waste much time on the incongruous pair, and, after wishing them a pleasant journey, hurried on.

Tramping forward for seven or eight miles, we breasted a rise, and there, on the banks of a river below, was the welcome sight of many horses at graze and smoke rising from among the trees to show that our long quest was

[449] PF: The town itself is 147 miles North East of Seven Weeks Gorge and the Dwyka River area.
Calvinia is a town in the Northern Cape province named after the French religious reformer Jean Calvin. Calvinia - Wikipedia

[450] Damned, (accursed), Englishman.

at an end. A mounted sentry rode out to see who we were, and, after shaking hands, galloped back to spread the news of our arrival. Soon the entire commando was running to meet us, and we were surrounded by a laughing, cheering crowd, all anxious to show their pleasure at our safe return. General Smuts was among the foremost to greet us. He said he had long given us up for lost, and warmly praised the way in which we had come through without losing a man.

Great indeed was our joy at getting back, but for me there was a fly in the ointment, for I found that the 'Rijk Section' was practically wiped out. First, there was Jack Borrius minus an eye, and still suffering from a swollen, festering hand; then there lay Ben Coetzee with a bullet in his leg; Nicolas Swart, with a shattered arm, the result of a revolver shot at close quarters; and Edgar Duncker, with a bullet through his thigh and three fingers of his right hand blown to pieces. In addition, my uncle Jan Mulder (he was really my step uncle)[451] and our inconspicuous but loyal companion, Jan van Zijl, had been captured, so that there was very little left of the old unit, Wyndell, Frits Balogh, and I being the only foundation members not incapacitated.

The total loss to the commando during our absence, apart from a dozen wounded who had come along, was not more than seven or eight men, in spite of heavy fighting all the way.

In the meanwhile, the other half of our force under van Deventer was not yet accounted for, but General Smuts fully expected that, under so experienced a leader, they would eventually turn up.

[451] Deneys' father Francis, a widower, had remarried. Deneys says very little about Francis' second family. Jan Mulder must be the brother of Francis' second spouse, Cornelia Maria Theresa Mulder.

South West Africa

This map and portions of it will prove useful in this portion of the book. While a contemporary map, many of the names are different. From Hillegas 1900 Pages 320 to 322 of the US Edition titled The Boers in War.[494]

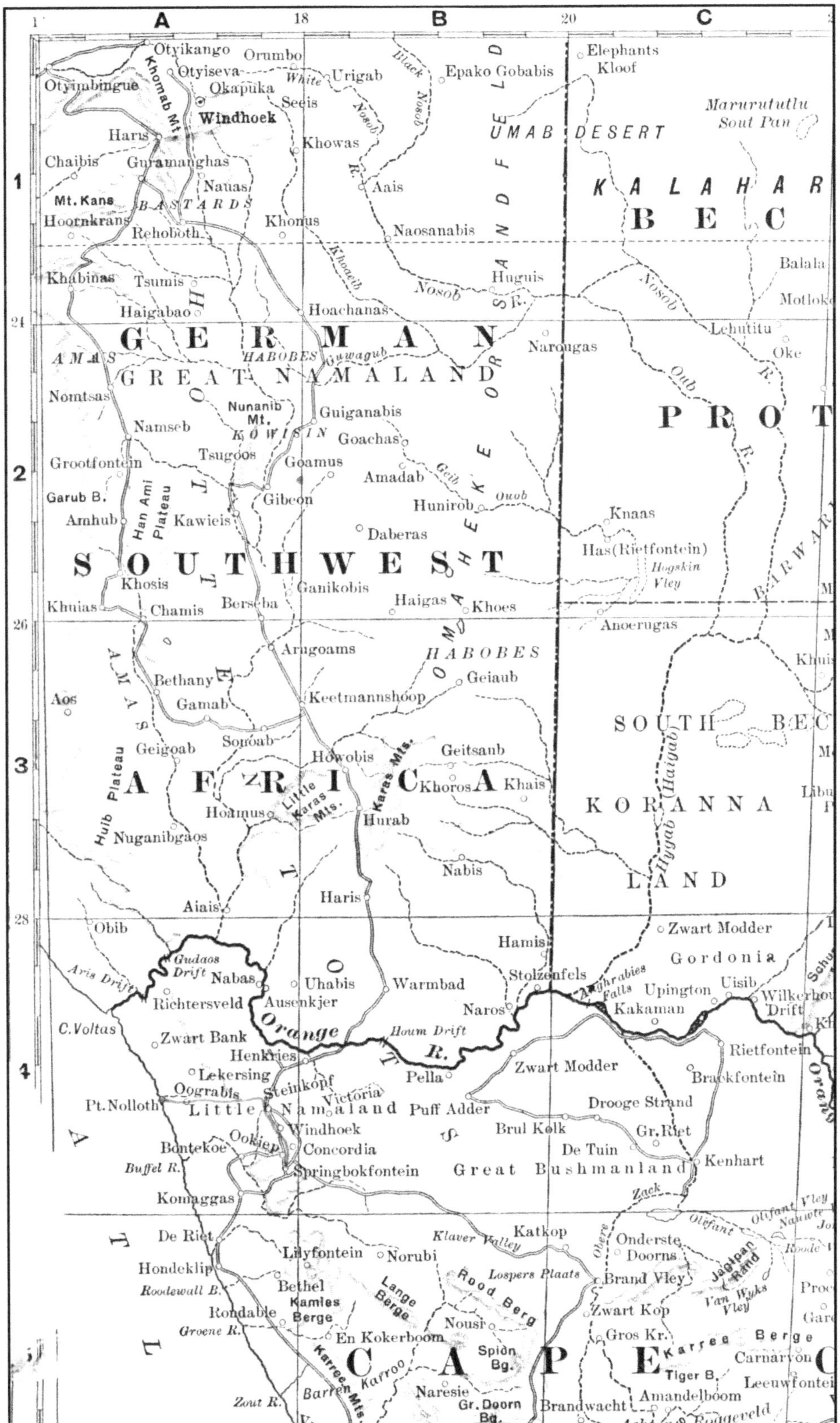

South West Africa.

XXIV. Calmer Waters

From now onward, the circumstances of our expedition into the Cape radically altered for the better. Here in the far-west there were no railways, arid the country was so difficult for large bodies of troops that we had reached comparative sanctuary. North, stretching towards the Orange River hundreds of miles away, lay a great territory practically free of the enemy, save for rare columns passing by, and a few garrisons scattered long distances apart so that we had the country almost to ourselves.

Small bands of local rebels had long been carrying on a desultory warfare of their own, between here and the coast, and General Smuts told us that he was going to reorganize these into larger commandos until he was strong enough to undertake big-scale operations, which he thought might ease the pressure in the two Republics. Thus we looked forward with fresh interest to the new stage of the war opening before us.

That same evening we moved off, still going west, our wounded now comfortably driven in carts, while those of us who had returned without horses were provided with temporary mounts, a welcome change after our long tramp. In a few days' time we reached Elandsvlei,[452] an oasis with waving palms and running water, and here we halted for two whole days. This was the first time since crossing the Orange River into the Cape that we had stayed in one place for even a day and a night, and, needless to say, both man and beast revelled in the unaccustomed holiday. In the hills near by ran a troop of mules, and some friends and I, managed to catch half a dozen of them, mine being a powerful black, who squealed and bit and threw me several times before I mastered him. Then he was quite gentle and I rode him for many hundreds of miles during the next few months.

From Elandsvlei we went north-west via Biddow to place called Kobbee, a deep valley that reminded me of Dreadful Hollow in Robbery Under Arms,[453] and here again we lay over for several days, feeding our animals on the plentiful crops. Thence we crossed the intervening mountains to the great plain that runs towards the Atlantic; sixty miles away.

At the foot of these mountains lies the village of van Rijnsdorp.[454] It had been recently garrisoned by the British, but Commandant [Manie] Maritz had swooped down and captured it. This Maritz was a policeman from Johannesburg, who, after many adventures, had established himself in these parts as a leader of various rebel bands.

He was a short, dark man, of enormous physical strength, crude and ruthless in his methods, but a splendid guerrilla leader, and according to his lights an ardent patriot, of whom I was to see more later on.

Manie Martiz on Commando.[455] Gen. Manie Maritz.[495] Salomon Gerhardus (Manie) Maritz.[496 497]

[452] Possibly this place: Ladismith is a town in the western Little Karoo region. In 1852 the farm Elandsvlei was set aside for the town. It was named after Lady Juana María Smith, the wife of Sir Harry Smith, one of two towns in South Africa named after her, the other being Ladysmith in KwaZulu-Natal. Ladismith - Wikipedia

[453] Robbery Under Arms is a bushranger novel by Thomas Alexander Browne, published under his pen name Rolf Boldrewood. It was first published in 1882. It is considered a classic of Australian colonial literature. Robbery Under Arms - Wikipedia

[454] Van Rhynsdorp (Afrikaans: Vanrhynsdorp) is a settlement in the Western Cape province Vanrhynsdorp - Wikipedia

[455] Salomon Gerhardus (Manie) Maritz was a Boer general and leader in the Rebellion of 1914. Manie Maritz - Wikipedia And Genl. Salomon Gerhardus Maritz (1876 - 1940) - Genealogy (geni.com)

He had sacked the village and disappeared with his retainers, and we found the original civil population in peaceful occupation, the English having apparently abandoned the place for good.

As the 'Rijk Section' was now practically defunct, General Smuts ordered William Conradi and me to join his staff. This was in recognition of our late exploit, and was tantamount to military promotion. The rest of the 'Rijk Section' was absorbed into Commandant Bouwer's commando, with the exception of Ben Coetzee, who rode away, his leg still in splints, in search of Maritz, for they were old friends. This, then, was the end of that small company with whom I had come into the Cape. Like Isaac Malherbe's Corporal ship and the A.C.C., it, too, had been destroyed, but I am glad to have served with three such bodies of men.

I was now what might correspond to a staff officer in a regular army, although none of us on General Smuts's headquarters bore any distinguishing title, beyond the fact that the rest of the men in friendly derision referred to us as 'Kripvreters'—a Kripvreter being a stall-fed horse, as distinguished from one having to scratch for its own living on the veld.

On assuming my new duties, however, I soon found that, so far from being staffed, the members of the staff, in addition to having to fight and forage for themselves like the rest, were employed as dispatch-riders, and during the next few weeks our life was one continual round of weary rides in search of one portion or another of the commando, for General Smuts had now divided his force into smaller groups, often stationed days apart, to provide more easily for grazing and food.

In December (1901) he left the commando scattered along the banks of the Olifants River[456] near van Rijnsdorp, and, with his staff, moved up the mountains to a spot called Willems River, where he began his work of collecting the various rebel bands into organized commandos. A large number of men from this area had either joined Maritz or were riding about on their own, sniping at British columns or waylaying convoys, and, in order to get into touch with these irregulars, we 'Kripvreters' were sent during the next few weeks on long rides, in the course of which we travelled unumbered miles from south of the Olifants River to far beyond Calvinia and back, until we knew every inch of the country.

The inhabitants sympathised with us, and looked upon us as their champions, so we were welcomed wherever we went and, despite the long gruelling journeys on mule-back, I enjoyed myself, for I was not above feeling a pleasant glow when the womenfolk waved from before the farmhouses, and the men shouted greetings from the road-side.

[456] The upper and main catchment area of the Olifants river is around Ceres and the Cederberg mountains. The Clanwilliam and Bulshoek dams are located on the river and provide water for the towns and farms along the watercourse. The river is approximately 285 km long with a catchment area of 46,220 km and flows into the Atlantic Ocean at Papendorp, 250 km north of Cape Town. Olifants River (Western Cape) - Wikipedia

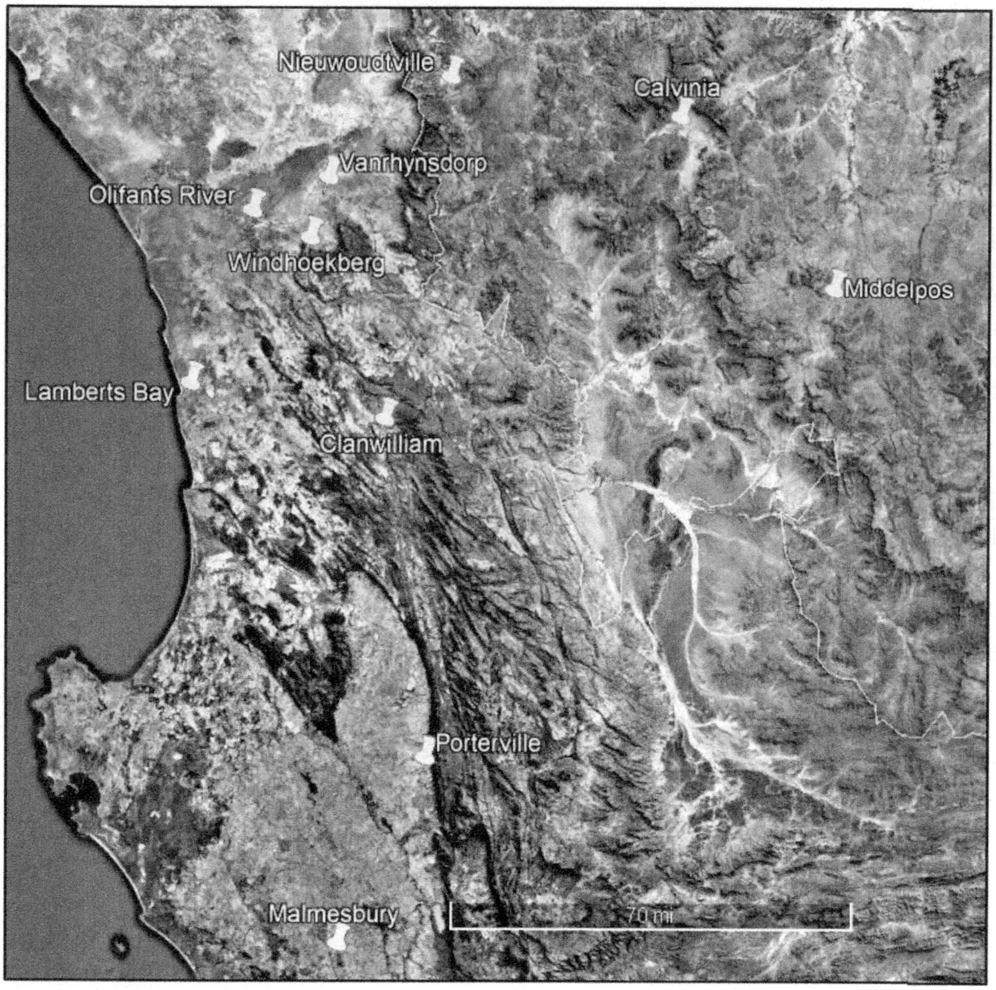

Map data: Google Image© 2024 Image Landsat/Copernicus.

On Christmas Day, [1901] returning from a hundred-mile errand beyond Calvinia, I saw mounted men camped in the distance, and making thither, found them to be Commandant van Deventer's missing column. My unexpected appearance amongst them was hailed with joy, for this was their first word of us since we had parted below the Zuurbergen, and for a long time I was busy telling them how we had fared and listening to their experiences. They had only just arrived here, having been driven out of their course for several weeks, but they had done well, for they had captured several convoys and many prisoners, and they had more horses, rifles, and ammunition than they could use.

After a pleasant evening's talk around the camp fires, I remained there for the night, and took the road again next day, getting back by the 28th to General Smuts, who was greatly pleased at my news.

My mule had unflinchingly trotted the better part of two hundred miles in five days, and the following morning I was off again in search of Maritz, whom I found eighty miles away in the neighbourhood of Tontelbos.[457] This place was an important grain growing centre, at which the, British had posted a force of men, to prevent the crops from being carried away.

Maritz attacked the garrison the day before I arrived, but was repulsed with heavy loss, he himself bring severely wounded. I found him seated on a chair in a farmhouse with two of his men dressing his wound, a terrible gash below the right armpit, exposing the lung, an injury that would have killed most men, but he was like a bull and seemed little the worse for it.

[457] Tinder Bush: Caputia tomentosa - Wikipedia

I saw the year 1902 in with them, and then started back, catching up General Smuts three days later at Nieuwoudsville[458] on the escarpment. During all this time our main commando under Commandant Bouwer was lying down on the plains to the south, having occasional brushes with the English, but on the whole passing a quiet time. Some of our patrols went beyond Porterville,[459] to within sight of Table Mountain, and my old companion [Petrus] Krige (the General's brother-in-law), with whom I had served in Isaac Malherbe's Corporalship, and who had been so badly wounded at Spion Kop, even penetrated as far as Malmesbury,[460] and brought back a large sum of money for the use of the commando from General Smuts's father, who lived there.

Jacobus and Catharina Smuts, 1893.[498]

During a visit which I paid at this time to our men along the Olifants River, I met my old Ladysmith tent-mate, Walter de Vos, who had likewise been wounded on Spion Kop, and whom I had last seen lying on the slope of that hill, on the day of the big fight two years before. He had latterly been in command of one of the local rebel bands, and we spent the morning talking over old times; but he was killed an hour after my going, in an outpost affair nearby.

Early in January [1902] General Smuts decided to go north to the Orange River, to organize the numerous rebel patrols that were under arms there. Our company consisted of himself and his staff only. It was a three-hundred-mile ride through desert country, and we went first of all to Tontelbos, now evacuated, as the crops were in. Maritz was here on a pallet of straw in an empty dwelling-house, but he made light of his wound and was well on the mend.

From Tontelbos we moved north through country thinly occupied by Nomad Boers (Trek Boers), who spend their lives going from one well to another with their flocks, like the old peoples in the Bible. They are a primitive patriarchal folk, knowing little of the outside world, but of a brave and sturdy stock, and many of them were under arms.

We travelled mostly at night to avoid the blazing heat of day, and at length reached Kakamas,[461] a small irrigation colony founded by the Dutch Church on the south bank of the Orange River.

458 The town lies on the Bokkeveld Escarpment, and was established in 1897. The Nieuwoudtville Falls on the Doring River are located a few kilometres north of the town. Nieuwoudtville - Wikipedia

459 The town is located at the foot of the Olifants River Mountains, 27 km south-east of Piketberg and 155 km north-east of Cape Town Porterville, South Africa - Wikipedia

460 Malmesbury is located about 65 km north of Cape Town. Malmesbury, South Africa - Wikipedia

461 Kakamas is located on the banks of the Orange River, Northern Cape. Kakamas - Wikipedia

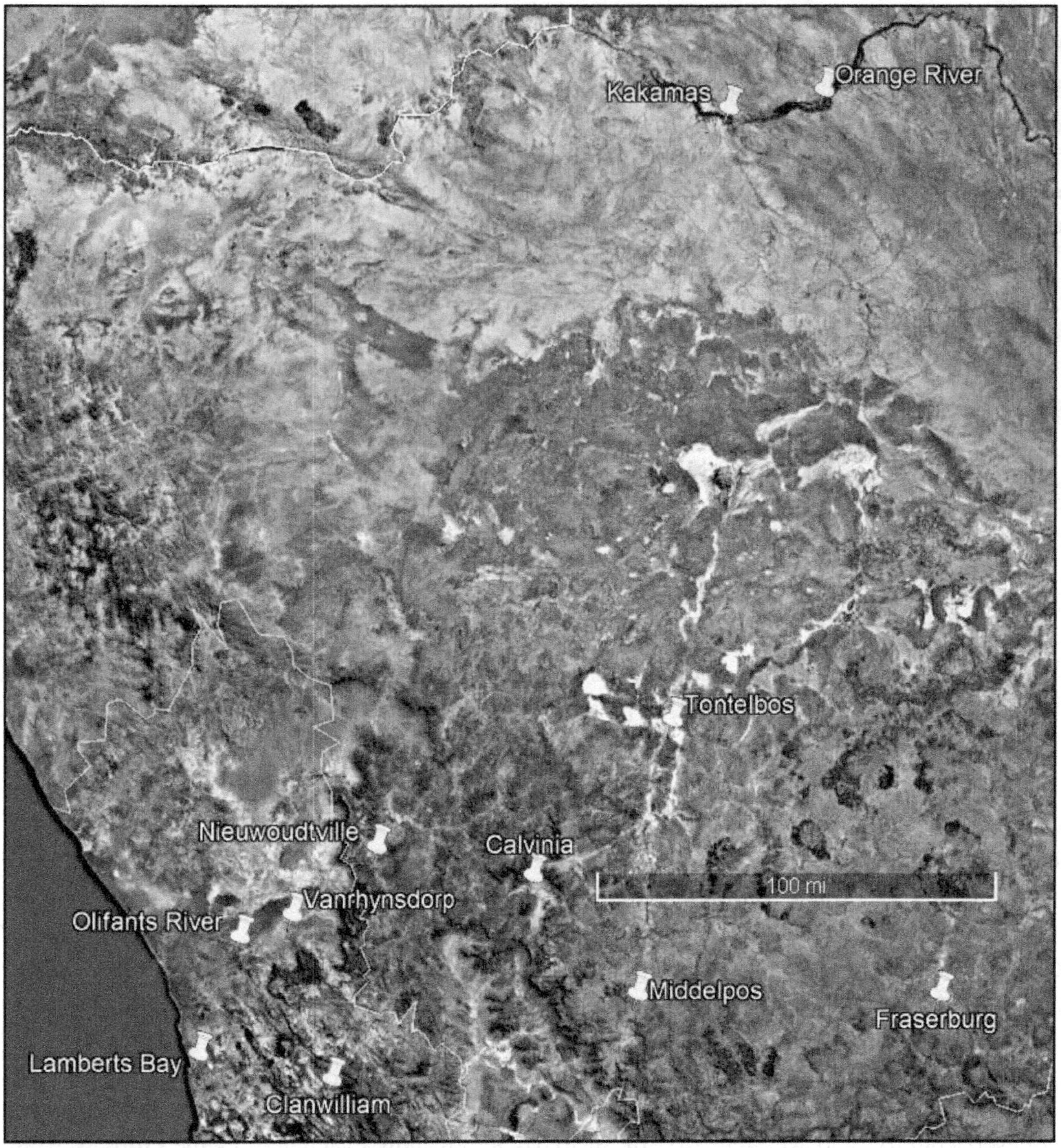

Map data: Google Image© 2024 Image Landsat/Copernicus.

Waterwheel on Orange River in Kakamas.[462]

[462] The waterwheel is situated in an irrigation canal fed from the Orange River to supply water to the vineyards in the background. File:Kakamas-Water wheel-001.jpg - Wikimedia Commons Author: NJR ZA.

The settlement was still in its infancy and the inhabitants lived in rude huts and shelters made of grass and reeds, but they had built a canal from the river, and had established fields and orchards so successfully that the place had become a supply depot for the surrounding districts. We spent a pleasant fortnight here, eating fruit and swimming in the river every day. As soon as General Smuts had completed his arrangements with the guerrilla bands, many of whom rode in from the desert to meet him, we returned south, reaching Tontelbos again towards the **second week in February**. Maritz was no longer here, but as the grazing was good in the cropped wheat-lands, we lay over a few days to rest our animals and ourselves.

General Smuts then decided to go eastwards in search of Commandant van Deventer. We did not know exactly where his commando was, but we travelled up along the Fish River, and after a day or two got word that he was thirty or forty miles away. We rode thither all that night, and towards daylight heard the sound of gun-fire and small-arms and saw a red glare in the sky. Quickening our pace, we reached a farmhouse called Middelpost[463] at dawn, and found two or three men here in charge of a dozen wounded.

They told us that van Deventer was fighting close by with an English column on its way to Calvinia, so, after a few hurried questions, we rode to where we could see his men, lining the crest of some small kopjes, their horses tethered below. On higher ground away to the left were small parties of English troops, and a single field-piece stood in full view but out of rifle-range. The men at the farmhouse warned us to ride for it, as they said that the gunners had the distance to a yard, so we set off at a gallop. They were right, for when we had got about halfway, there was a flash at the gun, and a shell came tearing at us.

A local schoolmaster named Hugo, who had joined a few weeks before, was riding beside me. The shell burst on us with a roar, but although I was nearer the gun, neither my mule nor I received a scratch. But when the smoke cleared I saw that my companion was badly hit. He was swaying in his saddle with blood streaming from his chest. His rifle dropped to the ground, and he fell forward on the neck of his animal. Then he recovered himself and said he was not going to give the gunners the satisfaction of knowing that they had hit anyone, so raising himself he rode for cover of the hill.

Other shells came after, but no one else was hit, and, having retrieved the fallen rifle, I rode on and found that Hugo had fainted and fallen from his horse, and that General Smuts and the others were trying to staunch his wound. It was at the base of his left lung, and I fished out the twisted buckle of his braces,[464] and a cartridge-clip with five rounds of ammunition, all of which had been driven into the cavity, in addition to the shell-fragment which I could not recover. I thought that he had not ten minutes to live, but two months later he was in the saddle once more. We made him comfortable, and climbed up to where Commandant van Deventer and his men were holding the ridge above. This was the first time that General Smuts had come among them since the parting in Somerset East, and there were cheers and shouts of greeting when they saw him. Van Deventer himself hastened forward to welcome us, and in a few seconds we were in the firing-line. Looking down the forward side of the hill, we saw an interesting sight.

Immediately below, on the level ground by the banks of a spruit, stood some hundred and twenty English convoy wagons, most of them burning fiercely to the crackle of exploding rifle ammunition, for every wagon seemed to carry several cases. Scattered among the blazing vehicles lay dead men and horses, and there were a large number of live troop-horses and mules, that had stampeded during the night, but had drifted back into the burring camp, where, in spite of the smoke and flames and the bursting cartridges, they were feeding on the seed oats and other fodder that our men had flung from the wagons during their hasty search for loot, before setting them on fire.[465]

[463] 40 Miles South East of Calvinia. 75 miles West of Fraserburg. Pin Located on Map. No other reference available.

[464] Suspenders.

[465] 2/4/1902 Capture and destruction of British convoy by Boers in Cape Colony. Major Crofton killed. (Two entries on ABW)

Crofton Henry William George Major 3rd Battalion East Surrey Regiment Demise: Killed in action 05 Feb 1902 Place: Uitspanfontein Source: In Memoriam by S Watt.

Crofton Henry William George Major He was killed in action in February 1902, in the attack on a convoy thirty miles from Fraserburg. He was transferred as a Captain from the 3rd Battalion East Surrey Regiment to the 3rd Royal West Surrey in February 1900, and, volunteering for active service, proceeded to South Africa. In May 1901, he was promoted Major in his old corps, the 3rd East Surrey, which had also been embodied, and was then in South Africa. In the Boer onslaught in which he was killed, the convoy, which was proceeding from Beaufort West, was attacked by Malan's Commando. The escort was overwhelmed, after a brave defence, and Major Crofton was killed. He had held the post of Station Commandant (graded as a Staff Capt), and had been awarded the medal and two clasps. Source: Donner

Van Deventer gave us a brief account of what had happened. A long convoy had approached the evening before, accompanied by a mounted column. He had disputed their way, whereupon the English troops parked their wagons beside the spruit, and took up covering positions, but during the night he and his men broke through the line on foot, and, entering the camp, set it on fire. In the dark the troops were unable effectively to hinder the work of destruction, and the position when we arrived was that van Deventer, having fired the wagons, had withdrawn before daylight, and the two sides were now facing one another with the convoy burning between them, neither side permitting the other to approach it. The bulk of the English soldiers had taken post at a farmhouse surrounded by a walled garden, about nine hundred yards away, from which they were maintaining a hot rifle-fire.

To the left, four hundred yards off, lay more of them on a stone hill, with their field-gun on a rise behind, and on the right in another kopje was an isolated detachment, so placed that they had anyone under short range who tried to enter the camp. Van Deventer told General Smuts that he was anxious to recover the animals feeding amongst the wagons, to which end he had a fewer minutes previously sent Field-Cornet Van der Berg with twenty-five men to clear the kopje overlooking the camp by a surprise attack from the rear. He had ordered them to ride round behind some other kopjes that screened the view.

I had now ridden my mule for upwards of a thousand miles. He was a willing animal, but with his shambling gait and long stride a mule at best makes a tiring mount, and I yearned for the easier seat of a horse. Another man in the staff, Martin Brink,[466] had also been on a mule for months past, and was equally anxious for a change, so we decided to overtake the attacking party in the hope of getting a troop-horse or two. We ran down, and, mounting, followed the tracks made by Van der Berg's men as fast as we could go. He had led them with skill, for nowhere was their route visible from the kopje, and when, after a breathless gallop, we raced round the corner of a ridge into the open we found that he had taken the soldiers by surprise and that he and his men had reached the foot, and were climbing up under a ragged rifle-fire, without having sustained any visible loss. By the time we joined them the affair was as good as over. A shot or two was loosed, but in a few seconds the last of the soldiers stood up to surrender. It had, however, been an expensive little fight. Alouin Weber, an ex-Transvaal artillery officer, and two more men lay dead and Field-Cornet van der Berg and another were badly wounded, while several soldiers were killed and three or four wounded, out of the dozen or so who had been holding the post.

In any case, know that the kopje was in our hands, it was possible to make our way down to the spruit, on the opposite bank of which stood the burning wagons, so, leaving their friends to attend to the wounded, the rest of us lost no time in descending the slope and jumping into the spruit. We ran along the sandy bottom until we could peer over at the camp, a stone's throw away. Then we climbed out and rushed for the horses that were nosing the fodder-strewn ground. When the troops from the distant farm-house saw us running amongst the wagons, they opened fire, but we were not to be denied. My first effort was to insure myself against further mule-riding, and in three successive raids I brought away three good horses with saddles and holsters complete. I hurried each into the shelter of the spruit, and ran out again for the next.

The other men were just as busy, and luckily no one was hit. As soon as I had secured my horses, I went back for other portable property, for several of the wagons and their loads were only half burnt, while some were scarcely damaged at all, and there was much useful loot still to be had. In dodging among the smouldering wagons I came on a fully laden scotch-cart[467] that had been overlooked in the dark. It was quite intact, and, as the firing from the farmhouse was increasing, I seized a large portmanteau and shovelled into it all that I could find in the way of books, papers, boots and clothing, including some Bank of England notes, and then dragged it over the ground to the spruit.

[466] Brink Marthinus Cornelis Petrus Kaptein Adjutant. Staff General Smuts. Awarded the DTD

[467] A small, stout, springless, two-wheeled tip cart which is horse- or ox-drawn and used mainly for transporting rough materials such as refuse, gravel, or manure.

Two Scotch Carts in Durban.[499]

I found that most of it belonged to Colonel Dorran,[468] who had commanded the convoy, and that amongst his papers were the records of the Court Martial of Commandant Scheepers, at which he had presided. We had heard already that this well-known guerrilla leader had been captured and executed in the Midlands some months before for alleged train-wrecking.

After a hurried inspection of my new property, I distributed my haul evenly on my three horses and my mule, and rode back to General Smuts very pleased with the morning's work, for, no longer a ragged muleteer, I was now better horsed, shod, and equipped than at any time of the war.

Soon after this the English troops sent off their gun and began to retire southward, abandoning a large number of horses and mules that had broken away, and were roaming about the veld.

General Smuts and Commandant van Deventer decided not to pursue the retreating column, for, even if we captured them, we should only have to let them go again, and we had done so well in horseflesh that it did not seem worth while to go after them. We were now free to revisit the camp unmolested, and, in addition to hundreds of animals, the men recovered a considerable quantity of ammunition, saddlery, etc., and, most valuable of all, many cases of horse-shoes and nails. Five or six soldiers lay dead in the camp, and, when some of us rode up to the farmhouse where their main body had been, we found twenty or thirty wounded who had been left there in charge of a military doctor, several of them very badly injured.

At the request of the Medical Officer I rode round to shoot the wounded horses and mules standing about the house, some with broken legs, others with blood dripping from their flanks, for, with no one to look after them, it was best to put them out of their misery.

One of the horses I had taken from the camp was a beautiful little dark-grey Arab mare with a coat like velvet and nimble as a goat. I was mounted on her when I rode to the farmhouse, and here her former owner, a wounded officer named Chapman, lying on a stretcher outside, recognized her, and offered to buy her back from me for £65. He said her name was Ninny and that she was the best horse in the country. As money was of no use to me, and I knew a good horse when I saw one, I refused to sell, but I promised to look after her and treat her well.

The commando spent the night at the other little farm-house, where we had first found van Deventer's wounded, and here we buried our dead at sunrise next day. The bodies had been placed ready on a wagon, and, not knowing this, I spent the night under it, and, waking in the morning, found myself dotted with blood that had oozed through the planking overhead.

At the funeral General Smuts made a moving speech. He pointed out that among the dead were a Transvaaler, a Freestater, and a Colonial, all parts of South Africa being thus represented in the common sacrifice for liberty. When the ceremony was over I was ordered to ride to the place twenty miles away to which our wounded had been taken, to see that all was well. I found most of the men fairly comfortable, although there were several bad cases. One was a Colonial who had been shot through the stomach, and the woman of the house asked me to have a look at him as his side was inflamed. While she and I were examining the wound, he gave a deep groan and died without speaking. A wagon-driver helped me to bury him. We dug a hole beside the threshing-floor, and as we knew no funeral service, we simply carried him by the shoulders and knees, laid him in the grave, covered him with earth, and left him.

[468] Three possibilities on ABW.com for Colonel Doran.

While I was at this farm we saw forty or fifty strange horsemen approaching from the north, and there was some alarm at first amongst the wounded, as we could not make out who they were. I fetched my horse and rifle and rode in their direction, until I was close enough to see that they were not British. They proved to be the survivors of a portion of the commando that General Smuts had left behind on his way through the Free State the year before, because their horses were too worn out to go on. He had placed Field Comet Dreyer in charge, with orders to follow later, when the condition of their animals permitted, and the faithful band had carried out these instructions to the letter.

T. F. J. Dreyer.[469]

As soon as possible they started south in our tracks, and, after many trials and dangers, this remnant had come through. Among them was the Reverend Mr. Kriel, with whom my brothers and I had quarrelled at Warm Baths in December, 1900.

In spite of his religious bigotry, he was a stout-hearted old man, whom I learned to respect. When they heard that General Smuts was in the district, they were so anxious to see him that they wanted to go off at once, but I told them to wait, as I knew that the General was coming our way, and he arrived that evening with van Deventer and his commando and there was great rejoicing on both sides.

Ever since General Smuts had gone to Kakamas in December, Commandant Bouwer with his commando had remained down on the Plains near the Olifants River beyond van Rijnsdorp, and I was now sent to find them. I gave away my mule, but took all three of my newly acquired horses, loaded with my loot from the camp. I reached van Rijnsdorp in three days, going via Nieuwoudsville, and thence down the mountain pass to the country below. I found Bouwer in van Rijnsdorp, and most of his men camped along the True-true River not far off. They were the more pleased when I told them of van Deventer's success, because they had suffered an unpleasant reverse the previous morning.

A week before a Colonial named Lemuel Colaine had turned up amongst them with a tale that the English had put him in prison at Clan-William[470] on a false charge of high treason. He said that he had escaped over the wall one night and had come in revenge to take up arms. Believing his story, they gave him a rifle and he joined the commando.

Colaine, however, was a spy in British pay, and, after collecting what information he could, he disappeared. No particular notice was taken of his absence, as the men were constantly riding off to visit farms, or look up friends at distant outposts, and it was thought that he had done the same; but the commando had a rude awakening when a body of English horse, with Colaine riding at their head, fell upon them at dawn, killing and wounding seventeen men, including my young friend Michael du Preez.

[469] Commandant T. F. J. Dreyer, as a civilian of the Potchefstroom commando, carried out the siege General Piet Cronjé's knowledge of Mafeking and all the fighting on the western border to escape lower to Magersfontein, participated in the battle of Abrahamskraal and was defeated the occupation of Pretoria was first elected field cornet and then Commandant. He as such fought several heavy battles in the districts of Rustenburg, Klerksdorp and Potchefstroom and followed with seventeen during the second half of 1901 year-old son of General J.C. Smuts on his journey through the Free State, as well as with a seventy Potchefstromers. However, Commandant Dreyer failed to reach the Cape Colony and lived to the age of 28 September 1901, in the south of the Free State, captured after fierce fighting and sent to India. Our [Transvaal] Military Officers: Dutch: Onze krijgs-officieren: album van portretten met levens-schetsen der. by Gustav Schoeman Preller, Frans Vredenrijk Engelenburg 1904.

[470] Clanwilliam is a town in the Olifants River valley in the Western Cape, Clanwilliam, South Africa - Wikipedia

The attacking force took our men so completely by surprise that the troopers rode through the camp using their swords, and got away safely on the other side before our men could recover their wits. All were fierce in their denunciation of Colaine's treachery, and hoped that he would fall into their hands. And later Nemesis[471] ran the right man to earth for once.

Meanwhile Bouwer was smarting under this setback, for not only had he lost good men, but the British were following up their success by an advance in force, with the object of retaking van Rijnsdorp, which we had come to regard as our headquarters, for it was the only town in South Africa still in Boer hands.

I remained with Bouwer overnight in the threatened village, [Vanrhynsdorp] and, as his scouts reported next morning that a strong column of English horsemen was pushing forward, he decided to retire northwards to the mountains until reinforcements could reach him.

I went out to watch the enemy movements with a party, among them being my old friends Nicolas Swart and Edgar Duncker, the former with his arm still in a sling, while the latter had his shattered hand in splints, and a pillow strapped to his saddle to ease his wounded thigh, for the sound of rifle-fire was an irresistible attraction to these two, and they refused to remain behind when they heard shots beyond the town. After going forward for a mile or two, we saw a long column of horsemen coming up from the direction of the Olifants River, their scouts thrown forward on a wide front, and we were soon engaged in a running fight, which continued until they pressed us back through the streets into the open country, where we took to our heels, to catch up with Bouwer's main body making for the mountains.

In the course of one of these skirmishes, Duncker, riding beside me, was shot through the chest. We plugged the bullet-holes with pieces of his shirt, and he rode on with us for the fifteen-odd miles that we had to go before we overtook the commando. He was then sent to a farm among the foothills, and completely recovered in a few weeks. The English contented themselves with reoccupying our little capital and came no farther, so Bouwer did not retire up the mountains after all, but, determined to recover his lost ground, he sent me hurrying up the pass to ask General Smuts for help.

After riding hard for two days I came up with him near Calvinia, sixty miles off, [to the East] and, when he heard that the troops were back in van Rijnsdorp, he ordered the commanders to gather. He sent word to van Deventer to bring his men to the head of the pass at Nieuwoudsville, where he would wait for him, while another messenger was sent to Bouwer, bidding him keep his men below until assistance came.

The various smaller local patrols were also ordered in, and General Smuts and his staff made for the appointed rendezvous at the edge of the berg.

The arrangements worked perfectly. In three days van Deventer arrived with his fighting-men, and we descended the mountains to Urion's Kraal on the plains, where Bouwer was eagerly awaiting us. This was the first time that our entire original commando had been reunited since parting under the Zuurbergen, and there was great cheering and handshaking when we rode up. That night our whole force marched out, intending to attack van Rijnsdorp at daybreak, but when it grew light we found that the English troops had been withdrawn to a place called Windhoek,[472] ten miles back, [Southwest] which was being turned into a fortified camp, so we lay over in the recovered village until dark. General Smuts had decided to attack Windhock at dawn next morning, but I missed the fight, for I was not told of what was pending, and was sent off at sunset with a message to a post stationed towards the Olifants [River]. I arrived after midnight, and spent the night with the picket. At dawn I was in the saddle on the return journey, and, as I rode towards van Rijnsdorp, I heard distant rifle-fire, and hurried towards it.

As I approached, the firing grew heavier for a while, and then died down altogether, so that I knew one side had been worsted. Then I came on Commandant van Deventer huddled on the ground before his horse, badly wounded and in great pain. Blood was pouring from a bullet-wound in his throat, and his tongue was so lacerated that he could not speak. Two men with him told me that the fight was over, and that the English camp at Windhoek had been captured. I galloped on, and met about a hundred disarmed soldiers, marching across the veld without their boots. They said our men had ordered them to find their way back to Clan-William, fifty miles away. [South]

[471] The Greek goddess of retribution. Nemesis - Wikipedia
[472] Now: Windhoekberg, South Africa.

Map data: Google Image© 2024 Image Landsat/Copernicus.

In a few seconds I reached the scene of action. General Smuts had surrounded the camp at daybreak, and, after a sharp fight, had overwhelmed it, killing and wounding many, and capturing the rest, about two hundred in number. He had not come off lightly either, having lost five men killed and sixteen wounded, but he had taken wagons, horses, arms and ammunition, and he head re-established his hold on these parts. As I rode through the camp I found Nicolas Swart lying on the ground apparently dead. A bullet had struck him in the chest and had traversed the length of his body, emerging at his left thigh, showing that he must have been bending forward when he was hit. His face was so pale that I thought him dead, so I went to one of the wagons in search of something to throw over his body, but when I came back his eyes were open, and he asked me in a whisper for a drink of water, which I gave him from my bottle. We carried him into the shade of a wagon, and roughly bandaged his wounds.

As we could do nothing further for the moment, I left him, in order to look around the rest of the captured Convoy now being ransacked by the men. It was parked around the dwelling-house in which the troops had made their last stand, and, seeing Wyndell of the 'Rijk Section,' I went to tell him about Nicolas. He had shared in the attack, but did not know that Nick was wounded, and he said we must search the house for pillow-slips or sheeting for better bandages. As we went through the rooms, strewn with upturned chairs, etc., in the hand-to-hand fighting, we saw a man in civilian clothing crouched under the arched fireplace in the kitchen. I thought it was the owner of the farm, not yet recovered from his fright, but when I drew Wyndell's attention to him he exclaimed, 'By God! it's Colaine!' I did not know Colaine, but Wyndell dragged him from the house, shouting to the men outside to come and see who was here, and soon dozens of angry men were muttering threats and curses at the wretched spy. He was a man of about forty-five, in appearance a typical back-veld Boer, with flowing beard and corduroys. He was brave enough now, for when the men fiercely assured him of his certain fate he shrugged his shoulders, and showed no sign of fear. Commandant Bouwer came up while we were crowding round, and ordered two men to guard him until General Smuts was notified.

Wyndell and I, having found some linen, went back to look for Nicolas, but found him gone, and were told that he had been loaded on a mule-wagon with other casualties, for removal to another farm.

As I was well found in horses and equipment since the Middelpost affair, the present convoy did not much interest me, but I collected some newspapers and books, and, leaving the men at their looting, I prepared to ride down to the farm known as Aties, belonging to old Isaac van Zijl,[473] the local member of Parliament, where General Smuts was said to be. But first I went to see who were killed, and was sad to find among them young Martin Wessels, a school-friend who had spent many of his holidays with my brothers and myself in the old Bloemfontein days. I had met him two days before, for the first time during the war, having come on him with one of the small rebel bands in the neighbourhood.

He had been wounded and captured by the British a year ago, but with Cornelius Vermaas, now also dead, he had leaped the train in the Hex River Mountains[474] to rejoin the commandos. When I entered the homestead at Aties, General Smuts was in the dining-room talking to the, owner, Isaac van Zijl, whose wife and daughters were there, too, and before long Colaine, the spy, was ushered in by his guards, who wanted to know what to do with their by prisoner.

General Smuts had heard the whole story of Colaine's treachery, and, after questioning the escort to make sure of the man's identity, he sentenced him to death without further formality. When the General said to the guards, 'Take him out and shoot him!' Colaine's nerve failed him, and, falling on his knees, he begged for mercy, while the women fled from the room in tears. General Smuts repeated his order, but as the condemned man was being led out, the Reverend Mr. Kriel came in, and asked leave to pray for the soul of this poor sinner. So Colaine was taken to a little smithy behind the dwelling-house, and when I looked in a little later, I saw him and the clergyman kneeling side by side against a plough-tail, deep in prayer.

After a while Andries de Wet of our staff was told to collect a firing party, and, as he disliked the job, he asked me to accompany him. We sent some Hottentot servants to dig a grave out of sight of the house, to spare the feelings of its inmates, and, ordering three men who had off-saddled in the garden to fetch their rifles, we went to the workshop door. Catching Mr. Kriel's eye, de Wet pointed to the prisoner, and the clergyman touched the kneeling man on the shoulder and said, 'Brother, be a man, your time has come.' Colaine took the news calmly; he rose from his knees, shook the parson by the hand, and bidding goodbye to the guards, said that he was ready. We led him to where the grave was being dug.

On the way he spoke to us. He said he knew he deserved to die, but he was a poor man and had taken blood-money to keep his wife and children from starving. The Hottentots were just completing the grave when we came up, and the unfortunate man blanched when he looked into the shallow pit. Perhaps he had still hoped for a reprieve, until he saw it. Even now he tried to gain time, by appealing to us to send for Mr. Kriel, to say a final prayer with him. Then he turned to me, and asked me to fetch General Smuts, but we felt that the sooner it was over the better, so de Wet blindfolded him, and placed him at the head of the grave. Realizing that this was the end, Colaine held up his hands, and in a low tone recited the Lord's Prayer, while the firing party silently ranged themselves. As he came to the final 'Amen,' they fired. With a convulsive jerk he pitched backward into the grave, and the frightened Hottentots quickly covered him with earth.

When we returned, we found that the wounded had been brought down from Windhoek, and were being placed in the main dwelling-house. Nicolas Swart was still alive, in fact the jolting seemed to have improved his condition, for he was conscious and able to speak. He was put in a room by himself, whilst the rest were laid on mattresses or on straw, wherever the mistress of the house and her daughters could find room for them. Nicolas was taken with a sick man's fancy that I should remain by his side. When I tried to leave him, he seized my hand and would not let me go, so General Smuts said I was to stay, and I sat by his side all that afternoon and all through the following night. At intervals I renewed the wad of damp cloth over the wound in his chest, doing this for close on twenty hours, and soon after daybreak he fell into an easy sleep. From then onward he began slowly to mend, and within a month was well again. Of the other wounded only one man died, the rest all making good progress, thanks to the care of the women and the wonderful climate.

The camp at Windhoek had cost us more men than it was worth, but the English were discouraged from further attempts to dislodge us and from the Olifants northwards we were left in possession of all area that we were beginning to regard as our peculiar property. So much was this the case that General Smuts once more broke up the commandos, and distributed the men in small patrols until he should need them for a fresh effort. This

473 Three possibilities for van Zijil on ABW.

474 The Hex River Mountains make up the second highest mountain range in the Western Cape province of South Africa and are located 120 kilometres (75 miles) north-east of Cape Town. Hex River Mountains - Wikipedia

entailed much work for the members of the staff, who were kept riding backwards and forwards from one detachment to another, in order to maintain contact.

I, however, stayed behind at the farm, as Nicolas Swart would not hear of my leaving.

Buffelshoek area of the Hex River Mountains in South Africa.[500]

While I was here I had time to read the English newspapers that I had found in the Windhoek camp. I gathered from certain letters and articles that there were many people in England who thought the war unfair. I cut out one poem and have kept it ever since. It ran:

Peace on Earth, Goodwill to Man
Christmas Day, 1901

The story is too old: no more it thrills.
Pity is dead; peace is a paltry art.
How can a glory on Judean hills
Make glad my heart?
The mighty splendours of our state shall show
A worthier creed than decalogue or love,
Let death and vengeance, launched on every foe
Our greatness prove.
Why mock us with the thoughts of Bethlehem foe
And glory humbled, and exalting grace?
Celestial music hits not with our theme
Our pride of race.
Dear God, forgive! Let other hearts be stone;
Christ's natal message shakes me like a reed.
Nor pride nor power nor country can condone.
The wild beast's creed?

At the end of ten days Nicolas was so much better that I was able to get away in search of General Smuts, whom I found on the banks of the Olifants River down towards the mouth. The sea lay only twenty-five miles from here and the day after my return he sent word to the units quartered within reach, that all who had never seen it were to be sent to him. Some sixty or seventy men arrived within the next forty-eight hours, and with these we set off for a small inlet on the coast called Fishwater. We rode via the Ebenezer Mission Station, and towards afternoon caught a glint of the sea through a gap in the dunes. It was amusing to watch the expression on the men's faces as the great expanse of ocean burst on their view, for few of them had seen anything bigger than the dam on their parents' farms, and, as we topped the last sand-hills, they looked in amazement on water that stretched beyond the horizon.

With one accord they reined in their horses in silence, and then, like the Greek soldiers, rushed forward in a body, crying, 'The sea! The sea!' each wanting to be first on the beach.

Soon they were throwing off their clothes, and our trouble was, not to get them to enter the waves, but to prevent them from venturing in too deep, for they were pitching down their saddles and riding barebacked into the surf, shouting and laughing whenever a rider and his mount were thrown headlong by the breakers.

After a while General Smuts ordered three of us to ride along the shore towards some huts in the distance, to inquire whether any troops had been here of late. In doing so we had an amusing encounter with a Hottentot fisherman. He stared open-mouthed at sight of armed Boers patrolling the water-line, and, seeing his surprise, I halted my horse and ordered him in a peremptory tone to show me where the road went through. He said, 'What road, Baas?' Pretending to be angry, I replied, 'The road to England, you fool, and show me the way at once, for we are crossing to-night to capture London.' He looked at me for a moment, and then exclaimed, 'My God, Baas, don't do it; the water is over your head here, and you will all be drowned.'

Soldiers of the Lothian and Border Horse Yeomanry.[501] With horses by the ocean. Caravan of camels approaching one of the South African ports.[502]

When next I met Maritz and told him this story, he said that two of his men had recently ridden on to the beach at Lambert's Bay,[475] where an English cruiser lay at anchor close in-shore. Dismounting, they opened fire. Their bullets pattered harmlessly against the armoured side of the warship, and when the crew turned a gun on them they made haste to disappear into the sand hills, but, on their return to their commando, they boasted that they had fought the only naval action of the war!

That night we camped in the dunes, sitting around great fires of driftwood, the men discussing what they had seen until far into the night, and telling each other of the things they would have to recount when they got home again.

We spent two more days here, boating on the estuary and helping the local fishermen to drag their nets. Then we returned along the Olifants River to our starting-place, proud of having ridden our horses into the sea.

One of the ports on the South African coast.[503] Local fishermen's dinghies sailing by one of the Royal Navy warships.[504]

[475] Lambert's Bay - Wikipedia

XXV. The Last Phase

General Smuts now made further plans.

To the north, one hundred and fifty miles away, lay the important copper-mining centre of O'Okiep,[476 477]with its subsidiary villages of Springbok[478] and Concordia.[479] These places were held by British garrisons, and he decided to look them up. So far as I could gather, his intention was not so much to capture the towns, as to lure a relief expedition thither, for he calculated that, if he threatened them, the British would be compelled to hurry a force round by the sea to their assistance. He would then break away south, and make for the old settled districts around the Cape. Thus at any rate was the rumour current among us, and the men were enthusiastic at the thought of raiding down towards Table Bay. They even talked of taking Cape town itself, and we on the staff were cheered, as we rode to tell the outlying patrols and corporalships that they were to gather once more.

After a few days the whole commando was assembled, and we faced north on a long journey through the barren rugged country of Namaqualand.[480]

Desert Road in Namaqualand.[505]

Spring Flowers Namaqualand.[506]

Namaqualand Terrain.[507]

Namaqualand.[508]

[476] Okiep is a small town in the Northern Cape. In the 1870s ranked as having the richest copper mine in the world. Okiep - Wikipedia

[477] Okiep's mine saw action on 4 April 1902 during the Anglo-Boer war when some 700 officers and men of the 3rd Battalion Queen's Royal Regiment, 5th Royal Warwickshire Regiment, Namaqualand Border Scouts, the Town Guard and the Cape Garrison Artillery, withstood a 30-day siege by Jan Smuts' forces. The village of Concordia with a garrison of 100 men, surrendered a day after the siege started. On 4 May 1902 a British relief column arrived from Port Nolloth and ended the siege. A ruined blockhouse is still visible on a hill north-east of the town. Okiep - Wikipedia

[478] Springbok, South Africa - Wikipedia

[479] Concordia was originally established as a Rhenish mission station in 1852 and copper mining began in 1853 through to 1983 in the area. Concordia, South Africa - Wikipedia

[480]Namaqualand (Meaning Nama Khoe people's land) is an arid region of Namibia and South Africa, Namaqualand - Wikipedia

Owing to difficulties of food-supply and water, we were, presently split into Smaller parties, each with instructions to make for a point in the Kamiesbergell.[481] General Smuts and his staff travelled by a separate route to the Leliefontein Mission Station,[482] which we reached in six days.

We lived in an atmosphere of rotting corpses for some days, for we had to wait here for news that our forces had arrived within striking distance of the copper-mills; then we moved nearer. At Silverfountains we found Bouwer with his men, as well as Maritz and a considerable number of local rebels, but, as van Deventer's commando was still absent, I was sent in search of him.

Map data: Google Image© 2024 Image Landsat/Copernicus

[481] Kamieskroon is a small town, lying in the foothills of the Kamiesberge at an elevation of approximately 800 m (2 600 ft.). The town is more or less in the centre of Namaqualand, about 70 km (43 mi) to the south of Springbok, Kamieskroon - Wikipedia

[482] Leliefontein is a village in the Kamiesberg Mtns. 18 miles (29 km) south-east of Kamieskroon, Leliefontein, South Africa - Wikipedia

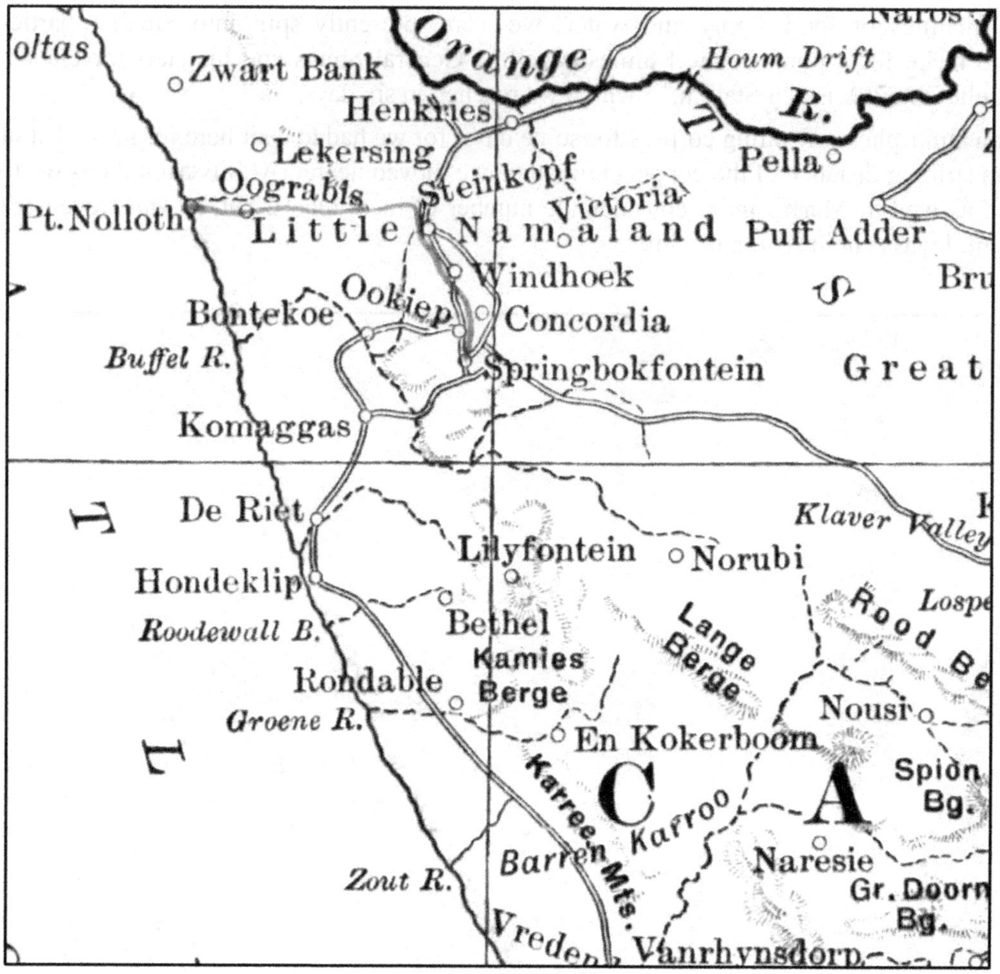

This is the same general area with some similar names assigned.

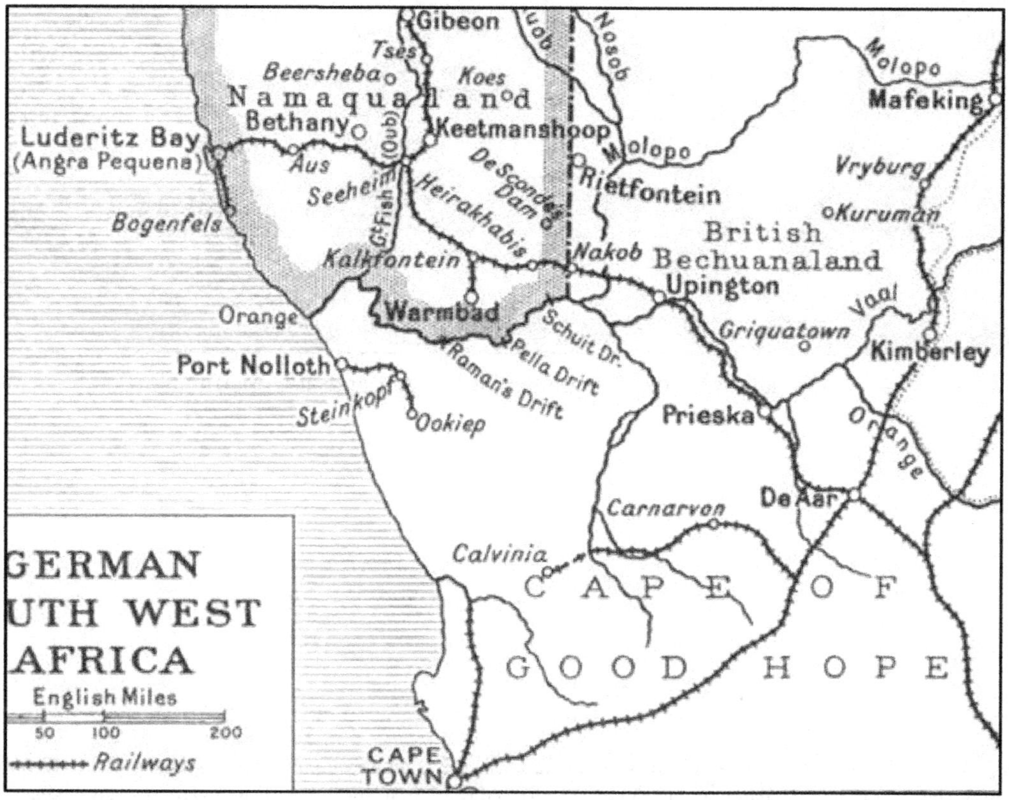

South West Africa.[509]

"Landscape near O'okiep."[510]

I started at daybreak one morning on what was the longest unbroken spell of riding and fighting that I had during the war, for I did not rest or sleep for eighty hours. I rode all that day, continually changing horses (I had my two spare mounts alongside), and making inquiries from farmers and shepherds. With the help of a guide I located van Deventer by midnight, and on receipt of my message he took the road at once. I had to travel back with him through that night and for most of the next day until, towards sunset, we reached General Smuts and the rest of our force at Silvermountains. Having now been in the saddles for thirty-six hours, I hoped for a rest, but at dusk the whistles blew, and the commando started off for the villages of Springbok, thirty miles away. Our course ran at first among rough hills, through which we made slow progress, and then across more open country, until, by four in the morning, we closed in on the village, the different sections moving round to prearranged posts, under the guidance of local farmers, who had volunteered to lead them.

Springbok lies about three miles from O'Okiep, which is the same distance from Concordia, and all three were held by mixed garrisons of British troops and Hottentot levies. Each place was to be attacked in turn, and Springbok came first. its defenders only numbered about one hundred and twenty, but they occupied three well-built forts on high ground, whose loopholes commanded all approaches, so that the disproportion of fighting strength (we had about four hundred men) was not so great as it seemed, particularly as we had to detach nearly half our men to watch Concordia and O'Okiep, in case of a sortie.

Whenever there was any fighting, the staff were re-absorbed into the ranks as ordinary privates, so I was now under Commandant Bouwer, with a detached party, whose instructions were to occupy a low neck over which the main road from O'Okiep crosses into Springbok.

Each contingent filed off quietly in the dark to its allotted post, our orders being to invest the place, but to make no frontal attack, as it was expected that the forts would surrender when they found themselves isolated.

No. 1 Fort, which we were to deal with, was a large round-house, standing on the slag-heap of a mine-shaft. It was heavily loopholed, and the approaches were obstructed by barbed-wire entanglements, so that although less than three dozen men were holding it, they had a clear sweep in all directions, and its capture would be by no means easy without our losing a number of men.

No. 2 Fort lay Some hundreds of yards farther away on a low hill, and No. 3 was built on rock overlooking the streets and houses at the far side of the village.[483]

[483] The village of Springbok 5 miles south of Okiep.

British Army piquet post.[511]

A Fort (Blockhouse at top of hill.) at Bushman's Hoek.[512]

Our party of about forty strong made its way to the neck under a guide, and halted to consult. The night was black, and, as none of us knew exactly where our fort lay, although the guide said it was close by, we decided to send a small patrol to investigate. There were two Irishmen with us, Lang and Gallagher, members of Bouwer's commando, and, with the Irish love of explosives, they had ferreted out a quantity of dynamite and fuses from some outlying mine the day before, with which they had made half a dozen hand-grenades. They were eager to try these, and volunteered to go forward with Edgar Duncker and myself. Leaving the rest behind the neck, we groped our way on until we could make out the dim outline of a wall, whereupon the Irishmen drew the blanket they had brought over their heads, and ignited the fuses of two of the bombs.

The moment the fuses took, the blanket was flung aside, and the projectiles went sizzling through the air, while we hugged the ground to await the result. The bombs blew up simultaneously with a crash and we rushed forward to find that they had exploded harmlessly in an empty cattle-kraal.

The noise, however, brought the enemy to life, for we heard a hoarse 'Halt! Who comes there?' out of the darkness ahead, followed by a crash of musketry, by the light of which we saw the fort, fifteen or twenty yards away, with every port-hole belching fire. One bullet struck the top of the kraal wall as I was looking over, and the fragments of lead and nickel splashed into my face. I thought at first that I was blinded, but I got off lightly, the punctures being only skin-deep, and I had the pieces of metal removed next day with a knife.

In the meanwhile, as the fire was heavy, we remained crouching under cover of the kraal, loosing an occasional shot, but making no attempt to advance on the fort, as we were already nearer than we liked. By now there was also fierce rifle-fire from the other forts, showing that there, too, our men had stirred up hornets' nests, and, when there was a slight lull on our front, we bolted back for the shelter of the neck behind which the others were halted.

General Smuts rode up out of the dark as we were discussing our next move, and, as it was clearing dawn, he ordered us to stay here and see that no one entered or left the fort during the coming day. Having given his instructions, he went off to visit the other posts, and our men distributed themselves amongst the rocks to watch for daylight, while I carried out a small operation of my own. Taking advantage of the still uncertain light, I slipped away, and crawled to a rise at the back of the fort. Here I searched out a convenient stone within forty yards of the enemy, and lay down to wait until the visibility was good enough for shooting. As soon as the sun was well up, I began putting bullets into the loopholes, until I had almost emptied my bandoliers.

The soldiers in the round-house tried hard to locate me, but there was a shrub screening my hiding-place, and, although an occasional bullet flattened itself close by, I lay undetected. When I had nearly finished my ammunition, I wormed my way from rock to rock until I was safe behind the slope and able to rejoin my companions, satisfied that I had given the occupants of the fort an unpleasant time. This was my birthday,[484] the third I had spent in the war, and we passed the rest of the day making more hand-grenades, as we were determined to try again that night.

General Smuts paid us another visit, and, after watching our efforts, told us to send dynamite round to our men at the other two places, which he thought should also be bombed. As soon as it was dark we made ready. Commandant Bouwer came up with some more men, and it was arranged that a smaller party should again go forward to throw the dynamite, after which the others would make a dash at the fort.

[484] April 2nd, 1902.

Commandant Bouwer, the two Irishmen, and Duncker and I formed the advance-guard. Carrying our boots in our hands for greater silence, we felt our way to the foot of the mound on which the fort stood. Quietly scaling the slope, we reached the outer circle of wire entanglements, without an alarm being raised, and, as we could get no farther, we crouched below the rim, and each lit and threw a bomb. Nearly all of them burst on the roof, and there followed a second or two of dead silence, which we took to mean that the men inside were dead or stunned, but, as we were preparing to climb the wires, a roll of fire flashed along the loopholes and sent us tumbling pell-mell down for cover.

The moment the explosions had taken place the balance of our party had rushed through the dark, in order to push home the attack, and they were scrambling up just as we came down, so that there was a collision, which brought us to the bottom in a heap, where we lay laughing helplessly, before we could disentangle ourselves. Despite the stream of bullets, General Smuts now came to us, and, after having climbed the embankment to look at the fort, told Bouwer to remain where he was, without making a direct assault, as the soldiers were bound to surrender sooner or later. Of this, however, there was no immediate sign, for, so far from surrendering, they were shouting strongly flavoured remarks at us, as an accompaniment to their rifle-fire, so we sat down below the rubble-heap to await developments.

During all this time there was firing from the other two forts, and after a while we heard a dull explosion at [Fort] No. 2, followed by cheers from our men. Soon word was shouted that Albert van Rooyen,[485] of our staff, had breached it with a single bomb that brought about the surrender of the defenders. From what we had seen of No. 3 in daylight, we judged it to be the most difficult of all, for it stood on a high rock, like a castle on the Rhine, and we had not much hope that it would be easily overcome. Before long, however, a bomb went off there too, and we heard the voice of Ben Coetzee calling out to the soldiers to give in. After a short interval there was more cheering, and No. 3 had also surrendered. We shouted to tell the men in No. 1 Fort, but they replied by jeers and volleys. We flung our remaining bombs at them, without apparent effect, and although, as we found afterwards, the steel girders of the roof were buckled, yet they braved it out.

A fortified Kopje.[513]

As our stock of grenades was exhausted, General Smuts told me to make my way round the outside of the village to No 3, to bring back dynamite from Maritz, who must have plenty left. I slipped away into the dark, escaping the bullets, but, instead of going wide, I ran as fast as I could through the streets, past the darkened houses.

When I reached No. 3, I saw light shining through the loopholes above, so I climbed the narrow sandbagged stairway and got through the steel trap-door that gave access to the inside, and here I found Maritz and some of his men sorting out ammunition and firearms by lantern light.

The English had hit on the same idea as ourselves for making hand-grenades, and there were several dozen home-made bombs very similar to ours, only smaller.[486]

[485] Van Rooyen A P Ass. Veldkornet Source: Roll of DTD recipients.

[486] Original Footnote: (The Times History of the War) 'A curious feature (of the operations around O'Okiep) was the use, first by the Boers, and then in retaliation by the defenders, of dynamite bombs.'

Filling a bag with as many as I could carry I accompanied Maritz to where his prisoners were collected at an hotel near by. There were about thirty including the Officer in command of the village. I asked him to give me a letter to No. 1 garrison advising them to lay down their arms. This he flatly refused to do. He told me that he had been obliged to surrender his own fort because part of it was built jutting over on wooden beams, and some of our men had got underneath it and were placing dynamite, so that resistance was useless, but he said that, if the men in No. 1 were able to hold out, then good luck to them.

The most I was able to get from him was a pencilled note, scribbled on the bar counter, and addressed to a Mr. Stewart who was in command there, to say that 2 and 3 had been taken, and that he must act according to circumstances.

With this and the bombs, I returned once more through the streets and reached No. 1 mound in safety. When I told general Smuts that I had a note for Mr. Stewart, he said I was to climb up and give it to him. This was more easily said than done, for, when I reached the top, the wire entanglements stopped me, and the soldiers, moreover, were still firing. However, I stood up, and called out, 'Mr. Stewart, Mr. Stewart, here is a letter for you.' At this there was silence, followed by a murmur of voices consulting within, and then a gruff voice asked what I wanted with Mr. Stewart. I said I wanted him to surrender, whereupon I was told to go to hell, and there was a renewed burst of firing, which I only escaped by leaping down.

As a matter of fact, although we did not know it, Mr. Stewart had been lying dead inside the round-house for many hours, and his men were holding out by themselves.

General Smuts, finding them resolute, ordered us to fling the bombs which I had brought. We did this, but they were too light, and only produced more sarcastic remarks and more firing, so he ordered Commandant Beuwer to withdraw his men to the neck, as thirst and hunger would subdue the garrison in the end. When the men slipped back I made off into the town again, for on my return journey through the streets I had heard the stamping of horses in a stable, and had decided to fetch them out.

After groping about the main street for some time, I found the stable, and, striking a match, saw two fine animals at a manger. While I was bringing them away I heard a horseman approaching in the dark. As he passed I seized his reins, and brought him to a standstill, at the same time digging the muzzle of my rifle into his side. He proved to be a British officer, a Lieutenant Mcintire, who told me, when he had got over his surprise, that he had been away on a long patrol towards the Orange River for over a week.

He had heard the sound of firing as he neared Springbok, but did not know that the place had been surrounded.

I relieved him of his horse, a sporting Lee-Metford, and a Webley revolver (a weapon which I had long coveted), and, as the hotel at which Maritz and his men were celebrating was brilliantly lit up, I directed my prisoner thither, for I did not wish to be bothered with him. He pretended to go there, but I discovered afterwards that he had doubled back, and made his way past our men to O'Okiep, where I had word of him a few days later, when I went in to demand its surrender.

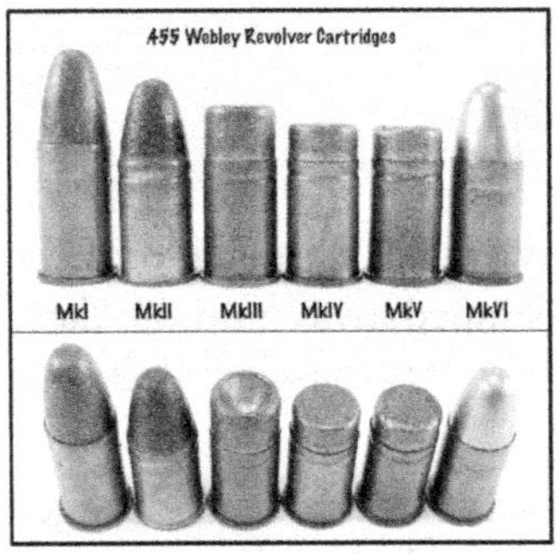

Webley Ammunition.[514]

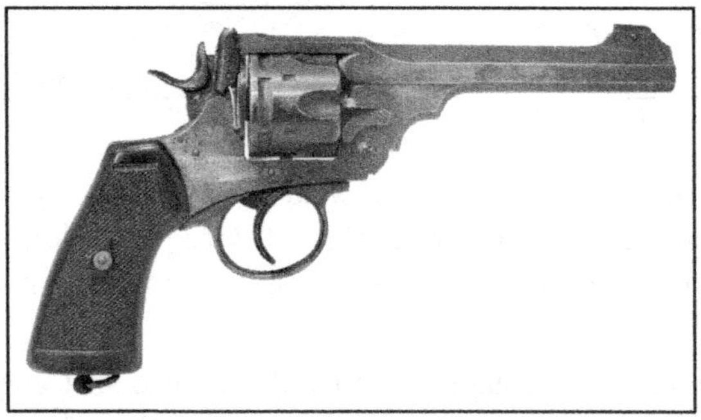

Webley Mk. VI top-break revolver.[515]

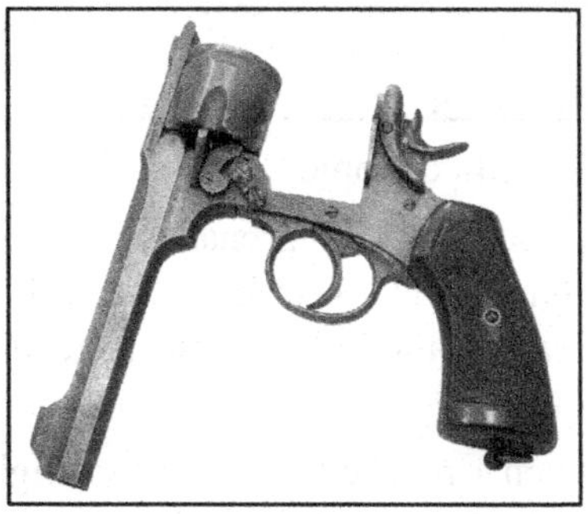

Webley Revolver.[516 517]

As for the rest of his patrol, I could hear by the jingling of bits and trampling of hoofs that they were waiting for their officer close by in the dark, so I yelled at them in Dutch and fired several shots, which sent them galloping back, and at sunrise our men rounded them up in a kopje a mile or two away, their horses being too done up to go farther.

I led my three captured horses along and as I went I stumbled on Edgar Duncker looking for a shop to loot. We joined hands and going along the street saw an open, lighted doorway. Here we found a room full of soldiers, with their rifles still in their hands. When we appeared some stood to attention, and one of them said that they were the defenders of No. 1 Fort that had so stubbornly denied us. They had held out for as long as they could, but their water-supply had long since given out, and they had been forced by thirst to vacate the block-house. Finding that our men had gone from below, they had marched silently into the town, hoping to find a dwelling-house provided with water-tanks where they could carry on the fight. When we appeared, they realized how forlorn was their hope of further resistance, and, after Duncker and I had helped them to find some drinking water in a yard behind, we took them to join their fellows at the hotel.

We were anxious to see the inside of No. 1 Fort, so we got a lantern, and, taking one of the soldiers to lead the way, went to inspect it. The entrance was a zig-zag passage, built high with sandbags, and the doorway was so low that we had to crawl inside. From floor to roof stood a huge iron water-cistern, occupying most of the cramped space. It had been pierced by so many bullets that the water had run out, and in the end the men had been obliged to leave. On a sort of firing-platform lay several dead men. One was Mr. Stewart, and another was a young local volunteer named van Couvorden, son of a doctor from Holland. Both had been shot through the head, and our guide said that they had been killed by a sniper from a rock, which he pointed out to me, when it grew light. It was the very rock from which I had fired the morning before, and this left little doubt in my mind that these two men had fallen to my rifle.

The whole village was now in our hands. Certainly we had far outnumbered the soldiers, but nevertheless it was good work done, for we had captured over a hundred prisoners, and a large stock of rifles, ammunition, and supplies, without the loss of a single man killed or wounded and this against fortified works.[487] General Smuts took up his quarters after daybreak in a large dwelling-house, and we members of the staff busied ourselves with collecting food-stuffs and stores for the use of our mess.

As I had gone for three full days and nights without sleep or rest, I now sought out a bed and turned in without removing my boots. I slept for twenty-four hours, and did not wake until the morning of the next day, when I found my friend Nicolas Swart sitting on the bed beside me. He was almost recovered from his wounds, and had just arrived from the south. He said that General Smuts had taken van Deventer's and Bouwer's men against the neighbouring town of Concordia, but had left special instructions that I was not to be disturbed, so they had gone away without me. Nicolas promised to take care of my spare horses and other belongings, so I saddled my mare Jinny, and rode after them.

When I reached Concordia, five miles away, it had just surrendered. About one hundred and fifty prisoners were taken, a motley collection of volunteers and Hottentot levies, with many rifles, and an abundance of clothing and other supplies.

In view of this success, General Smuts sent P. Mullar and myself to O'Okiep, the largest of the three mining places, with a letter demanding its surrender.

We set off at a gallop with a white cloth on a whipstick, and as the place lay only four miles away, we were soon approaching it. I saw at once that here was a harder nut to crack, for a ring of block-houses and wire entanglements stood all around the town, and inside the circle was a large central fort, flanked by a strong redoubt on a conical hill. As we rode up to the nearest block-house on the plain, half a dozen soldiers ran forward, and when I told them that we had a letter from our General demanding their surrender, one of them slapped the stock of his rifle and said, 'Surrender! Surrender be damned; we're Brummagem boys,[488] we're waiting for ye,' which also seemed unpromising.

[487] Friday, April 4, 1902, 902 Ookiep invested by Commandant Smuts.

[488] The Birmingham Boys (also known as the Brummagem Boys or the Brum Boys) were a street gang whose power extended from the North of England to London's underworld. Brummagem being slang for Birmingham. Birmingham Boys - Wikipedia

Two Boer fighters, one of them holding a white flag, coming up to a British soldier, probably with intention to negotiate.[518]

As we sat our horses, an infuriated officer rushed up from the next block-house, and violently abused us. He was an officer, but no gentleman, for he blustered and swore, and at the point of his revolver ordered us to put up our hands, while he went through our pockets. When I protested that we were under the flag of truce, he violently told me to hold my tongue, and, blindfolding us, marched us into town on foot, to a running fire of angry comment at our effrontery in daring to demand their surrender. I answered him at intervals, until he clapped his revolver to my forehead, and threatened to blow my brains out if I uttered another word, when I began to suspect that we had to do with a madman, and held my tongue.

The climax came when his eyes fell on my horse, being led behind us. I had taken the saddle from lieutenant McIntyre, whose name was marked on the holsters, and the fact that I had come into their lines on a British saddle, and a horse marked with the British broad-arrow, threw him into a fresh paroxysm, and, bawling obscene oaths, he hustled and jostled us along as if we were common malefactors.

He was the most disagreeable, in fact the only disagreeable, Englishman whom I met in the war, for, with this one exception, I had no unpleasant word from officer or private in all the time that we were out against them. At length we reached some sort of a camp, to judge by the sounds around us, and here we were left sulkily standing for about an hour, still subjected to abuse. Then a different stamp of man rode up, a superior officer, at whose approach our tormentor faded away, not to reappear. The newcomer was furious when he heard of our treatment, and he at once led us into a tent, and gave us each a cigar and a cooling, drink. When, after some time, a reply arrived from Colonel Shelton,[489] the officer in command of O'Okiep, our host personally helped us to mount our horses (for we were still blindfolded), and accompanied us to a point beyond the outer defences. There he uncovered our eyes and bade us a friendly good-bye.

The reply we brought from Colonel Shelton was couched in more elegant language than that received earlier in the morning from the Brummagem boys, but it was too much the same effect, for it said that he had plenty of men and munitions in the town, and that we could do as we liked.

When we handed this answer to General Smuts, he decided on a blockade. He said he did not very much care whether he took O'okiep or not, as he had all the arms, ammunition, and horses that he wanted, but he was going to make a show of besieging the place, until a relief force arrived, after which he would decide what to do next.

In order to judge for himself how the land lay, he took a patrol out within half an hour of our return. We rode forward until we reached a line of hills overlooking the town. From here we could see troops paraded on an

[489] Shelton Willington Augustus David, Major was born 14 May 1849, eldest son of Colonel J R W Shelton, of Bruree House, Limerick. He joined the 3rd Royal West Surrey Regiment; served in the South African War, 1900-2; was mentioned in Despatches twice; received the Queen's Medal with clasp; the King's Medal with two clasps, and was created a Companion of the Distinguished Service Order [London Gazette, 31 October 1902]: "Willington Augustus David Shelton, Major, 3rd Battalion Royal West Surrey Regiment. In recognition of services during the operations in South Africa." Lieutenant Colonel W A D Shelton married (1st), in 1877, May daughter of Henry Goodlake; (secondly), Victoria, daughter of John Chancellor, and (thirdly), in 1889, Florence, daughter of Edward Campbell, of Stapleton, Gloucester. He died 14 June 1909

open space close to the main fort, and it seemed that they were being addressed by an officer. This was probably Colonel Shelton, for heliograph messages from him to the relief force, intercepted later, showed that he was given to oratory, and high-flown phrases about England's far-flung Empire and the upholding of the flag.

We opened fire at fifteen hundred yards, and, although we could not see that we had hit anyone, we cleared the parade ground in a few seconds, the soldiers running to their stations, and returning our fire from the blockhouses. If we did no damage below, I drew blood nearer at hand. A large flock of goats was coming up the slope of the hill from the direction of the town, in charge of a Hottentot herdboy. When the firing started, he turned the animals back towards the English lines, so I ran down to within shouting distance, and beckoned him to drive the goats up our way, but, instead of obeying, he, bravely enough, urged them back the faster. As it would have been foolish to allow so valuable a meat-supply to escape, I was obliged to shoot, but I aimed low, and brought him to earth with a bullet through his leg. Then I chased the flock to our side of the hill, with shots from the town spattering about. The wounded herdboy crawled down to the block-house fence, and was fetched in on a stretcher, and I daresay he recovered. As a result of his inspection General Smuts decided that it would be worth his while to bomb some of the defences. He said that by pressing we should force the authorities to dispatch reliefs round by sea from Cape town. Already there was a heliograph winking away towards Port Nolloth[490] on the coast, no doubt asking for help, and his forecast ultimately proved correct in every particular, for in a few days we had word that ships were arriving with a considerable force.

Using a Heliograph.[519]

Soldiers of the 3rd Warwickshire Regiment on Parade.[520]

We now returned to our headquarters at Concordia, and, as there was a supply of dynamite and fuses at the copper-mine, we once more set about making bombs.

That night a party of Maritz's men was sent against two block-houses that had been marked down during the day. One of them was the redoubt on a high sugar-loaf kopje, the other was to the right on a somewhat lower ridge.

The assaulting party consisted of twenty men, and as they were under Ben Coetzee, my old friend, I joined them when they met after dark. We started from Concordia, and, when we reached the spot from which we had watched the town that afternoon, we left our horses behind, and worked our way on foot over broken ground to the ridge on which the smaller block-house was situated. We crawled stealthily upward until we were challenged by a sentry, with his 'Halt ! Who goes there?' Ben immediately lit and flung a bomb, timing it so well that it brought down a section of the block-house wall, out of which the garrison of ten men came tumbling. They belonged to the Warwickshire Regiment,[491] and, although none of them were killed, all were shaken and dazed. We sent the prisoners, rifles and ammunition to the rear under escort, and then set out to look for the second work on the hill. It was so dark that we lost our bearings and presently found ourselves in the town's cemetery.

After yet more floundering we were challenged and fired at from Fort Shelton, the main stronghold of O'Okiep, situated in the village. We retreated, intending to make back for our horses, but as we went we were halted for the third time, a voice calling to us from somewhere high above our heads. Rifle shots followed, in the glare of

[490] Port Nolloth is a town and small domestic seaport 144 kilometres (89 mi) northwest of Springbok. Port Nolloth - Wikipedia

[491] The 5th Battallion of the Royal Warwickshire Regiment. Royal Warwickshire Regiment - Wikipedia

which we made out the block-house for which we were looking. Ben Coetzee said we must climb up to it, and we did so unharmed, for the shooting was wild. As soon as we got under cover of the rock-capped summit, we lobbed our remaining bombs on to the roof and against the loopholed wall without making any visible impression, the soldiers within continuing to fire. Then we wormed our way down to our horses, and home to Concordia by midnight.

When General Smuts got our report in the morning, he ordered another raid against the same block-house for the next night. This time Commandant Bouwer supplied the bombers, and I went with them. We left our horses at the same place after dark, and again climbed up. We threw more than a dozen bombs, but in vain; the men within maintained an unceasing fire; the dynamite seemed ineffective, and we had to return empty-handed. Our efforts must have been anxiously followed by the rest of the O'Okiep garrison, for, as we scrambled down, there came a hail through the darkness, 'No. 4 Blockhouse, how are you, lads?' to which a voice replied, 'No. 4 Blockhouse is a-a-all right' and there was much cheering in the town below.

Next morning General Smuts said that the block-house was of no value to us, but that, as we had committed ourselves to taking it, we must carry the work, and he ordered Maritz to go in person. With him went the Marquis de Kersauson,[492] a young French adventurer who had been his constant associate since the war began. I went too, and, as before, we were challenged, but reached the ledge in safety. The first thing Maritz did was to stand on another man's shoulders in order to calculate the throwing distance.

He then climbed down and fastened three bombs together, weighing about twenty pounds. No other man could have hoped to throw so heavy a missile that length, but standing precariously on the shoulders of one of his men, he lit the fuse and hurled the triple grenade right on the roof. The fuse flared and sizzled for a second or two, lighting up the scene for many yards around, and then there was a tremendous roar, and stones and sandbags went flying in all directions. Silence followed, and realizing that the defenders were dead or stunned, we helped each other on to the rocky platform and, crawling over and under the wire entanglements, rushed the entrance passage.

From within we heard groans and a muffled voice, 'Stop throwing; stop throwing,' so we crowded in. Striking matches, we saw that the roof was blown down upon the soldiers. A few were dead, and the rest lay on the ground stunned. About half were Warwicks and the rest Hottentot levies. The sergeant in command, on recovering, told us they were the original garrison who had been there since our first attack, having specially asked to remain, and certainly they had acquitted themselves well. The dead and injured men were removed, also the arms and ammunition, and then we placed the remaining sticks of dynamite in the loopholes, and blew No. 4 Block-house into still further ruin before we returned to Concordia.

As the other works and block-houses around O'Okiep could only be approached over level ground bare of cover, General Smuts forbade any further bombing expeditions, and contented himself with investing the town, until such time as the relief force arrived. We had captured some two hundred prisoners in the other two villages and a large stock of supplies, so we could afford to rest for the present, and the next two weeks were spent quietly.[493]

I thoroughly enjoyed the lull, and commandeered a small cottage in Concordia to set up housekeeping with Duncker and Nicolas Swart, who was not yet completely recovered from his wounds and required attention. We pressed several Hottentot prisoners into our service as grooms and cook-boys, and as there was plenty of food in the military and mining depots, and sheep and goats in the hills to be had for the fetching, we lived well. General Smuts and the rest of the staff took another building in the village, while Bouwer and Maritz, with their men, lay camped in small parties in the hills around O'Okiep, and van Deventer's commando was posted twenty miles west astride the rail way line that comes up from Port Nolloth, to watch the progress of the relief expedition which was now assembling. And so we waited quietly for the order to break away south, on what would have been the most dramatic stroke of the war.[494]

[492] De Kersauson De Pennedreff Robert. Lt. Commando: General Maritz. Awarded the DTD (Dekoratie voor Trouwe Dienst) and received the Wound Ribbon (Lint voor Verwonding). See: French Volunteers: 274 records: https://www.angloboerwar.com/?option=com_content&view=article&id=3244

[493] ~April 5th to 19th 1902

[494] PF: He is alluding to a drive towards Cape town. Clan-William is only 120 miles from territory in control of the Boer forces. Paternoster is only 60 miles further on and Cape town is another 80.

Gen. Jan Smuts and his staff in Concordia, Namaqualand, May 1902.[521] Smuts stands fourth from left and Deneys Reitz sat on the left. Deneys Reitz on commando during the Anglo-Boer War.[522]

Smuts' establishment of an area of operations on the West coast of Africa, directly South of German South West Africa: German South West Africa - Wikipedia was a strategic rather than tactical threat to the British Empire. The German Government was quite expansionist and also in strong sympathy with the Boers. It only took a few weeks for the British Foreign Ministry to ascertain this and, while other factors were at work, to decide to call for peace negotiations. This military accomplishment is what would guarantee that Jan Smuts' opinions and recommendations would carry great weight in the upcoming negotiations.

XXVI. The Lost Cause

On the surface things looked prosperous. Five months ago we had come into this western country hunted like outlaws, and to-day we practically held the whole area from the Olifants to the Orange River four hundred miles away, save for small garrison towns here and there, whose occupants could not show themselves beyond the range of their forts without the risk of instant capture at the hands of the rebel patrols told off to watch them, while we roamed all the territory at will we had enjoyed a number of successes which the British probably regarded as minor incidents, but which our men looked upon as important victories and all this had greatly raised their spirits. Unfortunately, while matters stood thus well with us, the situation in the two Republics up north was far otherwise. Lord Kitchener's relentless policy of attrition was slowly breaking the hearts of the commandos. We had been out of touch with them for so long that we did not realize the desperate straits to which they had come, and our men judged the position from our own more favourable circumstances. Personally, I was not quite so sanguine, for, from such English newspapers as had come my way, I had learned something of the true state of affairs, but I hoped that all would yet be well and I kept my thoughts to myself.

Towards the end of April [~April 25th, 1902] I rode out one afternoon with Duncker and Nicolas Swart to snipe at the English posts on the other side of O'Okiep, and, as we were returning to our horses, we saw a cart coming along the road from the south with a white flag waving over the hood. Galloping up, we found two British officers insides who said that they were the bearers of a dispatch from Lord Kitchener.[495]

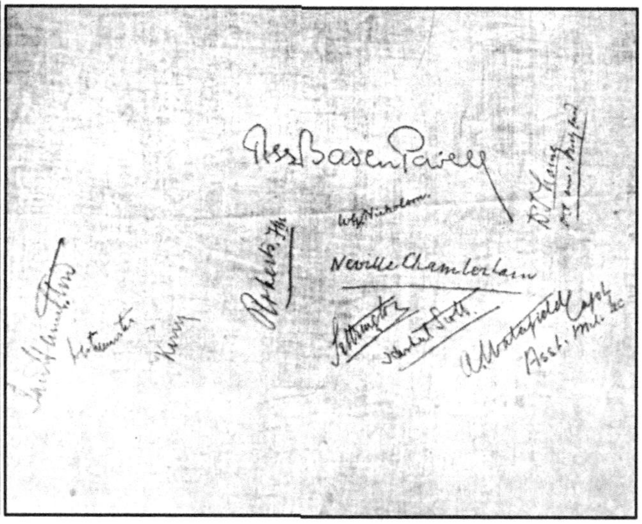

Cape Cart, with flag of truce. This one was requisitioned by Roberts for peace negotiations with Louis Botha,[523] A flag of truce.[524] Inscribed with the following signatures for verification: Roberts, (Vertical in Center), Baden Powell, Roberts' Secretary Neville Chamberlain. Kerry, Nicholson, Scott, and others.

We took them to Concordia, our pickets amongst the hills riding down from all sides to hear what it was about, but the officers professed ignorance of the contents of their message, although I had an uneasy suspicion of the truth.

When we reached Concordia, General Smuts took them inside his house and remained closeted with them for some time, after which he came out and walked away into the veld by himself in deep thought. We knew then that there was grave news.

That evening he showed me the dispatch. It was a communication from Lord Kitchener to say that a meeting between the English and Boer leaders was to be held at Vereeniging, on the banks of the Vaal River, with a view to discussing peace terms, and he was summoned to attend. A safe-conduct was enclosed, under which he was to proceed through the English lines to Port Nolloth, where he would be taken by sea to Cape town, and from there by rail to the Transvaal.

495 4/9/1902 Conference between Transvaal and Orange Free State leaders at Klerksdorp in regard to negotiations for peace.
4/11/1902 Meeting of Boer representatives at Klerksdorp in relation to Peace movement.

All this was ominous, and he spoke forebodingly of the future, but, in spite of the shadow that hung over us, one item almost made me forget the darker side, for the safe-conduct provided for a Secretary and an Orderly, and he said that I was to go with him as one of these. I was so delighted at the prospect of going on a journey like this that for the time being I gave thought to little else.

The men were the real tragedy. They had endured against great odds, facing years of peril and hardship without pay or reward, and they still had so much faith in the cause for which they were fighting that, when the news trickled through next day that General Smuts was to go to a peace conference, they were convinced that the British were suing for terms and were ready to restore our country.

It was pitiful to listen to their talk, and to see their faces light up when they spoke of having won through at last, and I, for one, had not the heart to disillusion them, or even to hint at a result other than favourable, so steadfast was their trust.

General Smuts set to work at once. Next morning a messenger was sent into O'Okiep, to advise the garrison that both sides were to refrain front active military operations while the Congress lasted, and the two British officers went on ahead to Steinkopf,[496] to warn the relief force collecting there that we were shortly passing through their-pickets.

The day after that the commando came in from the outlying posts to say good-bye to their leader. The men paraded before the Court House, each man sitting his horse, rifle at thigh, while General Smuts addressed them. He briefly told them of the object of his going, and asked them to be prepared for disappointment if need be, but there were only cheers and shouts of courage, as they pressed from all sides to wish him farewell.

Boer Commando assembled.[525]

I steered through the throng to shake hands with such as I could reach, waving to others beyond, and in this way I saw the last of many good friends and companions.

We set off next day, escorted by a small patrol. I left my spare horses, rifle, and gear with Nicolas Swart and Edgar Duncker, my best friends, whom I have not met again. We reached van Deventer's commando that afternoon where they were watching the troops that had come up from the sea, and for the last time we spent a night around camp fires. In the morning we made ready to pass into the English lines. As we started, General Smuts told me that his brothers law, [Petrus] Krige, was the other man to accompany him, and he said we were to arrange amongst ourselves which was to be Secretary and which Orderly, so I chose to be Orderly, as I thought it meant an aide-de-camp, and left my companion to be Secretary. Soon after that we saddled our

[496] Steinkopf is a town in Namakwa District. The town is located about 45 km north-north-west of Springbok. Formerly known as Kookfontein. Steinkopf, South Africa - Wikipedia

horses, said good-bye to van Deventer and his men, and rode down the valley towards the English lines. Far down we were met by Colonel Collins,[497] who commanded the relief expedition.

Here our escort took over our horses, and, after singing our Commando Hymn, and firing a farewell volley into the air, they wheeled round and galloped cheering away towards their own side, to the manifest interest of the English officers and troopers lined up beside the road. With them went the last of our free life and all that it had meant to us.

A cart was brought, in which General Smuts, [Petrus] Krige, and I were driven to a large camp standing beside the railway line, where a guard of honour was drawn up to receive us, behind which crowds of soldiers had gathered to see the Boer Emissaries. I now discovered that I had made a mistake, and that an Orderly was an officer's batman, whereas a Secretaryship carried commissioned rank. Krige was invited into Colonel Collins's tent with General Smuts, while I was led to the servants' mess, and when, an hour later, a train stood ready to take us to Port Nolloth he and General Smuts were ceremoniously ushered into a first-class compartment, whilst I was put aboard an open cattle-truck with the luggage.[498]

However, being in an enemy camp and travelling by rail for the first time for nearly two years was so exciting that it made no difference to me where I was, and Krige and I appreciated the humorous side of our respective positions. Whenever he looked out of the carriage window and saw me sitting in the truck behind, he roared with laughter, and so did I, at his having become an officer, and I a servant.[499]

When the train drew up at the next station there was another guard of honour for General Smuts and his Secretary, who were taken in to a grand luncheon, whilst I forgathered with the batmen in the kitchen behind. At the next halting-place I underwent record promotion. There was an officer of the Hussars, Captain Barclay,[500] who had been deputed to accompany us to the coast, and having seen me standing about, he asked General Smuts who I was. The General explained that on commando there were no social distinctions, but that he had brought me along because he thought my father might be at the Conference. Captain Barclay telephoned up the line to Colonel Collins, to say that a son of the Transvaal State Secretary was of the party as an orderly, and he presently came to me and said, 'Young man, you are Chief-of-Staff to General Smuts; come along and join us.' He jokingly assured me that the promotion from batman to field rank in the course of one morning was the quickest known in any army.

Towards evening we reached Port Nolloth, a dreary little seaport, where many troopships lay at anchor. One of these, the Lake Erie, was under steam, and even as the train ran into the station, a boat set out to fetch us.

This was the end of our long roving. We stood on the quayside, silently looking back on the way we had come, each busy with his own thoughts. I do not know what was in the minds of my companions, but perhaps they, too, were thinking of the long road we had travelled, of camp fires on mountains and plains, and of the good men and splendid horses that were dead.

With heavy hearts we got into the boat that was to take us to the ship, and the moment we were on board they weighed anchor, and we sailed southward.

[497] Collins Ernest Rokeby Captain was born 12 October 1870, fourth son of Brigade Surgeon Francis Collins, late 5th Fusiliers, of Lyme Regis. He was gazetted to the East Lancashire Regiment 19 November 1892; became Lieutenant 1 July 1894, and Captain 4 June, 1900. He served in the South African War, 1900-2, employed with Mounted Infantry, and took part, in operations in the Transvaal, December 1900 to January 1902, and April to 31 May 1902; operations in Orange River Colony, July to September 1901, and January to February 1902; operations in Cape Colony, May and September 1901, and May 1902. He was mentioned in Despatches [London Gazette, 10 September and 15 November 1901]; received the Queen's Medal with three clasps; the King's Medal with three clasps, and was created a Companion of the Distinguished Service Order [London Gazette, 27 September 1901]: "Ernest Rokeby Collins, Captain, East Lancashire Regiment. In recognition of services during operations in South Africa." Captain Collins was invested by the King 18 December 1902. He was promoted to Major 14 May 1913, and served in the European War, 1914–18. Major Collins married Margaret Alice, second daughter of the Reverend E W Sotheron-Estcourt, MA, of Estcourt, Tetbury, Gloucestersliire, and of his wife, Eleanor Lucy, daughter of the Reverend W J Bucknall Estcourt. Source: DSO recipients (VC and DSO Book) East Lancashire Regimen.

[498] "Batmen" A batman or an orderly is a soldier assigned to a commissioned officer as a personal servant. Before the advent of motorized transport, an officer's batman was also in charge of the officer's "bat-horse" that carried the pack saddle with his officer's kit during a campaign. Batman (military) - Wikipedia

[499] "Batmen" A batman or an orderly is a soldier assigned to an officer as a personal servant. Before the advent of motorized transport, an officer's batman was also in charge of the officer's "bat-horse" that carried the pack saddle with his officer's kit during a campaign. Batman (military) - Wikipedia

[500] Barclay Cameron Captain QSA (3). Source: QSA medal rolls. (The Prince of Wales's Own Royal) 10th Hussars.

In spite of our mission, the voyage was one of great pleasure to me. After years of rough fare and hard living, we had luxurious cabins, with soft beds to lie on; a steward with coffee in the morning, a bath ready prepared and food such as I had almost forgotten the existence of. All this seemed like a dream, and I enjoyed every moment of it.

We reached Cape town in five days, and were transhipped to Simonstown on board the battleship Monarch, Captain Parkes,[501] and here again we spent a week in comfort, for officers and men vied with each other in their efforts to welcome us.

The British, with all their faults, are a generous nation, and not only on the man-of-war, but throughout the time that we were amongst them, there was no word said that could hurt our feelings or offend our pride, although they knew that we were on an errand of defeat.

Monarch after her 1872 conversion to barque rig.[526] [502]

HMS Monarch by William Frederick Mitchell[527]

Cape Town, Devils Peak, Table Mountain and Lion's Head from Table Bay.[528] Cape Town– Seat of Government, Cape Colony.[529]

Panorama of Cape Town.[530]

[501] Parkes George Kale Serapis Captain Source: Nominal roll in WO127 Served with Gorringe's Flying Column. Which was raised by Lt Col G F Gorringe in 1901 and served in Cape Colony against numerous Boer Commandos.

[502] HMS Monarch (1868) was the first seagoing British warship to carry her guns in turrets, and the first British warship to carry guns of 12-inch (300 mm) calibre. Broken up, 1905

Cape town docks.[503]

The Squadron at the Cape.[531]

Simonstown – Headquarters of the Cape Squadron.[532] H.M. Ships Powerful, Terrible, Doris, Monarch and Penelope at Simons Bay. X.[533]

At length orders came for us to go north. We were rowed ashore after dark to a landing-stage below the Simonstown railway station, and taken to a train that was standing ready. We were hurried through the suburbs of Cape town and then switched on to the main line at Salt River Junction, to find ourselves at Matjesfontein[504] in the Karroo next day.

[503] Cape Town docks. Somerset Hospital in centre. File:The Second Boer War, 1899-1902 Q71990.jpg - Wikimedia Commons

[504] Matjiesfontein is a settlement in Central Karoo District Municipality in the Western Cape province. Matjiesfontein - Wikipedia

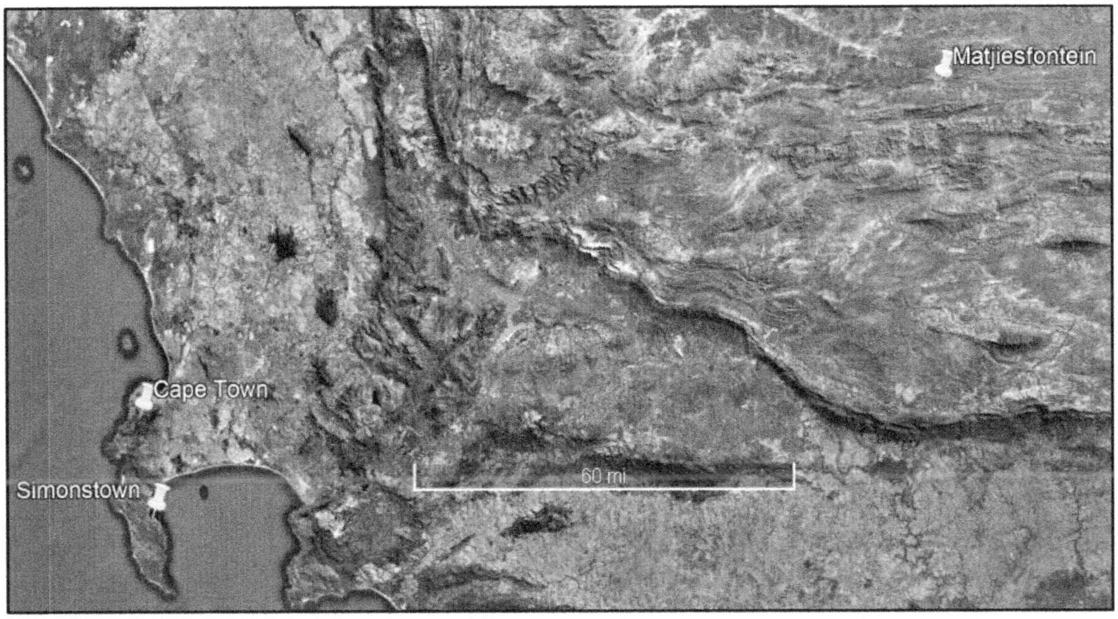

Map data: Google Image© 2024 Image Landsat/Copernicus.

Here General French came to see us, a squat, ill-tempered man, whom we did not like, although he tried to be friendly. He sat talking to us for an hour or more trying to draw General Smuts, who had no difficulty in parrying his clumsy questions. When he made no headway, he became more natural and spoke of his experiences during the war, in the course of which he told us how narrowly we had missed capturing him that night below the Stormbergen.

General John French.[505] View of Standerton.[534]

From Matjesfontein we continued our journey, travelling at night only, an armoured train puffing ahead all the way, its searchlight sweeping the veld. Each day we were side-tracked at some lonely spot till dark, and thus made slow progress. I have been told that we were purposely delayed lest, coming from the Cape where the outlook was brighter, we might persuade the Transvaalers that things were not so bad as they seemed. For this reason Lord Kitchener did not wish us to appear amongst them until matters had gone too far for them to turn back. However that may be, it took us the better part of a week to reach Kroonstad in the Northern Free State, where Lord Kitchener was to meet us.

[505] Lieutenant-General Sir John French Commanding Cavalry Div. 1899–1900 Commanding Cape Colony, 1901–1902. Times History War in S.A Vol 3 Photo from MacMillan & Co. from "With General French and the Cavalry in South Africa."

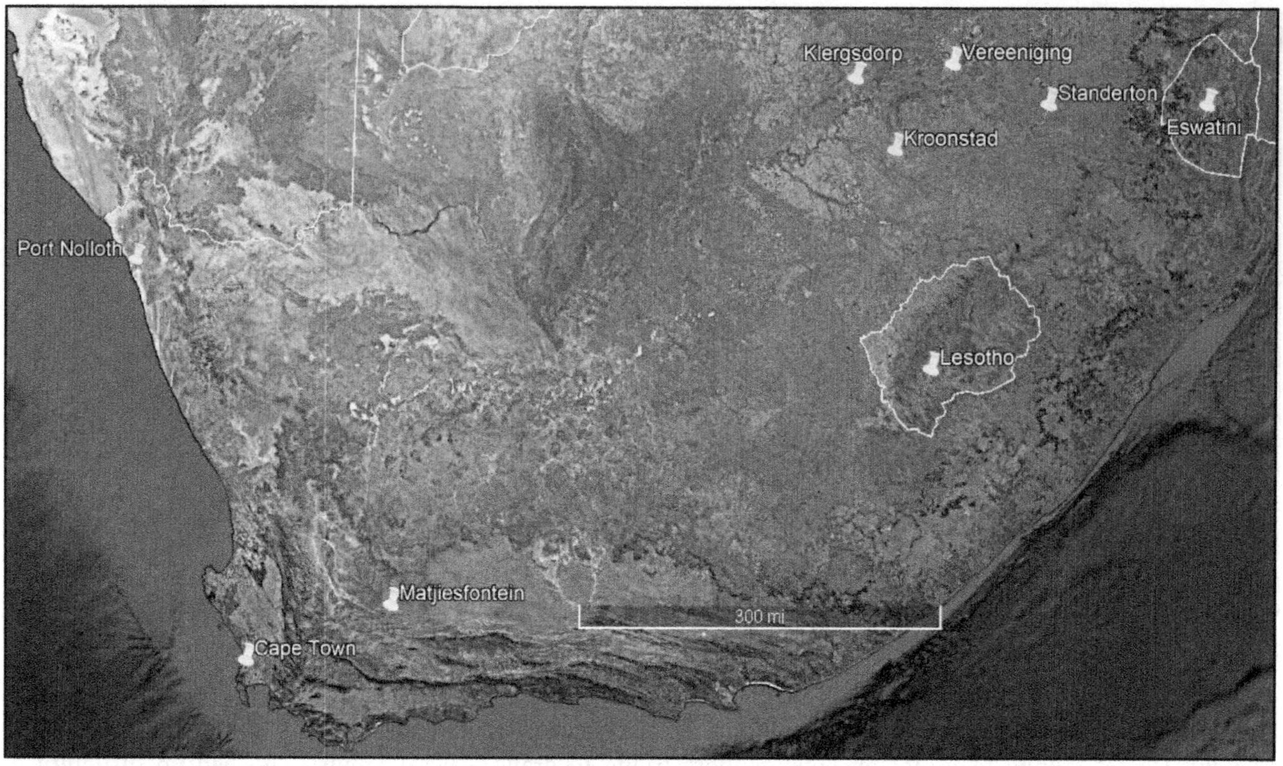

Map data: Google Image© 2024 Image Landsat/Copernicus.

Soon after our arrival he rode up to the station on a magnificent black charger, followed by a numerous suite, including turbaned Pathans,[506] in Eastern costume with gold-mounted scimitars.

A Group of Bengal Lancers at the remount [horses] depot at Kroonstad.[535] Note the vertical held sabers.

His retinue waited outside while he came into our compartment to talk. He was anxious to bring the war to a close, for he referred again and again to the hopelessness of our struggle, telling us that he had four hundred thousand troops in South Africa against our eighteen thousand. He said that he was prepared to let the burghers retain their horses and saddles in recognition of the fight that they had made, and that the British Government would help to rebuild the destroyed farmhouses, the burning of which he defended on military grounds.

[506] Pashtuns are an Iranian ethnic group native to the region of Afghanistan and Pakistan. Pashtuns - Wikipedia

General (the taller) and Gen. Pole-Carew at the railway station, Pretoria, June 5th.[536] General Viscount Kitchener, Commander-in-Chief, South Africa, Nov., 1900 – June, 1902.[537]

General Smuts taxed him with having unfairly executed our men in the Cape, and this, too, he justified, on the plea that we had used khaki uniforms to decoy his soldiers.

Before going he told us that we were to proceed to the Eastern Transvaal, to find General Botha, and that the conference at Vereeniging[507] would only take place after that.

Accordingly, from Kroonstad, still escorted by an armoured train, we crossed the Vaal River into the Transvaal. We went through Johannesburg at night, and here they turned us east on to the Natal line, until we came to the town of Standerton,[508] where we left the train and travelled by cart along a block-house line that ran straight over the high veld. At intervals there were small English camps, at each of which the troops turned out and treated us with courtesy.

Standerton concentration camp.[538]

We journeyed for a day and-a half, until we reached a point where a party of horsemen sent by General Botha was awaiting us. They had brought spare horses, so we left the cart with the troopers, and, striking across country, travelled for two days over bare and deserted plains, to the place where the Commandant-General was expecting us. Here about three hundred men were assembled. They were delegates from every commando in the Eastern Transvaal, come to elect representatives to the Peace Congress to be held at Vereeniging, and nothing could have proved more clearly how nearly the Boer cause was spent than these starving, ragged men, clad in skins or sacking, their bodies covered with sores, from lack of salt and food, and their appearance was a great shock to us, who came from the better-conditioned forces in the Cape. Their spirit was undaunted, but they had reached the limit of physical endurance, and we realized that, if these haggard, emaciated men were the pick of the Transvaal commandos, then the war must be irretrievably lost.

Food was so scarce that General Botha himself had only a five strips of leathery biltong to offer us, and he said that, but for the lucky chance of having raided a small herd of cattle from the British a fortnight before, he would have been unable to hold the meeting at all.

[507] Vereeniging - Wikipedia

[508] Standerton is a large commercial and agricultural town lying on the banks of the Vaal River in Mpumalanga, South Africa, which specialises in cattle, dairy, maize and poultry farming. General Jan Smuts won this seat during elections and went on to assist in setting up the League of Nations. Standerton - Wikipedia

I inquired at once for news of my father and my three brothers. General Botha gave me word of my father. He told me that he was with one of the northern commandos, and would in all probability be at the coming Conference. He could tell me nothing of my brothers, but by asking among the men, I learned that my eldest brother, Hjalmar, had been captured by the Australians more than a year before, and that my second brother, Joubert, was taken prisoner whilst lying ill of malarial fever in the low country, apparently not long after I had last seen him at the Warm Baths, towards the end of 1900, I could find out nothing about my youngest brother Arnt.

Next day the elections were held. Even in adversity the Boer instinct for speeches and wordy wrangling asserted itself, and the time was passed in oratory, and with nominations and re-nominations of candidates, but by evening the complicated balloting was finished, and some thirty delegates elected.

Next morning the gathering dispersed, the men riding off on their hungry-looking horses to rejoin their distant units, while General Botha and the successful deputies started back for the English block-house line.

We arrived here by the following evening. The troops supplied us with food, for we were famishing, and we now returned along the block-houses to Standerton, the soldiers everywhere standing respectfully to attention as our tattered cavalcade went by. At Standerten we entrained for Vereeniging. This is a small mining village on the banks of the Vaal River, where, nearly two years before, I had watched the Irishmen burning the railway stores during the retreat from the south.

Vereeniging, first station in the Transvaal from Cape Town.[539] General L. Botha [With 'Aussie' hat, right side pinned up.] in conversation with General Lyttelton and Staff at Newcastle Station.[509]

Vereeniging, first station in the Transvaal from Cape Town.[540]

[509] Navy & Army Vol 14 p. 309 June 14th, 1902 Photo Copyright by "Kodak"

General Christiaan de Wet [Far Right] on his way to peace talks at Vereeniging, 13 April 1920.[541] Boer general Christiaan de Wet, photographed at the end of the Second Boer War in Bloemfontein on 6.6.1902.[542]

The British had prepared a large tented camp for our reception, and almost the first man I saw as we entered was my father, shaggy and unkempt, but strong and well, and our greeting after so long a parting was deep and heartfelt.

And now the delegates came in from the rest of the Transvaal and from the Free State. Every leader of note was there. General de la Rey, Christian de Wet, President Steyn, Beyers, Kemp, and many others, the best of the Boer fighting-men.[510]

We learned from General de Wet that my younger brother [Arnt] had been serving under him for more than a year, and that he was still safe and sound, so we were all accounted for. Although two were prisoners of war, we had been luckier than the majority of families, most of whom were mourning their dead, whereas all five of us were still alive.

I know little of the actual Peace Conference as I was not a delegate, but the outcome was a foregone conclusion. Every representative had the same disastrous tale to tell of starvation, lack of ammunitions horses, and clothing, and of how the great block-house system was strangling their efforts to carry on the war. Added to this was the heavy death-roll among the women and children, of whom twenty-five thousand had already died in the concentration camps, and the universal ruin that had overtaken the country. Every homestead. was burned, all crops and live-stock destroyed, and there was nothing left but to bow to the inevitable.

After prolonged debates the Conference suspended its sittings for a day, whilst General Botha, my father, general de la Rey and others went to Pretoria to conclude the final treaty with Lord Kitchener and Lord Milner.

On their return peace was an accomplished fact.[511]

Of the sting of defeat—I shall not speak, but there was no whining or irresponsible talk. All present accepted the verdict stoically, and the delegates returned quietly to their respective commandos, to make known the terms of surrender.

I was spared the ordeal of returning to break the news to our men in the Cape, for my father insisted upon my remaining with him when General Smuts went south. When he came to take leave of me, he said that he dreaded the task of telling the men, and our hearts were heavy at the thought of the disappointment awaiting them. We shook hands for the last time, and then he, too, was gone.

My father was sent into the low country, to arrange for the bringing in of the commando with which he had been serving. We travelled by rail to Balmoral Station on the Delagoa Bay line, and from there on horseback into the wild country through which my brother and I had ridden in search of General Beyers, earlier in the war. After two days hard going we found the camp, and my father had the unpleasant duty of telling the men that all was over. Most of them took it calmly, but some cursed and vowed that they would never surrender.

[510] 5/27/1902 Malan mortally wounded and captured by Jansenville District Mounted Horse (under Major Collett), and Lovat's Scouts.

[511] Signed just before midnight on May 30th, 1902.

Commandos

Vieldcornet Thring's Commandos near Ladysmith, Natal, South Africa, 1900.[543]

Group of Boer leaders in Kimberley area. General Koos de la Rey is the man with black beard in right foreground.[544] Per Album of 300 Commandant Wessels is the "old burgher with a white beard" [white shirt]

"General Louis Botha with his commandants. Botha is seated in the centre."[545]

Infantry wing of the Johannesburg Vrijwilliger Corps.[546]

Boer family of sharpshooters at Johannesburg.[547]

Three Generations under arms.[548] (L to R) P.J. Lemmer (65) J.D.I. Botha (15) and C.J. Pretorius (43) of the Takhaar and Penkop Commando's that fought side by side. "Young Botha in Trench."[549] [Third from right.]

Pretoria town burghers at the front.[550]

A section of the Penkop Regiment.[551]

My father, although he had himself voted against peace at the Conference, pointed out to them that they should either submit to what had been done, or leave the country, as he intended doing. This quieted the more turbulent, and we started back next day for Balmoral, where the men were to hand in their rifles. This depressing ceremony was presided over by an English officer, seated at a table beneath the trees, with a regiment of troops in reserve close by. Despite his protests, our men fired away their ammunition into the air, smashed their rifle-butts and sullenly flung the broken weapons down, before putting their names to the undertaking which each man was called upon to sign, that he would abide by the peace terms.

British troops at assembly point during probably surrender ceremony of General Piet Cronjé's units.[552] British soldier standing in the assembly line, possibly during the surrender of the Boer fighters ceremony.[553]

Surrender of Prinsloo's Boer forces in Brandwater basin, 1900.[554] Stacks of rifles taken from Cronje's Laager at Paardeberg.[555]

A heap of Mausers with martini rifles surrendered by Cronje's army.[556] Captured Boer Rifles at Cape town.[557]

British soldiers collecting rifles of the Boer fighters.[558] Ending in Smoke, Destroying Captured Boer Rifles.[559]

When my father's turn came, he handed over his rifle to the officer in charge, but refused to sign. He said that although he was one of the signatories to the Peace Treaty, he had told Lord Milner at the time that he was setting his hand to the document in his official capacity as State Secretary of the Transvaal and not as a private individual, and Lord Milner had accepted his signature on that basis.

The officer pointed out that he would not be allowed to remain in the country, and my father agreed. I had no very strong convictions on the subject, but I had to stand by him, so I also refused to sign, and was told that I would be put across the border, which troubled me little, as I was eager to see more of the world.

When all was over, the men rode off on their different ways, to search for what remained of their families and ruined homes.

My father and I went to Balmoral Station, where a message had been received from Lord Milner confirming the order that we were to be deported, but qualifying it to the extent of allowing my father a fortnight in which to settle his affairs in Pretoria. And so we returned, after more than two years of wandering. We found our home in the possession of a British General with sentries outside who forbade approach. Our household goods had disappeared, and had it not been for the hospitality of a friend we should have gone roofless. During this time my younger brother came [Arnt] riding in from the Free State, six inches taller than when I last saw him, and none the worse for his long adventure. He, too, decided to go, and so, towards the end of June, we went into self-imposed exile.

As we were waiting on the border at Komati Poort, before passing into Portuguese territory, my father wrote on a piece of paper a verse which he gave me.

It ran:

South Africa

Whatever foreign shores my feet must tread,
My hopes for thee are not yet dead.
Thy freedom's sun may for a while be set,
- But not for ever, God does not forget.

And he said that until liberty came to his country he would not return.

Komatie Poort Railway Bridge.[560]

He is now in America and my brother and I are under the French flag in Madagascar.

We have heard of my other two brothers. The eldest [Hjalmar] has reached Holland from his Prison camp in India, and the other [Joubert] is still in Bermuda awaiting release.

Maritz and Robert de Kersauson[512] are with us in Madagascar. We have been on an expedition far down into the Sakalave country, to see whether we could settle there.

General Gallieni[513] provided us with riding-mules and a contingent of Senegalese soldiers, as those parts are still in a state of unrest. It was like going to war again, but all went quietly, and we saw much that was of interest—lakes and forests; swamps teeming with crocodiles, and great open plains grazed by herds of wild cattle. But for all its beauty the island repels one in some intangible manner, and in the end we shall not stay.

At present we are eking out a living convoying goods by ox-transport between Mahatsara on the East Coast and Antananarive, hard work in dank fever-stricken forests, and across mountains sodden with eternal rain; and in my spare time I have written this book.

Antanarive,[514]

Madagascar, 1903.

[512] Marquis de Kersauson, the young French officer previously mentioned.

[513] Joseph Simon Gallieni (24 April 1849 – 27 May 1916) was a French soldier, active for most of his career as a military commander and administrator in the French colonies. Gallieni is infamous in Madagascar as the French military leader who exiled Queen Ranavalona III and abolished the 350-year-old monarchy on the island. Joseph Gallieni - Wikipedia

[514] Antananarivo, is the capital and largest city of Madagascar. Antananarivo - Wikipedia

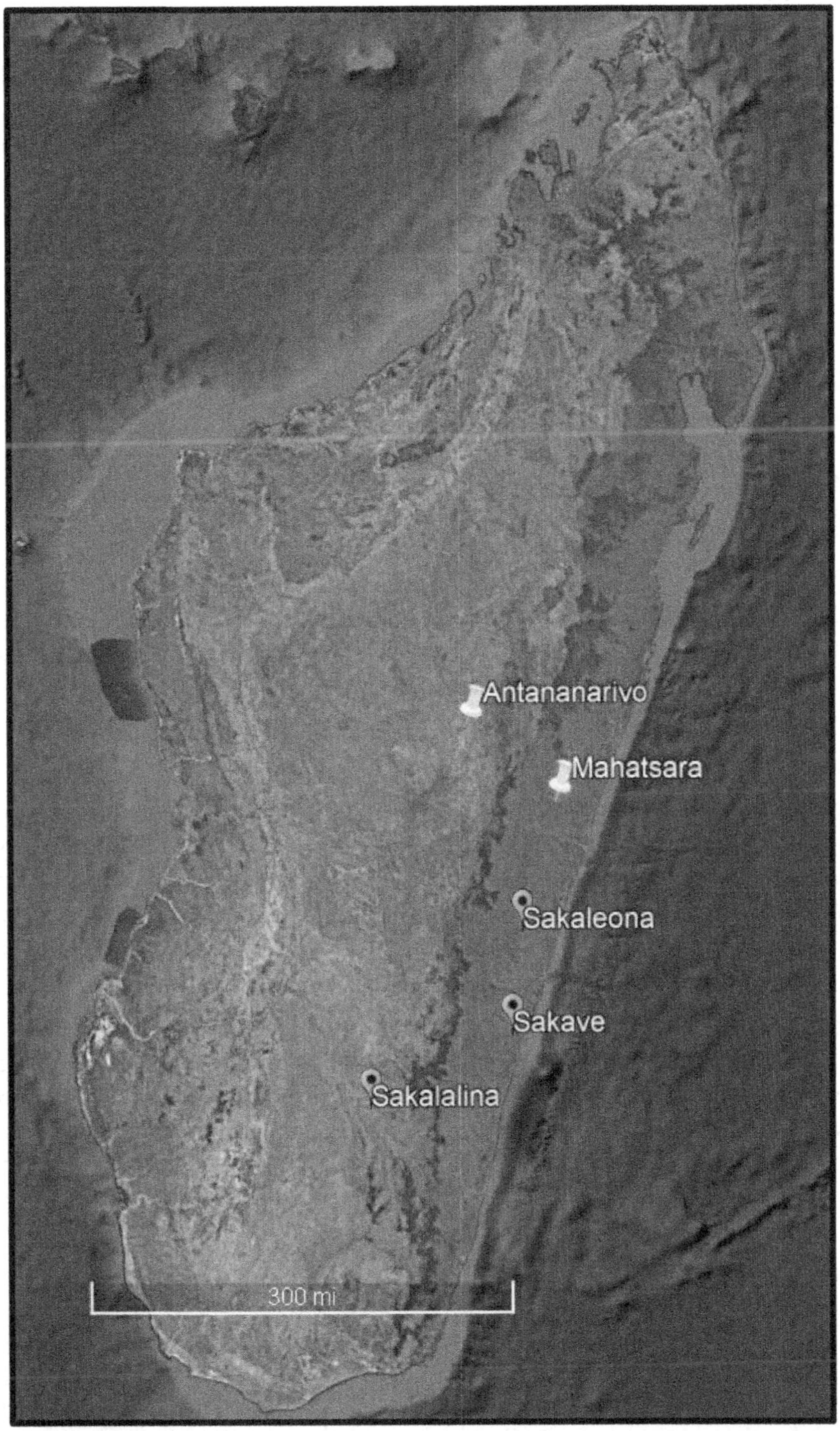

Madagascar - Map data: Google Image© 2024 Image Landsat/Copernicus.

Deneys continues his story, starting with his sojourn in Madagascar, in his next work, 'Afrikander'.[515]

[515] Deneys Reitz, Afrikander, New York, Minton, Balch & Company, 1933. A narrative of the author's life in exile following the Boer War, his work upon his return to South Africa, and his part in the European War, first in South Africa and later in Europe. Catalog Record: Afrikander | HathiTrust Digital Library U. Cal. Scan: #7 - Afrikander, by Deneys Reitz ... - Full View | HathiTrust Digital Library https://babel.hathitrust.org/cgi/pt?id=uc1.$b582073&seq=7 Direct Link to Book: #7 - Afrikander, by Deneys Reitz ... - Full View | HathiTrust Digital Library

O'kiep & Cape Copper

O'kiep

Jan Smuts accompanied by Deneys Reitz, and [Petrus] Krige departed O'kiep on April 26th, 1902. Deneys does not mention the episode of running a locomotive and train car loaded with dynamite into town on May 1st, since he had already departed. Also, Jan Smuts had left instructions that no further military operations were to be carried out while peace negotiations were occurring.

Okiep - Port Nolloth narrow gauge railway.[561] Okiep - Port Nolloth narrow gauge railway. Circa 1885. Senior officers' coach by mules on its way down the mountain.[562]

By the time Smuts departed with a British safe-conduct to the deliberations the siege became something resembling a good-humoured blockade.

On 1 May 1902, the commandos launched an attack on O'okiep, using the commandeered locomotive "Pioneer" of Concordia's Namaqua United Copper Company to propel a mobile bomb in the form of a wagon-load of dynamite into the besieged town. The protective defences at O'okiep consisted of a barbed wire fence, which was erected across the railway line at Braakpits Junction, just north of the town. The points at the junction were rigged to the fence, with the result that when the dynamite laden wagon breached the fence, it derailed at the points and spilled its load of dynamite on the ground, where it burned out harmlessly without exploding.

According to Jan Smuts, the railway between the two towns was still intact, but since there were women and children in O'okiep town, all the commando [Gallagher] was allowed to do was to give O'okiep a tremendous fright with a harmless explosion. Boer General Ben Bouwer had inspected the trainload before it was sent hurtling into the besieged town, to make sure that there were no caps in the dynamite.

Namaqua Copper Company 0-4-2ST Pioneer, derailed outside O'okiep while being run into the town with a truckload of dynamite during the South African War.[563 564] [Note small mining 'truck' off the rails.] 'Pioneer' locomotive used to send a truck loaded with dynamite into O'Kiep.[565]

Cape Copper Article

In the Colonies and Boer Republics which became the Union of South Africa in 1910, several unofficial military decorations and medals were instituted and awarded during the nineteenth and early twentieth century. The Cape Copper Company Medal for the Defence of O'okiep is a private campaign medal which was instituted by the Cape Copper Company Limited in 1902.[566] The medal was awarded to members of the O'okiep Garrison who defended the town while it was besieged by Boer Commandos from 4 April to 4 May 1902, near the end of the Second Boer War.

The garrison of O'okiep, commanded by Lieutenant Colonel W.A.D. Shelton DSO, consisted of some 900 men and included local miners who assisted Shelton's troops. The civilian defenders were mostly employees of the Cape Copper Company, three-quarters of whom were Coloured.

After the war ended, the British government awarded the Queen's South Africa Medal to all those regular troops who had seen active service in Namaqualand during the war, as well as the King's South Africa Medal to those who had seen service on or after 1 January 1902 and completed 18 months service before 1 June 1902.

When it transpired that the Coloured members of the O'okiep garrison were precluded from receiving these medals, the Cape Copper Company decided to strike a medal of its own and to present it to all the defenders of the town, regardless of race.

Cape Copper Company defense medal.[567]

Silver Medal.[568]

Two versions of the Cape Copper Company Medal for the Defence of O'okiep were struck. The bulk were struck in bronze and were awarded to the rank and file of the military and civilian defenders of the town, most of whom were mine workers who had taken part in the defence of O'okiep. A silver medal was awarded to officers, higher ranking mine officials and at least one non-commissioned officer, Paymaster Sergeant Harry Rodda, who was also the paymaster of the O'okiep Copper Company.

Times History Quote:

Latterly the siege [of O'kiep] had degenerated into a mere blockade, conducted with so much mutual good humour that on one occasion a challenge to a football match was carefully considered by the garrison and eventually declined. On April 26 Smuts had been summoned to Pretoria to join in the final deliberations, and thenceforward even the blockade was little more than nominal.[516] [517]

For forty-five miles eastward from Port Nolloth the railway crosses a sandy and waterless desert. Then it climbs a rampart of steep and rugged hills and, bending sharply to the south, reaches the high plateau where the mining settlements stand. ......The viaducts being intact, Klipfontein, the summit, was reached without difficulty and the march continued on foot. A series of very strong positions were abandoned one after the other by the rebels, and on May 3 [Colonels] Cooper and Callwell, ... deserve all credit for their difficult march, entered Ookiep.

[Operationally] It was a case of stale-mate. In seven months Smuts had succeeded in raising the number of men in arms to 3,000, six sevenths of whom were rebels and one-seventh Boers. Tactically he had inflicted far more loss than he had suffered none of his bands had met the fate of Lotter's; none, unhappily, had ever received a sound beating.

[516] Times History of the War Vol 5 pl 552–553

[517] Times History Vol 5 Re: Peace Chapter p. 564–606 Attendees: Page 583.

Commandant Lotter being brought into Graaff Reinet on 5 September 1901 after being captured by the Coldstream Guards after the Battle of Groenkloof.[569 518 519] Commandant Hans Lotter in Graaff-Reinet prison.[570]

The death sentence is pronounced on Commandant Lötter at Middleburg[571]

Reading the sentence of Lotter after the court martial.[572] Lotter fainting on hearing his death warrant announced.[573]

Strategically, by means of an insignificant abstraction from the fighting strength of the two Republics, he had kept one of the best British generals and eight or nine thousand Imperial troops actively employed in defending British territory. There was no reason why he should not have persisted for an indefinite period in the same sort of campaign.

But to what end? Smuts knew very well that the main purpose of the campaign, a general rebellion, had failed, and it was in this spirit that he travelled to Pretoria in May.

[518] 9/7/1901: Lotter and his entire commando captured in Cape Colony.
In the Battle of Groenkloof on 5 September 1901, a British column under Colonel Harry Scobell defeated and captured a small Boer commando led by Commandant Lötter in the Cape Colony during the Second Boer War.

[519] 10/11/1901: Commandant Lotter sentenced to death. Death sentence on five members of his commando was commuted to penal servitude for life. The Times History simply recites that Lotter was condemned to death under martial law. Vol 5 p. 320. As a rebel and possibly for train wrecking. Battle of Groenkloof - Wikipedia

Executions

Number	Place	Date	Reason
		1901	
2	De Aar	March 19	Train wrecking.
2	Pretoria	June 11	Boers breaking oath of neutrality.
1	Middelburg	July 10	Fighting.
1	Cape Town........	" 13	"
1	Cradock...	" 13	"
2	Middelburg.......	" 24	"
2	Kenhardt	" 25	"
1	Pretoria	Aug. 22	Boer spy.
3	Colesburg.	Sept. 4	Fighting.
1	Middelburg.	Oct. 10	"
1	Middelburg.	" 11	"
1	Vryburg (hanged).	" 12	"
Several	Tarkastad.........	" 12	"
1	Tarkastad.........	" 14	"
1	Middelburg	" 15	"
2	Cradock (1 hanged, 1 shot)..........	" 17	Train-wrecking and murdering native.
2	Vryburg	" 29	Fighting.
1	Mafeking.........	Nov. 11	Shooting a native.
1	Colesburg	" 12	Fighting, marauding, and assaulting, etc.
1	Johannesburg	" 23	Persuading surrendered burghers to break oath.
1	Aliwal North.....	" 26	Cape Police deserter.
1	Krugersdorp......	Dec. 26	Shooting wounded.
2	Mafeking.........	" 27	Kaffir murder.

Doyle, Arthur Conan, 1859-1930 The war in South Africa, its cause & conduct..[520]

Conan Doyle Summary

In the last two months of the year 1901,[521] a new system was introduced into the Cape Colony campaign by placing the Colonial and district troops immediately under the command of Colonial officers and of the Colonial Government. It had long been felt that some devolution was necessary, and the change was justified by the result.

Without any dramatic incident, an inexorable process of attrition, caused by continual pursuit and hardship, wore out the commandos. Large bands had become small ones, and small ones had vanished. Only by the union of several bodies could any enterprise higher than the looting of a farmhouse be successfully attempted.

Such a union occurred, however, in the early days of February 1902, when Smuts, Malan, and several other Boer leaders showed great activity in the country round Calvinia. Their commandos seem to have included a proportion of veteran Republicans from the north, who were more formidable fighting material than the raw Colonial rebels. It happened that several dangerously weak British columns were operating within reach at that time, and it was only owing to the really admirable conduct of the troops that a serious disaster was averted.

Two separate actions, each of them severe, were fought on the same date, and in each case the Boers were able to bring very superior numbers into the field.

The first of these was the fight in which Colonel Doran's column extricated itself with severe loss from a most perilous plight. The whole force under Doran consisted of 350 men with two guns, and this handful was divided by an expedition which he, with 150 men, undertook in order to search a distant farm.

[520] Doyle, Arthur Conan, 1859-1930 The war in South Africa, its cause & conduct; 1902 McClure, Phillips & Co. U. Cal. Copy: Page 115 of the tract. https://archive.org/details/warinsouthafrica00doylrich

[521] Doyle, Arthur Conan. The Great Boer War 17th Edition, 1902 (Chapter 35) From Project Gutenberg: The Great Boer War by Arthur Conan Doyle - Free Ebook (gutenberg.org) Posting Date: February 1, 2009 [EBook #3069] Produced by Robert Laing, and Sue Asscher

The remaining two hundred men, under Captain Saunders, were left upon February 5th with the guns and the convoy at a place called Middlepost, which lies about fifty miles south-west of Calvinia. These men were of the 11th, 23rd, and 24th Imperial Yeomanry, with a troop of Cape Police.

The Boer Intelligence was excellent, as might be expected in a country which is dotted with farms. The weakened force at Middlepost was instantly attacked by Smuts's commando. Saunders evacuated the camp and abandoned the convoy, which was the only thing he could do, but he concentrated all his efforts upon preserving his guns.

The night was illuminated by the blazing wagons, and made hideous by the whoops of the drunken rebels who caroused among the captured stores. With the first light of dawn the small British force was fiercely assailed on all sides, but held its own in a manner which would have done credit to any troops. The much criticised Yeomen fought like veterans.

A considerable position had to be covered, and only a handful of men were available at the most important points. One ridge, from which the guns would be enfiladed, was committed to the charge of Lieutenants Tabor and Chichester with eleven men of the 11th Imperial Yeomanry, their instructions being 'to hold it to the death.'

The order was obeyed with the utmost heroism. After a desperate defence the ridge was only taken by the Boers when both officers had been killed and nine out of eleven men were on the ground. In spite of the loss of this position the fight was still sustained until shortly after midday, when Doran with the patrol returned.

The position was still most dangerous, the losses had been severe, and the Boers were increasing in strength. An immediate retreat was ordered, and the small column, after ten days of hardship and anxiety, reached the railway line in safety. The wounded were left to the care of Smuts, who behaved with chivalry and humanity.

At about the same date a convoy proceeding from Beaufort West to Fraserburg was attacked by Malan's commando. The escort, which consisted of sixty Colonial Mounted Rifles and 100 of the West Yorkshire militia, was overwhelmed after a good defence, in which Major Crofton, their commander, was killed. The wagons were destroyed, but the Boers were driven off by the arrival of Crabbe's column, followed by those of Capper and Lund. The total losses of the British in these two actions amounted to twenty-three killed and sixty-five wounded.

The re-establishment of settled law and order was becoming more marked every week in those south-western districts, which had long been most disturbed. Colonel Crewe in this region, and Colonel Lukin upon the other side of the line, acting entirely with Colonial troops, were pushing back the rebels, and holding, by a well-devised system of district defence, all that they had gained.

By the end of February there were none of the enemy south of the Beaufort West and Clanwilliam line. These results were not obtained without much hard marching and a little hard fighting. Small columns under Crabbe, Capper, Wyndham, Nickall, and Lund, were continually on the move, with little to show for it save an ever-widening area of settled country in their rear.

In a skirmish on February 20th Judge Hugo, a well-known Boer leader, was killed, and Vanheerden, a notorious rebel, was captured. At the end of this month Fouche's tranquil occupation of the north-east was at last disturbed, and he was driven out of it into the midlands, where he took refuge with the remains of his commando in the Camdeboo Mountains. Malan's men had already sought shelter in the same natural fortress. Malan was wounded and taken in a skirmish near Somerset East a few days before the general Boer surrender. Fouche gave himself up at Cradock on June 2nd.

The last incident of this scattered, scrambling, unsatisfactory campaign in the Cape peninsula was the raid made by Smuts, the Transvaal leader, into the Port Nolloth district of Namaqualand, best known for its copper mines. A small railroad has been constructed from the coast at this point, the terminus being the township of Ookiep.

The length of the line is about seventy miles. It is difficult to imagine what the Boers expected to gain in this remote corner of the seat of war, unless they had conceived the idea that they might actually obtain possession of Port Nolloth itself, and so restore the communications with their sympathisers and allies.

At the end of March the Boer horsemen appeared suddenly out of the desert, drove in the British outposts, and summoned Ookiep to surrender. Colonel Shelton, who commanded the small garrison, sent an uncompromising reply, but he was unable to protect the railway in his rear, which was wrecked, together with some of the blockhouses which had been erected to guard it. The loyal population of the surrounding country had flocked into Ookiep, and the Commandant found himself burdened with the care of six thousand people.

The enemy had succeeded in taking the small post of Springbok, and Concordia, the mining centre, was surrendered into their hands without resistance, giving them welcome supplies of arms, ammunition, and dynamite. The latter was used by the Boers in the shape of hand-bombs, and proved to be a very efficient weapon when employed against blockhouses. Several of the British defences were wrecked by them, with considerable loss to the garrison; but in the course of a month's siege, in spite of several attacks, the Boers were never able to carry the frail works which guarded the town.

Once more, at the end of the war as at the beginning of it, there was shown the impotence of the Dutch riflemen against a British defence. A relief column, under Colonel Cooper, was quickly organised at Port Nolloth, and advanced along the railway line, forcing Smuts to raise the siege in the first week of May.

Immediately afterwards came the news of the negotiations for peace, and the Boer general presented himself at Port Nolloth, whence he was conveyed by ship to Cape Town, and so north again to take part in the deliberations of his fellow-countrymen. Throughout the war he had played a manly and honourable part. It may be hoped that with youth and remarkable experience, both of diplomacy and of war, he may now find a long and brilliant career awaiting him in a wider arena than that for which he strove.

German West Africa

This map shows the successful invasion route taken by the British in 1915 after the suppression of the Martiz rebellion[522] Notice how close Port Nolloth is to the German colony's frontier. The port was small. It could not accommodate large warships. The British would be at a disadvantage. Germany, a very expansionist power, had decided to assist the Boers on this side of the continent.

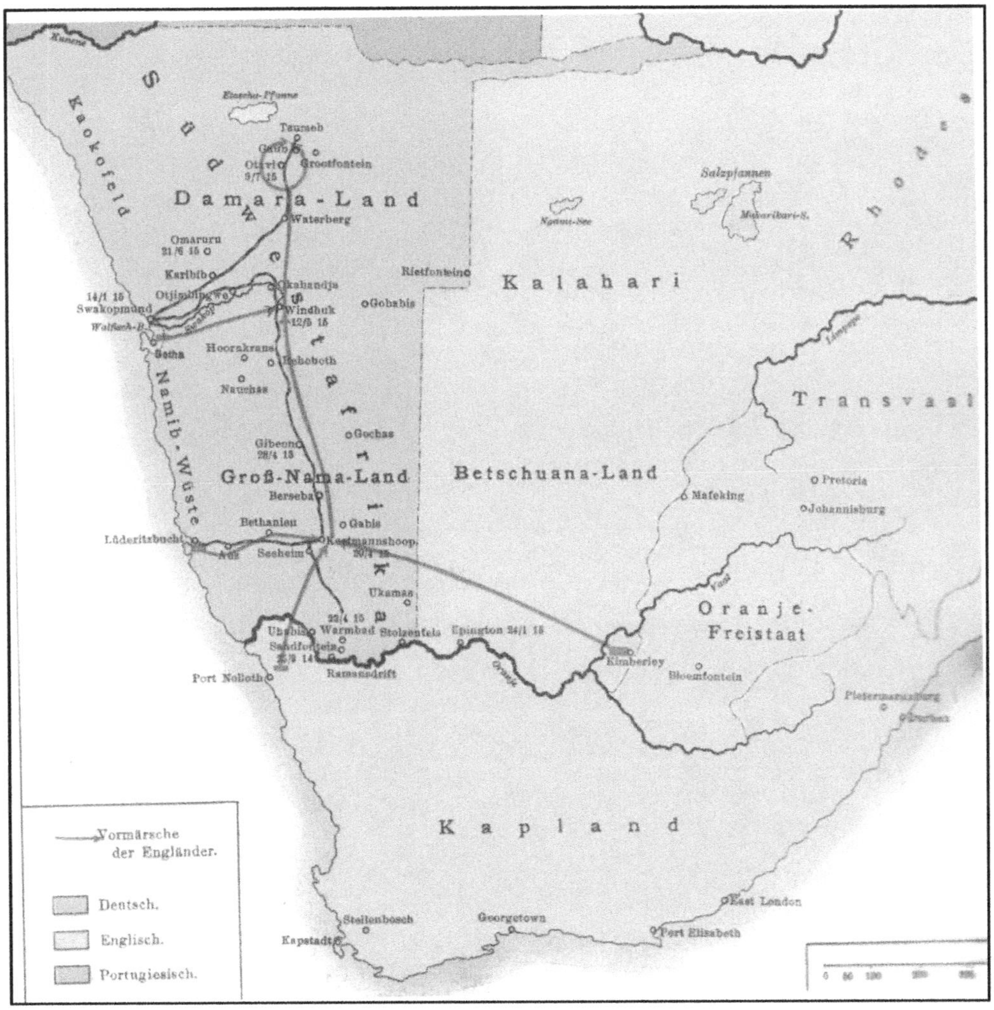

German West Africa.[574]

[522] (see Martiz Rebellion infra). The insurrectionists wanted to align with Germany at the start of the First World War and evict the British. Jan Smuts, Louis Botha and other Boer leaders sided with the British and were forced to confront and defeat their former comrades in arms, like Martiz and de La Rey.

Partial Chronology

Sunday, March 23, 1902 Arrival at Pretoria of Acting Transvaal Government to discuss the terms of peace.

Saturday, March 29, 1902: Total defeat of Beyers and dispersal of investing commando.

Monday, March 31, 1902: De le Rey defeated in engagement with Colonels Keir and Cookson.

Friday, April 4, 1902: O'okiep invested by Commandant Smuts.

Tuesday, April 8, 1902: Successful attack on Beyers' laager near Pietersburg.

Wednesday, April 9, 1902: Conference between Transvaal and Orange Free State leaders at Klerksdorp in regard to negotiations for peace.

Friday, April 11, 1902: Meeting of Boer representatives at Klerksdorp in relation to Peace movement.

Thursday, May 1, 1902: Relief of O'okiep by British troops under Colonels Cooper and Caldwell.

Concluding Material

Times History Discussion

From the 12th of May [1902] onwards the tension of war relaxed. From far and wide the delegates elected by the commandos travelled to the meeting-place.

Though the rules governing the armistice did not, of course, apply to Cape Colony, General Smuts, …was transported by sea from Port Nolloth to Cape Town and thence sent to Vereeniging. Nine hundred miles from Ookiep, Beyers left his ragged bands in the Waterberg to travel to the same destination. Chris Botha came from the Swazi Border, Van Heerden from the Zwartruggens, Meuwhoudt and Brand from the banks of the Orange, Wessels from the rushing rivers and wild ravines of the Drakensberg. From north, south, east and west of the vast theatre of war they came to decide their nation's destiny.

On the 15th of May, 1902, a cold, misty morning, the sixty delegates and the members of the two Governments came together at Vereeniging. Two camps, one for each [Republic].

In the appointment of the sub-committee the military element was eliminated, and the negotiations were left in the hands of Smuts and Hertzog, the legal advisers of the two Republics, and, on the British side, to Lord Milner and his legal adviser, Sir Richard Solomon.[523]

Even so, agreement was only secured by the suggestion that the acceptance of the British terms should be accompanied by a formal statement which made clear that the acceptance was under protest and due to the hopelessness of continuing the struggle.
The statement was drawn up by Hertzog and Smuts, and was moved as a substantive proposal at the afternoon session by Commandant H. P. Pretorius and seconded by General C. [Christian] Botha.[524] [525]

General Christian Botha.[575]

[523] The Times' History of the War in South Africa 1899–1902 by Amery, L. S. (Leopold Stennett), Volume 5 p. 592.

[524] Christiaan Botha, 6 October 1864 – 29 October 1902) was a younger brother of Louis Botha and likewise a Boer general in the Second Boer War who fought the British to the end. Chris Botha - Wikipedia

[525] The Times' History of the War in South Africa 1899–1902 by Amery, L. S. (Leopold Stennett), Volume 5 pages 604–606.

Boer Declaration

The Boer Declaration prior to signing the peace treaty in May of 1902.[526]

"We, the national representatives of both the South African Republic and the Orange Free State, at the meeting held at Vereeniging, from the 15th of May till the 31st of May, 1902, have with grief considered the proposal made by His Majesty's Government in connexion with the conclusion of the existing hostilities, and their communication that this proposal had to be accepted, or rejected, unaltered. We are sorry that His Majesty's Government has absolutely declined to negotiate with the Government of the Republics on the basis of their independence, or to allow our Government to enter into communication with our deputation. Our people, however, have always been under the impression that not only on the grounds of justice, but also taking into consideration the great material and personal sacrifices made for their independence, that it had a well-founded claim for that independence.

"We have seriously considered the future of our country, and have specially observed the following facts:

"Firstly, that the military policy pursued by the British military authorities has led to the general devastation of the territory of both Republics by the burning down of farms and towns, by the destruction of all means of existence, and by the exhausting of all resources required for the maintenance of our families, the existence of our armies, and the continuation of the war."

"Secondly, that the placing of our families in the concentration camps has brought on an unheard-of condition of suffering and sickness, so that in a comparatively short time about twenty thousand of our beloved ones have died there, and that the horrid probability has arisen that, by continuing the war, our whole nation may die out in this way."

"Thirdly, that the Kaffir tribe, within and without the frontiers of the territory of the two Republics, are mostly armed and are taking part in the war against us, and through the committing of murders and all sorts of cruelties have caused an unbearable condition of affairs in many districts of both Republics. An instance of this happened not long ago in the district of Vryheid,[527] where fifty-six burghers on one occasion were murdered and mutilated in a fearful manner."[528]

"Fourthly, that by the proclamations of the enemy the burghers still fighting are threatened with the loss of all their movable and landed property—and thus with utter ruin—which proclamations have already been enforced."

"Fifthly, that it has already, through the circumstances of the war, become quite impossible for us to keep the many thousand prisoners of war taken by our forces, and that we have thus been unable to inflict much damage on the British forces (whereas the burghers who are taken prisoners by the British armies are sent out of the country), and that, after war has raged for nearly three years, there only remains an insignificant part of the fighting forces with which we began."

"Sixthly, that this fighting remainder, which is only a small minority of our whole nation, has to fight against an overpowering force of the enemy, and besides is reduced to a condition of starvation, and is destitute of all necessaries, and that notwithstanding our utmost efforts, and the sacrifice of everything that is dear and precious to us, we cannot foresee an eventual victory."

"We are therefore of opinion that there is no justifiable ground for expecting that by continuing the war the nation will retain its independence, and that, under these circumstances, the nation is not justified in continuing the war, because this can only lead to social and material ruin, not for us alone, but also for our posterity. Compelled by the above-named circumstances and motives, we commission both Governments to accept the proposal of His Majesty's Government, and to sign it in the name of the people of both Republics."

[526] The Times' History of the War in South Africa 1899–1902 by Amery, L. S. (Leopold Stennett), Volume 5 pages 604–606.

[527] Vryheid - Wikipedia And:

[528] The Battle of Holkrans or Holkrantz (6 May 1902) took place at Holkrans, near Vryheid. To stop the raids operated by the Boers on their cattle and crops as a punishment for helping the British, the leader of the abaQulusi, Sikhobobho, responded by assembling his forces and attacked the commando camp at the foot of Holkrans hill. The Zulus attacked at night, and in a mutual bloodbath, the Boers lost 56 killed and 3 wounded, while the Zulus suffered 52 killed and 48 wounded. This battle was the last victory of the abaQulusi to date Battle of Holkrans - Wikipedia

"We, the representative delegates, express our confidence that the present circumstances will, by accepting the proposal of His Majesty's Government, be speedily ameliorated in such a way that our nation will be placed in a position to enjoy the privileges to which they think they have a just claim, on the ground not only of their past sacrifices, but also of those made in this war."

"We have with great satisfaction taken note of the decision of His Majesty's Government to grant a large measure of amnesty to the British subjects who have taken up arms on our behalf, and to whom we are united by bonds of love and honour; and express our wish that it may please His Majesty to still further extend this amnesty."

The motion was carried by fifty-four votes to six. The irrevocable decision had been made. The time for speech making had gone, but Schalk Burger closed the meeting in a few sentences of simple eloquence.

"We are standing here at the grave of the two Republics. Much yet remains to be done, although we shall not be able to do it in the official capacities which we have formerly occupied.

Let us not draw our hands back from the work which it is our duty to accomplish. Let us ask God to guide us, and to show us how we shall be enabled to keep our nation together. We must be ready to forgive and forget whenever we meet our brethren.

That part of our nation which has proved unfaithful we must not reject."

The Commissioners travelled to Pretoria, and on the same night, at five minutes past eleven, only an hour before the expiry of the term of grace, the Treaty was signed.

Let us Shake Hands

"We are good friends now," said Kitchener, as he shook hands with the representatives of the late Republics.[529]

Treaty of Vereeniging Terms

Quote from The Great Boer War by Arthur Conan Doyle.[530] [531] [532]

On May 31st this resolution was notified to Lord Kitchener, and at half-past ten of the same night the delegates arrived at Pretoria and set their names to the treaty of peace. ....

The following were in brief the terms of surrender:—

1. That the burghers lay down their arms and acknowledge themselves subjects of Edward VII.
2. That all prisoners taking the oath of allegiance be returned.
3. That their liberty and property be inviolate.
4. That an amnesty be granted—save in special cases.
5. That the Dutch language be allowed in schools and law-courts.
6. That rifles be allowed if registered.
7. That self-government be granted as soon as possible.
8. That no franchise be granted for natives until after self-government.
9. That no special land tax be levied.
10. That the people be helped to reoccupy the farms.
11. That 3,000,000 pounds be given to help the farmers.
12. That the rebels be disfranchised and their leaders tried, on condition that no death penalty be inflicted.

These terms were practically the same as those which had been refused by Botha in March 1901. [At the Middleburg Conference] Thirteen months of useless warfare had left the situation as it was."

Peace Treaty of Vereeniging original documents.[533] And.[534]

[529] The Times' History of the War in South Africa 1899–1902 by Amery, L. S. (Leopold Stennett), Volume 5 page 606.

[530] Doyle, Arthur Conan. The Great Boer War 17th Edition, 1902 with the two volumes as one document: The Great Boer War by Arthur Conan Doyle - Free Ebook (gutenberg.org) Chapter 39. Pages 232–234 of the Word document.

[531] General Article: Treaty of Vereeniging - Wikipedia

[532] Text: Peace Treaty of Vereeniging - transcript | South African History Online (sahistory.org.za)

[533] Peace Treaty of Vereeniging - original document | South African History Online (sahistory.org.za)

Kitchener's Final Despatch

Despatch by General H. Kitchener, dated 23rd June, 1902, relative to Military Operations in South Africa.[535]

From Lord Kitchener to the Secretary of State for War. Cape Town, 23rd June, 1902.

Sir, I have the honour to submit a final despatch which covers the period between the signature of the terms of peace at Pretoria and my departure from Cape Town. ....

2. In my last despatch dated 1st June, I had the pleasure of reporting the conclusion of peace. Directly the terms were signed I appointed the following three Peace Commissioners to receive the surrenders in the territories of the new Colonies:

General Bruce Hamilton for the Eastern Transvaal.

General W. [Walter] Kitchener for the Western Transvaal.

General Elliot[536] for the Orange River Colony.

I also directed General French to make the necessary arrangements for the same purpose in the Cape Colony.

3. The reports of General French and the three Peace Commissioners are herewith submitted. .... The Boer leaders concerned, played their part with loyalty and in a thoroughly straightforward manner, whilst the burghers and rebels on commando also accepted the situation in a manly spirit which augurs well for the future of South Africa. So strongly did I feel this that I sent the following telegram to Generals Louis Botha, De la Rey, and Christian de Wet:

> "Now that the surrender of the arms in the hands of the burghers in the Transvaal and Orange River Colony has been completed, I desire to place on record my high appreciation of the unflagging energy and unfailing tact with which you have facilitated the work of my Commissioners. I recognize how much it is due to your exertions that the burghers have displayed everywhere such a loyal spirit in accepting the change of Government that has taken place, and I can inform you that the manner in which the burghers have come in has greatly pleased His Majesty, the King, and deeply impressed the British people, who are heartily anxious to welcome them as fellow citizens. I feel confident that a new era of complete reconciliation between all races has now dawned in South Africa."[537]

The total numbers of the surrenders of armed burghers [were]

in the Transvaal were	11,166
in the Orange River Colony	6,455
in the Cape Colony	3,635
Total	21,256

4. I well understand that the accompanying list, ... who have in my opinion rendered exceptionally good service during the campaign, is incomplete, ... there must necessarily be a proportion of Officers omitted, who have rendered equal or even better service than those who have been mentioned. This, however, is the fortune of war, and will, I am sure, be well understood by the Army I have been so proud to command. What follows are 21 pages of lists of individuals and units and their service.]

[534] Treaty Page 1 File:Vereeniging1.jpg - Wikimedia Commons
Treaty Page 2 File:Vereeniging2.jpg - Wikimedia Commons
Treaty Page 3 File:Vereeniging3.jpg - Wikimedia Commons
Treaty Signatures 2 File:Vereeniging4.jpg - Wikimedia Commons

[535] Proclamations Issued In South Africa: Vol II p.757 https://babel.hathitrust.org/cgi/pt?id=hvd.32044089026009&seq=757&view=1up

[536] Lieutenant General Sir Edward Locke Elliot, KCB, KCIE, DSO (28 January 1850 – 12 December 1938) was a British Army General during the Second Boer War and in British India. Edward Locke Elliot - Wikipedia

[537] At this time the Dutch and English were thought of and considered different 'races.' It was also British policy and thought that the conclusion of the conflict would materially improve the condition of all people in South Africa.

Crowds in Johannesburg celebrating peace. Dated 26 June 1902.[576] Note the white Arch.

Repatriating St. Helena Prisoners

Partial Chronology

5th June 1902: The St Helena Guardian leads with the headline 'Peace, Perfect Peace,' following the end of the Boer War.

8th June 1902: Churches across the island hold special services to celebrate the end of the War.

26th June 1902: The first Boer Pow's leave St Helena for Cape Town aboard the Canada.

3rd October 1902: 95 Boer Pow's are reported to have died while on St Helena, mostly from enteric fever.

21st October 1902: The last of the pardoned Boer Pow's leave St. Helena for Cape Town aboard the Golconda.

Prisoners Departing St. Helena.[538]

Graves of Prisoners on St. Helena.[539]

[538] Text and Image Source: Boer PoWs | Saint Helena Island Info: All about St Helena, in the South Atlantic Ocean

[539] The original monuments can be seen at the bottom. Image source: Boer PoWs | Saint Helena Island Info: All about St Helena, in the South Atlantic Ocean

Costs of Second Boer War

The total cost of the War was approximately £211 million. It is not possible to give an exact figure because the costs were borne from many sources. For example, the colonial units were initially funded by their source governments. However, the figures that follow give a reasonable estimation.[540 541]

Costs of Second Boer War		GBP	1900 US Dollars	USD
		2022 Value	£ * 5 = USD	2022 Value
1899 - 1900	£23,000,000	£3,516,046,041	$115,000,000	$4,594,065,757
1900 - 1901	£63,737,000	£9,743,575,066	$318,685,000	$12,730,955,181
1901 - 1902	£67,670,000	£10,344,818,939	$338,350,000	$13,516,540,426
1902 - 1903	£47,500,000	£7,261,399,433	$237,500,000	$9,487,744,499
Sub Total	£201,907,000	£30,865,839,478	$1,009,535,000	$40,329,305,862
Interest	£9,249,000	£1,413,909,123	$46,245,000	$1,847,413,660
Total	£211,156,000	£32,279,748,601	$1,055,780,000	£41,963,673,181

Repatriation - Times History Vol 6 Chapter II

The cessation of hostilities [In June of 1902] was, in one sense, merely an incident in the progress of the work of reconstruction.[542]...........................

The actual terms, … of the treaty of surrender contained nothing that involved the slightest deviation from the settled policy of the British Government. But the fact that there were terms-terms formally asseverated by the British representatives and formally agreed to by the Boer leaders-created an entirely new situation.

The most obvious feature of that situation was the explicit and definite acceptance of the British flag by the Boer leaders. The surrender of Vereeniging gave to the annexations, hitherto resting on force alone, the validity of a mutual compact, however reluctantly that compact may have been signed. It meant not only an earlier peace but an enduring peace. For who could henceforward dare to summon the Boers to a renewal of the struggle for a cause which its most devoted and stubborn champions had formally abandoned? Vereeniging was the grave not only of the two Republics but of the Republican idea. ....

The surrender definitely established the commandos left in the field as the predominant section, to which the rest of the Boer population would inevitably gravitate; it established the Boer generals as the recognized political leaders of their people.

While making an end of the opposition in arms, it created and endowed with leaders and with organization, a new political opposition with which the work of reconstruction would have to reckon. ....the Boer leaders devoted themselves whole-heartedly to carrying out their obligations under the treaty of surrender.

Even before they left Vereeniging on May 31, Botha and Schalk Burger issued an open letter to their burghers, thanking them for their heroism and enjoining on them to acquiesce in the peace and to obey and respect the new Government.

The same letter also announced the formation of a head committee, representing both the late Republics, entrusted with the task of collecting money to provide for the needs of the widows and orphans, and concluded with an earnest appeal for unity and mutual forbearance in the great struggle for the spiritual and social welfare of their people which lay before them.

540 Adapted from various sources but primarily from: https://www.angloboerwar.com/other-information/16-other-information/1845-cost-of-the-war And: Second Boer War - Wikipedia For a detailed list of expenses by type see the Appendix Volume.

541 United States Dollar Converter: https://www.usinflationcalculator.com/ This calculator runs from the creation of the Federal Reserve System in the US in 1913. Results from 1913 for Inflation to 2022 = 2856.10% Multiplicand: 29.562.
GB Pounds Converter: UK Values from: https://www.thisismoney.co.uk/money/bills/article-1633409/Historic-inflation-calculator-value-money-changed-1900.html Values from 1900 to 2022 use a multiplicand of: 152.87157.
2024 Exchange Rate for GBP to USD: 1USD = .77 GBP And: 1 GBP = 1.33

542 This material is from Times History, Vol 6 Chapter 2 Pages 39-61. Some material has been omitted as noted in the text.

On Monday, June 2, Lord Kitchener and the Boer leaders went down to Vereeniging. Crowded together in a large tent the Boer delegates listened with eager and friendly curiosity to …[him].

Briefly eloquent General Beyers translating sentence for sentence-Kitchener praised their courage and endurance, dwelt on the prospects of the future, and explained to them their immediate duty in connexion with the surrender of the commandos.

A complete programme of surrenders was then drawn up in consultation with the Boer leaders. A few hours later leaders and delegates dispersed to their districts. During the next three weeks the surrenders proceeded steadily and without a hitch, thanks largely to the loyal co-operation of the Boer leaders.

There were practically no absentees, in the new colonies at any rate, though not a few Cape rebels managed to evade surrender, while others made their way across into German territory.

In all, 11,166 surrendered in the Transvaal, 6,455 in the Orange River Colony and 3,635 in Cape Colony, or a grand total of 21,256. The surrenders formed a fitting conclusion to a struggle unique, no less for its good nature and informality than for its stubbornness.

A commando or group of commandos would assemble at some selected spot, where it would be met by a British force. After speeches from their own leaders, and from the British general deputed to meet them, the burghers proceeded to hand over their rifles-mostly captured British Lee Metfords, it must be confessed-signed the oath of allegiance or the equivalent declaration, and then passed on to draw their rations from the Army Service Corps and to fraternize with their old adversaries.

Columns and commandos which had hunted and eluded each other for months met like friends long separated, and many an engagement was fought out again over the camp fire.

Throughout the Empire the tidings of peace was received with profound emotion. Only when the war was really at an end could men realize how long and heavily it had weighed upon their minds. In other countries the news met with vague and puzzled acquiescence.

The generosity of the British terms, the friendliness of the Boers towards their conquerors, were equally unintelligible to those who had pictured a war of conquest waged with brutal severity and resisted with unquenchable hate.

There was one at least to whom the news must have been bitter. At Utrecht, in Holland, old President Kruger learnt that the struggle to which his life had been given was over, and that his voice had ceased to count in the councils of his people.

On June 8, a solemn military peace celebration was held at Pretoria and Bloemfontein.

On the 20th, Lord Kitchener[543] resigned the South African command into the hands of General Lyttelton, and left Pretoria to spend a few months of well-earned rest in England before taking up the command which had been waiting for him in India.

On the following day the now Crown Colony Government of the Transvaal was formally inaugurated in the old Raadzaal by the issue of the Letters Patent[544] and the swearing-in of the Governor now Viscount Milner-and his Executive Council. On the 23rd the same ceremony took place at Bloemfontein.

[543] Lord Kitchener received a. viscounty and II grant of £50,000 for his services in concluding the war.

[544] Letters patent (always in the plural), in the United Kingdom, are legal instruments generally issued by the monarch granting an office, right, title or status to a person (and sometimes in regards to corporations and cities). Letters patent have also been used for the creation of corporations or offices, for granting city status, for granting coat of arms, and for granting royal assent. https://en.wikipedia.org/wiki/Letters_patent_(United_Kingdom)

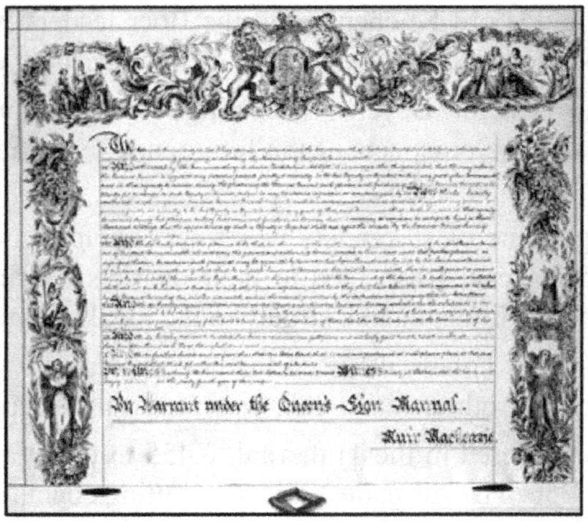

Letters patent issued by Queen Victoria in 1900, creating the office of Governor-General of Australia as part of the process of federation.[577]

The 25th was fixed for the celebration throughout the Empire of the coronation of King Edward VII, but a sudden and dangerous illness postponed the great Imperial ceremony to August 9.

The surrenders were immediately followed by the establishment of the framework of British administration in the country districts.

Everywhere the new resident magistrates hurried out to their posts, and from the ruin or the tent which served them for an office proceeded to administer justice and maintain order. In this task they were admirably seconded by the South African Constabulary,[545] who were released from military duty in the middle of June, and distributed in small detachments all over the country. One of their first tasks was the disarmament of the natives, many of whom had possessed themselves of rifles, while some had been directly given arms by the British authorities for the defence of their districts.

Contrary to expectation, the task proved an easy one. More difficult and never satisfactorily ·solved was the question of restoring to their rightful owners the stock looted by natives from Boers, and by Boers from natives. Even after the peace stock-thieving proved by far the most serious crime the constabulary had to deal with. Otherwise its efforts were less required for the prevention of crime than for giving general assistance in the work of repatriation, which, for the moment, thrust all other issues into the background.

It was, in truth, an extraordinary and unparalleled task which now confronted the British administration. On its hands were over 200,000 men, women, and children whom it had undertaken to bring back to their homes, and for whose subsequent welfare it was morally responsible.

Of these not more than 30,000, including women and children, represented the remnant of the Boer nation still at large at the end of the war. Of the rest, 24,000 were prisoners in camps overseas, 7,000 prisoners in South Africa or on parole, 4,500 serving with the British forces in one capacity or another, 110,000 inmates of concentration camps, and the balance living on their own means or on Government relief in the various towns.

The whole of this population was to be planted out and helped till it could support itself in a country which the war had converted into an absolute wilderness. From end to end of the two colonies there was hardly a farmhouse left standing. In large areas the live stock had been almost exterminated. Except for a few dams and

[545] The South African Constabulary (SAC) was a paramilitary force set up in 1900 under British Army control to police areas captured from the two independent Boer republics of Transvaal and Orange Free State during the Second Boer War.
Once hostilities were over, each troop took up its assigned position in the two new colonies as soon as possible. In this way a network of posts and patrols was established in a very short space of time. With the SAC patrolling in every direction, including the Portuguese and Tongaland frontiers, they also visited all the farms at least once a week. By the beginning of August 1902, 28 districts, 64 sub-districts and 210 stations across South Africa were occupied and the force had over 10,000 men.
In November 1902, the size of the force was reduced to 6 000 men and after further reductions it was down to 4,000 men in 1906. During this time, a commission of enquiry was appointed to look into the administration and organisation of the SAC. The commission divided the force into two divisions in the Transvaal and the Orange River Colony and changed the military ranks into civilian ones, for example captains became inspectors and lieutenants became sub-inspectors. By May 1908, after further reductions, there were 1,068 police of European ancestry in the Transvaal and 674 in the Orange River Colony. https://en.wikipedia.org/wiki/South_African_Constabulary

occasional fences which had escaped destruction the whole apparatus of rural civilization had practically been wiped out. The first stage in the process of repatriation was one of sorting out and preparatory concentration.

For some time before the peace transfers had been going on between the different concentration camps in order to bring each family into the district to which it belonged.

The moment peace was signed the camps became the fixed bases and organizing centres of the whole work of repatriation. Into them flocked the surrendering burghers, well content to enjoy the luxury of tents and regular meals, delighted to rejoin their families, and surprised to find them so well cared for and on such excellent terms with the British.[546].

Into them also flocked, to rest and recover strength, the starving remnants of the women and children left on the veld. Later on, in a steady stream, came the prisoners. Thus, in spite of the progress of the work of repatriation, the total population of the Transvaal and Orange River Colony camps showed,

As a purely military measure the camps were no doubt a mistake. But they proved invaluable for the work of repatriation-in fact, if they had not existed already they would have had to be invented for that purpose. The return of the prisoners was a task of considerable difficulty. Those in South Africa were released as soon as the surrender of the commandos was completed, on signing the oath or equivalent declaration.

Thanks to the exertions of Mr. Schalk Burger, who visited the Natal camps, and of other influential leaders, any hesitation that was at first felt about signing was dispelled.

In the over-sea camps, it was very different. With characteristic suspiciousness the majority at first believed that the news of peace, and even the telegrams from the Boer generals, were merely cunning British devices to entrap them into signing the oath, and their fears were played upon by a small section of genuine irreconcilables. It was not till the middle of August, in fact, that the over-sea prisoners began to sign in any numbers.

The first transports, consequently, brought back very few, except those who had already before the peace openly accepted the situation.[547]

These were entitled to priority in any case, the order of return for prisoners being fixed by their disposition towards the new Government, with a further preference to married men and owners of property. On arrival in South Africa the prisoners went to the rest camps at Simonstown or at Umbilo, near Durban, and were fitted out with clothes, blankets, and other personal necessaries, before being sent off by train to the concentration camps in their respective districts.

By the beginning of 1903, all were brought back except some 900, mostly in India, who still refused to sign.

A visit from two leading Boers satisfied many of these, but there still remained some 500 who continued unconvinced, until General De la Rey visited them in person at the end of the year.

The task of repatriation necessitated the improvisation in each colony of an enormous temporary organization, only the rudiments of which were in existence before the peace.

In each colony the organization consisted of a central department which was responsible for general policy and for the purchase and distribution of supplies and transport of all kinds, and of local commissions, presided over by the resident magistrates, and comprising the Commandant of the Constabulary and three or four leading residents of each district, which estimated the requirements of the burghers and supervised the work of relief.

The actual restoration of the people to their homes was carried out by a gigantic field transport based on a vast network of depots. ....

In the Transvaal, the Central Board, after settling some of the more general problems connected with repatriation, fell into abeyance, and the work of the department was carried on by Captain A. M. Hughes, under the general supervision of Mr. Duncan, the Treasurer. Again, in the Transvaal, a representative of the

[546] "The attitude of the surrendering burghers was one of surprise and relief to find their families so well cared for ill the camps. The most extravagant and false rumours had prevailed among the commandos in the field as to the mortality and suffering among the inmates of the burgher camps, and when they learnt the truth large numbers hastened to express their gratitude to the local superintendents.

The excellent feeling which was manifested between the women and children of the camps and the camp staffs, more especially the matrons and nurses; was a revelation to the husbands and brothers from the veld. "Report of Director of Transvaal Burgher Camps, Cd. 1553, p. 5.

In many cases surrendering burghers and others were rationed from the camps without actually becoming inmates. no appreciable diminution for fully three months after the peace.

[547] Several hundred of these had expressed their willingness to fight on the British side.

surrendered Boers or of the National Scouts was attached to each of the local commissions, whereas in the Orange River Colony the repatriation of those Boers who had espoused the British cause was conducted by a separate branch of the department.

On the whole, the local commissions proved a distinct success, especially in the Orange River Colony, where they were given much greater latitude in fixing prices and settling other details of the work. The Boer members were of the very greatest practical assistance, quite apart from the excellent political effect of enlisting their active co-operation with the new Government at the outset. The working staff of the Repatriation Departments, composed mainly of soldiers and ex-irregulars, was of very varying efficiency, as was inevitable considering the circumstances under which it was improvised and the purely provisional character of the appointment offered.

On the whole, the repatriation was much more efficiently and economically conducted in the Orange River Colony than in the Transvaal.

In Colonel MacLaughlin the department had an exceptionally energetic and capable chief, while, in the absence of the many pressing problems which distracted the Transvaal authorities, Sir H. Goold-Adams was able to give himself whole-heartedly to the task of supervising the work and seeing that the fullest use was made of all available resources.

In the Transvaal Mr. Duncan, in the heavy rush of other administrative work, was unable to give sufficiently close and continuous attention to repatriation, with the result that there undoubtedly was an enormous amount of wasteful expenditure which might have been kept down by more effective supervision and greater business experience on the part of some of the officers concerned. But repatriation must not be judged as an ordinary business operation.

It was a gigantic campaign waged for two whole years against famine, drought, locust and murrain,[548] and like other campaigns begun under conditions of great urgency and prolonged by unforeseen developments, was inevitably costly.

The fact remains that the Repatriation Departments in both colonies carried through an immense task in the face of extraordinary difficulties, and carried it through with success.

According to Article X of the Vereeniging terms the local commissions would have at their disposal for repatriation the free grant of £3,000,000, in the distribution of which they were to take into account the evidence of war losses afforded by Boer notes or receipts.- They were also to make advances for the same purpose, free of interest for two years, and afterwards paying 3 per cent., or less than half the current South African rate. The purport of the article was by no means clear, and a pedantic interpretation of it might have paralyzed the whole work of repatriation by postponing the distribution of relief to the assessment of war losses, or might have excluded altogether those who could produce no evidence of war losses, and had no security to offer for a loan.

Milner decided to carry out the task of repatriation first, limiting its scope by no other consideration than the needs of the population, and to fit in the interpretation of Article X afterwards.

Every ex-burgher was to be set on his feet again. Subsequent investigation should determine whether the assistance given was to be reckoned against the free grant of £3,000,000, or treated as a loan, or simply written off' owing to the recipient's inability to repay. This decision opened the door to a vast additional expenditure, and subsequently furnished a handle for no small amount of misrepresentation. But its wisdom and generosity cannot be questioned.

The moral responsibility of the new administration could not be limited by the terms of the treaty. A literal fulfilment of its terms would have created widespread distress, if not starvation, and would have aroused a bitter resentment throughout the people, compared with which the vigorous abuse of the repatriation work by the Boer leaders was a mere harmless and conventional expression of political disagreement.

It was the same sense of moral responsibility which led Supplying to the further decision that the Repatriation Departments the people. should not only convey the people to their homes and provide them with money for the purchase of necessaries, but should also undertake the task of purchasing and directly supplying all that the people required. Here, again, the right thing was done with complete indifference to the difficulties and the risk of misinterpretation involved. Except for a certain amount of stock and seed corn which the civil administration

[548] The noun is derived from Late Middle English morein, morine, moreyn ("(widespread) death; widespread sickness, plague; fatal disease; carnage; carrion" cattle disease.)

had been able to save for repatriation, practically the whole of the supplies, stock and transport in the two colonies was in the hands of the Army.

The railways were almost monopolized by the troop movements consequent on the general demobilization, and getting through supplies from the coast was a matter of no small difficulty. Kitchener was determined, in the interests of military finance, to exercise his monopoly to the fullest. Such contractors as could have got supplies and stock up country in spite of him would have been no less interested in combining to force up all prices to the most exorbitant rates. To have left the Boers at the mercy of the "open market" at such a time would have been a criminal violation of the whole spirit, though not of the letter, of the British covenant.

Milner alone was in a position to make even a tolerable bargain with Kitchener and the War Office. After weeks of desperate haggling he secured some 5,500 vehicles, [Wagons] including 47 traction trains, 80,000 oxen, mules, and donkeys, and the whole of the blockhouses and wire entanglements for £1,391,000, a first instalment of a total of £3,521,000 of Army stores eventually taken over for repatriation purposes.

Even so some of the charges, more particularly for the first lot of mules and oxen, were inordinately high, considering the condition of the animals. In the matter of food-stuffs, indeed, the military demands at that moment were so exorbitant· that the bulk of the supplies required were imported. Enormous quantities' of implements, building materials, transport animals, and live stock of all kinds were also bought by skilled buyers for the Repatriation Departments in the other colonies or imported from oversea.

There was no conceivable requirement of the returning population, from ox-wagons to wooden legs, which the departments did not manage to supply. Here, again, the whole weight of the civil administration had continually to be thrown into the scale in order to secure railway facilities. All these articles were offered to the Boers at cost price. There was no profit, no charge for administration, or for the enormous expenses involved in conveying the bulk of these articles out to scattered farms, often over a hundred miles from the railway. There was no charge for the keep of animals before they were disposed of, though in some instances a proportion was added to the price to cover casualties.

In some cases even, where the cost price was excessive, as, for instance, in the case of the first lot of mules purchased from the Army, the issue price was at a heavy reduction. There was no compulsion upon the Boers to purchase from the department. They could always buy for themselves, taking the money on loan from their local commission, and eyen receiving free railway passes to Cape Colony and Natal for the purpose. But except in the case of the more prosperous farmers, who wished to select a picked team of oxen or a particularly choice breed of sheep, the vast majority found it more convenient to buy from their repatriation depot.

Repatriation prices were, in fact, on the whole, below the then current market prices. They were, without a doubt, far below the level which those prices would have reached if matters had been left to themselves, and an unorganized mass of ignorant peasants, filled with impatience to get back to their homes, had been handed over to the tender mercies of the inevitable ring of speculators. It is true the prices of transport animals were considerably above the average of ordinary years of small demand and abundant supply. But then the end of the war was not a normal period. All these considerations, however, could hardly be expected to appeal to the ordinary Boer, who only knew that some of the prices charged were more than he had been accustomed to, and naturally grumbled. And the legend that the Boers were robbed of a large share of the promised relief by the extortions and extravagance of the Repatriation Departments was much too useful politically not to receive sedulous propagation from the Boer leaders.

The restoration of the people to their homes was undertaken in the face of almost inconceivably adverse circumstances. was midwinter, to begin with, and the veld was bare of all sustenance for the transport oxen, which in the main were the worn-out surplus of the Anny transport. The mules were in a no less wretched condition, and the state of the transport animals was equaled by that of the wagons. As a climax to all these difficulties glanders,[549] which had been kept within limits by the exertions of the Army Veterinary Department

[549] An infectious disease of horses, mules and donkeys caused by the bacterium Burkholderia, one species of which may be transmitted to humans. https://en.wikipedia.org/wiki/Glanders

From the Middle Ages to the 1900s, glanders was a significant threat to armies.[8][1] Before the Battle of Blenheim in 1704, glanders may have afflicted and greatly diminished the horses of Marshal Tallard's cavalry, helping the Duke of Marlborough win the battle.[9] Glanders was a significant problem for civilian use of horses, as well. In the 18th-century veterinary hospital at the École Nationale Vétérinaire d'Alfort, glanders was the most common disease among their equine patients and the one most likely to cause death.[10]

during the war, broke out with disastrous effect among the mules, and crippled the whole work for months, till the small veterinary staff of the Repatriation Departments could cope with it.[550]· ....

It is easy to say that the danger should have been foreseen and provided against. It could have been done no doubt, but only at the cost of serious delay, and at the moment the one supreme need seemed to be that of getting the people established on their farms before the ploughing season. In spite of all difficulties, however, the work went on. As transport was available each family was sent out supplied from the camp with tent, bedding, and a month's rations[551] and from the adjoining repatriation depot with tools, agricultural implements, seeds, and such vehicles and live stock as could be furnished. A repatriation officer[552] has given a vivid account of the proceedings:

"Day by day, and all day long for many weeks, clouds of dust were seen to rise as the long spans of oxen and mules, dragging wagon after wagon laden with food, tent, and a strange assortment of domestic articles, crawled snake-like out of the canvas town on to the illimitable veld, wending their way to some distant farmstead. The loading has not been so easy a task as might be supposed. The conductor knows that his wagons are not of the best, and his animals none of the strongest. He cannot contemplate with equanimity a breakdown on the open veld.

The camp superintendent, ever mindful of the welfare of his protégés, insists upon the absolute necessity of taking the whole of the first equipment-the first month's rations, bedding, and a tent against the restoration of the homestead. But the old' vrouw' has a host of broken furniture, in addition to the food-stuffs she has managed to save from liberal camp rations. Her man has his curios, collected or manufactured during his imprisonment oversea.

The children are loth to leave behind their books and toys, for which they have acquired a taste in the camp school. There have been hot debates as to essentials and non-essentials. "But eventually, by hook or by crook, and by dint of much rearranging, room has been found for all. The stodgy, expressionless, sallow-complexioned, but withal hospitable and kindly women crawl, not without difficulty and danger of upsetting, to the top of the motley pile, and at length a start is made. The crack of the long whip and the familiar shouts of the native driver as he urges forward his cattle ring out clearly on the sunlit air." On arrival at the homestead the goods were dumped down, the repatriation wagons lumbered away for their next job, and the family found themselves alone with the heap of ruins which represented their old home.

For the first time the Boers realized the full meaning of the guerilla war, and passed through the depths of despondency when faced with its results. But, for the most part, they pulled themselves together manfully, and, with great energy and resource, set to work patching up habitable rooms out of such walls as were still standing, repairing fences, dams, and kraals, and getting ready for the ploughing. From time to time they visited the nearest repatriation depot to draw a fresh supply of rations, and to inquire anxiously when ploughing cattle would be available.

Every effort was made to distribute these. But the enormous demand for transport purposes stood in the way, and in spite of every effort to reduce transport requirements, in spite of large additional purchases from the military and from outside the new colonies, a serious shortage was inevitable. The difficulty was met by the bold but simple device of sending round Government teams to plough, at a very low charge, a certain number of acres on every farm which had no team of its own, including those farms whose owners had not yet returned.

In this way enough was ploughed by the end of the year to provide a crop which might be calculated to keep the people going for the greater part of 1903. Nothing that the British Government had done hitherto produced so deep an impression on the Boer mind, and inspired a genuine a feeling of gratitude and confidence as the thoughtfulness shown in this matter.

Due to the high mortality rate in humans and the small number of organisms required to establish infection, B. mallei is regarded as a potential biological warfare or bioterrorism agent, as is the closely related organism, B. pseudomallei, the causative agent of melioidosis. During World War I, glanders was believed to have been spread deliberately by German agents to infect large numbers of Russian horses and mules on the Eastern Front.[

550 "The district reports for many months were merely repetitions of the sa.me story-plans frustrated, hopes blighted, supplies scarce, transport disorganized, ploughing impossible. Repatriation was frequently a thankless task; disease made it appear at times It. hopeless one."-G. E. Beak: "The Aftermath of War."

551 These were charged to camp funds, and thus came out of the pocket of the British taxpayer and not out of repatriation.

552 G. B. Beak: "The Aftermath of War."

Throughout, even when the outlook was blackest, the attitude of the people towards the Government had not been otherwise than friendly. But it was not till after the ploughing that the despondency of the first few months began to pass away, and that they began to look more happily towards the future. Closely connected with the repatriation of the white population and, indeed, from the South African point of view, an essential part of it, was the repatriation of the farm natives, the greater part of whom had been collected into the native camps or had taken service with the British. It was at first feared that there might be trouble over this. The natives had in many instances become insolent, owing to unduly high wages and to the familiarity with which the soldiers had treated them.

They expected the Boers to be treated as a conquered race, to whom they would no longer stand in a dependent relation. But they soon discovered that the British conquest, though it might give the black man greater security against oppression and more clearly defined rights, involved no essential alteration in the superior status of the white man, whether Briton or Boer. ....

The most difficult problem, in some ways, connected with the general re-settlement was that of those ex-burghers who had espoused the British side during the war. The position of these men and their families was a pleasant one. The feeling of the "bitter enders" against them was intense. In most of the camps measures of segregation had been necessary before the peace, and special precautions were taken afterwards to prevent acts of violence.

Few such acts, indeed, took place-a remarkable testimony, it may be said in passing, to the lack of vindictiveness which is one of the most favourable traits of the Boer character. But it was impossible to prevent Insulting language, boycotting, or the religious ostracism enforced by the predikants.[553]

The British Government was bound by the strongest ties of obligation and honour to the men who had taken the momentous step of fighting for it against their own countrymen. To have treated them on exactly the same footing as those who fought against it, and to have left them exposed, not only to the hostility, but also to the contempt of their neighbours, would have been as foolish as it was dishonourable. Yet there had been more than one precedent for such dishonourable folly in the conduct of the British Government in the past, and among the bitterest enemies of England during the war were the sons of those Boers who fought on the British side in 1881. ....

To the military they were auxiliaries whose services were no longer needed; to the civil administration they were just ordinary Boers requiring repatriation. Fortunately for the honour of the British Government, the ex-military burghers at this critical juncture found a devoted and resourceful champion. Major E. H. M. Leggett, R.E., one of the many able junior officers engaged on the work of the railways, had acted as staff officer at headquarters for the National Scouts and Orange River Colony Volunteers.

Realizing the disastrous effect of a policy of neglect, he now decided to stay on and devote himself to the interests of the men he had largely been instrumental in enlisting. As a first measure he secured permission from Kitchener to keep up the staff and depots of the two corps as long as might be necessary to see the men duly settled, and to extend the advantages of this organization to all ex-burghers who had served with the British forces in any capacity, as well as to those oversea prisoners who had offered to do so. For all the men who had served he further secured their pay for a month after peace, a pony, saddle and bridle, and, in the case of a considerable number, a permit for a rifle and ammunition.

In addition to this he wheedled out of the Commander-in-Chief a large quantity of wagons and teams, tents and rations, which were sold at moderate prices to those of the men who could afford to buy them. The proceeds were used as a general fund for purchasing necessaries at wholesale prices, and eventually applied in various ways to benefit the men. This fund was administered by their own leaders and everything was done to promote the sense of solidarity among the men, and to make them feel that they were being looked after. All this help was in addition to the ordinary repatriation on which the interests of the ex-military burghers were provided for either by representation on the district commissions as in the Transvaal, or by a separate branch of the department, as in the Orange River Colony.

The repatriation of the ex-military burghers was from the outset complicated by a difficult problem. A very large proportion of them had been bywoners,[554] i.e., squatters on the lands of farmers who now made their

[553] Preacher, minister (used in the Dutch Protestant church)

conduct during the war an excuse for refusing to take them back.[555] This refusal was most marked in the case of the bywoners who had served the British, but was by no means confined to them.

The problem, indeed, was no new one. The steady growth of a population of hangers-on, mostly poor relatives, was far from being appreciated by the farmers before the war, and after the peace many of them showed a marked reluctance to take back their former quota of bywoners.

For a certain number of the ex-military bywoners Leggett found posts in the Constabulary, on the railways, and in other public departments.

For the majority it was necessary to devise special measures The first thing was to appeal to the wealthier men in his organization to help their comrades by finding them land. The second was to make it worth their while doing so. The real evil of the bywoner system, so Leggett urged, was the absence of any clear arrangement between landlord and squatter. The squatter had no security of tenure to encourage him to cultivate seriously; the landlord had nothing to gain from an idle and starveling dependent.

The remedy lay in substituting a system under which the squatter, on his side, should be guaranteed a certain fixity of tenure, and compensation for betterments, the landlord in return receiving a fixed proportion of the crop. On this basis a number of leading ex-military burghers formed a Farmers' Association and rented land near Standerton, on which bywoners were settled. The experiment promised so well at first that other associations were rapidly constituted by various private individuals, the most important at Potchefstroom. Applications poured in from bywoners of every shade of political sympathy, and Leggett and his fellow-workers wisely decided to accept all who were ready to live on a friendly footing with their neighbours.

To facilitate the work, Leggett was made "Director of Burgher Land Settlement" in October, and in November Director of the Transvaal Burgher Camps, whose inhabitants were by then reduced mainly to members of the bywoner class.

Over 3,000 ·persons were thus settled by the beginning of 1903, and the number was more than doubled eventually. The subsequent history of the Burgher Land Settlements was, unfortunately, not satisfactory. The associations were private ventures, and suffered from bad management and, in some cases, from the actual dishonesty of their members or employees. The tenants were, for the most part, shiftless and easily discouraged by adversity. In the end, most of the settlements came to grief, and the advances for implements, stock, etc., made by the Repatriation Department had to be written off.

The fact is, that the experiment, though based on an absolutely sound conception, was in some respects premature, in view of the very primitive conditions of agriculture in the Transvaal, and, in any case, required far more effective supervision than it ever received. The best that can be said is that the loss was better incurred in this way than in keeping the men on in the camps, and that the experiment solved the immediate difficulty of disposing of the bywoners who were not wanted by other farmers.

In the Orange River Colony Burgher Land Settlement was not carried out on any extensive scale. The bywoner problem was here attacked from an entirely different quarter by the provision of relief works. As early as July a special department was created under Lieutenant B. H. O. Armstrong, R.E., for this purpose. Special camps were formed, and in the course of the next two years a number of irrigation dams, several short sections of railway work, and some public buildings were all constructed by white manual labour. The men received 4 [shillings]. 6d [pence] a day, and their work was, on the whole, far more satisfactory than had been expected.

The actual results were not, indeed, profitable from a purely business point of view, partly owing to the cost of the labour, and, in some instances, owing to the hasty selection of unsuitable sites for the works. But the true test of the success of the experiment lies, not in treating it as an ordinary commercial venture, but in a comparison with the alternative of feeding the men in idleness, and in a consideration of the ultimate social and economic consequences of familiarizing the Boers with the conception of white manual labour.

Amid all the urgently pressing problems of the first few months after the peace, the work of establishing British settlers on the land was inevitably relegated to a secondary position. In the Orange River Colony, where other problems were less complex and where the agricultural value of the country was better known, the work pursued

[554] Literally 'Tenant'. A poor tenant farmer who labours for the owner and does some farming of his own.

[555] To accept their farm hands and other dependents back onto their farms.

a course of quiet and steady progress under the capable direction of Major K. P. Apthorp, the Secretary of the Land Board.

By the end of 1902 some 360 settlers, selected from nearly 5,000 applicants, were established, partly on land previously belonging to the Government, partly on land purchased for the purpose.

In the Transvaal the history of Land Settlement was somewhat chequered. There was a first stage, immediately after the peace, when impatient crowds of ex-irregulars, anxious to settle on the land, besieged a cautious and timid Land Board. The direct intervention of Lord Milner, through his energetic private secretary, Mr. John Buchan, galvanized the Board into a spasmodic activity.

Some 300 settlers were provisionally planted out, and large quantities of land were bought, not always with judgment or discretion. The Land Settlement Departments, especially the Transvaal one, were freely criticized afterwards for gross extravagance in their purchase of lands. In some instances there was undoubtedly justification for the charge. In the main, however, the prices paid were simply those prevailing everywhere after the peace, and only seemed unreasonably high at a later time when all prices had fallen.

For the first few months after the peace the one unceasing complaint against those responsible for land settlement was not their extravagance, but their slowness in purchasing land and their reluctance to pay what then seemed reasonable prices. At the end of 1902, a Department of Lands was instituted in the Transvaal, and Dr. A. Jameson, Minister of Lands in Western Australia, was appointed Commissioner. Eventually some 700 or more settlers, the survivors of a considerably larger number, were successfully established in each colony, and after many vicissitudes, were in a fair way to prosperity when self government was introduced. Considering the scarcity of really capable settlers and the extraordinary difficulties in the way of agriculture in the first few years, before the scientific study of the problems and the development of the railway system had cleared the ground, it is doubtful whether the experiment could really have been carried out with advantage on a much larger scale.

Milner, indeed, had originally hoped for a much larger number to produce any real influence on the character of the country population. But there can be no doubt that even the small number actually established, together with the settlers introduced by private agencies, has already exercised a very considerable moral and economic effect in bringing the races together, and in stimulating agricultural progress.

The repatriation of the Rand population was already in a large measure accomplished before the peace. By September, 1902, it had reached its old maximum and was still growing steadily. A large part of this increase was furnished by disbanded irregulars who came in to Johannesburg hoping for employment, either permanently, or till land should be available for settlement. But the revival of the mining industry was checked by the shortage of native labour, and employment was scarce.

As a temporary measure of relief, not only to the men but also to some extent to the labour situation, some 1,200 of these were taken on by the Rand [Gold] Mines Limited to do unskilled work at a rate of five shillings a day and their keep.

Another new element in the Rand population, not very large at first, but very desirable, was contributed by the women emigrants who were being sent out by the South African Expansion Committee of the British Women's Emigration Association, a body organized as a result of conversations between certain ladies and Lord Milner in 1901.

Soon after the peace, Milner started a Women's Immigration Department in the Transvaal to cooperate with the Committee in providing suitable employment for the immigrants, contributing to the cost of the voyage out, receiving them at Government hostels on arrival, and generally watching over their interests. In spite of all difficulties, due to the subsequent economic crisis, some 4,000 women in all were helped out to South Africa by the Women's Emigration Association between 1902 and 1908.

At the headquarters of the new administration the pressure of work during these first few months of peace was enormous. Except for a flying visit in August to Lorenzo Marques to secure the assistance and goodwill of the Portuguese authorities in the recruiting of labour, Lord Milner was kept close to his desk or at the council chamber by the overwhelming rush of problems that came to him for decision.

The arrival in September of Sir A. Lawley, as Lieutenant-Governor of the Transvaal, relieved him of no small part of the current administrative work, and set him free for the more important and more congenial task of initiation and supervision. ....

The need of Milner's restraining influence over Johannesburg was, indeed, apparent from the very first.

Within a month of peace, signs of discontent with the Crown Colony government began to show themselves. Johannesburg expected the peace to be immediately followed by the millennium, and in its sanguine mood deplored the peddling caution of a policy of development which it was ere long to denounce for reckless optimism and wild extravagance.

With no practical experience of government it criticized the slowness and incapacity of an administration whose like for ability and persistent energy South Africa had never seen. Forgetful of its old state of helotry to the Boers, it chafed at being governed by an" imported bureaucracy" of Cape Colonists and Englishmen. At the end of June a movement, of which the leaders were Mr. Dale Lace and Mr. R. J. Pakeman, the editor of the Transvaal Leader, was initiated for forming a "Transvaal Political Association" which was, according to its promoters, to support the Government by criticizing it. Overtures were even made to some of the Boer leaders, who politely declined them with the just observation that the less politics were discussed at that moment the better.

Fortunately the real leaders of Johannesburg, the men of the Imperial Light Horse, as well as the representatives of the mining houses, were of the same mind, and in face of their strong opposition the agitation soon collapsed. In so far as anything could be done to remove the causes of discontent by greater publicity and by the introduction of a representative element into the Government, Milner was anxious to do it. As early as July 4, he had telegraphed to Chamberlain urging the immediate enlargement of the Legislative Council by the addition of nominated but representative unofficial members. But the chief conclusion he drew from the incident was the necessity of hurrying on the revival of the industry and pushing forward the work of reconstruction and development at its utmost speed. ....

Little has been said in the foregoing pages about military affairs, yet the departure or disbandment of nearly 200,000 men and the redistribution and reorganization of the troops still left in the country was no small event, and for some months imposed a heavy task on the capacity of the railways. The troops left behind at first remained scattered over a large number of garrisons; gradually as their numbers grew less, and as any lingering doubts as to the possibility of disturbance were dispelled, they were drawn together into the stations destined to be permanently occupied, of which the chief were Pretoria, Potchefstroom, Middelburg, and Standerton in the Transvaal, Newcastle in Natal, Bloemfontein and Harrismith in the Orange River Colony, and Middelburg in Cape Colony.

At one time, indeed, Mr. Brodrick had announced that the whole South African garrison was to be reduced to a strength of 15,000 men. This decision was eventually reversed, largely owing to the representations of Lord Milner and Mr. Chamberlain, not so much from fear of the inadequacy of the numbers proposed-though the unwisdom of leaving a force insufficient to cope with even a purely local rising was obvious-as on broad grounds of training and strategy South Africa offered an ideal training ground, with a climate admirably suited to young soldiers, and was moreover a halfway house to India. In spite of the higher cost of maintaining troops in South Africa as compared with England, and of other objections arising out of the rigidity of the linked battalion system, these arguments prevailed, and the permanent strength of the forces in South Africa was in July, 1903, fixed at 25,000 men.[556]

Meanwhile in July, 1902, a Volunteer Force was formally inaugurated by ordinance in the Transvaal. This force was very popular, serving as it did to keep alive the organization and the name of more than one corps that had become famous in the war. At first mainly confined to the towns, the volunteer organization was gradually extended over the country districts, and was keenly taken up by many of the Boer farmers within very few years of the war. The development of cadet corps was also a prominent feature of the organization. No volunteer force was created in the Orange River Colony.

In November, 1902, the state of the new colonies was considered sufficiently settled to permit of the reduction of the South African Constabulary from 10,000 to 6,000 men.[557]

[556] The force at that time was still over 80,000. The figure then fixed has been gradually reduced at intervals since, more especially in the last three years. Its present strength (1909) is about 10,500.

[557] Reduced progressively by nearly a thousand a year during the next four years, and still further reduced after the establishment of self-government.

Martial law, which had continued in force, though no recourse to it was actually made, was at the same time withdrawn, and an ordinance issued granting the usual indemnity [see examples [558] [559] and making provision for the public safety in case of need.

The revival of industrial activity on the Rand, due to the solution of the labour crisis, had already been preceded by an improvement in the agricultural situation. A plentiful rainfall in the closing weeks of 1903 not only promised a good crop, but also encouraged the Repatriation Departments to sell a considerable proportion of their transport animals to the farmers for ploughing, thus rendering possible a large increase in the area under cultivation. By the end of June, 1904, the Orange River Colony [Repatriation] department was finally closed down.[560]

[558] The Indemnity and Oblivion Act 1660 (12 Cha. 2. c. 11) was an act of the Parliament of England, the long title of which is "An Act of Free and General Pardon, Indemnity, and Oblivion". This act was a general pardon for everyone who had committed crimes during the English Civil War and subsequent Commonwealth period, with the exception of certain crimes such as murder (without a licence granted by King or Parliament), piracy, buggery, rape and witchcraft, and people named in the act such as those involved in the regicide of Charles I. It also said that no action was to be taken against those involved at any later time, and that the Interregnum was to be legally forgotten. https://en.wikipedia.org/wiki/Indemnity_and_Oblivion_Act

[559] The Indemnity Ordinance, 1975 was a controversial law enacted by the martial law administration of Bangladesh on 26 September 1975. It provided legal immunity to all persons involved in the assassination of Sheikh Mujibur Rahman, who was killed with most of his family on 15 August 1975. Immunity meant the assassins were immune from any legal action. https://en.wikipedia.org/wiki/Indemnity_Ordinance,_1975

[560] This quote from the beginning of Chapter 6.

Epilogue

Koos De la Rey

After the war De la Rey travelled to Europe with Louis Botha and Christiaan de Wet to raise funds for the impoverished Boers whose families and farms had been devastated. In 1903 he was in India and Ceylon, persuading the prisoners of war interned there to take the oath of allegiance and return to South Africa.[561]

General J. H. De La Rey.[578]

On 15 September 1914 Christian Frederick Beyers, Commandant-General of the armed forces and an old comrade of De la Rey, resigned his commission and sent his car to fetch the latter from Johannesburg to Pretoria as he wished to consult with him. The two generals then set out that evening for Potchefstroom military camp where General Jan C.G. Kemp had also resigned.

They encountered several police roadblocks but refused to stop; the roadblocks had in fact been set to capture the Foster gang. At Langlaagte the police fired on the speeding car and a bullet struck De la Rey's back, ending his life;[562] his last words were dit is raak ('It hit').

He returned to his Lichtenburg farm as van Rensburg had predicted. Many Boers were convinced he had been deliberately assassinated, while others could not believe that he would have joined a rebellion, breaking his oath. According to Beyers, the plan was to co-ordinate the simultaneous resignation of all the senior officers in protest at the attack on South West Africa. The theory of a government assassination holds sway to this day.

Not long after De la Rey's funeral the short-lived Maritz rebellion broke out and De Wet, Beyers, General Maritz, commander of a force on the border of the German colony, Jan Kemp, and other Boer veterans took up arms again but most of the army remained loyal and the rebellion was swiftly put down by Botha and Smuts.

The rebels were pardoned just two years later by Botha in the interests of national reconciliation. While De la Rey would probably have been quite capable of taking to the field again at 67, it seems unlikely he would have gone against his word, especially as he had played such a leading role in bringing about the peace of Vereeniging.

De la Rey was buried in the Lichtenburg graveyard, where a bronze bust by sculptor Fanie Eloff adorns his grave.

[561] See: Koos de la Rey - Wikipedia

[562] Hon. Mr. Justice Reinhold Gregorowski (1914). Judicial Commission of Inquiry into the circumstances leading up to and attending upon the deaths of Senator General the Honourable J.H. de la Rey and Dr. G. Grace. Cape Times Limited. pp. 1–4, 7–11. Retrieved 8 September 2009.

Maritz Rebellion

The Maritz Rebellion (also known as The Rebellion of 1914 and the Huilo War) was a South African uprising that took place in 1914.[563]

Potchefstroom students march against government policy during the Rebellion.[579]

Despite the formation of the Union of South Africa, considerable embitterment due to the Second Boer War continued to live on in the hearts of many Afrikaners. When Great Britain declared war on Germany in 1914, many former Republicans did not see their way to fighting on the side of the Allies. Fueled by narratives of visions of Siener van Rensburg, who predicted the demise of the British Empire, opposition to any military obligation began to take shape.

The matter came to a head in September 1914 when the Prime Minister, Gen. Louis Botha, announced that South Africa intended to invade German South-West Africa. Counter-voices rang out in many parts of the country, but there was little coordination between the groups led by Gens. Manie Maritz in the North West, Christiaan Frederik Beyers, commander of the Uniale Army in the Transvaal, Captain Jopie Fourie and Gen. Christiaan de Wet. And Jan Kemp.[564]

Concerted efforts by the Dutch Reformed Church, President M.T. Steyn and other powerful forces to prevent hostility failed, and fighting broke out almost simultaneously in the Transvaal, in the Cape Province near Upington and in the Free State. Gen. Louis Botha called up the newly formed Union Defence Force and called on loyal citizens to support the government.

Gen. Christiaan de Wet was defeated in the Free State and took refuge in Bechuanaland (Botswana), where he was later captured. Gen Beyers drowned while trying to cross the Vaal River and Jopie Fourie was arrested and killed by firing squad in terms of Martial Law.

The last Rebels surrendered in early 1915, many of whom were sentenced to prison. Attempts were made by German rulers in South West Africa to foment the Rebellion by concluding a so-called treaty with Gen. Manie Maritz, which guaranteed him the establishment of a republic in the Union.

Background

General Manie Maritz

The Cabinet, constituted by Gen. Louis Botha at the request of the Governor-General at the formation of the Union of South Africa in 1910, contained representatives of both the Afrikaans and English-speaking political groups in the country. This was a reflection of Botha's desire to reconcile the Afrikaans and English elements in

[563] Translation of Afrikaans Wikipedia Article: https://af.wikipedia.org/wiki/Maritz-rebellie And: Maritz Rebellion - Wikipedia

[564] Jan Christoffel Greyling Kemp (10 June 1872 – 31 December 1946) was a South African Boer officer, rebel general, and politician. Jan Kemp (general) - Wikipedia

the country and unite them into a united South African nation that would maintain imperial ties with the British Empire.

The formation of the South African Party (SAP) in 1911 was a further step towards Botha's so-called "conciliation." Most Afrikaners and many English speakers sought their political salvation from the party. In practice, it soon became apparent that many Afrikaners did not agree with the policy of conciliation. With Gen. J.B.M. Hertzog as leader, they broke away from the SAP and formed the National Party (NP) in January 1914.

The NP campaigned for the maintenance of Afrikaner culture parallel to that of the English speakers. They accused Botha of neglecting South Africa's interests in favour of Britain and advocated, among other things, the republican ideal, which lived on among the Afrikaners after the Second Boer War. On 4 August 1914, Britain declared war on Germany and requested that the South African Government actively assist them by conquering the radio stations in German South West Africa.

The British request put the government in a difficult position. For many Afrikaners the Anglo-Boer War was still too fresh in the memory and they would rather take up arms against than for Britain. Their distrust was exacerbated by the spread of rumours that the government intended to use civilian commandos for the conquest of South West Africa. They saw the war in Europe as a golden opportunity to regain their independence by force.

On the other hand, there was a possibility that the English-speaking section of the population, especially Natal and Eastern Cape, would refuse to be part of the Union further if the British request was not met. Botha and his deputy, Gen. J.C. Smuts, favored active participation in the war. They saw that South West Africa would in any case be conquered by Britain, probably with the help of Australian or Indian troops, and that South Africa would thereby lose the initiative in southern Africa.

They also saw it as South Africa's moral duty to assist Britain. However, the case had to be submitted to a special session of Parliament for decision (September 1914). The position of the disgruntled section of the Africans was stated in the Volksraad by Gen. Hertzog. He acknowledged that South Africa was at war with Germany as part of the British Empire, but argued that this did not yet oblige him to resort to military action.

The conquest of South West Africa, he said, would be accompanied by unnecessary expense and loss of life. However, both in the House of Assembly and in the Senate, it was voted in favor of participation by an overwhelming majority.

Consequences

The 1914 Rebellion had far-reaching consequences. Many Afrikaners became opposed to the Botha government and gave support to the National Party. The change in the political climate was already evident with the elections of 1915 when, barely a year after its formation, the National Party drew 78,000 votes against the 94,000 of the SAP and won 26 seats in the House of Assembly against the SAP's 54.

In order to stay in power, the Botha government was increasingly reliant on the support of the Unionists. The real growth of the National Party and the national sense of unity of the Afrikaners date from the Rebellion. Less than ten years later in 1924, the National Party took over the government.

Women's March on Reddersburg in 1914 to demand the release of Gen. C. R. de Wet.[580]

Manie Maritz

Salomon Gerhardus (Manie) Maritz was a Boer general and leader in the Rebellion of 1914.[565 566]

Manie Martiz on Commando.[567] Gen. Manie Maritz.[581] Salomon Gerhardus (Manie) Maritz.[582 583]

Smuts' Commando in April of 1902. Martiz seated next to Smuts.[568]

Maritz was born on 26 July 1876 in Kimberley in the Cape Colony. So he was a born British national. His parents were Johannes Stephanus Maritz and Anna Coetzee, and he claimed he was a direct descendant of the voortrekker Gerrit Maritz. He writes that opportunity for education was slim: He had to take turns herding livestock and attending school.

He left for Johannesburg at the age of 19 where he helped his uncle drive a "keb" with a hooded car. During the Jameson Raid, he volunteered to look after the fort in Johannesburg. This entitled him to citizenship of the South African Republic. This entitled him to become a member of the South African Republic of Politie, the ZARPs. After completing his education, he focused on barring liquor and smuggling trade. However, he was later dismissed from the police service after he turned himself in at the Johannesburg police station because his conscience was bothering him due to improprieties.

565 Salomon Gerhardus (Manie) Maritz was a Boer general and leader in the Rebellion of 1914. Manie Maritz - Wikipedia And Genl. Salomon Gerhardus Maritz (1876 - 1940) - Genealogy (geni.com)

566 The Maritz Rebellion (also known as The Rebellion of 1914 and the Huilo War) was a South African uprising that took place in 1914. Maritz Rebellion - Wikipedia

567 Salomon Gerhardus (Manie) Maritz was a Boer general and leader in the Rebellion of 1914. Manie Maritz - Wikipedia And Genl. Salomon Gerhardus Maritz (1876 - 1940) - Genealogy (geni.com)

568 Jan Smuts is seated. Viewer left standing is General Jaap van Deventer and standing viewer right with his leg up is General Manie Maritz. General Smuts Vol 1 by Millin Sarah Gertrude (1936) p.159. File:Smuts Boereoorlog.jpg - Wikimedia Commons

Second Anglo-Boer War

In 1899, at the outbreak of war, he participated in battles in Natal with the Boksburg commando. England began reinforcing its troops in Natal and the Cape Colony and in September 1899 the South African Rebupliek, including Maritz, called up the Boksburg commando. The commands marched to Sandspruit, between Standerton and Volksrust. On October 12, 1899, the commando Natal went binn. Later he joined the Johannesburg Police Command and they were sent to Colesberg. Later he joined Danie Theron's scout corps. Then he sailed into the Cape Colony under Wynand Malan. He then went to the Northwest of the colony. He writes that he was elected commander in chief there. But he did not seem to accept it. Then he writes that Jan Smuts appointed him fighting general. Shortly afterwards, the massacre took place at Leliefontein.

Origin and youth

Lily Fountain

Leliefontein is a farm in the north-west of the Northern Cape. Centuries ago, it was awarded to the local Namaqua tribe. In 1816 a Wesleyan mission station was established there. In January 1902, Manie Maritz and his commando passed by. Maritz wanted to read a proclamation to the Hottentotte (today Khoi) that read that they were not allowed to be employed by English employers. One of the Namas then cursed him, saying that the English were their government. He then hit Maritz over the head with a cane.

A fight then broke out. Maritz later had to summon reinforcements to help calm the fight. A few of the armed Hottentotte trapped some of Maritz's men at the ring wall of the church. Maritz's instruction to the remaining crew was to go to the nearest hill to fire from a distance. The next morning they returned to the mission station on a counterattack, shooting 8 and then later 30 Namas. After the incident, Maritz and his crew loaded all the wounded onto a horse-drawn carriage and took them to Garies for care. Shortly before the incident, they brutally murdered two of Maritz's crew members by putting sand in their mouths and pushing sticks down their throats. Deneys Reitz writes that he visited with General Jan Smuts at Leliefontein. The place was ruined and burned down.

The bodies of the Namas were still lying there as well as some front loaders. Except for the blow over the head, his version matches that of Maritz. According to Reitz, Maritz's actions were reckless and unjustifiable. General Smuts said nothing, but walked into the field alone. Upon his return, Smuts was cranky and short-tempered because he was displeased.

After that

A few days after the Leliefontein battle, Maritz received orders from General Smuts to go and protect the Calvinia district and that he who is Smuts would have given the Boland protection. Due to little fodder for the horses, Maritz could not stay there for long and Smuts had to move into Namaqualand with a portion of his commando. Maritz moved ahead to get enough feed for the horses and to provide for the surrounding commandos as well.

Maritz was aware of the fact that Britain was planning to send a convoy from Garies to Springbok. With the arrival of the British at the Kamiesberg Mountains, General Smuts, van Deventer and Bouwer were late and by the second day of the battle the British had returned to Garies. Upon Smuts' arrival, they decided to attack Springbok. The next day, 28 members of the crew decided to go explore Concordia. Upon arrival, they sent in two members of the crew carrying a white flag. The next day they went to Okiep to explore the position there. General Smuts did not want Maritz to move in front and suggested that the leaders cover the rearguard.

The self-made bombs they used consisted of a syrup can that was filled with dynamite and then soldered again. Maritz was well disposed towards the British at a distance and one day they even called a truce for a football match at Concordia.

In April 1902, Smuts left for Vereeniging for peace negotiation. Smuts' words to Maritz were: "Old Mania, do you see a chance to hold your commando together for though the Transvaal and the Free State make peace, can we fight still further?" After many battles, Maritz and the commando went to Calvinia to wait for Smuts. Maritz received a letter from Smuts telling them to lay down their weapons. Everyone could sign a declension stipulating that their private property still belonged to them and that they would return home.

When Maritz was in Europe, he received a letter from Piet de Wet, saying that they were held in camps by the and treated badly.

Madagascar

Reitz encountered Maritz in Madagascar after the war. The latter arrived there with piles of money in March 1903. He relates that after the war he slipped across the German West border and eventually arrived in Paris, France. There he met a millionaire, Jacques Lebaudy, who sent him to Madagascar to buy land. Reitz and Maritz then explored the interior of the island, eventually arriving at Lake Itasy. Maritz was so impressed with the place that he just wanted to stay there. However, they all contracted malaria and when he became healthy, Maritz went back to the capital. Reitz and Maritz left Madagascar together, Reitz returned to South Africa and Maritz to Europe.

Back in South Africa

Maritz returned to South Africa and farmed horses at Paarl. He also went to help the Germans during the Herero genocide. He then went to the Transvaal, but was arrested for not wanting to sign the oath of allegiance. He then went back to Cape Town. When the Free State achieved self-government, he went to farm there and married Emma Nel.

Betrayal

When the Union of South Africa's Defence Force was formed, Maritz was offered a commission in the militia and he joined. At the beginning of August 1914 he was promoted to lieutenant colonel and appointed commander of the Uniale forces towards Upington and Kakamas. The appointment was made jointly by Smuts (the minister concerned) and the head of the militia, General Beyers. Maritz began to haggle with the Germans even before war broke out.

On September 23, Maritz was ordered, in support of Lieutenant Colonel Lukin's invasion of German West, to move with a large force toward the border. Maritz refused to do so. He refused to relinquish his order. He refused to go to Pretoria. He finally decided to rebel on 9 October 1914. That was the day he informed his commando of his plans. On 11 October, he occupied Keimoes. Maritz was wounded in the knee on 22 October and he was taken to German West.

Back to South Africa

He then eventually defected to Angola. The Portuguese sent him to Luanda in 1916 and later to Portugal. Four years later, he went to Germany. That same year, after many obstacles, he arrived in Lourenço Marques (Maputo) and was arrested for treason when he entered the Union.

In Cape Town, he was charged with treason in the High Court. He was convicted and sentenced to three years in prison. However, General Hertzog's National Party came to power and Maritz was let go after three months.

His anti-Jewish statements in his book, My Life and Pursuit,[569] made him clash with the law again. He was convicted on a charge of clouding race relations and fueling racial hatred. He had to pay a fine of £75.

He died on 20 December 1940 in Pretoria at the age of 64.

[569] Manie Maritz, "My Life and Aspiration," or My Life and Pursuit 1938. PF: Unable to locate this work.

Family Biographical Sketches

Francis Reitz

Because Deneys father, Francis, plays such a large role in his life and in the life of the Republics during the Second Boer War, I am including this brief family summary.

Francis William Reitz, 1890.[584]

Francis William Reitz Jr.[570] (October 5th, 1844 – March 27th, 1934) was a South African lawyer, politician, statesman, poet and member of the Parliament of the Cape Colony, Chief Justice and fifth State President of the Orange Free State, State Secretary of the South African Republic at the time of the Second Boer War and the first president of the Senate of the Union of South Africa.

Reitz had a very diverse political and legal career that spanned more than forty years and played out in four different political entities: the Cape Colony, the Orange Free State, the South African Republic and the Union of South Africa. Reitz. After graduating at the South African College in Cape Town he finished his studies at Edinburgh University, and afterwards at the Inner Temple, where he was called to the Bar in 1868. He then returned to the Cape, and, after practising as a barrister in the Cape courts for six years, was appointed Chief Justice of the Orange Free State, a post which he held for fifteen years.

When State President Brand died suddenly in 1888, Reitz won the presidential election unopposed. After being re-elected in 1895 and making a trip to Europe, he became very ill and had to retire. In 1898, after recovering, he was appointed State Secretary of the South African Republic and became a leading political figure during the Second Boer War. After the war, he voluntarily went into exile because he was unwilling to show allegiance to the British. Several years later, he returned to South Africa and opened another law firm in Pretoria. In the early 1900s he became involved in politics again and with the declaration of the Union of South Africa in 1910 he was elected the first President of the Senate.

Reitz was an important figure in the cultural life of the Afrikaner throughout his life, especially through his poems and other publications.

With respect to his son Deneys and his extended family and upbringing his father Francis was born in Swellendam, the son of Francis William Reitz, Sr., farmer and agronomist and Cornelia Magdalena Deneys. He was the seventh child in a family of twelve. He grew up on his father's farm, Rhenosterfontein, on the border of the Breede River in the Cape Colony.

Francis married twice. His first marriage was to Blanka Thesen took place on June 24th, 1874 in Cape Town. Blanka was born in Stavanger, Norway on October 15th, 1854. She was the daughter of Arnt Leonard Thesen, a

[570] Francis William Reitz, Jr. Francis William Reitz - Wikipedia Francis William Reitz, Jr. (5 October 1844 – 27 March 1934) was a South African lawyer, politician, statesman, publicist, and poet who was a member of parliament of the Cape Colony, Chief Justice and fifth State President of the Orange Free State, State Secretary of the South African Republic at the time of the Second Boer War, and the first president of the Senate of the Union of South Africa. See Also: African Wikipedia Article: Francis William Reitz - Wikipedia

merchant, and Anne Catherina Margarethe Brandt.[571] Francis and Blanka had seven sons and one daughter, before her death on October 5th, 1887 in Bloemfontein.

Deneys' Brothers

Blanka Reitz gave birth to Deneys on April 2nd 1882. In the first chapter he mentions that there are a total of five brothers from this marriage. Deneys is the middle child with two older and two younger brothers.

His oldest brother is Hjalmar[572] born in 1877 who was five years older. Hjalmar is captured by the Australians and sent to India for the duration of the conflict. He later goes on to a distinguished career as a jurist. See infra.

His next oldest brother is Joubert who is one year older than Deneys. Born in 1881. He served in many engagements detailed by Deneys and then was captured and sent to Bermuda.[573]

The younger brother who is mentioned most is Arnt Leonard Reitz, referenced as 16 in 1899 having been born in 1883.[574]

Another brother Jack is referenced as being 12 in 1899, having been born in 1887. At the beginning of the war he was sent home when he was caught at the front because he was too young.[575]

After the brothers' mother Blanka died in 1887; Francis married Cornelia Maria Theresa Mulder on December 11th, 1889.[576] At the time of her marriage Mulder was acting director of the Eunice Ladies' Institute at Bloemfontein. With his second wife, he had six sons and one daughter. Deneys mentions his stepmother several times and also other family members.

With the fall of Johannesburg and Pretoria in June of 1900, Blanka departed South Africa and went to live in Holland for the duration of the war.

"We only heard later that my stepmother Cornelia Maria Theresa Mulder[577] and the younger children had been sent to Delagoa Bay and thence by sea up the East Coast of Africa to Holland, where they still are." From June of 1900.[578]

Hjalmar Reitz Bio

Hjalmar Reitz[579] (Bloemfontein, 25 February 1877 – Johannesburg, 10 January 1946) was a South African politician, writer and the eldest son of Free State President F.W. Reitz.[580]

He married Elzine Mulder of Delft in 1906.[581]

Training Reitz studied law in London and Amsterdam and returned to the Free State in 1899 to take part in the Anglo-Boer War, but was captured by Australian forces while ill with malaria and exiled to India. After the war, he resumed his studies in the Netherlands and obtained a doctorate at the University of Amsterdam in law. He returned in 1904 and became registrar for Judge J. S. Curlewis and the following year he commenced practising as a member of the bar in Pretoria.

[571] The Thesen family took up residence in Knysnain the Cape Colony in 1869.

[572] Reitz Hjalmar, Burger Received the Wound Ribbon (Lint voor Verwonding). Pretoria Commando

[573] Joubert Reitz Prisoner number: 21388 Captured: Majatje 02 Jun 1901 Sent to: Bermuda, Burts Island Age: 20 Source: Anglo Boer War Museum 2016. This is also cited in Deneys' narrative where he departs the story.

[574] Reitz Arnt Leonard Asst Veldkornet Commando: Commandant Malan.

[575] "**Arnt and Jack**, aged sixteen and twelve respectively, had been at school up to now; but they had at last persuaded my father to let them come to the war. Jack only remained for a few days, as old [General] Maroola on one of his visits to our laager caught sight of him and ordered him to be sent home."

[576] Born in Delft, Netherlands 25 December 1863; died in Cape Town on 2 January 1935.

[577] Cornelia Maria Theresa Mulder on December 11th, 1889. At the time of her marriage Mulder was acting director of the Eunice Ladies' Institute at Bloemfontein. Born in Delft, Netherlands 25 December 1863; died in Cape Town on 2 January 1935 With his second wife, he had six sons and one daughter.

[578] PF: I suspect that she was on the same ship as Kruger? If not, then some other sailing shortly thereafter.

[579] Deneys' quote "My eldest brother (named Hjalmar, after a Norwegian uncle Hjalmar Thesen)"

[580] Hjalmar Reitz - Wikipedia

[581] Potgieter, D.J. (red.) 1972. Standard Encyclopaedia of Southern Africa. Cape Town:

Political career He was elected to the Transvaal Provincial Council in 1915 as a supporter of Gen. J.B.M. Hertzog and remained a member of the Council until he became National Party member of the Union House of Assembly for British in 1924. He accompanied Gen. Hertzog and Dr. D. F. Malanin 1919 as a member of the republican delegation to the Paris Peace Conference. An avid debater, he was elected to the Union House of Assembly as a member of Tielman Roos' party (the Roosites) for Jeppesin 1933 when he defeated the Labour Party's Henry Sampson by 1,222 votes after Sampson had represented the constituency since 1920. He followed Gen. Hertzog and entered the United Party in December 1934 and in 1936 he was appointed chairman of the Central Road Transport Board.

Author With Tielman Rooshe he co-authored Principles of Roman-Dutch law in 1909 and was an editor along with Harm Oost on a national yearbook, in 1931–32. He was also author of a war novel "De dochter van den handsopper (1903)." It is written in Dutch, but the dialogue is in Afrikaans. His autobiography, The conversion of a South African nationalist, was published in 1946.

Hjalmar Reitz.[582]

Conrad Hjalmar Reitz.

Conrad H. Reitz was the Head of the Acquisitions Department at the university of Windsor, Ontario, Canada. He received his B.A. at the University or Pretoria, South Africa. in 1958, and t hen worked for three years as a probation officer before going to library school. He was awarded the Diploma of Librarianship by the University of Cape Town School of librarian ship in 1962, and was regional librarian in the Cape Provincial Library Service for several months before emigrating to Canada in 1963.

He was a cataloger at the University of Toronto Library for two years, and then became Head of the Acquisitions Department, University of Waterloo (Ontario) Library, In 1966 he moved to his present position, and in the same year received his M.L.S. degree from the University of Toronto School of Library Science. He has also published The Reitz Family; an Annotated Bibliography (1964) in the University of Cape Town School of Librarianship Bibliographic Series.[583] And.[584]

[582] Hjalmar Reitz - Hjalmar Reitz - Wikipedia

[583] Reitz, Conrad Hjalmar. The Reitz Family: an Annotated Bibliography. [Cape Town]: University of Cape Town, School of Librarianship, 1964. 62 Pages. Catalog Record: The Reitz family; an annotated bibliography | HathiTrust Digital Library

[584] Reitz, Conrad Hjalmar, South African bibliography. 1967 University of Illinois Graduate School of Library Science Occasional Papers #90. 22 pages. South African bibliography: Reitz, Conrad Hjalmar: Free Download, Borrow, and Streaming: Internet Archive

Deneys Reitz Bio

Deneys Reitz[585] [586] [587] (Bloemfontein, April 2nd, 1882 – London, October 19th, 1944) was a South African soldier, politician, cabinet minister, writer and the third eldest of Free State Pres. Francis William Reitz's five sons. His mother, Blanka Thesen, a member of the Norwegian family from Knysna, was the first wife of the Free State president. Deneys Reitz took his first name from the family name of his paternal grandmother, Cornelia Magdalena Deneys.

Deneys Reitz in 1923.[585]

Deneys Reitz as an older man.[586]

At Aldershot, 1939.[587]

Deneys 1940.[588]

Home Fleet Visit 1943.[589]

Reitz received his education at Grey College in Bloemfontein and joined the Boer forces in the Second Anglo-Boer War when he was seventeen.

During the initial phase of the War, he fought in several battles, including the engagement at Surprise Hill (Vaalkop) and in the Boer victory at Spionkop. After a string of Boer defeats and the British capture of Pretoria, Reitz was one of the fighters who remained in the field eventually joining up with Jan's Smuts' commando.

On 17 September 1901, Smuts' commando encountered the 17th Lancers in the vicinity of Tarkastad. Smuts realised that the Lancers' camp was their one opportunity to re-equip themselves with horses, food and clothing. A fierce fight, subsequently to be known as the Battle of Elands River, took place, with the Lancers being caught in a cross-fire and suffering heavy casualties. Stunned by the onslaught, the remaining Lancers put up a white flag. Reitz encountered Captain Sandeman, the Lancers' commander, and his lieutenant Lord Vivian among the wounded.

At the end of the war, Smuts' commando was finding success in the west of the Cape Colony and was besieging the garrison of Okiep, Northern Cape.

[585] Cite: Afrikaans Wikipedia article: Deneys Reitz - Wikipedia
[586] Reitz, Deneys. *Dictionary of South African Biography*. (1981). Human Sciences Research Council. 670. ISBN 0-409-09183-9.
[587] English Wikipedia: Deneys Reitz - Wikipedia https://en.wikipedia.org/wiki/Deneys_Reitz

In April of 1902 when Smuts was summoned to participate in peace negotiations, Reitz formed part of the negotiating delegation from his commando, given passage to meet the delegates from the other commandos still in the field.

At the peace conference, Reitz's father was among the signatories of the peace treaty, but as he related to Kitchener, only in his official capacity; since he refused to personally submit to British rule. He was given two weeks to settle his affairs in Pretoria before leaving the country. Deneys felt that he had to stand by his father and accordingly refused to take the oath of allegiance in order to remain in the country.

He left for Madagascar with his brother, where they eked out a living convoying goods by ox-transport "hard work in dank fever-stricken forests and across mountains sodden with eternal rain." In his spare time there he wrote Commando, dated 1903 but not published until 1929.[588]

After three difficult years and at the persistent requests of Isie Smuts,[589] Gen. Smuts' wife, he returned to South Africa in 1904 or 05 where he was able to regain his health. He then completed his studies and in 1908 in Heilbron began his successful career as a lawyer.

In World War I, he first helped Smuts suppress the Maritz Rebellion in the Orange Free State; then he served on Smuts' army staff in the "German West campaign" (in the German colony of South-West Africa) and in the "German East campaign" (in German East Africa) he commanded the mounted regiment Fourth South African Horse. In 1917 he went overseas to the western front where he was wounded in late 1917 while serving with 6/7th Battalion Royal Scots Fusiliers. In early 1918 he was again wounded and subsequently became colonel in command of a battalion of the First Royal Scots Fusilliers. He led his men to the Rhine after the Armistice, as detailed in his book Trekking On.[590]

After the war, he entered politics as a supporter of Gen. Smuts' South African Party (S.A.P.) becoming the member of the House of Assembly of South Africa for Bloemfontein South, defeating Dr. Colin Steyn (a son of a former Free State president, M.T. Steyn) of the National Party by 101 votes in the first of their three contests for this seat. His principles during his political career included loyalty to General Smuts, loyalty to the British Empire as guarantor of South African freedom, and harmony between Dutch and English South Africans. He opposed the Ossewa Brandwag organisation, which planned to take control of South Africa as soon as Britain had been crushed.

In 1920 he married Leila Agnes Buissiné Wright (Cape Town, 13 December 1887–Cape Town, 27 December 1959). She was a social reformer, an outspoken advocate of women's rights and elected representative in her own right becoming the first woman member of the Assembly (representative for Parktown in Johannesburg, 1933–1944). [See infra]

On 3 August 1920, Reitz and Steyn contested the same constituency in a by-election. (It's not known why, unless Steyn successfully contested the result.) Reitz won again, this time with a majority of 141, but Steyn stood in a by-election on 26 August 1920 in the safe National Party (N.P.) seat of Vredefort against the South African Party's candidate, S. P. van W. Botha, winning with a majority of 520 through which he also became Volksraadslid.

In the 1921 general election, which Gen. Smuts wanted to use so that he could strengthen his position in the Volksraad after the Unionist Party was incorporated into his South African Party, Reitz and Steyn faced off against each other a third time. This time Steyn was returned with a majority of 47.

Dr. Steyn stood in both the Bloemfontein seat and Vredefort and won both. However, he preferred to represent Bloemfontein South, probably because he wanted to prevent the S.A.P. from winning a by-election in the capital and thus regaining a hold in the Free State, because in the general election the N.P. again won 16 out of the 17 seats. An exception was Bloemfontein North, where the Labour Party's Arthur Barlow won by 316 votes against N.P.'s J.W.G. Steyn.

As was the custom at the time, a member of a safe seat retired so that a serving minister, Reitz, could once again take his place in the Volksraad. In this case, it was E. H. Walton who comfortably won Port Elizabeth Central in the 1921 general election. Reitz was elected unopposed at the by-election on 15 March 1921.

[588] No Outspan. Deneys Reitz. Faber and Faber, London, 1943. No ISBN And Footnote 4 Deneys Reitz - Wikipedia
[589] Isie Smuts - Wikipedia
[590] Trekking On: This work is a UK republication of Afrikander along with an index. Since it was first published in the US in 1933. Publisher Faber & Faber Ltd.

In the early general elections of 1924, Reitz did not again risk a seat in his hometown, and only well, because the N.P.'s Colin Steyn convincingly won Bloemfontein South with almost two-thirds of the vote. In the then still predominantly English port city, Reitz once again prevailed when he won Port Elizabeth Central with a majority of 1,044 against the Labour Party's R.L. Weir's 661 votes.

However, it was the election in which the Smuts government fell; therefore, it marked the end of Reitz's ministership, although he remained as representative of the Bay Area seat until April 1929.

When the Smuts government fell in 1924, Reitz returned to his law practice.

He undertook extensive private and business trips during which he visited Northern and Southern Rhodesia, the Belgian Congo, Angola and especially the Kalahari and Kaokoveld in South West Africa. His last book, No outspan (1943), describes this period in his life.

Reitz won back the Barberton constituency in the 1929 general election for the South African Party after W.H. Rood won it in 1924 with a majority of 24 for the N.P. The S.A.P. candidate's majority is 363 votes against the National Party's C.L. Worral.

Reitz was elected unopposed in Barberton in 1933, after the N.P. and S.A.P. formed a coalition. He retained the seat in 1938 for the United Party with a majority of 1,381 votes against the National Party's W.C.J. Brink.

With the formation of the coalition between the National Party and the South African Party in 1933, which would culminate in the formation of the United Party on 5 December 1934, he accepted the office of Minister of Lands and Irrigation and in 1935 became the Minister of Agriculture and Forestry and in 1938 Minister of Mines.

In 1939, following the split in the U.P. when Gen. J.B.M. Hertzog resigned as prime minister due to the Union House of Assembly's decision to participate in World War II, Smuts appointed him as Minister of Native Affairs and Deputy Prime Minister until 1943, when he was appointed as South African High Commissioner to London.

In South Africa he was regarded as an enterprising cabinet minister and in Britain as an outstanding representative of the Union of South Africa.

As Minister of Lands in the Smuts Cabinet, he took the lead at a number of important undertakings, most notably the legislation relating to the establishment of the Kruger National Park.

The South African Academy for Science and Arts elected Reitz as a member when he was still minister of countries in the Smuts cabinet. He was the son of a founding member of the institution and a leader in the First Language Movement. At the time, membership was still limited to 50 members and it was therefore a great honour when someone was elected to this select group. However, Reitz declined membership because he saw the Academy as a "racist" organisation that stood in the way of good relations between Afrikaners and English because he campaigned for Afrikaans. At the same time, "racism" pertained to the relations between Afrikaans and English speakers in South Africa and not to different ethnic groups.

Reitz spoke only English in Parliament and in Cape Town, but Afrikaans on the farm with his labourers.

The Free State town of Deneysville is named after Reitz. His brother Hjalmar Reitz was likewise a writer and politician, but an ardent supporter of the National Party.

Published works:

"Commando: A Boer Journal Of The Boer War," first published in Great Britain in 1929, ISBN 0-571-08778-7

Trekking On (1933), dealing with the Boer War through World War I, and

No Outspan (1943), which covers life in South African politics between the wars and concludes with him as Deputy Prime Minister of South Africa.

He is buried south of Mariepskop, approximately 10 km (6.2 mi) east of the Blyde River Canyon in Mpumalanga.

Leila Reitz Bio

Leila Agnes Buissinné Reitz,[591] [592] (née Wright; 13 December 1887 – 27 December 1959) of Wynberg, Cape Town was a South African politician who served as the first woman elected to Parliament in South Africa. She represented Parktown in the House of Assembly of South Africa from 1933 until 1943.

Reitz was the daughter of Claude Wright, a physician from Wynberg. After completing her matric in the first class in 1905 at the High School for Girls (now Rustenburg Girls High School), in Rondebosch, she matriculated studying history in the South African College, Cape Town, and received a B.S. degree with distinction from the University of the Cape of Good Hope.

From 1910 to 1913 she studied as a bursary[593] holder for Cape University at Newnham College, Cambridge in England. After obtaining the tripos[594] in history and returning to South Africa, she taught history at the University of Cape Town until 1920 when she married Deneys Reitz.

Leila's association with the South African Party dates from early in her married life, firstly while supporting her husband in his political career and later in her own right as an executive member of the women's section of the Party, the Witwatersrand executive committee and the national steering committee. Along with Bertha Solomon,[595] among others, she helped secure passage of the Women's Enfranchisement Act of 1930.

Subsequently she contested the 1933 South African general election for Parktown. Her success in this election made her the first woman to become a member of the Union House of Assembly and she continued to represent Parktown in Johannesburg from 1933–1944, initially as a South African Party candidate in the coalition elections of 1933 and then as a United Party candidate in 1938.

During her parliamentary career she was active in the field of social welfare and played a major part in the passage of the Children's Act in 1937.

She left politics in 1943 to accompany her husband, Deneys Reitz, to London after he was appointed South African High Commissioner to the United Kingdom. She retired to Cape Town and died in 1959 aged 72.

Leila Agnes Reitz's tombstone[590] at Pilgrim's Rest.[596]
Humanitas et sapientia Humanity and wisdom

[591] Afrikaans Wikipedia: Leila Reitz - Wikipedia
[592] English Wikipedia: Leila Reitz - Wikipedia
[593] A monetary award to university students that allows them to continue their studies.
[594] Any of the final examinations for a BA honours degree. And: The list of successful candidates in such an examination.
[595] Bertha Solomon - Wikipedia
[596] Pilgrim's Rest - Wikipedia

Publishers' Credit

C O M M A N D O

☆ DENEYS REITZ ☆

A BOER JOURNAL OF THE BOER WAR
WITH A PREFACE BY
GENERAL J. C. SMUTS

19 30

Charles Boni PAPER BOOKS *New York*

The PAPER BOOKS was founded in nineteen twenty-nine by Charles Boni, to place good books, well designed and carefully made, within the reach of any reader; and is under the direction of Editors: Horace M. Kallen, Lincoln Colcord, Padraic Colum, Everett Dean Martin, Louis Untermeyer, Rockwell Kent, Elmer Adler.

A new book is published and mailed to subscribers on the 25th of each month. The first book in the series was published in September 1929. In May, THE BRIDGE OF SAN LUIS REY by Thornton Wilder was set up and printed as an example of the format.

DATE	TITLE	AUTHOR
FOR FORMAT	*The Bridge of San Luis Rey*	Thornton Wilder
SEPT 25	*The Golden Wind*	Takashi Ohta and Margaret Sperry
OCT 25	*Frederick the Great*	Margaret Goldsmith
NOV 25	*Dewer Rides*	L. A. G. Strong
DEC 26	*Prosperity: Fact or Myth*	Stuart Chase
JAN 25	*Commando*	Deneys Reitz

Subscribers have the privilege of purchasing extra copies of any Paper Books as long as they are in print. They will be sent postpaid to any name and address supplied by a subscriber on receipt of 50 cents each

Charles Boni PAPER BOOKS *New York*
Number 80 *Fifth Avenue*

FIRST PUBLISHED IN OCTOBER MCMXXIX
BY FABER AND FABER LIMITED
24 RUSSELL SQUARE LONDON W.C.1
SECOND IMPRESSION NOVEMBER MCMXXIX
THIRD IMPRESSION DECEMBER MCMXXIX
FOURTH IMPRESSION JANUARY MCMXXX
SECOND EDITION REVISED MARCH MCMXXXI
RESET AND FIRST REPRINTED
IN THIS NEW EDITION OCTOBER MCMXXXII
REPRINTED NOVEMBER MCMXXXIII
FEBRUARY MCMXXXV
AND MARCH MCMXXXVIII
PRINTED IN GREAT BRITAIN
BY LATIMER TREND AND CO LTD PLYMOUTH

Bibliography

Many works have been footnoted in the text, enthusiastically if also somewhat idiosyncratically as to form. Restating those references again here would extend this work unnecessarily.

Anyone who has read this work will easily come across the major historical works which chronicle this era and conflict as they follow the leads I have cited. I would rather cite material directly in the location where it is most relevant than bury it at the end of the work or in the endnotes where I primarily include citations to images.

The full and equally idiomatic bibliography for this series is over sixty pages long and will be included in an Appendix volume to be published at the end of the series.

For this volume an abbreviated bibliography is as follows:

Deneys Reitz. Commando: A Boer Journal of The Boer War. London Faber & Faber, 1929. With A Preface By General the Right Honourable J. C. Smuts. First Published In October MCMXXIX [1929] By Faber and Faber Limited 24 Russell Square London W.C I.
Reset and Reprinted October MCMXXXIII [1933] Printed In Great Britain by Latimer Trend and Co Ltd Plymouth. Commando A Boer Journal Of The Boer War: Reitz Deneys: Free Download, Borrow, and Streaming: Internet Archive And: https://archive.org/details/in.ernet.dli.2015.207576
Charles Boni Edition: Deneys Reitz, Commando; a Boer journal of the Boer war, with a preface by General J.C. Smuts New York: C. Boni, 1930. HathiTrust: University of Michigan Scan. Catalog Record: Commando | HathiTrust Digital Library Direct Link: #25 - Commando / [by] Deneys Reitz; a Boer journal of the Boer war, ... - Full View | HathiTrust Digital Library

Also: Anglo Boer War - Reitz: Commando And: https://www.angloboerwar.com/books/35-reitz-commando

See Also: https://www.angloboerwar.com/books/35-reitz-commando

And: Commando: A Boer Journal of the Boer War - Deneys Reitz - Google Books https://books.google.com/books?id=x7kxEb_wEZYC

Also available from Commando: A Boer Journal Of The Boer War, And: http://www.ourcivilisation.com/smartboard/shop/reitzd/commando/index

Our [Transvaal] Military Officers: Dutch: Onze krijgs-officieren: album van portretten met levens-schetsen der. by Gustav Schoeman Preller, Frans Vredenrijk Engelenburg 1904 Catalogue Record: https://archive.org/search?query=title%3A%28Onze%20Krijgs-Officieren%29

Our Military Officers. Album of portraits with life sketches. Original Publication: Onze krijgs-officieren: album van portretten met levens-schetsen der. by Gustav Schoeman Preller, Frans Vredenrijk Engelenburg 1904 Publisher "Volksstem" kantore. In Dutch with a preface by Louis Botha. Transcribed and published by: Mr. Pieter van der Merwe see infra. Original Publication : Onze krijgs-officieren: album van portretten met levens-schetsen der. by Gustav Schoeman Preller, Frans Vredenrijk Engelenburg 1904 Publisher "Volksstem" kantore Catalogue Record: https://archive.org/search?query=title%3A%28Onze%20Krijgs-Officieren%29

Direct Link: https://archive.org/details/onzekrijgsoffic00engegoog

Reproduction: of Our Military Officers. Album of portraits with life sketches of the Transvaal Generals and Commandants), Volksstem, Pretoria 1904. Transcribed and published by: Mr. Pieter van der Merwe..

File:Onze Krijgs-Officieren. Album van portretten met levens-schetsen, Pretoria, Volksstem, 1904.pdf

Download Access: https://en.wikipedia.org/wiki/File:Onze_Krijgs-Officieren._Album_van_portretten_met_levens-schetsen,_Pretoria,_Volksstem,_1904.pdf

Direct Link to File: https://upload.wikimedia.org/wikipedia/commons/5/52/Onze_Krijgs-Officieren._Album_van_portretten_met_levens-schetsen%2C_Pretoria%2C_Volksstem%2C_1904.pdf

Source Page: https://en.wikipedia.org/wiki/Annie_Botha#cite_note-4

License Links

My many thanks to all the authors, libraries, conservators, and scanners/uploaders who have allowed me to populate, illustrate and annotate this compilation. Responsible efforts have been made to ensure that material in this work public domain material and otherwise is governed by the following licenses.

Copyright Discussion

[597] Creative Commons License: Creative Commons license - Wikipedia And About CC Licenses - Creative Commons

[598] If the work was commercially published within 75 years of the end of the calendar year in which it was made, Crown copyright will last for 50 years from the end of the calendar year in which it was published.
https://copyrightservice.co.uk/copyright/p01_uk_copyright_law Flow Chart: Crown copyright flowchart 2015

[599] post mortem auctoris or "pma" [post mortem author]

Otherwise, with respect to the underlying texts and images recast in a more permanently accessible form I choose to pay forward my creative energy to the people who lived through these experiences and it has been a pleasure to find and assemble their collective words and images in a new synoptic and synergistic arrangement.

Except where noted otherwise, every effort has been made to ensure that no material has been used without an attributed source. However, if any non-attributed source, incorrect source, or copyrighted material used without permission or justification, has been included then I would be grateful for information that would enable any errors or inappropriate inclusions to be corrected or omitted in subsequent impressions.

Starting from creation, to publication, through to purchase, archiving, and digitizing, this annotated, illustrated and expanded edition is the sixth step in the journey of these works. My thanks go to all who have helped preserve and perpetuate these important historical records.

'Courage reader, we near the end.'[600]

[600] Per, Will Durant, in his eleven-volume work 'The Story of Civilization,' Simon & Schuster, 21st Edition. 1972 ISBN: 0-671-41800-9

Endnotes

[1] Deneys Reitz op kommando - Deneys Reitz - Wikipedia Made Available By: Morné van Rooyen Source: Newhistory.co.za
[2] Deneys Reitz as minister of native affairs visit British army camp Aldershot 22 November 1939. File:Reitz at Aldershot.jpg - Wikimedia Commons Author: Gthvanbrakel
[3] Genl Jan Smuts en sy staf te Concordia, Mei 1902 - Deneys Reitz - Wikipedia
[4] South African High Commissioner Visits the Home Fleet. 10 April 1943, Scapa Flow, Colonel, The Hon Deneys Rietz, High Commissioner for South Africa, During His Recent Visit To The Home Fleet. File:SOUTH AFRICAN HIGH COMMISSIONER VISITS THE HOME FLEET. 10 APRIL 1943, SCAPA FLOW, COLONEL, THE HON DENEYS RIETZ, HIGH COMMISSIONER FOR SOUTH AFRICA, DURING HIS RECENT VISIT TO THE HOME FLEET. A15880.Jpg - Wikimedia Commons Author Davies, F A
[5] Description: Deneys Reitz. circa 1940. Source: Newhistory.co.za. Made available by: Morné van Rooyen Author: Unknown File:Deneys Reitz parent.jpg - Wikipedia
[6] From C. Boni NY Publication, 1930 page 330 of scan. #15 - Commando / [by] Deneys Reitz; a Boer journal of the Boer war, ... - Full View | HathiTrust Digital Library
[7] Deneys Reitz. Commando: A Boer Journal of The Boer War. London Faber & Faber, 1929
[8] With the flag to Pretoria Vol 2 This is in May 23. This is only 125kb Larger one is available.
[9] Heroes of the Boer War. Review of Reviews, 1903.
[10] Davitt, Boer Fight for Freedom, 3rd ed. 1902. (Duke Scan)
[11] Richard Harding Davis With Both Armies in South Africa. Charles Scribner's Sons, New York 1900
[12] Davitt, Boer Fight for Freedom, 3rd ed. 1902. (Duke Scan)
[13] File:SouthAfrica1885.svg - Wikimedia Commons
[14] Basutoland - Wikipedia (Lesotho) Bechuanaland Protectorate - Wikipedia (Botswana) Cape Colony - Wikipedia German South West Africa - Wikipedia (Namibia) Griqualand East - Wikipedia Griqualand West - Wikipedia Matabeleland - Wikipedia Natal: Colony of Natal - Wikipedia Orange Free State - Wikipedia Portuguese Mozambique - Wikipedia Stellaland - Wikipedia Swaziland: Eswatini - Wikipedia South African Republic - Wikipedia Zulu Kingdom - Wikipedia File:SouthAfrica1885.svg - Wikimedia Commons
[15] South African Railway System circa 1900: File:SouthernAfricaRailwayMap 1901.png - Wikipedia
[16] SA Map Roberts, Churchills in Africa p. 120
[17] Veld - Wikipedia
[18] Geology of South Africa - Wikipedia
[19] History of the War in SA, Birch, 1899.
[20] File: The Escarpment and the Drakensberg.jpg - Wikimedia Commons Author: Oggmus See: Great Escarpment, Southern Africa - Wikipedia
[21] File:GiantsCastlePanoramaSmall.jpg - Wikimedia Commons
[22] Navy & Army Illustrated Vol 9, October 7th 1899 and Creswicke, Louis South Africa and the Transvaal War, Vol. 1
[23] Doyle, Arthur Conan. The Great Boer War: Vol 1 Smith and Elder Co. Stanfords.
[24] Cunliffe, History of the Boer War Vol 1 (NY Library Scan)
[25] Image Source: London to Ladysmith, map by G. Philip & son, London & Liverpool. UVA Scan.
[26] Ian Hamilton's March 1900 The Copp, Clark Co. Toronto. And Frontiers & Wars.
[27] Reports on military operations in South Africa and China. U.S. Adjutant-General's Office. 1901
[28] File:10 of 'History of the Boers in South Africa ... with three maps' (11189449154), crop.jpg - Wikimedia Commons Author: The British Library
[29] File:11 of 'History of the Boers in South Africa ... with three maps' (11191958004), crop2.jpg - Wikimedia Commons
[30] Doyle, Arthur Conan. The Great Boer War 17th Edition, 1902
[31] Doyle, Arthur Conan. The Great Boer War 17th Edition, 1902
[32] Doyle, Arthur Conan. The Great Boer War 17th Edition, 1902
[33] Davitt, Boer Fight for Freedom, 3rd ed. 1902. (Duke Scan)
[34] Davitt, Boer Fight for Freedom, 1902 (U. Cal. Scan)
[35] With Roberts to the Transvaal. C.N. Robinson [1900]
[36] Davitt, Boer Fight for Freedom, 1902 (U. Cal. Scan)
[37] #25 - Commando / [by] Deneys Reitz; a Boer journal of the Boer war, ... - Full View | HathiTrust Digital Library C. Boni, NY 1930.

[38] General Botha, the Career and the Man, By Harold Spender, 1916 Google Books, UCAL-Davis: https://books.google.com/books?id=dxw1AQAAIAAJ
[39] General Botha, the Career and the Man, By Harold Spender, 1916 Google Books, UCAL-Davis: https://books.google.com/books?id=dxw1AQAAIAAJ
[40] Heroes of the Boer War. Review of Reviews, 1903. and File:FWReitz VA0947.jpg - Wikimedia Commons
[41] Hillegas, Boers in War, 1900.
[42] Jan and Isie Smuts with one of their daughters, Santa, Source: One Man and his Time. File:Smutsgesin.jpg - Wikipedia
[43] Smuts kabinet 1923 - Deneys Reitz - Wikipedia
[44] General Smuts of the Union of South Africa. Date Created/Published: 1942 Digital ID: (digital file from intermediary roll film) fsa 8e00860 http://hdl.loc.gov/loc.pnp/fsa.8e00860 Reproduction Number: LC-USW33-019083-C (b&w film neg.) Repository: Library of Congress Prints and Photographs Division Washington, D.C. 20540 USA http://hdl.loc.gov/loc.pnp/pp.print https://www.loc.gov/pictures/resource/fsa.8e00860/
[45] With the Flag to Pretoria, H. W. Wilson Vol 2 Dec., 1899.
[46] Black & White Weekly Vol 1-2 1899-1900
[47] War with the Boers, H. Brown, London [1902?] Vol 1
[48] Black & White Weekly Vol 1-2 1899-1900
[49] Anglo-Boer War Album, Cape Town, D. Edwards & co. [1902?]
[50] Creswicke, Louis South Africa and the Transvaal War, Vol. 1 Photo by Wilson, Aberdeen.
[51] Black & White Weekly Vol 1-2 1899-1900
[52] Navy & Army Illustrated Vol 9 Photo by G. W. Wilson, Aberden.
[53] Creswicke, Louis South Africa and the Transvaal War, Vol. 1 Photo by Wilson, Aberdeen.
[54] File:The Second Boer War, 1899-1902 Q71990.jpg - Wikimedia Commons
[55] Black & White Weekly Vol 1-2 1899-1900
[56] Black & White Weekly Vol 1-2 1899-1900
[57] Black & White Weekly Vol 1-2 1899-1900
[58] Black & White Weekly Vol 1-2 1899-1900
[59] Black & White Weekly Vol 1-2 1899-1900 Photo by D. Barnett
[60] History of War SA, Birch, 1899
[61] War in SA, Harding 1899
[62] Navy & Army Illustrated Vol 9
[63] Creswicke, Louis South Africa and the Transvaal War, Vol. 2 1900 Photo by Elliott & Fry, London
[64] Blake, John Y. Fillmore, A West Pointer with the Boers: (Angel Guardian Press, 1903), U. Cal Scan: https://archive.org/details/westpointerwithb00blakrich
[65] Richard Harding Davis With Both Armies in South Africa. Charles Scribner's Sons, New York 1900
[66] Arthur Mann The Boer in Peace and War, 1900
[67] Hillegas, Boers in War, 1900.
[68] Davitt, Boer Fight for Freedom, 3rd ed. 1902. (Duke Scan)
[69] War in SA, Harding 1899
[70] File:The Second Boer War, 1899-1902 Q72365.jpg - Wikimedia Commons
[71] Citizens of the Transvaal https://www.angloboerwar.com/unit-information/boer-units/1953-boer-forces?start=3
[72] Pretoria. Train with troops bound for Norvalspont. - Atom site for DRISA P2301_02.jpg (1400×1024) (drisa.co.za)
[73] https://www.angloboerwar.com/unit-information/boer-units/1953-boer-forces?start=3
[74] Heroes of the Boer War. Review of Reviews, 1903.
[75] Heroes of the Boer War. Review of Reviews, 1903.
[76] Navy & Army Illustrated Vol 11 p. 123 October 27th, 1900.
[77] South Africa; its History, Mackenzie 1899
[78] Basuto-Pony - Basuto pony - Wikipedia
[79] File:Howard C. Hillegas.jpg - Wikimedia Commons Source: With the Boer Forces Howard C. Hillegas, Methuen & Co. 36 Essex Street W.C., London, 1900. Author: Thomas Franklin Fairfax Millard for Howard C. Hillegas.
[80] File:Rifle, bolt action (AM 1930.61-17).jpg - Wikimedia Commons
[81] With the Flag to Pretoria, H. W. Wilson Vol 1
[82] Arthur Mann The Boer in Peace and War, 1900
[83] Anglo-Boer War Album, Cape Town, D. Edwards & co. [1902?]

[84] File:The Second Boer War, 1899-1902 Q68471.jpg - Wikimedia Commons Author Unknown author
[85] Hillegas, Boers in War, 1900. This is below Majuba as a backdrop.
[86] Anonymous (1910) A Handbook of the Boer War - With General Map of South Africa and 18 Sketch Maps and Plans, London & Aldershot: Gale & Polden Ltd File:NorthNatalWarTheatre, colour, crop.png - Wikimedia Commons
Black and White Original: NorthNatalWarTheatre - File:NorthNatalWarTheatre, colour, crop.png - Wikimedia Commons
[87]
https://commons.wikimedia.org/wiki/File:Boer_generals_Lucas_Meijer,_Louis_Botha_and_Daniel_Erasmus,_before_1903.jpg
and Black & White Weekly Vol 1-2 1899-1900
[88] Neville Constantine 3rd: Pretoria_Commando_Crop_3.jpg
[89] File:GiantsCastlePanoramaSmall.jpg - Wikimedia Commons
[90] File:Battle of Glencoe Map.jpg - Wikimedia Commons
[91] Cunliffe, History of the Boer War Vol 1 (NY Library Scan)
[92] Hillegas, With the Boer forces, 1901 (Photograph by Reginald Sheppard, Pretoria.)
[93] Photograph of one of the Transvaal Staatsartillerie 75 mm QF Creusot guns (Thanks to MC Heunis) Anglo Boer War - Artillery and Ammunition - Page 17 - Boer War Forum
[94] Hillegas, With the Boer forces, 1901 (Photograph by Reginald Sheppard.)
[95] . https://commons.wikimedia.org/wiki/File:Light_horse_walers.jpg
[96] Hillegas, With the Boer forces, 1901 (Photograph by Reginald Sheppard.)
[97] File:Battle of Elandslaagte Map.png - Wikimedia Commons
[98] Cunliffe, History of the Boer War Vol 1 (NY Library Scan)
[99] With Roberts to the Transvaal. C.N. Robinson [1900]
[100] BritishBattles.com https://www.britishbattles.com/
[101] BritishBattles.com Map by John Fawkes BritishBattles.com analysing and documenting British Battles from the previous centuries
[102] Imperial Light Horse at Elandslaagte by William Barnes Wolllen.
[103] Image by R. Canton Woodville File:Elandslaagte 1.jpg - Wikimedia Commons
[104] BritishBattles.com https://www.britishbattles.com/
[105] Heroes of the Boer War. Review of Reviews, 1903.
[106] Black & White Weekly Vol 1-2 1899-1900 Photo by D. Barnett. This must have been early in the battle or the next day since the battle ended at night.
[107] File:Boer War; searching for the wounded from the battlefield at Wellcome V0015550.jpg - Wikimedia Commons Watercolour sketch by F. Craig, 1899.
[108] File:Boer War; removing the wounded after the Battle of Elandslaa Wellcome V0015512.jpg - Wikimedia Commons
[109] Image included to show R.A.M.C. personnel working with Indian bearers. “Removing the wounded after the battle of Elandslaagte. This operation had to be performed in a pouring rain and in almost total darkness.” Image Source: With the Flag to Pretoria, H. W. Wilson Vol 1 1900 Sketch by W. Dewdr
[110] Heroes of the Boer War. Review of Reviews, 1903.
[111] Heroes of the Boer War. Review of Reviews, 1903.
[112] File:Hollanderkorps Memorial at Elandslaagte.jpg - Wikimedia Commons Author: David Biggins
[113] The Dutch Corps Monument is a provincial heritage site in Kliprivier in the KwaZulu-Natal district of South Africa. It was designed by Gerard Moerdijk. The monument was vandalized in 2014. In 1982, it was described in the Government Gazette as
This memorial, which was designed by G. Moerdyk of Pretoria, was built by the firm Barker and Nel of Volksrust and was erected by the citizens of The Netherlands in memory of members of the Dutch Corps who died during the battle of Elandslaagte on 21 October 1899. Dutch Corps Monument - Wikipedia. Credit: David Biggins.
[114] File:Boer War Photograph Taken by B W Caney Q115353.jpg - Wikimedia Commons Author::B.W. Caney
[115] The Mounted Police of Natal by H. P. Holt, published by John Murray, London, 1913 The Mounted Police of Natal : H. P. Holt : Free Download, Borrow, and Streaming : Internet Archive
[116] File:Positions round Ladysmith - November 1899 - Project Gutenberg eText 16466.jpg - Wikimedia Commons
[117] Anglo-Boer War Album, Cape Town, D. Edwards & co. [1902?]
[118] Heroes of the Boer War. Review of Reviews, 1903.

[119] Creswicke, Louis South Africa and the Transvaal War, Vol. 2 1900 Photo by Wilson, Aberdeen. And Haldane, How We Escaped from Pretoria. Photo by Stereoscopic Co.
[120]SA Boer-British War Vol 1. Hopkins, 1900
[121] Hillegas, Boers in War, 1900.
[122] Image source: With the Flag to Pretoria, H. W. Wilson Vol 1 1900 and britishbattles.com. RML 2.5-inch mountain gun - Wikipedia
[123] File:Devonshire Regiment-Ladysmith.JPG - Wikimedia Commons
[124] 155 mm Creusot Long Tom - Wikipedia Image Source: 155 mm Creusot Long Tom - Wikipedia File:Long tom on platform at Mafeking.jpg - Wikimedia Commons
[125] Anglo-Boer War Album, Cape Town, D. Edwards & co. [1902?]
[126] Black & White Weekly Vol 1-2 1899-1900 Photo by Rene' Bull.
[127] Creswicke, Louis South Africa and the Transvaal War, Vol. 2 1900 Photo by Cribb, Southsea.
[128] File:The Second Boer War, 1899-1902 Q72305.jpg - Wikimedia Commons Author: William Skeoch Cumming
[129] Image source: britishbattles.com
[130] File:Termite Magnetic DSC03613.jpg - Wikimedia Commons
[131] BritishBattles.com Map by John Fawkes.
[132] Image Source: With the Flag to Pretoria, H. W. Wilson Vol 1
[133] File:Boer War; Indians carrying a wounded soldier in a dhoolie. P Wellcome V0015530.jpg - Wikimedia Commons
[134] Navy & Army Illustrated Vol 10 May 26th, 1900 File:British Army, Boer War, Q 71941.jpg - Wikimedia Commons
[135] Times History War in S.A Vol 4 (1906)
[136] Image Source: BritishBattles.com
[137] Webster Davis, John Bull's crime; or, Assaults on Republics. 1901 New York The Abbey Press
[138] Webster Davis, John Bull's crime; or, Assaults on Republics. 1901 New York The Abbey Press
[139] Davitt, Boer Fight for Freedom, 3rd ed. 1902. (Duke Scan)
[140] Image source: BritishBattles.com
[141] Black & White Weekly Vol 1-2 1899-1900
[142] Image source: britishbattles.com
[143] Navy & Army Illustrated Vol 10
[144] Navy & Army Illustrated Vol 10 and With the Flag to Pretoria, H. W. Wilson Vol 1 1900 Photo by Knight.
[145] Black & White Weekly Vol 1-2 1899-1900
[146] "Our special Correspondent who was locked up in Ladysmith sends an illustration of the balloon being sent up to observe the Boer positions." Navy & Army Illustrated Vol 10
[147] Richard Harding Davis With Both Armies in South Africa. Charles Scribner's Sons, New York 1900
[148] The Record of a Regiment of the Line Being a Regimental History of the 1st Battalion Devonshire Regiment during the Boer War 1899-1902, by Mainwaring George Jacson. Hutchinson and Co., London, 1908. Harvard Copy: #13 - The record of a regiment of the line: being a regimental history ... - Full View | HathiTrust Digital Library And: Siege of Ladysmith - Wikipedia File:Ladysmith Town Hall 1900 - Project Gutenberg eText 15972.png - Wikimedia Commons
[149] After Pretoria: The Guerilla War Vol 3 Feb., 1901
[150] The Staats Model School Officers Prison. Image credit: Haldane and National Library of Scotland
[151] Haldane, How We Escaped from Pretoria.
[152] Neville C 2nd Batch
[153] From a Photograph by Mackintosh, Kelso. From: Winston Spencer Churchill by A. MacCallum Scott, Methuen, London 1905. Lord Randolph Churchill - Wikipedia
[154] Anglo Boer War - Small Arms and Ammunition - Page 15 - Boer War Forum Credit: Neville Constantine
[155] Rinderpest 1896-CN - Rinderpest - Wikipedia
[156] Arthur Mann The Boer in Peace and War, 1900.
[157] Richard Harding Davis With Both Armies in South Africa. Charles Scribner's Sons, New York 1900
[158] Courtesy of Neville Constantine: Anglo Boer War - Artillery and Ammunition - Page 18 - Boer War Forum
[159] Per Neville Constantine: Anglo Boer War - Artillery and Ammunition - Page 15 - Boer War Forum
[160] With the Flag to Pretoria, H. W. Wilson Vol 2 Dec., 1899. PF: The end of the barrel was damaged and it was cut down and returned to action.
[161] File:Photograph; 'Long Tom,' RAMC; 801-22-3 Wellcome L0021295.jpg - Wikimedia Commons

[162] Creswicke, Louis. South Africa and the Transvaal War. Vol 2. 1900 And: Battle of Colenso Map - Battle of Colenso - Wikipedia
[163]With the Flag to Pretoria, H. W. Wilson Vol 1, 1900
[164] Benett Burleigh The Natal Campaign Chapman Hall Ltd. 1900
[165] Navy & Army Illustrated Vol 10
[166] Transvaal Staatsartillerie 120 mm Krupp Howitzer No. 1 at Ladysmith. Per Neville C. Anglo Boer War - Artillery and Ammunition - Page 18 - Boer War Forum
[167] Creswicke SA War Vol 2
[168] Courtesy of Neville Constantine. See: Anglo Boer War - Artillery and Ammunition - Page 4 - Boer War Forum
[169] Hillegas, With the Boer forces, 1901 (Photograph by R. Steger, Pretoria.)
[170] Image and text credit Neville Constantine: Anglo Boer War - Artillery and Ammunition - Page 16 - Boer War Forum From the album of Majoor Friedrich Wilhelm von Wichmann, Transvaal Staatsartillerie.
[171] Anglo-Boer War Album, Cape Town, D. Edwards & co. [1902?]
[172] Original lent by 'Navy and Army Illustrated' File:QF 4.7 inch gun Colenso.jpg - Wikipedia
[173] Frontispiece to "Surgical Experiences in South Africa: File:Varieties of Ammunition collected at Ladysmith - Project Gutenberg etext 21280.jpg - Wikimedia Commons and Courtesy of Neville Constantine. Anglo Boer War - Artillery and Ammunition - Page 5 - Boer War Forum
[174] Credit Rob Davidson: Anglo Boer War - Artillery and Ammunition - Page 7 - Boer War Forum
[175] Credit Rob Davidson: Anglo Boer War - Artillery and Ammunition - Page 7 - Boer War Forum
[176] The leaguer of Ladysmith, by Captain C.M. Dixon. London, Eyre and Spottiswoode [1900] Catalog Record: The leaguer of Ladysmith | HathiTrust Digital Library See Also: Clive MacDonnell Dixon - Wikipedia
[177] The leaguer of Ladysmith, by Captain C.M. Dixon. London, Eyre and Spottiswoode [1900] Catalog Record: The leaguer of Ladysmith | HathiTrust Digital Library See Also: Clive MacDonnell Dixon - Wikipedia
[178] Clive Dixon's Ladysmith Lyre Image Credit: Rob Davidson
[179] War with the Boers, H. Brown, London [1902?] Vol 2
[180] Creswicke SA War Vol 2
[181] The Sphere - Saturday 24 November 1900 p. 13 https://www.britishnewspaperarchive.co.uk/viewer/bl/0001861/19001124/016/0013
[182] With the Flag to Pretoria, H. W. Wilson Vol 1
[183] File:Spioenkop01.jpg - Wikipedia User: Vberger~commonswiki Notation: I, the copyright holder of this work, release this work into the public domain. This applies worldwide. I grant anyone the right to use this work for any purpose.
[184] Cunliffe, History of the Boer War Vol 1 (NY Library Scan)
[185] Image source: With the Flag to Pretoria, H. W. Wilson Vol 1 1900 mage Credit: Rob Davidson Anglo Boer War - Small Arms and Ammunition - Page 6 - Boer War Forum
[186] Image source: britishbattles.com
[187] Image source: With the Flag to Pretoria, H. W. Wilson Vol 1 1900 and britishbattles.com
[188] Image source: britishbattles.com
[189] Webster Davis, John Bull's crime; or, Assaults on Republics. 1901 New York The Abbey Press
[190] Image source: With the Flag to Pretoria, H. W. Wilson Vol 1 1900 and britishbattles.com
[191] October 21 Folder Anglo Boer War - Small Arms and Ammunition - Page 4 - Boer War Forum
[192] One of the 10th Mountain Battery's 2.5-inch Mountain Guns captured from the British at Nicholson's Nek on 30 October 1899. Anglo Boer War - Artillery and Ammunition - Page 25 - Boer War Forum
[193] https://www.rijksmuseum.nl/nl/zoeken/objecten?set=RP-F-F01113#/RP-F-F01113-M,13
[194] File:Spioenkop2.jpg - Wikipedia "A killed Boer sniper after the Battle of Spioenkop, near Ladysmith, Natal, South Africa. "Courtesy of the Cape Archive, J5262
[195] Black & White Weekly Vol 1-2 1899-1900
[196] With Roberts to the Transvaal. C.N. Robinson [1900] Navy and Army.
[197] Image source: britishbattles.com
[198] Image Source: With the Flag to Pretoria, H. W. Wilson Vol 1 1900 By A. Forestier after a sketch by Fred Villers.
PF: Indians are only sparingly depicted in the documentation of battles. The sketch was made contemporaneously with the event and since Indian bearers are in the middle distance and background it leads credence to the conclusion that they were assigned a role on the summit of the Kop. Also, Indian bearers were not perceived as threatening to the Boers since they were known to not carry weapons, were clearly carrying stretchers and regarded as noncombatants. Since they were not trained soldiers or R.A.M.C. personnel, they were more expendable in the event they were detained or captured by Boers

who were working details in the same vicinity and in military control of the Kop. It is a shame that this picture is marred by the one vertical line.
Whether Gandhi participated in this detail is uncertain. In the context of his entire article recalling his first interactions with Gandhi Mr. Vere Stent is not specific enough about which morning he spoke to Gandhi after a grueling night's work. He was near the action the day before in order to receive General Woodgate, a task that would not have been entrusted to someone not able to provide due care for so important a charge. I can only presume that his skill as a bearer and leader were known to the medical staff from his prior work at Colenso.

[199] Process print after R. Caton Woodville. Halftone Co. Wellcome Collection. https://www.jstor.org/stable/community.24840258

[200] With the Flag to Pretoria, H. W. Wilson Vol 2 March 1900.

[201] Navy & Army Illustrated Vol 9 Photo by 'Navy & Army.'

[202] Black & White Weekly Vol 1-2 1899-1900

[203] Image source: britishbattles.com

[204] Often reproduced reversed. In this case the negative is correct, the buttons are on the right and button holes are on the left Champ de bataille de Spion-Kopje - Battle of Spion Kop - Wikipedia

[205] Frontiers and Wars https://archive.org/details/frontierswarshis0000chur

[206] Richard Harding Davis With Both Armies in South Africa. Charles Scribner's Sons, New York 1900

[207] File:The Second Boer War, 1899-1902 Q71942.jpg - Wikimedia Commons

[208] Creswicke, Louis South Africa and the Transvaal War, Vol. 2 1900 Photo by Cribb, Southsea.

[209] Repository: Library of Congress Prints and Photographs Division Washington, D.C. 20540 USA https://www.loc.gov/pictures/item/2006684574/resource/

[210] Image source: Navy–Army Illustrated & britishbattles.com

[211] Navy & Army Illustrated Vol 9

[212] Frontiers and Wars https://archive.org/details/frontierswarshis0000chur

[213] Buller's Campaign with the Natal Field Force Map by A.I. Ltd.

[214] Navy & Army Illustrated Vol 10

[215] Navy & Army Illustrated Vol 10

[216] Times History War in S.A Vol 3 Photo Perceval Landon. File:The Second Boer War, 1899-1902 Q71654.jpg - Wikimedia Commons

[217] Frontiers and Wars four early books, 1962. Credit Radio, Times Hulton Picture Library.

[218] With the Flag to Pretoria, H. W. Wilson Vol 2 Feb., 1900

[219] Hillegas, Boers in War, 1900.

[220] Anglo-Boer War Album, Cape Town, D. Edwards & co. [1902?]

[221] File:Resa del bacino del Brandewater.jpg - Wikimedia Commons

[222] Anglo-Boer War Album, Cape Town, D. Edwards & co. [1902?] T.J. Hall Photo.

[223] Blake, John Y. Fillmore, A West Pointer with the Boers: A Personal Narrative of Colonel J Y F Blake, Commander of the Irish Brigade (Angel Guardian Press, 1903, and: John Blake (soldier) - Wikipedia And Irish commandos - Wikipedia

[224] Hillegas, With the Boer forces, 1901 (Photograph by Leo Weinthal.)

[225] File:Officers of the Irish Brigade.jpg - Wikimedia Commons Jennifer Bosch 2019.

[226] Howard Hillegas, The Boers in War Chapter 9. https://www.angloboerwar.com/books/68-hillegas-with-the-boer-forces/1465-hillegas-chapter-9-foreigners-in-the-war

[227] Black & White Weekly Vol 1-2 1899-1900

[228] Black & White Weekly Vol 1-2 1899-1900

[229] Johannesburg, 24 April 1900. Explosion accident at the Begbies munition works. - Atom site for DRISA P1880.jpg (1400×1007) (drisa.co.za)

[230] Anglo-Boer War Album, Cape Town, D. Edwards & co. [1902?]

[231] File:The Second Boer War, 1899-1902 Q68477.jpg - Wikimedia Commons

[232] File:The Second Boer War, 1899-1902 Q68484.jpg - Wikimedia Commons

[233] Image: 10 of 'History of the Boers in South Africa ... with three maps' (11189449154), crop - Vet River - Wikipedia

[234] With the Flag to Pretoria, H. W. Wilson Vol 2 July, 1900 File:Koos de la Rey.jpg - Wikimedia Commons

[235] Hiley and Hassell, The Mobile Boer, Grafton Press, New York, (c. 1902)

[236] Second Anglo-Boer War - 1899 - 1902 | South African History Online (sahistory.org.za)

[237] File:The Second Boer War, 1899-1902 Q71945.jpg - Wikimedia Commons Author: William Skeoch Cumming

[238] Members of the Staffordshire Regiment with their Vickers-Maxim 1-pounder. Note the belts of 25 rounds, coiled on top of their ammunition boxes, to left and right. Courtesy of Neville C: Anglo Boer War - Artillery and Ammunition - Page 22 - Boer War Forum

[239] File:The Second Boer War, 1899-1902 Q72153.jpg - Wikimedia Commons

[240] SA National Archives: http://www.nationalarchives.gov.za/node/1236

[241] Vet River, circa 1900. Bridge damaged by dynamite. (Van Hoepen, Pretoria) (Donated by Mr. GJ Krister, Pretoria) - Atom site for DRISA

[242] Reports on military operations in South Africa and China. U.S. Adjutant-General's Office. 1901

[243] Louis Botha with the Italian Volunteer Legion Italian Volunteer Legion - Wikipedia File:Louis Botha - Project Gutenberg eText 16462.jpg - Wikimedia Commons

[244] Black & White Weekly Vol 1-2 1899-1900

[245] Black & White Weekly Vol 1-2 1899-1900

[246] Heroes of the Boer War. Review of Reviews, 1903.

[247] Anglo Boer War - Small Arms and Ammunition - Page 14 - Boer War Forum Credit: Rob Davidson

[248] Heroes of the Boer War. Review of Reviews, 1903.

[249] Ian Hamilton's March 1900 The Copp, Clark Co. Toronto and online sources. See Bibliography,

[250] Ian Hamilton's March 1900 The Copp, Clark Co. Toronto and online sources. See Bibliography,

[251] File:Gezicht op een rivieroever met paardenkarren bij Kroonstad, Zuid-Afrika History Recorders and History Makers, Correspondents' Carts waiting Lord Roberts' entry to Kroonstadt, S.A., crop.jpg - Wikimedia Commons

[252] After Pretoria: The Guerilla War Vol 3 July, 1900

[253] Doyle, Arthur Conan. The Great Boer War 17th Edition, 1902

[254]Daniel (Danie) Johannes Stephanus Theron File:Danie Theron.jpg - Wikipedia

[255] Heroes of the Boer War. Review of Reviews, 1903.

[256] File:The Second Anglo - Boer War, South Africa 1899 - 1902 Q72318.jpg - Wikimedia Commons

[257] File:Boer War Q71976.jpg - Wikimedia Commons

[258] File:Yeomanry South Africa.jpg - Wikimedia Commons

[259] Rhenoster River, South Africa. 1900. Members of the Royal Engineers construct a temporary railway bridge beside that destroyed by the retreating Boers. (Original stereocard housed in AWM Archive Store) See Rhenoster River, South Africa. 1900. Members of the Royal Engineers construct a temporary railway ... | Australian War Memorial (awm.gov.au)

[260] British troops guarding the Rhenoster River bridge whilst repair work is being done after Boers dynamited it ABW British troops guarding the Rhenoster River bridge whilst repair work is being done after Boers dynamited it ....ABW | Troops, War, Armed conflict (pinterest.com)

[261] File:Viljoensdrif station.jpg - Wikipedia

[262] Anglo-Boer War Album, Cape Town, D. Edwards & co. [1902?]

[263] Anglo-Boer War Album, Cape Town, D. Edwards & co. [1902?]

[264] ABW Source

[265] War in SA, Harding 1899

[266] Heroes of the Boer War. Review of Reviews, 1903.

[267] File:South Africa; English tourists at the battlefield of Doornkop Wellcome V0037949.jpg - Wikimedia Commons File from: Wellcome Images.

[268] File:Doornkop Battle.jpg - Wikimedia Commons Author: AxelRavera

[269] South Africa; its History, Heroes and Wars. Mackenzie [1899]

[270] Navy & Army Illustrated Vol 9 Photo by W. P. Edwards

[271] Hillegas, Boers in War, 1900.

[272] Richard Harding Davis With Both Armies in South Africa. Charles Scribner's Sons, New York 1900

[273] File:The Second Boer War, 1899-1902 Q72488.jpg - Wikimedia Commons

[274] Richard Harding Davis With Both Armies in South Africa. Charles Scribner's Sons, New York 1900

[275] Black & White Weekly Vol 3 1900

[276] With the Flag to Pretoria, H. W. Wilson Vol 2 June, 1900.

[277] After Pretoria: The Guerilla War Vol 3 August, 1900

[278] Rosslyn, James, 5th Earl of. Twice Captured: A Record Of Adventure During The Boer War William Black Wood & Sons, Edinburgh 1900

[279] Volume 14 at HathiTrust June 1901 to November 1901: Catalog Record: Catalog Record: The Windsor magazine : an illustrated monthly... | HathiTrust Digital Library Michigan Scan: #7 - The Windsor magazine : an illustrated monthly for ... v.14. - Full View | HathiTrust Digital Library
[280] WSC, London to Ladysmith.
[281] Lord Rosslyn, stationed at a window with his camera, managed to get some capital snapshots. At the head of one body of troops rode the Duke of Marlborough and Mr. Winston Churchill, and they recognised Lord Rosslyn's head at an upper storey window with "Halloo, Harry!" The Sphere - Saturday 21 July 1900 P.13. https://www.britishnewspaperarchive.co.uk/viewer/bl/0001861/19000721/014/0013
[282] File:The Second Boer War, 1899-1902 Q72136.jpg - Wikimedia Commons Author: William Skeoch Cumming
[283] File:The Second Boer War, 1899-1902 Q72151.jpg - Wikimedia Commons William Skeoch Cumming
[284] With the Flag to Pretoria, H. W. Wilson Vol 2 June, 1900.
[285] Richard Harding Davis With Both Armies in South Africa. Charles Scribner's Sons, New York 1900
[286] With the Flag to Pretoria, H. W. Wilson Vol 2 June, 1900.
[287] Richard Harding Davis With Both Armies in South Africa. Charles Scribner's Sons, New York 1900
[288] Library of Congress Prints and Photographs Division Washington, D.C. https://www.loc.gov/pictures/item/00651051/resource/
[289] To the Bitter End: a photographic history of the Boer War, 1899-1902, by Lee, Emanoel C. G. Harmondsworth: Penguin, 1986. https://archive.org/details/tobitterend00eman_0
[290] Crop of map from, My Early Life.
[291] Haldane, How We Escaped from Pretoria. Image credit: Neville Constantine, Neville C. at Angloboerwar.com Anglo Boer War - Armored Train Ambush 11-15-99 Chuchill & Haldane - Page 13 - Boer War Forum
[292] File:Termine cremagliera Saline-Volterra.jpg-Wikimedia Commons
[293] Rack railway-Wikipedia
[294] Waterval-Boven, circa 1901. Armoured train during Anglo-Boer War at railway station. - Atom site for DRISA N78689.jpg (1400×1045) (drisa.co.za)
[295] Our Military Officers. Album of portraits with life sketches. Original Publication: Onze krijgs-officieren: album van portretten met levens-schetsen der. by Gustav Schoeman Preller, Frans Vredenrijk Engelenburg 1904 Publisher "Volksstem" kantore
Catalogue Record: https://archive.org/search?query=title%3A%28Onze%20Krijgs-Officieren%29 Direct Link: https://archive.org/details/onzekrijgsoffic00engegoog
[296] File:South Africa-Mpumalanga-Long Tom01.jpg - Wikimedia Commons Author: NJR ZA
[297] File:Kaapsehoop-berg.jpg - Wikimedia Commons Author: Karen Wiebe.
[298] Courtesy of Neville Constantine: Anglo Boer War - Artillery and Ammunition - Boer War Forum Image courtesy of MC Heunis.
[299] Courtesy Neville C: Anglo Boer War - Artillery and Ammunition - Page 24 - Boer War Forum
[300] Anglo-Boer War Album, Cape Town, D. Edwards & co. [1902?]
[301] Wildebeest - Wikipedia And: File:Black wildebeest, or white-tailed gnu, Connochaetes gnou at Krugersdorp Game Reserve, Gauteng, South Africa (30570451563).jpg - Wikimedia Commons Credit: Derek Keats
[302] Sable antelope - Wikipedia And: File:Sable antelope (Hippotragus niger) adult male.jpg - Wikipedia Credit Charles J. Sharp
[303] Blake, John Y. Fillmore, A West Pointer with the Boers: (Angel Guardian Press, 1903), U. Cal Scan: https://archive.org/details/westpointerwithb00blakrich
[304] Blake, John Y. Fillmore, A West Pointer with the Boers: (Angel Guardian Press, 1903), U. Cal Scan: https://archive.org/details/westpointerwithb00blakrich
[305] Blake, John Y. Fillmore, A West Pointer with the Boers: (Angel Guardian Press, 1903), U. Cal Scan: https://archive.org/details/westpointerwithb00blakrich
[306] File:CFbeyers.jpg - Wikimedia Commons
[307] https://commons.wikimedia.org/wiki/File:Genl_Christiaan_Frederik_Beyers.jpg
[308] File:The Second Boer War, 1899-1902 Q72006.jpg - Wikimedia Commons
[309] Image Source: Boer PoWs | Saint Helena Island Info: All about St Helena, in the South Atlantic Ocean
[310] The prisoners' camp on Hawkin's Island. Navy & Army Vol 13, Feb. 15th, 1902
[311] Black & White Weekly Vol 1-2 1899-1900
[312] File:Magaliesberg08.jpg - Wikimedia Commons Credit: Androstachys
[313] File:Magaliesberg te Skeerpoort, a,jpg.jpg - Wikimedia Commons Author: JMK

[314] Battle of Nooitgedacht. https://commons.wikimedia.org/wiki/File:Plan_of_the_battle_of_Nooitgedacht_(Battles_of_the_nineteenth_century,_1901).jpg (Battles of the nineteenth century, 1901) Author:Archibald Forbes

[315] With the Flag to Pretoria, H. W. Wilson Vol 2 Nov, 1899.

[316] Library of Congress Prints and Photographs Division Washington, D.C. 20540 USA https://www.loc.gov/pictures/resource/cph.3a40944/

[317] Davitt, Boer Fight for Freedom, 1902 (U. Cal. Scan)

[318] Davitt, Boer Fight for Freedom, 3rd ed. 1902. (Duke Scan)

[319] Per Neville Constantine Anglo Boer War - Armored Train Ambush 11-15-99 Chuchill & Haldane - Page 21 - Boer War Forum

[320] Courtesy of Neville Constantine.

[321] Per Neville Constantine: Here is a photograph of a similar setup for the lead truck. These are H.M.S. Philomel men with a Hotchkiss 3-pdr, opposed to Tartar men with a 9-pdr, and the truck benefits from the addition of armour plate to its sides. This train was photographed in Durban, before proceeding to the front. Anglo Boer War - Armored Train Ambush 11-15-99 Chuchill & Haldane - Page 6 - Boer War Forum

[322] BoerWarBlockHouse SouthAfrica - Second Boer War - Wikipedia Credit: Danie van der Merwe.

[323] A gun carrying carriage named “Cock of the North” of No. 3 Artillery Train at the railway station at Bronkhorstspruit, Gauteng Province. File:The Second Boer War, 1899-1902 Q72437.jpg - Wikimedia Commons Author: William Skeoch Cumming

[324] Image Source: With the Flag to Pretoria, H. W. Wilson Vol 1

[325] Anglo-Boer War Album, Cape Town, D. Edwards & co. [1902?]

[326] File:The Second Boer War, 1899-1902 Q72217.jpg - Wikimedia Commons Author: William Skeoch Cumming

[327] After Pretoria: The Guerilla War Vol 3 Jan-April 1901

[328] Navy & Army Vol 14 p. 285, June 7th 1902. By the time of publication the hostilities had ended.

[329] Navy & Army Vol 13 p. 59 Oct. 5th, 1901

[330] Navy & Army Vol 13 p. 59 Oct. 5th, 1901

[331] SA National Archives: http://www.nationalarchives.gov.za/node/1236

[332] Navy & Army Vol 13 p. 59 Oct. 5th, 1901

[333] Navy & Army Illustrated Vol 11 p. 123 October 27th, 1900.

[334] Photographical Collection Anglo-Boer War Museum, Bloemfontein SA. Also Available Here: File:VerskroeideAarde1.jpg - Wikimedia Commons

[335] File:The Second Boer War, 1899-1902 Q72284.jpg - Wikimedia Commons Author: William Skeoch Cumming

[336] File:The Second Boer War, 1899-1902 Q72160.jpg - Wikimedia Commons Author: William Skeoch Cumming.

[337] After Pretoria: The Guerilla War Vol August 1900.

[338] De Camp Courant Herdenkings-uitgawe vir Slagoffers van die Tweede Vryheidsoorlog 1899-1902 ** De Camp Courant: A Memorial Publication for Victims of the Anglo-Boer War 1899-1902 (arkpark.net) And: https://decampcourant.arkpark.net/

[339] Concentration Camps where the Boer woman, their children and the elderly were held during the Anglo-Boer War 1899-1902 (arkpark.net)

[340] Photographical Collection Anglo-Boer War Museum, Bloemfontein SA. Image: Concentration camps in the South African War? Here are the real facts by Fransjohan Pretorius, (The Conversation), February 18, 2019 | South African History Online (sahistory.org.za)

[341] SA National Archives: http://www.nationalarchives.gov.za/node/1236

[342] File:The Second Boer War, 1899-1902 Q72453.jpg - Wikimedia Commons

[343] File:Boercamp1.jpg - Wikimedia Commons and Women and Children in White Concentration Camps during the Anglo-Boer War, 1900-1902 | South African History Online (sahistory.org.za)

[344] File:Norvals Pont Camp, c.1901. (22988974995).jpg - Wikimedia Commons

[345] Photographical Collection Anglo-Boer War Museum, Bloemfontein SA. Image: Concentration camps in the South African War? Here are the real facts by Fransjohan Pretorius, (The Conversation), February 18, 2019 | South African History Online (sahistory.org.za)

[346] Navy & Army Illustrated Vol 9

[347] Boer family with a wagon on a farm in South Africa circa 1900. Titled: Father and son going to the front by Boer War Colourised Photographs File:Boer Family with a wagon.jpg - Wikimedia Commons Tinus le Roux (Photos Redux - Colourising History)

[348] Dordrecht, 1902. Engine buried 25 feet deep for 8 days before dug out by 100 boys. Dordrecht, 1902. Engine buried 25 feet deep for 8 days before dug out by 100 boys. - Atom site for DRISA P2899_01.jpg (1400×1008) (drisa.co.za)
[349] SA Political. Source: https://www.warmuseum.ca/cwm/exhibitions/boer/boerwarmap-lrg_e.html
[350] File:Schoemansville panorama from cable-way hill.jpg - Wikimedia Commons Author: Farawayman
[351] The Queen's Empire. Volume 1. Cassell & Co. London Eland crossing QE1 48 - Elands River (Mpumalanga) - Wikipedia
[352] Hillegas, Boers in War, 1900.
[353] "Aliwal North district, 1978. Derelict cart." - Atom site for DRISA N85729.jpg (1400×1080) (drisa.co.za)
[354] Drakensberg. Men fixing overturned oxwagon, mountain in background. - Atom site for DRISA N08606.jpg (1400×961) (drisa.co.za)
[355] File:The Second Boer War, 1899-1902 Q72211.jpg - Wikimedia Commons
[356] File:The Second Boer War, 1899-1902 Q72102.jpg - Wikimedia Commons
[357] Black & White Weekly Vol 1-2 1899-1900
[358] Anglo-Boer War Album, Cape Town, D. Edwards & co. [1902?]
[359] Nicolaas Pieter Johannes Janse van Rensburg (30 August1864 – 11 March1926) Seer van Rensburg - Wikipedia See Also ABW.com for multiple references with identical names, some of whom were captured prior to 1901.
[360] File:Seer van Rensburg.jpg - Wikipedia
[361] Black & White Weekly Vol 1-2 1899-1900
[362] Original Source: Edwin Sharpe Grew: Field-Marshal Lord Kitchener. His Life and Work for the Empire. London 1916. Volume 2. File:Kitchener at the Peace Conference that ended the Second Boer War.png - Wikimedia Commons On 28 February 1901 British commander Lord Kitchener met Boer commander Louis Botha at Middelburg, Transvaal. The talks failed to bring peace and Kitchener commenced the barbed wire and block house campaign to thwart attacks by Boer guerilla fighters. Category:Middelburg Peace Conference - Wikimedia Commons Also: File:The Second Boer War, 1899-1902 Q68497.jpg - Wikimedia Commons
[363] File:Kitchener at the Peace Conference that ended the Second Boer War.png - Wikimedia Commons
[364] Image: Hillegas, With the Boer forces, 1901 (Photograph by Leo Weinthal, Pretoria.) and quote: With the Flag, After Pretoria: The Guerilla War Vol 3 p. 91 1900.
[365] Page 13 https://www.britishnewspaperarchive.co.uk/viewer/bl/0001684/19010309/091/0013
[366] Southport Guardian - Saturday 23 March 1901 page 5. Col 7 https://www.britishnewspaperarchive.co.uk/viewer/bl/0001783/19010323/121/0005
[367] Southport Guardian - Saturday 23 March 1901 page 5. Col 7 https://www.britishnewspaperarchive.co.uk/viewer/bl/0001783/19010323/121/0005
[368] Southport Guardian - Saturday 23 March 1901 page 5. Col 7 https://www.britishnewspaperarchive.co.uk/viewer/bl/0001783/19010323/121/0005
[369] Southport Guardian - Saturday 23 March 1901 page 5. Col 7. https://www.britishnewspaperarchive.co.uk/viewer/bl/0001783/19010323/121/0005
[370] File:Philip Rudolph Botha (1851-1901), Boer War general.jpg - Wikimedia Commons
[371] Davitt, Boer Fight for Freedom, 1902 (U. Cal. Scan)
[372] Hillegas, Boers in War, 1900.
[373] Tuesday, March 26, 1901 Victory by General Babington over Delarey at Ventersdorp. Nine Boer guns captured.
[374] File:The National Archives UK - CO 1069-215-66.jpg - Wikimedia Commons
[375] Navy & Army Illustrated Vol 10
[376] With the Flag to Pretoria, H. W. Wilson Vol 2 April, 1900.
[377] Repository: Library of Congress Prints and Photographs Division Washington, D.C. 20540 USA http://hdl.loc.gov/loc.pnp/pp.prin
https://www.loc.gov/pictures/item/2016645958/resource/
[378] Anglo-Boer War Album, Cape Town, D. Edwards & co. [1902?]
[379] Navy & Army Illustrated Vol 10
[380] After Pretoria: The Guerilla War Vol 3 April, 1901
[381] File:10 of 'History of the Boers in South Africa ... with three maps' (11189449154), crop.jpg - Wikimedia Commons
[382] Map of Stellaland, from Scottish Geographical Magazine of 1885, cropped and enhanced. https://commons.wikimedia.org/wiki/File:Map_Stellaland.jpg
[383] South African Railway System circa 1900: File:SouthernAfricaRailwayMap 1901.png - Wikipedia
[384] South Africa; its History, Mackenzie 1899

[385] Davitt, Boer Fight for Freedom, 1902 (U. Cal. Scan)
[386] File:The Second Boer War, 1899-1902 Q72267.jpg - Wikimedia Commons Author: William Skeoch Cumming
[387] Black & White Weekly Vol 1-2 1899-1900
[388] Ox wagon stuck in Crocodile River - Atom site for DRISA N00061.jpg (1400×925) (drisa.co.za)
[389] File:The Second Boer War, 1899-1902 Q72297.jpg - Wikimedia Commons Author: William Skeoch Cumming
[390] SA National Archives: http://www.nationalarchives.gov.za/node/1236
[391] File:State artillery corps, Orange Free State - Korps van die Staatsartillerie, Oranje-Vrystaat, near Colesberg.jpg - Wikimedia Commons
[392] History of the Boers in South Africa, Theal, George McCall. Sonnenschein & Co. London 1887 British Library HMNTS 9061.eee.18. Identifier: 003607007 https://access.bl.uk/item/pdf/lsidyv2afa3205
[393] File:The Second Boer War, 1899-1902 Q72085.jpg - Wikimedia Commons Author: William Skeoch Cumming
[394] Author: Montagu Brownlow Parker Blockhouse 1901 - Category:Blockhouses of the Second Boer War - Wikimedia Commons
[395] Black & White Weekly Vol 1-2 1899-1900
[396] Anglo-Boer War Album, Cape Town, D. Edwards & co. [1902?]
[397] Anglo-Boer War Album, Cape Town, D. Edwards & co. [1902?]
[398] Anglo-Boer War Album, Cape Town, D. Edwards & co. [1902?]
[399] Black & White Weekly Vol 1-2 1899-1900
[400] SA War Vol 3 1900 Photo by Vickers Sons & Maxim, Limited.
[401] File:Modder River pontoon bridge.jpg - Wikimedia Commons
[402] File:The Second Boer War, 1899-1902 Q72248.jpg - Wikimedia Commons Author: William Skeoch Cumming
[403] File:Boer War Photograph Taken by B W Caney Q115360.jpg - Wikimedia Commons Author: B.W. Caney
[404] Anglo-Boer War Album, Cape Town, D. Edwards & co. [1902?]
[405] The leaguer of Ladysmith, by Captain C.M. Dixon. London, Eyre and Spottiswoode [1900] Catalog Record: The leaguer of Ladysmith | HathiTrust Digital Library See Also: Clive MacDonnell Dixon - Wikipedia
[406] Anglo-Boer War Album, Cape Town, D. Edwards & co. [1902?]
[407] File:The Second Boer War, 1899-1902 Q72417.jpg - Wikimedia Commons
[408] Pocket tinderbox with firesteel and flint. This type was used during the Boer War due to a scarcity of matches. Author: Theo V. Bresler, File:Tinderbox.jpg - Wikimedia Commons
[409] Cunliffe, History of the Boer War Vol 2 (Composite of Mich. St. & U. Cal. Scans)
[410] Times History, Vol 5 page 600.
[411] Heroes of the Boer War. Review of Reviews, 1903.
[412] File:The Second Boer War, 1899-1902 Q72521.jpg - Wikimedia Commons Author: William Skeoch Cumming
[413] Author: Andrew Hall File:Blockhouse, Prieska.JPG - Wikimedia Commons
[414] Davitt, Boer Fight for Freedom, 1902 (U. Cal. Scan)
[415] Cunliffe, History of the Boer War Vol 1 (NY Library Scan)
[416] Navy & Army Illustrated Vol 11 One of a very few pictures of a large body of Boer soldiers. This was a Scene In The Boer Lines Before Joubert's surrender.
[417] Our Military Officers. Album of portraits with life sketches.
Original Publication: Onze krijgs-officieren: album van portretten met levens-schetsen der. by Gustav Schoeman Preller, Frans Vredenrijk Engelenburg 1904 Publisher "Volksstem" kantore
Catalogue Record: https://archive.org/search?query=title%3A%28Onze%20Krijgs-Officieren%29
File:Onze Krijgs-Officieren. Album van portretten met levens-schetsen, Pretoria, Volksstem, 1904.pdf
Download Access: https://en.wikipedia.org/wiki/File:Onze_Krijgs-Officieren._Album_van_portretten_met_levens-schetsen,_Pretoria,_Volksstem,_1904.pdf
Direct Link to File: https://upload.wikimedia.org/wikipedia/commons/5/52/Onze_Krijgs-Officieren._Album_van_portretten_met_levens-schetsen%2C_Pretoria%2C_Volksstem%2C_1904.pdf
Source Page: https://en.wikipedia.org/wiki/Annie_Botha#cite_note-4
[418] Titled: Commando at Dundee https://www.angloboerwar.com/unit-information/boer-units/1953-boer-forces?start=3
[419] File:The Second Boer War, 1899-1902 Q72215.jpg - Wikimedia Commons
[420] File:The Second Boer War, 1899-1902 Q72219.jpg - Wikimedia Commons
[421] SA National Archives: http://www.nationalarchives.gov.za/node/1236

[422] To The Bitter End: a photographic history of the Boer War, 1899-1902, by Lee, Emanoel C. G. Harmondsworth: Penguin, 1986. https://archive.org/details/tobitterend00eman_0

[423] Black & White Weekly Vol 5 1901 Original Caption: "Boer Snipers at work on a British Convoy"

[424] To The Bitter End: a photographic history of the Boer War, 1899-1902, by Lee, Emanoel C. G. Harmondsworth: Penguin, 1986. https://archive.org/details/tobitterend00eman_0

[425] Military Officer Album.

[426] Ben Bouwer (Barend Daniël Bouwer), 1875-1938.
https://commons.wikimedia.org/wiki/File:Ben_Bouwer_(Barend_Dani%C3%ABl_Bouwer),_1875-1938.jpg

[427] File:The Second Boer War, 1899-1902 Q72179.jpg - Wikimedia Commons Author: William Skeoch Cumming

[428] File:The Second Boer War, 1899-1902 Q72039.jpg - Wikimedia Commons Author: William Skeoch Cumming

[429] Anglo-Boer War Album, Cape Town, D. Edwards & co. [1902?]

[430] Anglo-Boer War Album, Cape Town, D. Edwards & co. [1902?]

[431] File:The Second Boer War, 1899-1902 Q72291.jpg - Wikimedia Commons Author: William Skeoch Cumming\

[432] Anglo-Boer War Album, Cape Town, D. Edwards & co. [1902?]

[433] File:View south to Stormberg - panoramio.jpg - Wikimedia Commons Author: Hendrik van den Berg

[434] Landscape Anysberg Nature Reserve, Kannaland Local Municipality, Western Cape, South Africa. Author: Winfried Bruenken (Amrum) File:Anysberg P1020029.JPG - Wikimedia Commons

[435] Source: Under the Union Jack 208-Pumping lead www.angloboerwar.com/forum/19-ephemera/3...-jack?start=54#93200 Credit David Biggins.

[436] Black & White Weekly Vol 5 1901

[437] Engineers repairing a Railway destroyed by the Boers near Kroonstadt. File:Reparaties aan de spoorlijn naar Kroonstad, Zuid-Afrika Royal Engineers repairing a Railway destroyed by the Boers - Kroonstadt, South africa (titel op object), RP-F-F09002.jpg - Wikimedia Commons

[438] Near Greylingstad, February 13th, 1901 After Pretoria: The Guerilla War Vol 3 April, 1901

[439] Pattern 1 Gun Truck and Train . Circa 1901. First pattern gun truck No 1 armoured train. (Publication on armoured trains in the Anglo Boer War) - Atom site for DRISA
PB3032_005.jpg (1240×971) (drisa.co.za)

[440] File:The Second Boer War, 1899-1902 Q72436.jpg - Wikimedia Commons Author: William Skeoch Cumming

[441] Pattern 2 train. Circa 1901. New pattern bogie on No 12 armoured train. (Publication on armoured trains in the Anglo Boer War) - Atom site for DRISA
PB3032_007.jpg (1237×969) (drisa.co.za)

[442] Anglo-Boer War Album, Cape Town, D. Edwards & co. [1902?]

[443] Anglo-Boer War Album, Cape Town, D. Edwards & co. [1902?]

[444] Anglo-Boer War Album, Cape Town, D. Edwards & co. [1902?]

[445] Anglo-Boer War Album, Cape Town, D. Edwards & co. [1902?]

[446] Anglo-Boer War Album, Cape Town, D. Edwards & co. [1902?]

[447] Anglo-Boer War Album, Cape Town, D. Edwards & co. [1902?]

[448] File:The Second Boer War, 1899-1902 Q72071.jpg - Wikimedia Commons Author: William Skeoch Cumming

[449] File:The Second Boer War, 1899-1902 Q72207.jpg - Wikimedia Commons Author: William Skeoch Cumming

[450] The Biograph in battle: its story in the South African War related with personal experiences. by William Kennedy Laurie Dickson, London: T. Fisher Unwin, 1901. Catalog Search:
https://archive.org/search?query=title%3A%28The+Biograph+in+Battle%29

[451] File:Camps-british-armyRFA.jpg - Wikimedia Commons Unknown Author

[452] File:The Second Boer War, 1899-1902 Q71993.jpg - Wikimedia Commons Author: William Skeoch Cumming

[453] Anglo-Boer War Album, Cape Town, D. Edwards & co. [1902?]

[454] Navy & Army Illustrated Vol 10 May 26th, 1900 File:British Army, Boer War, Q 71941.jpg - Wikimedia Commons

[455] A Section with a Maxim Gun in Action Navy & Army Illustrated Vol 11

[456] Navy & Army Illustrated Vol 11 At Beaufort West. A " Sangar " on the Line of Communications.

[457] Navy & Army Illustrated Vol 10 Taken under fire. The 2nd Queens in action

[458] Navy & Army Illustrated Vol 10 May 26th, 1900

[459] 17th Lancers at the Battle of Modderfontein Farm. Painting by Richard Caton Woodville, Jr. File:All That Was Left of Them (17th Lancers at Moddersfontein).jpg - Wikimedia Commons Credit: Alonso de Mendoza.

[460] War with the Boers, H. Brown, London [1902?] Vol 1

[461] File:The Second Boer War, 1899-1902 Q72296.jpg - Wikimedia Commons Author: William Skeoch Cumming
[462] File:Visrivier.jpg - Wikimedia Commons Author: Jcwf
[463] File:Canadian troops in armoured train South Africa Jan 1902 LAC 3407086.jpg - Wikimedia Commons Author Roberts, George F.
[464] File:Encephalartos longifolius05.jpg - Wikimedia Commons Author: Herselman Werner and Encephalartos longifolius - Wikipedia
[465] After Pretoria: The Guerilla War Vol 3 p. 493 1901
[466] After Pretoria: The Guerilla War Vol 3 p. 493 1901
[467] Anglo Boer War - Artillery and Ammunition - Page 31 - Boer War Forum_Credit: Neville Constantine
[468] Cunliffe, History of the Boer War Vol 1 (NY Library Scan)
[469] File:Encephalartos Altensteinii in Lednice Greenhouse.jpg - Wikimedia Commons
[470] File:Syncerus caffer nanus 001.JPG - Wikimedia Commons
[471] The African buffalo (Syncerus caffer) is a large sub-Saharan African bovine. Syncerus caffer caffer, the Cape buffalo, is the typical subspecies. S. c. nanus (the forest buffalo) is the smallest subspecies, common in forest areas of Central and West Africa, African buffalo - Wikipedia
[472] File:17th Lancers - cap badge, original, antique.jpg - Wikimedia Commons Author: GMJ
[473] File:The Second Boer War, 1899-1902 Q72100.jpg - Wikimedia Commons Author: William Skeoch Cumming Note the shadow of the photographer, probably of William Skeoch Cumming himself.
[474] File:Opuntia littoralis var vaseyi 4.jpg - Wikimedia Commons Author: Stan Shebs
[475] Bysing Wood is full of bluebells at the beginning of May - geograph.org.uk - 786661 - Coppicing - Wikipedia
[476] File:Coppice2.png - Wikimedia Commons
[477] File:The Second Boer War, 1899-1902 Q72098.jpg - Wikimedia Commons Author: William Skeoch Cumming
[478] Heroes of the Boer War. Review of Reviews, 1903.
[479] File:Swartberg, 2001.jpg - Wikimedia Commons Author: Andresdewet
[480] File:Karoo P1030683.JPG - Wikimedia Commons Author: Winfried Bruenken (Amrum)
[481] File:The Second Boer War, 1899-1902 Q72442.jpg - Wikimedia Commons Author: William Skeoch Cumming
[482] File:The Second Boer War, 1899-1902 Q72008.jpg - Wikimedia Commons
[483] South Africa; its history, heroes and wars by Mackenzie, William Douglas (William Douglas), 1859-1936; Assisted by Alfred Stead, 1877-1933, joint author. Published 1899 by The Co-Operative Publishing Company, Chicago, Ill. Page 160 South Africa; its history, heroes and wars: Mackenzie, W. Douglas (William Douglas), 1859-1936: Free Download, Borrow, and Streaming: Internet Archive
[484] File:RoadThroughMeiringspoort.jpg - Wikimedia Commons Author: Martinvl
[485] File:Great Swartberg Pass soon after being cut - Cape Colony.jpg - Wikimedia Commons Author: Cape Archives Anon.
[486] The over 2000 m high Greater Swartberg range in the distance, viewed from farmlands near the town of Oudtshoorn in the relatively well-watered south. File:Swartberg01.jpg - Wikimedia Commons Author: Vberger
[487] File:The Second Boer War, 1899-1902 Q72019.jpg - Wikimedia Commons
[488] File:The Second Boer War, 1899-1902 Q72020.jpg - Wikimedia Commons Author: William Skeoch Cumming
[489] File:Pai e filho, garanhões premiados da raça Crioula..jpg - Wikimedia Commons Author: Kc.sminsights
[490] File:Yeguacriolla.jpg - Wikimedia Commons
[491] File:TradeCardMasonAndHamlinOrganCoCirca1880s.jpg - Wikimedia Commons
[492] Lower Karoo File:Karoo National Park.jpg - Wikimedia Commons Author: Tjeerd Wiersma
[493] File:The Second Boer War, 1899-1902 Q72197.jpg - Wikimedia Commons
[494] From Hillegas 1900 Pages 320 to 322 of the US Edition titled The Boers in War.
[495] Photos in Photos of Genl. Manie Maritz (geni.com)
[496] Photos in Photos of Genl. Manie Maritz (geni.com)
[497] File:Gen. Manie Maritz.jpg - Wikipedia
[498] File:JacobusenCatharinaSmuts1893.jpg - Wikimedia Commons Attribution: Levy, Naphtali
[499] Durban. Old NGR station at Point with scotch cart in the foreground. - Atom site for DRISA PB1222.jpg (1400×917) (drisa.co.za)
[500] Buffelshoek area of the Hex River Mountains in South Africa. Author: Andres de Wet. File:Buffelshoek-pano.jpg - Wikimedia Commons
[501] File:The Second Boer War, 1899-1902 Q72030.jpg - Wikimedia Commons Author: William Skeoch Cumming
[502] File:The Second Boer War, 1899-1902 Q72362.jpg - Wikimedia Commons Author: William Skeoch Cumming

[503] File:The Second Boer War, 1899-1902 Q72367.jpg - Wikimedia Commons Author: William Skeoch Cumming
[504] File:The Second Boer War, 1899-1902 Q72385.jpg - Wikimedia Commons Author: William Skeoch Cumming
[505] File:Namaqualand Desert Road.jpg - Wikimedia Commons Source: South African Tourism
[506]File:Spring flowers in Namaqualand.JPG - Wikimedia Commons Author: Chen Hualin
[507] File:Namaqualand, Goegap 0035.jpg - Wikimedia Commons Author: Winfried Bruenken (Amrum)
[508] File:Namaqualand, Goegap 1035.jpg - Wikimedia Commons Author: Winfried Bruenken (Amrum)
[509] General Botha, the Career and the Man, By Harold Spender, 1916 Google Books, UCAL-Davis: https://books.google.com/books?id=dxw1AQAAIAAJ
[510] File:Orbicule Koppie, Concordia.JPG - Wikimedia Commons_Author: Andrew Hall
[511] File:The Second Boer War, 1899-1902 Q72521.jpg - Wikimedia Commons Author: William Skeoch Cumming
[512] Cunliffe, History of the Boer War Vol 1 (NY Library Scan)
[513] Black & White Weekly Vol 1-2 1899-1900
[514] Gun: .455 Webley - Wikipedia Image: File:455 Webley Revolver.jpg - Wikimedia Commons Credit Drake00
[515] Webley & Scott Mk VI. Caliber .455 Collection Paul Regnier, Lausanne, Switzerland Image: File:Webley IMG 6789.jpg - Wikimedia Commons
[516] File:Webley IMG 6793.jpg - Wikimedia Commons Author: Rama
[517] Cartridge .455 Webley Mk II Calibre .455 (11.6 mm) Action: Single or double action. Rate of fire, 20–30 rounds/minute. Muzzle velocity 620 ft./s (190 m/s) Feed system 6-round cylinder Webley Revolver - Wikipedia
[518] File:The Second Boer War, 1899-1902 Q71974.jpg - Wikimedia Commons Author: William Skeoch Cumming
[519] Anglo-Boer War Album, Cape Town, D. Edwards & co. [1902?]
[520] File:Prospect Camp, Bermuda - Parade in Camp.jpg - Wikimedia Commons Source: An old postcard. No author indicated. Image of: 3rd Battalion on parade at Prospect Camp, Bermuda, circa 1902.
[521] Genl Jan Smuts en sy staf te Concordia, Mei 1902 - Deneys Reitz - Wikipedia
[522] Deneys Reitz op kommando - Deneys Reitz - Wikipedia
[523] The Biograph in battle: its story in the South African War related with personal experiences. by William Kennedy Laurie Dickson, London: T. Fisher Unwin, 1901. Catalog Search: https://archive.org/search?query=title%3A%28The+Biograph+in+Battle%29
[524] The Biograph in battle: its story in the South African War related with personal experiences. by William Kennedy Laurie Dickson, London: T. Fisher Unwin, 1901. Catalog Search: https://archive.org/search?query=title%3A%28The+Biograph+in+Battle%29
[525] Boer Commando assembled from Pretoria. 1899. File:1899 Boer Commando - Pretoria.jpg - Wikimedia Commons
[526] File:HMS Monarch (1868).jpg - Wikimedia Commons unattributed; probably Royal Nav
[527] File:HMS Monarch (1868) William Frederick Mitchell.jpg - Wikimedia Commons Author: William Frederick Mitchell (1845–1914)
[528] Creswicke, Louis South Africa and the Transvaal War, Vol. 1 Photo by Wilson, Aberdeen.
[529] "Cape Town– Seat of Government, Cape Colony" Navy & Army Illustrated Vol 9 Photo by G. W. Wilson, Aberden.
[530] Black & White Weekly Vol 1-2 1899-1900
[531] "The Bulwarks of South Africa – The Squadron at the Cape." Navy & Army Illustrated Vol 9
[532] Navy & Army Illustrated Vol 9
[533] War with the Boers, H. Brown, London [1902?] Vol 2
[534] After Pretoria: The Guerilla War Vol 3 Dec. 1900
[535] After Pretoria: The Guerilla War Vol 3 Jan. 1901
[536] General (the taller) and Gen. Pole-Carew at the railway station, Pretoria, June 5th—S. Africa
Digital ID: (b&w film copy neg.) cph 3c02310 http://hdl.loc.gov/loc.pnp/cph.3c02310
Reproduction Number: LC-USZ62-102310 (b&w film copy neg.)
https://www.loc.gov/pictures/item/91705267/resource/
[537] Times History War in S.A Vol 3 Photo by Bassano.
[538] "Under British Protection. The great concentration camp near Standerton." Navy & Army Vol 13 March 15th, 1902
[539] Black & White Weekly Vol 1-2 1899-1900
[540] War in SA, Harding 1899
[541] Description: Boer general Christiaan de Wet on his way to peace talks at Vereeniging. 13 April 1920, Source Scan of a publicly available post card from 1920. Author Unknown. File:De Wet by Ferreirastasie oppad na Vereeniging.jpg - Wikimedia Commons

[542] Boer general Christiaan de Wet, photographed at the end of the Second Boer War in Bloemfontein on 6.6.1902, before embarking for Europe [To raise funds for relief.] Source: S. King File:Generaal Christiaan de Wet - Bloemfontein 6.6.1902.jpg - Wikimedia Commons
[543] File:The Second Boer War, 1899-1902 Q68470.jpg - Wikimedia Commons Author Unknown author
[544] Group of Boer leaders in Kimberley area. General Koos de la Rey is the man with black beard in right foreground File:The Second Boer War, 1899-1902 Q68490.jpg - Wikimedia Commons Author Unknown author
[545] Heroes of the Boer War. Review of Reviews, 1903.
[546] Anglo Boer War - Pictures from Under the Union Jack - Page 3 - Boer War Forum Credit: Neville Constantine
[547] War in SA, Harding 1899
[548] (L to R) P.J. Lemmer (65) J.D.I. Botha (15) and C.J. Pretorius (43) Hillegas, Boers in War, 1900.
[549] Anglo-Boer War Album, Cape Town, D. Edwards & co. [1902?]
[550] Davitt, Boer Fight for Freedom, 3rd ed. 1902. (Duke Scan)
[551] Hillegas, Boers in War, 1900.
[552] File:The Second Boer War, 1899-1902 Q72252.jpg - Wikimedia Commons Author: William Skeoch Cumming
[553] File:The Second Boer War, 1899-1902 Q72174.jpg - Wikimedia Commons
[554] File:Resa del bacino del Brandewater.jpg - Wikimedia Commons
[555] Black & White Weekly Vol 1-2 1899-1900
[556] With the Flag to Pretoria, H. W. Wilson Vol 2 Feb. 1900
[557] After Pretoria: The Guerilla War Vol 3 Feb., 1901
[558] British soldiers collecting captured rifles of the Boer fighters. File:The Second Boer War, 1899-1902 Q72023.jpg - Wikimedia Commons
[559] Navy & Army Illustrated Vol 11 Sept. 29th, 1900
[560] Anglo-Boer War Album, Cape Town, D. Edwards & co. [1902?]
[561] Okiep - Port Nolloth narrow gauge railway. Circa 1884. Senior officers' coach by mules on its way down the mountain. - Atom site for DRISA P1262_21.jpg (1400×1056) (drisa.co.za)
[562] Okiep - Port Nolloth narrow gauge railway. Circa 1885. Senior officers' coach by mules on its way down the mountain. - Atom site for DRISA P1262_29.jpg (1400×921) (drisa.co.za)
[563] Namaqua Copper Company 0-4-2ST Pioneer, derailed outside O'okiep while being run into the town with a truckload of dynamite during the South African War File:NCC Pioneer 0-4-0ST b.jpg - Wikimedia Commons Source: Transnet Heritage Foundation Library, via Peter Bagshawe. See: Bagshawe, Peter (2012). Locomotives of the Namaqualand Railway and Copper Mines (1st ed.). Stenvalls. pp. 35–39. ISBN 978-91-7266-179-0. And: See: Cape Copper Company Medal for the Defence of O'okiep - Wikipedia
[564] Okiep - Port Nolloth narrow gauge railway. Derailed locomotive. - Atom site for DRISA P1262_17.jpg (1400×1078) (drisa.co.za) And Port Nolloth district, 1898. Locomotive breakdown on the Okiep - Port Nolloth line. - Atom site for DRISA N56232.jpg (1400×886) (drisa.co.za)
[565] Springbok district. 'Pioneer' locomotive used during Anglo-Boer War to send a truck loaded with dynamite into O'Kiep. - Atom site for DRISA N75973.jpg (1400×1153) (drisa.co.za)
[566] Cape Copper Company Medal for the Defence of O'okiep - Wikipedia
[567] File:Okiep00.jpg - Wikimedia Commons Source: McGregor Museum, Kimberley
[568] File:Cape Copper Company Medal, Silver.jpg - Wikipedia
[569] Navy & Army Vol 13 p. 58 March 1st, 1902 From a photo. supplied by Mr. W. P. Watkins, Bexhill-on-Sea.
[570] Commandant Hans Lotter in Graaff-Reinet prison. Blood spatter can be seen on his shirt. Text in white: "Commandant / J.C. Lotter. / The Spots on shirt / are blood stains / Ivie H. Allan / Photo / Graaff-Rt." https://commons.wikimedia.org/wiki/File:1875-1901_Johannes_Cornelius_L%C3%B6tter.jpg
[571] Lotter Death Penalty - Battle of Groenkloof - Wikipedia Unknown author
[572] Navy & Army Vol 13 p. 58 March 1st, 1902 From a photo. supplied by Mr. W. P. Watkins, Bexhill-on-Sea.
[573] Navy & Army Vol 13 p. 58 March 1st, 1902 From a photo. supplied by Mr. W. P. Watkins, Bexhill-on-Sea.
[574] File:Südwestafrika 1915.jpg - Wikimedia Commons Source Eduard Rothert, Karten und Skizzen zum Weltkrieg, Druck und Verlag von A. Bagel, Düsseldorf, 1916.
[575] South Africa; its History, Heroes and Wars. [1899]
[576] File:The Second Boer War, 1899-1902 Q68483.jpg - Wikimedia Commons
[577] Letters patent issued by Queen Victoria in 1900, creating the office of Governor-General of Australia as part of the process of federation. https://commons.wikimedia.org/wiki/File:Letters_Patent_Australia.jpg
[578] Times History War in S.A Vol 5 (1907) Photo by Duffus Bros., Cape Town

[579] File:Potchefstroom students march against government policy during the Rebellion.jpg - Wikipedia

[580] File:Women's march on Reddersburg in 1914 to demand the release of Gen. CR de Wet.jpg - Wikipedia

[581] Photos in Photos of Genl. Manie Maritz (geni.com)

[582] Photos in Photos of Genl. Manie Maritz (geni.com)

[583] File:Gen. Manie Maritz.jpg - Wikipedia

[584] File:FWReitz VA0947.jpg - Wikimedia Commons

[585] Smuts kabinet 1923 - Deneys Reitz - Wikipedia

[586] File:Deneys Reitz.jpg - Wikimedia Commons

[587] Deneys Reitz as minister of native affairs visit British army camp Aldershot 22 November 1939. File:Reitz at Aldershot.jpg - Wikimedia Commons Author: Gthvanbrakel

[588] Description: Deneys Reitz. circa 1940. Source: Newhistory.co.za. Made available by: Morné van Rooyen Author: Unknown File:Deneys Reitz parent.jpg - Wikipedia

[589] South African High Commissioner Visits the Home Fleet. 10 April 1943, Scapa Flow, Colonel, The Hon Deneys Rietz, High Commissioner for South Africa, During His Recent Visit To The Home Fleet. File:SOUTH AFRICAN HIGH COMMISSIONER VISITS THE HOME FLEET. 10 APRIL 1943, SCAPA FLOW, COLONEL, THE HON DENEYS RIETZ, HIGH COMMISSIONER FOR SOUTH AFRICA, DURING HIS RECENT VISIT TO THE HOME FLEET. A15880.Jpg - Wikimedia Commons Author Davies, F A

[590] File:Leila Reitz's tombstone on Pilgrim's Rest, photo by Rehan Opperman, July 28, 2009.JPG - Wikipedia Made available by: Morné van Rooyen. Author, Rehan Opperman

Made in the USA
Monee, IL
06 November 2026

67383122R00201